Springer
Proceedings in Physics 71

Springer Proceedings in Physics

Managing Editor: H. K. V. Lotsch

Volumes 1–43 are listed on the back inside cover

C. Y. Yang M. M. Rahman G. L. Harris (Eds.)

Amorphous and Crystalline Silicon Carbide IV

Proceedings of the 4th International Conference,
Santa Clara, CA, October 9–11, 1991

With 336 Figures

Springer-Verlag
Berlin Heidelberg New York
London Paris Tokyo
Hong Kong Barcelona
Budapest

Professor Dr. Cary Y. Yang
Professor Dr. M. Mahmudur Rahman

Microelectronics Laboratory, Santa Clara University,
Santa Clara, CA 95053, USA

Professor Dr. Gary L. Harris

School of Engineering, Howard University,
Washington, DC 20059, USA

ISBN-13:978-3-642-84806-3 e-ISBN-13:978-3-642-84804-9
DOI: 10.1007/978-3-642-84804-9

Softcover reprint of the hardcover 1st edition 1992

Typesetting: Camera ready by authors/editors
54/3140 - 5 4 3 2 1 0 - Printed on acid-free paper

Preface

This volume contains written versions of papers which were presented at the Fourth International Conference on Amorphous and Crystalline Silicon Carbide and Other IV-IV Materials (ICACSC '91), held at Santa Clara University on October 10 and 11, 1991. Over 150 participants from nine countries attended the conference. This attendance was the largest of all four ICACSCs held thus far.

Of the 7 invited and 86 contributed papers, 64 were accepted for publication in this volume after a peer review process. These numbers far exceed those in the previous three conferences and they confirm the continuing revival of interest in this technologically important field.

The volume is divided into four parts according to growth, characterization, and applications. Some exciting and technologically promising results are reported for the first time; for example, HBTs using microcrystalline Si and a-SiC, SiC hetero-emitter and epitaxial base transistors, optoelectronic applications of 6H-SiC, formation of SiGe/Si heterostructures by low-temperature germanium ion implantation, growth of 40 mm diameter 6H-SiC single crystal boules, and growth of silicon carbon alloy on Si(100) superlattices, to name only a few.

ICACSC '91 was made possible by the generous support of the many industrial and government sponsors, and by the time and effort contributed by conference committee members. We are immensely grateful for the financial support from the following organizations: Fujitsu Laboratories, Ltd., Furukawa Electric Technologies, Inc., Hewlett-Packard Circuit Technology R & D, Hitachi Research Laboratory, National Science Foundation, Office of Naval Research, Sanyo Electric Co., Ltd., Toshiba America Electronic Components, Inc., and Westinghouse Electric Corporation. We acknowledge additional support from Santa Clara University Microelectronics Laboratory and Howard University Materials Science Research Center of Excellence. Finally, we express our heartfelt appreciation to the students and staff of Santa Clara University Microelectronics Laboratory, especially Dr. Jianmin Qiao, for their tireless efforts in making this ICACSC the best ever. This volume is dedicated to them.

Santa Clara, CA
May 1992

Cary Y. Yang
M. Mahmudur Rahman
Gary L. Harris

Preface

Contents

Part III Growth and Characterization of Polycrystalline, Microcrystalline, and Amorphous Silicon Carbide

Part IV Applications

Part I

Growth of Crystalline Silicon Carbide

Recent Progress in Epitaxial Growth of SiC

H. Matsunami

Department of Electrical Engineering, Kyoto University,
Yoshidahonmachi, Sakyo, Kyoto 606, Japan

Abstract. Recent progress in epitaxial growth of SiC in the author's group is surveyed. Growth of 3C-SiC on a Si substrate, a typical success in heteroepitaxial growth with large lattice mismatch, is briefly reviewed. Step-controlled epitaxy of α-SiC, which the author's group named and is a key technology for future development, is discussed in detail. Growth mechanism and effects of off-direction and off-angle are described. As an advanced epitaxial growth, atomic level control in SiC crystal growth by gas source MBE is given.

1. Introduction

Strong desire has been given for developing silicon carbide(SiC) devices, e.g. blue light-emitting diodes(LEDs) and transistors for high-temperature and/or harsh environments[1,2]. The material has been ordinarily produced by the Acheson method above 2500℃ using silica and coke, which inevitably brings about the problem of purity for semiconductor use. Through many efforts, high-purity single crystals are grown using sublimation(Lely method). However, difficulties in device processing technology at high temperatures have prevented the development of SiC devices.

Recent progress in epitaxial growth of SiC at lower temperatures has given superior single crystalline layers and easy control of in-situ impurity doping[3,4]. Break-through is now appearing in SiC processing technology. In this report, growth of 3C-SiC on a Si substrate, a typical success in heteroepitaxial growth with large lattice mismatch, is briefly reviewed. Then, step-controlled epitaxial growth of α-SiC, which the author's group named and is a key technology for future development of devices, is discussed in detail. Homoepitaxial growth on 6H-SiC $(01\bar{1}4)$ substrates prepared by polytype conversion

Editors: C.Y. Yang · M.M. Rahman · G.L. Harris

of 3C-SiC grown on Si is given to discuss on the growth mechanism in step-controlled epitaxy. Atomic level control in SiC crystal growth is given as an advanced study on epitaxial growth.

2. Heteroepitaxial Growth of 3C-SiC on Si

Since single crystals of 3C-SiC produced by sublimation methods are small, heteroepitaxial growth on Si has been studied extensively using vapor phase epitaxy(VPE). However, reproducible growth has been difficult, which is caused by the growth on Si at high temperatures and large lattice mismatch of 20% between 3C-SiC and Si (a_{SiC} = 0.4358nm, a_{Si} = 0.5430nm). One of the approaches to overcoming this difficulty is to produce a very thin layer of 3C-SiC by carbonization before chemical vapor deposition (CVD)[3,4], which was proposed by the author's group[5].

Heteroepitaxial growth of 3C-SiC was carried out under atmospheric pressure in a horizontal quartz reaction tube using SiH_4, C_3H_8 and H_2[6-8]. A Si(001) substrate placed on a SiC-coated graphite susceptor was heated by rf induction at 200 kHz. Single crystal growth procedure for 3C-SiC consists of etching, carbonization and CVD. A Si substrate was etched at 1180℃ using a HCl flow (HCl:10 sccm, H_2:1 slm). Then, in a flow of C_3H_8 (C_3H_8:1.2 sccm, H_2:1 slm) the substrate was heated up to 1360℃ in short time(~2 min) for carbonization. Immediately after, CVD growth of 3C-SiC was carried out at around 1330℃ with the standard gas composition (SiH_4: 0.3 sccm, C_3H_8:0.25 sccm). During CVD, the flow rate of H_2 carrier gas was kept at 3 slm, corresponding to a linear velocity of 5 cm/s. The growth rate of 3C-SiC was about 1.8μm/h. At present, a diameter of 50-75mm for single crystalline 3C-SiC can be obtained.

Reflection high-energy electron diffraction(RHEED) patterns changed gradually from a spot pattern to a streaked one with increasing layer thickness. Kikuchi lines are clearly observed in a thick layer of 5.4 μm as shown in Fig.1. In a cross-sectional transmission electron microscope(XTEM) photograph shown in Fig.2, the interface between 3C-SiC and Si is rather sharp. Stripes spread along the <111> directions in 3C-SiC and Si. Since the lattice spacings differ in 3C-SiC and Si, one out of 4 or 5 lines of the stripes is disconnected at the interface showing the existence of misfit dislocations, which agrees well with the lattice mismatch of 20 %. Dark and

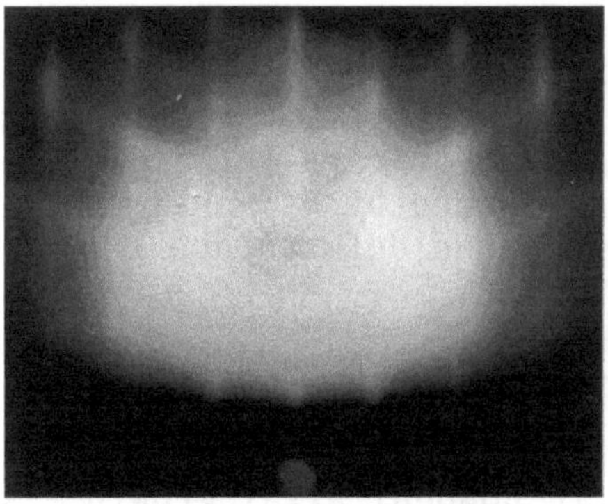

Fig.1 RHEED pattern of 3C-SiC on Si(100).

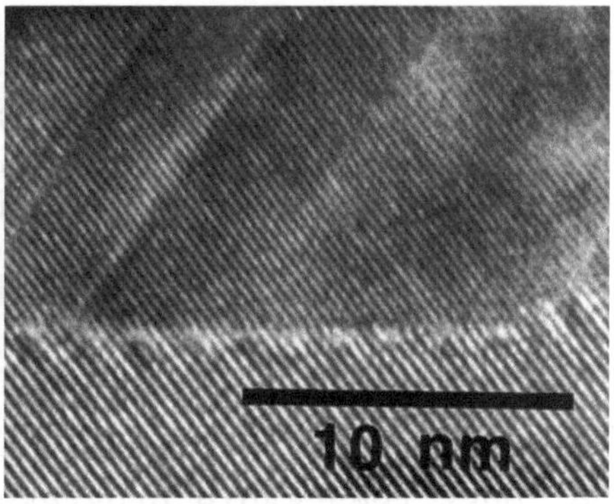

Fig.2 XTEM photograph of 3C-SiC on Si(100).

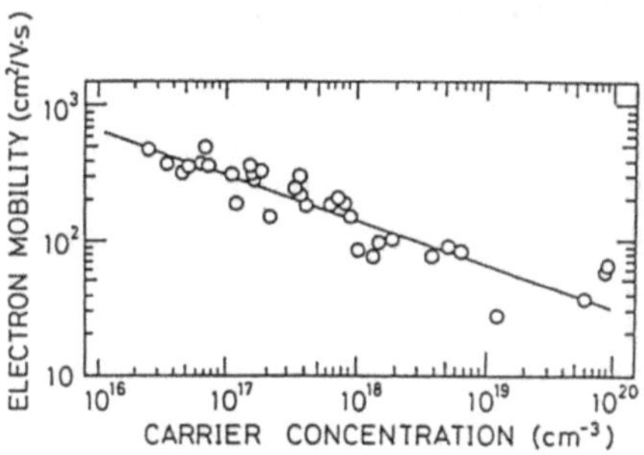

Fig.3 Electron mobility vs. carrier concentration of 3C-SiC.

white stripes along the <111> directions in the 3C-SiC layer indicate the inclusion of plane defects, probably stacking faults or microtwins.

Single crystals of 3C-SiC on Si(001) can be used for device application. However, antiphase domains(APDs) exist owing to the growth of polar crystals (3C-SiC) on nonpolar substrates(Si)[9]. APDs give texture-like morphology consisting of wedge shapes intersecting to each other. The boundaries were confirmed as random grooves by etching with molten KOH at 600℃. APDs in epitaxial layers should be eliminated, since they cause surface roughness and may work as scattering centers or recombination centers. Thorough detailed growth experiments on spherically polished Si(100) substrates, APD-free growth was realized by using a Si(001) wafer tilted towards [110] [10].

Control of conduction type (n- or p-type) and carrier concentration is possible by in-situ doping. Electron mobilities of the grown layers are shown in Fig.3[11].

Standard Hall measurements using long sheet-shaped samples were carried out for 3C-SiC grown on 2° -, 4° - and 6° -off Si (100) substrates. There exists significant difference in electron mobilities for parallel and perpendicular to atomic steps introduced by off-angle. The difference may be the change in plane defects due to the introduction of off-angle[11].

3. Homoepitaxial Growth of α-SiC (Step-Controlled Epitaxy)

The most remarkable progress in homoepitaxial growth is step-controlled epitaxy proposed by the author's group [12,13]. High-quality single crystals of 6H-SiC can be grown on angle-lapped 6H-SiC {0001} at 1500℃, which is more than 300℃ lower compared with ordinary growth temperatures. The density of surface steps is increased by angle lapping. Utilizing step-flow growth (lateral growth from the steps), single crystalline 6H-SiC with a very smooth surface is grown at lower temperatures. Hereby, the author's group named "step-controlled epitaxy", in which the polytype-controlled epitaxial growth at low temperatures can be realized by the aid of a number of steps.

Single crystals of 6H-SiC grown by the Acheson method were used as substrates. The low-index basal planes of the crystals are either (0001)Si or $(000\bar{1})$C faces. To control the step density, the basal planes were angle-lapped with carborundum and polished with diamond paste. The off-angle and off-direction were determined by X-ray diffraction.

Crystal growth was carried out mainly at 1500℃ under atmospheric pressure in a horizontal quartz reaction tube using SiH_4, C_3H_8, and H_2. Substrates put on a SiC-coated graphite susceptor was heated by rf induction. First the substrates were etched with HCl gas at 1500℃. Then CVD growth was carried out with a standard gas composition (SiH_4:0.3 sccm, C_3H_8:0.2 sccm, H_2:3 slm). The growth rate was 2 μm/h.

3.1 Effects of Off-Orientation

On a well-oriented (0001)Si face, the grown layer shows a mosaic pattern as in Fig.4(a). The grown layer is separated by step- or groove-like boundaries. Triangular etch pits indicating 3-fold symmetry are rotated by 180° each

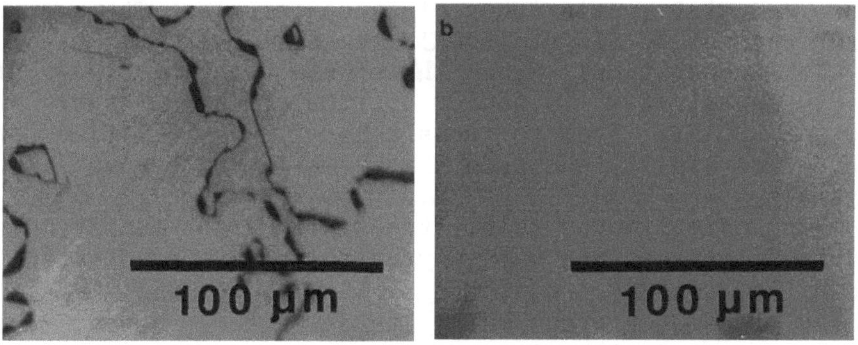

Fig.4 Nomarski microphotographs of layers grown on 6H-SiC(0001); (a) well-oriented, and (b) off-oriented (3-6° off toward [11$\bar{2}$0]).

other across the boundaries. The RHEED pattern shows the twin relation of 3C-SiC against c-axis. Both results indicate the layers grown on well-oriented (0001)Si faces to be twin crystalline 3C-SiC. Similar results are obtained on (000$\bar{1}$)C faces.

On an off-oriented (0001)Si face inclined by 3-6° toward [11$\bar{2}$0], the grown layer shows a very smooth mirror-like surface without boundary structure as shown in Fig.4(b). Etch pits of rounded hexagonal-shape are observed. The RHEED pattern shows that of 6H-SiC. Hence, on an off-oriented (0001)Si faces inclined toward [11$\bar{2}$0], single crystalline 6H-SiC is homoepitaxially grown even at 1500℃. The introduction of off-orientation brings about changes in growth mechanism, lateral growth from steps. Similar results are obtained on (000$\bar{1}$)C faces. Step-controlled epitaxy on {0001} faces can be realized even at 1200℃[14].

3.2 Growth Mechanism

Crystal growth is mostly initiated from steps (or kinks), rather than by two-dimensional nucleation on terraces, if the migration length on the surface is much larger than the step width. The migration length is changed by growth condition, mainly by temperature. At high temperatures above 1700℃, layers grown on well-oriented {0001} faces become 6H-SiC as in previous reports, probably because the migration length is increased. At 1500℃, the migration length seems to be sufficiently short compared with the terrace width on the well-oriented {0001} faces.

Hence, crystal growth takes place on terraces. The grown layer becomes 3C-SiC at 1500℃, because 3C-SiC is stable at low temperatures. Since the atomic stacking order of 6H-SiC substrates is ABCACB..., the stacking order of growing 3C-SiC from terraces takes 2 possibilities of ABC... and ACB..., and thus the grown layers become twin crystals.

On off-oriented {0001} faces, the step density is high and terraces are not very vast (calculated terrace-width is around 2.4 nm for 6° off-angle), and lateral growth from the steps mainly takes place. Hence, the growth proceeds inheriting the atomic stacking order of substrates, and 6H-SiC grows at 1500℃.

3.3 Effects of Off-Direction and Off-Angle[15]

On (0001)Si faces inclined toward [1$\bar{1}$00] by 3° , stripe-type morphology with some textured features are observed after 120 min growth as shown in Fig.5(a). The stripes are parallel to one of edges (crystal facets) of the hexagonal shape substrates and perpendicular to the off-direction. Triangular etch-pits are observed together with rounded hexagonal-shape etch pits. In further long-time(300 min) growth, the stripe-type morphology changes to grain-like structures as shown in Fig.5(b). Thus, 3C-SiC components appear in the layer. However, the surface of layers grown on (0001)Si faces inclined toward [11$\bar{2}$0] remains very smooth even for long-time growth.

The feature of the growth on off-oriented (0001)Si faces varies with the off-direction, because the configu-

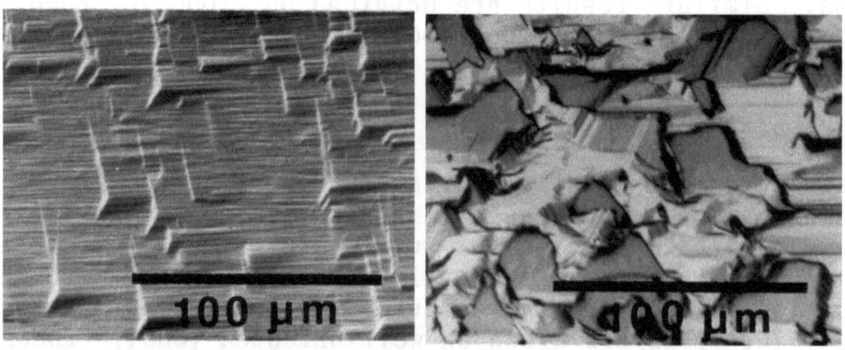

Fig.5 Nomarski microphotographs of layers grown on off-oriented 6H-SiC(0001) (3° off toward [1$\bar{1}$00]); (a) 120 min, and (b) 300 min growth.

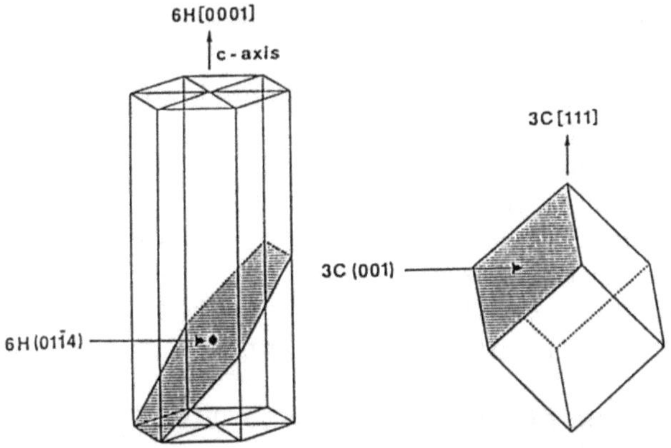

Fig.6 Relation between (01$\bar{1}$4) plane and hexagonal unit cell of 6H-SiC.

ration of atomic steps may be different. The steps are supposed to be aligned in parallel to the crystal facets of {0001} substrates. On the substrates inclined toward [11$\bar{2}$0], "zigzag steps" may exist, and on the substrates inclined toward [1$\bar{1}$00], "parallel steps" may exist. The zigzag steps are not easily bunched, because those are composed of steps developed to two different directions. On the other hand, the parallel steps are easily bunched and make terraces vast, since they are aligned in parallel. In the early stage of the growth on off-oriented (0001)Si faces inclined toward [1$\bar{1}$00], 6H-SiC is grown by lateral growth from the steps. However, after the terraces become vast by step bunching, 3C-SiC components begin to grow on terraces through two-dimensional nucleation.

Since the high-index crystal plane of 6H-SiC(01$\bar{1}$4) corresponds to the 54.74° -inclined plane from the {0001} basal plane toward [1$\bar{1}$00] as shown in Fig.6, the reduction of growth temperature by step-controlled epitaxy is expected. 6H-SiC (01$\bar{1}$4) ingots were prepared by sublimation on a 6H-SiC (01$\bar{1}$4) seed, which was polytype-converted from 3C-SiC(100) on Si[16]. Similar results as on {0001} faces inclined toward [11$\bar{2}$0] are obtained, although the wafer is inclined toward [1$\bar{1}$00]. This is mainly due to the high off-angle of 54.74° . Homoepitaxial growth of 6H-SiC can be obtained even at 1100℃[17].

The similar step-controlled epitaxy was realized even for other polytypes, for examples, 4H-SiC and 21R-SiC.

4. Atomic-Level Control in SiC Crystal Growth

The most advanced subject in epitaxial growth of SiC is "atomic layer epitaxy". Growth of 3C-SiC was tried by gas source MBE utilizing the change in surface superstructures by alternate supply of Si_2H_6 and C_2H_2. As a substrate, antiphase-domain free 3C-SiC (001) grown by CVD on Si(001) 2° off-oriented toward [110] was used. Source gases were introduced into the growth chamber through individual nozzles each located 10 cm distant from the substrate.

The substrate was first heated up to 1100℃ in a high vacuum (below $3x10^{-8}$ Torr) for oxide removal, and then the substrate temperature was lowered to 1000℃ at which temperature experiments were carried out. The starting 3C-SiC surface shows centered(c)(2x2) superstructure in its RHEED patterns. As the substrate surface is exposed to the Si_2H_6 beam, the surface reconstruction changes to (3x2) superstructure. The superstructure is thought to be formed by adsorbed Si atoms. Those Si atoms adsorbed on the surface react with C_2H_2. As the C_2H_2 is supplied to the (3x2) surface, the RHEED pattern returns to c(2x2). By repeating the alternate supply of Si_2H_6 and C_2H_2 many times as shown in Fig.7, epitaxial growth of 3C-SiC can be obtained[18,19].

By dynamic RHEED observation for the transition of 3C-SiC (001) surface superstructure under Si_2H_6 gas molecular beam irradiation, the surface superstructure changed in the order of c(2x2)→(2x1)→(5x2)→(3x2). By the detailed analysis, the relative amount of the constituent Si atoms of the superstructures was determined, and an additional dimer model was confirmed[20].

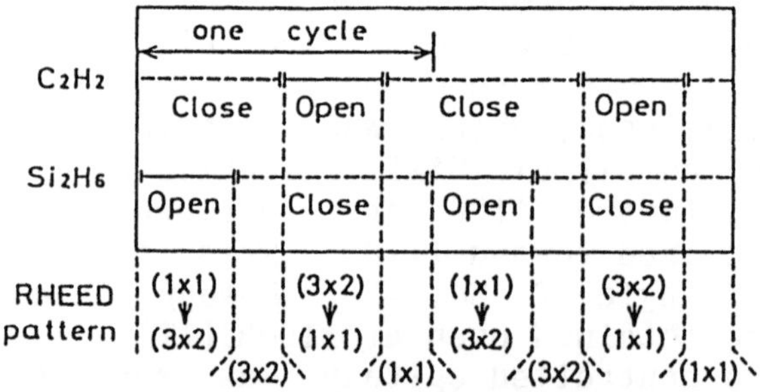

Fig.7 Sequence of source supply and change of surface superstructure.

Summary

Recent progress of epitaxial growth of SiC in the author's group was surveyed. Heteroepitaxial growth of 3C-SiC on Si has reached a certain level, and future research should be oriented to grow layers with low defect density. To know much about 3C-SiC, growth on α-SiC is one of challenges. Step-controlled epitaxy opened the way to utilize SiC for active devices. Still, there are many interesting phenomena in this subject both from scientific and engineering points of view. Gas source MBE with the technique of alternate gas supply will give much information on atomic layer epitaxy.

References

1. R.B.Campbell and H.C.Chang, Semiconductors and Semimetals 7 (Academic Press, 1971) p.625.
2. E.Pettenpaul, W.von Muench and G.Ziegler, Inst. Phys. Conf. Ser. No.53 (The Inst. Physics, 1980) p.21.
3. H.Matsunami, Thin Films from Free Atoms and Particles (Academic Press, 1985) p.301.
4. H.Matsunami, Optoelectronics-Devices and Technologies, 2 (1987) 29.
5. H.Matsunami, S.Nishino and H.Ono, IEEE Trans. Electron Devices ED-28 (1981) 1235.
6. H.Matsunami, Mat. Res. Symp. Proc. 116 (1988) 325.
7. H.Matsunami, Heterostructures on Silicon; One Step Further with Silicon (Kluwer Academic Publisher, 1989) p.311.
8. S.Nishino, H.Suhara, H.Ono and H.Matsunami, J. Appl. Phys., 61 (1987) 4889.
9. K.Shibahara, S.Nishino and H.Matsunami, J. Crystal Growth, 78 (1986) 538.
10. K.Shibahara, S.Nishino and H.Matsunami, Appl. Phys. Lett., 50 (1987) 1888.
11. K.Shibahara, T.Takeuchi, S.Nishino and H.Matsunami, Jpn. J. Appl. Phys., 28 (1989) 1341.
12. N.Kuroda, K.Shibahara, W.S.Yoo, S.Nishino and H. Matsunami, Ext. Abstr. 19th Conf. Solid State Devices and Materials (Tokyo, 1987) p.227.
13. H.Matsunami, T.Ueda and H.Nishino, Proc. Mat. Res. Soc. Symposium 162 (1990) 397 .
14. T.Kimoto, H.Nishino and H.Matsunami, to be submitted.
15. T.Ueda, H.Nishino and H.Matsunami, J.Cryst. Growth 104 (1990) 695.

16. W.S.Yoo, S.Nishino and H.Matsunami, J. Crystal Growth 99 (1990) 278.
17. A.Yamashita, W.S.Yoo, T.Kimoto and H.Matsunami, to be submitted.
18. T.Fuyuki, M.Nakayama, T.Yoshinobu, H.Shiomi and H. Matsunami, J.Cryst. Growth 95 (1989) 461.
19. T.Yoshinobu, M.Nakayama, H.Shiomi, T.Fuyuki and H. Matsunami, J. Cryst. Growth 99 (1990) 520.
20. T.Yoshinobu, I.Izumikawa, H.Mitsui, T.Fuyuki and H. Matsunami, Appl. Phys. Letters, to be published.

$Si_{1-y}C_y$ Alloys – Extending Si-Based Heterostructure Engineering

S.S. Iyer, K. Eberl, M.S. Goorsky, J.C. Tsang, F.K. LeGoues, F. Cardone, and B.A. Ek

IBM Research Division, T.J. Watson Research Center,
P.O. Box 218, Yorktown Heights, NY 10598, USA

Abstract: $Si_{1-y}C_y$ alloy single layers and strained layer superlattices on Si (100) with C concentrations of up to a few percent have been synthesized using Solid Source Molecular Beam Epitaxy. While the presence of even small surface contaminants containing carbon disrupts Si epitaxy, we show that co-evaporated Si and C can yield defect-free epitaxial films with C concentrations up to 5 atomic %. Critical parameters include growth temperature, the total Si growth rate and the atom flux ratio. Outside the acceptable process window, the films are highly twinned, and in some cases amorphous. In addition, growth temperature also plays a significant role in preventing twinning and islanding. Low growth temperature also suppresses the precipitation of β-SiC and leads to the formation of pseudomorphic $Si_{1-y}C_y$ random alloys. Secondary Ion Mass Spectroscopy, X-ray Diffraction, Raman Spectroscopy, and electron microscopy have been used to verify that the layers are indeed substitutional alloys without silicon carbide precipitation. We have also studied the thermal stability of these layers and find that the layers degrade by qualitatively different mechanisms: interdiffusion at low temperatures, and silicon carbide precipitation at high temperatures. The ability to synthesize and process these alloy layers in combination with Si and $Si_{1-x}Ge_x$ layers, allows for exploitation of the ternary system and the possibility of more flexible bandgap engineering in a Si-based technology.

1. Introduction

Heterostructures have been used to demonstrate significant leverage in a silicon-based technology. $Si_{1-x}Ge_x$ alloys and Strained Layer Superlattices (SLS), have been used to fabricate Heterojunction Bipolar Transistors (HBTs) [1, 2] as well as other devices such as MOS devices resonant tunneling diodes, strained layer superlattice quantum well detectors for long wavelength and IR applications. The $Si_{1-x}Ge_x$ system allows one to integrate a smaller bandgap semiconductor with Si. In addition, one can engineer both Ge composition and strain independently to obtain a variety of bandgaps and band alignments. However, it is not possible to exceed the bandgap of pure unstrained silicon

Springer Proceedings in Physics, Vol. 71 Amorphous and Crystalline Silicon Carbide IV
Editors: C.Y. Yang · M.M. Rahman · G.L. Harris

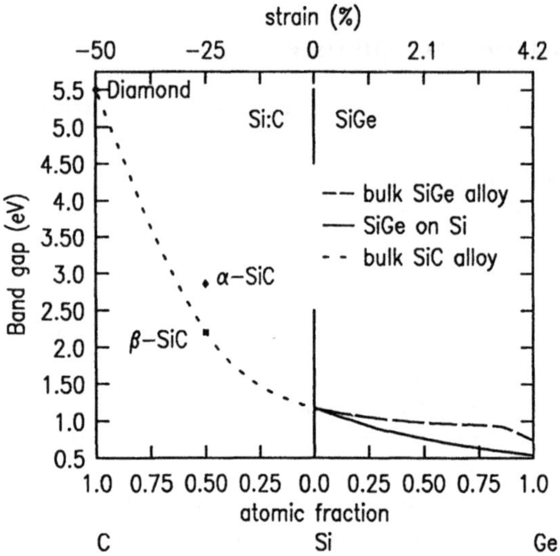

Fig. 1: Measured bandgap of $Si_{1-x}Ge_x$ alloys as a function of Ge content for bulk (upper curve) alloys and layers pseudomorphic to Si (lower curve), and the estimated bandgap of $Si_{1-y}C_y$ alloys as a function of C content. The estimate is based on a spline fit through the gaps for Si, β-Si and diamond. Also shown is the gap for α-SiC. The top scale shows the strain in the film for growth on Si.

(1.17 eV at 300K) using this system. To have a similar flexibility in integrating a Si-compatible wide bandgap semiconductor, a number of approaches have been attempted over the past few years: GaP [3], Semi-Insulating POlycrystalline-Silicon (SIPOS) [4-6], Oxygen-Doped Si Epitaxial Films (OXSEF) [7], β-SiC [8, 9], and amorphous and microcrystalline silicon (α-Si and μc-Si) [10, 11]. These approaches are severely limited because the material is not always epitaxial and high quality Si cannot be grown on these materials. In addition, many of these approaches require the use of high temperature growth techniques that may be incompatible with advanced VLSI processing trends. A more comprehensive treatment of wide-bandgap materials may be found in other papers in this proceedings.

The Si-C alloy system offers the possibility of a wide-bandgap material compatible with Si. Fig. 1 shows the bandgap for elements in column IV of the periodic table. We also show the projected bandgap for the Si-C system. The Si-C layer is assumed to be strain-free in this heuristic projection. Note that diamond has a bandgap of about 5.5 eV, and β-SiC 2.2 eV. However, unlike the Si-Ge system, Si and C are not miscible at all. In fact the solubility of C in Si even at 1420°C is barely 10^{-6}. Furthermore, stoichiometric silicon carbide is a stable compound phase and exists in several phases and polytypes, the most notable being β-SiC (zinc blende structure) and α-SiC (hexagonal structure) with bandgap of 2.9 eV.

Also shown in Fig. 1 is the lattice constant for the unstrained alloy. Note that while the misfit for pure Ge grown on Si is about 4.2 %, diamond has a misfit of about 50%. Thus both phase separation and high mismatch are two challenges posed by this system. At the outset we recognize that we will be limited to relatively small percentages of C incorporation. In terms of strain, when grown on silicon, a pseudomorphic $Si_{1-y}C_y$ (y = 0.01) layer has a comparable magnitude of strain to a $Si_{1-x}Ge_x$ (x = 0.12) layer. The $Si_{1-y}C_y$layer is under tension while the $Si_{1-x}Ge_x$ layer is in compression.

The bandgap of such a strained $Si_{1-y}C_y$ layer is not easily determined. By analogy to the Si/SiGe system, both composition and tetragonal strain will play a role. Diamond has a band structure quite similar to Si. Thus we do not expect the X-L crossing as seen in the SiGe system. However, the effect of strain will be to lift the 6-fold conduction band degeneracy. Strain, therefore, is expected to reduce net bandgap increase.

Posthill *et al.* [12, 13] have reported the use of Remote PECVD to grow a fairly thick alloy of over 7 μm with total C concentrations of about 3.5%. We estimate from their x-ray diffraction data that approximately 0.2 % C is substitutional. Other attempts to grow such alloys from the gas phase invariably result in the deposition of the carbide phase, or very rough films, though very good morphology of β-SiC Strained Layer Superlattices (SLS) have been reported by Sugii *et al.* [14] on Si (111) using gas source MBE at 900°C.

We have used solid source Molecular Beam Epitaxy (MBE) to synthesize $Si_{1-y}C_y$ layers and SLS with C concentrations of a few percent. We have used elemental C sources from resistively heated graphite filaments, or by electron beam evaporation. Si was evaporated from an electron beam evaporator. The two dimensional growth of Si in the presence of surface carbon is a challenge and strides in Si epitaxy have, in fact, been a consequence of the reduction of the C background in the reactor both for MBE and CVD techniques. Early work [15] shows that C on the surface especially at moderate to high temperatures results in the formation of SiC particles that pin the motion of steps. This in turn leads to the disruption of epitaxy and in severe cases to 3-dimensional nucleation of Si on the growing surface. The reduction of C background in the reactor has been a prime consideration for growing high quality epitaxial films both by CVD and MBE. We show in this work that the use of elemental beams combined with low temperature growth can overcome these difficulties to a large extent.

2. Growth of $Si_{1-y}C_y$ alloys and SLS

The tendency to island or nucleate in three dimensions depends on several factors including interface and surface free-energy considerations and strain. The presence of nucleating centers such as C or SiC particles on the growing surface will enhance this tendency. Low temperature growth inhibits the tendency to island by reducing the surface mobility. Fig. 2 shows the Raman spectra of a series of samples grown at different temperatures between 450 and 700°C. Raman spectroscopy has been a valuable tool in the study of Si/SiGe structures. In the Si-C system, it can be used as a quick probe to determine the bonding configuration of C in the Si matrix. C that is tetrahedrally bonded to Si as in silicon carbide would have a a phonon mode at about 850 cm^{-1}. C that is substitutional in the Si matrix, would have a phonon mode near 600 cm^{-1}, close to the substitutional B phonon mode at about 620 cm^{-1}. In all cases we see no evidence for the carbide mode. The relative intensities of the substitutional Si-C mode indicate a maximum at a growth temperature of about 525°C. We have shown in earlier work [16] that growth at higher tempertaure leads to island growth. In that case, the Raman probe (488nm)samples a smaller volume of the layer since it samples only a fixed depth. This accounts for the reduced Raman intensity of this mode at high growth temperatures. At low temperatures, we have observed a progressive disruption of epitaxy,

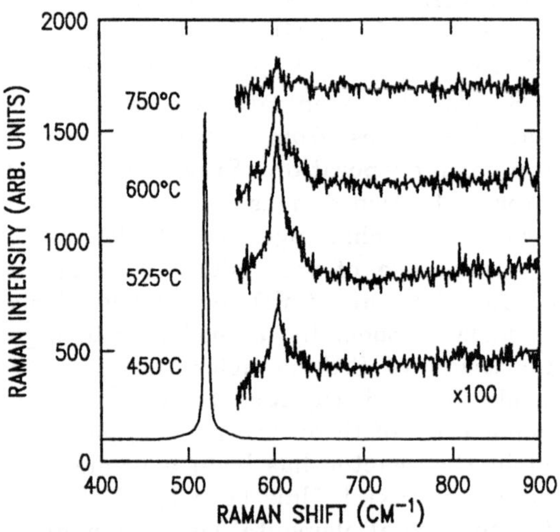

Fig. 2: Raman spectra for 50 nm $Si_{1-y}C_y$ (y=0.02) grown on Si (100) at temperatures between 450 and 750°C. Also shown is Si substrate peak for comparison. The Si-C vibration at 600 cm^{-1} is from substitutional C in Si. No features from C tetrahedr bonded to Si is seen at 850 cm^{-1}.

1nm $Si_{0.95}C_{0.05}$ / Si cap layer

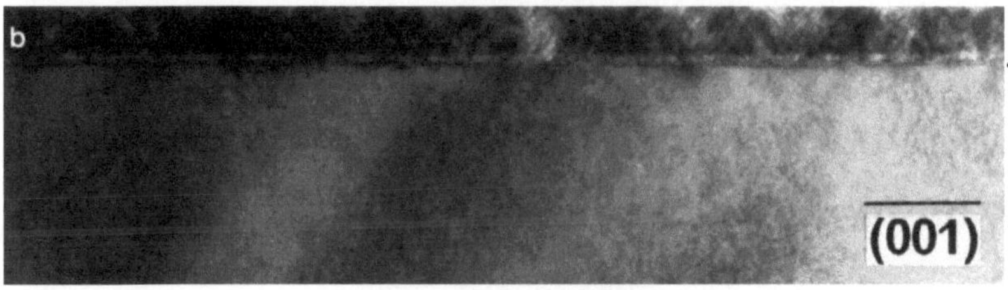

6nm $Si_{0.95}C_{0.05}$ / Si cap layer

Fig. 3: Cross sectional micrographs of two $Si_{1-y}C_y$ (y = 0.05) layers (a) 1 nm and (b) 6 nm. The layers wre capped with 25 nm of Si. The thicker film shows evidence of misfit dislocations but the thinner is free of misfit defects.

first via twin formation and then to amorphous growth. TEM observations confirm that breakdown of epitaxy has occurred in the layer grown at 425°C.

Fig. 3 shows cross sectional Transmission Electron Micrographs (TEM) of two $Si/Si_{1-y}C_y$ (y = 0.05) layers grown at 550°C for thicknesses of (a) 1 nm and (b) 6 nm. The Si cap thickness in both cases was 25 nm. The mismatch at this C concentration corresponds to about 2.5%. The Mathew-Blakeslee critical thickness[17] for this mismatch is about 1-2 nm. Clearly the 1 nm film is within this limit and shows no evidence of misfit dislocations from TEM. The thicker 6 nm film shows several misfit dislocations that propagate into the cap layer as well. In both cases however, no evidence of silicon carbide precipitates is observed. We show later that such precipitates can be observed after annealing at high temperatures, confirming the sensitivity of TEM to detect the microcrystallites.

Fig. 4 shows the cross sectional TEM for a $Si(26nm)/Si_{1-y}C_y(4nm)$ SLS. Fig. 5 shows a double crystal X-ray rocking curve for the superlattice. The first 10 layers were grown with y = 0.0086 and the subsequent layers were grown with y = 0.015. The X-ray diffraction also shows two peaks labelled a and b in Fig. 5 that correspond to these two concentrations, which were verified by SIMS. The contrast in the TEM follows this C concentration profile. The oxygen concentration

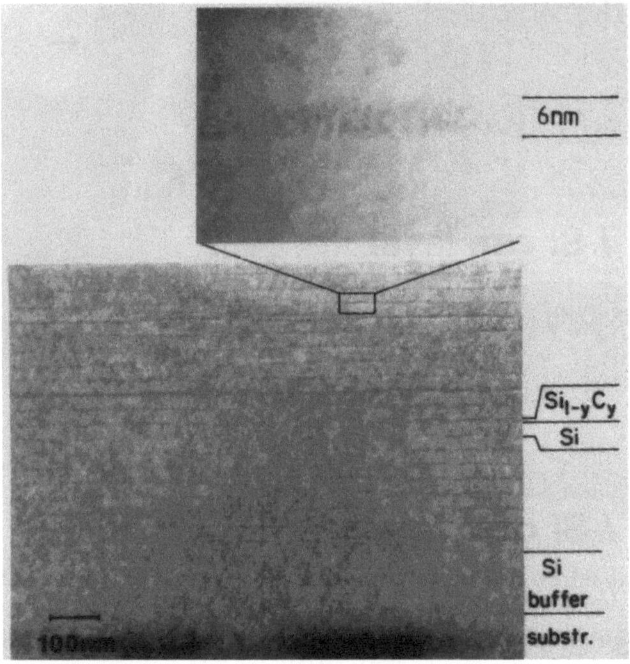

Fig. 4: TEM cross section for a Si(24 nm)/$Si_{1-y}C_y$(y = 0.009/0.015, 6 nm) strained layer superlattice. Inset shows a high resolution image.

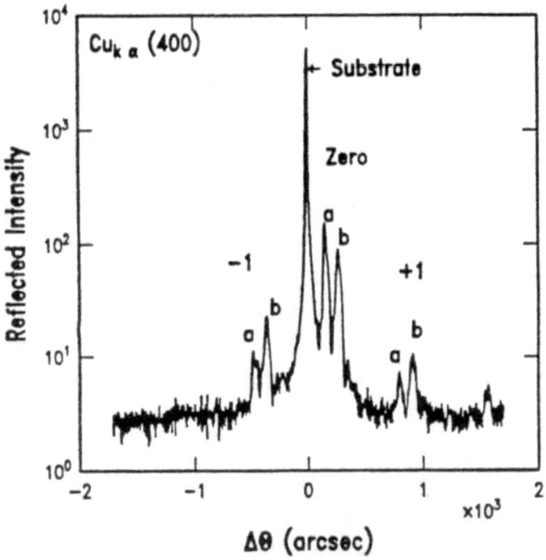

Fig. 5. Double crystal X-Ray diffraction (400) of the superlattice shown in Fig. 4. The peaks for the Si substrate (400) reflections and for the two distinct $Si_{1-y}C_y$ (a) y = 0.0086 and (b) y = 0.015 epitaxial layers corresponding to SIMS observations (not shown).

in the films is below that in the substrate and comparable to that in Si films grown under similar conditions. The defect density in the lower concentration part of the film is below TEM detection levels. Some threading dislocations have been observed to originate in the layer with the highest C concentration (layer 12) and these propagate to the surface. The growth temperature in this case was about 500°C and the growth rate was 0.4nm/sec.

Increasing the C concentration to even higher levels results in a highly twinned crystal growth. We have found that if the C flux is increased beyond a critical value without increasing the Si flux a transition to amorphous growth can occur even at 500°C. Thus the Si flux, the C/Si flux ratio, and the growth temperature define an acceptable process window where good quality $Si_{1-y}C_y$ films may be grown. We also expect that this window is larger when single alloy layers rather than superlattices are to be grown.

III. Thermal stability of $Si/Si_{1-y}C_y$ SLS

Thermal stability of these layers is important especially for VLSI processing. We have annealed as-grown $Si/Si_{1-y}C_y$ SLS at temperatures from 600° C to 1100° C. We consider here the case of a 20 period $Si(25\ nm)/Si_{1-y}C_y(5\ nm)$ ($y = 0.003$) SLS. Fig. 6 shows X-ray diffraction results for the as grown sample and those annealed at 900°C and 1100°C for one hour. Fig. 7 shows the SIMS profiles for these conditions. The x-ray results show quite distinctly, both the (400) Si peak (mainly from the substrate) and the peak corresponding to the average

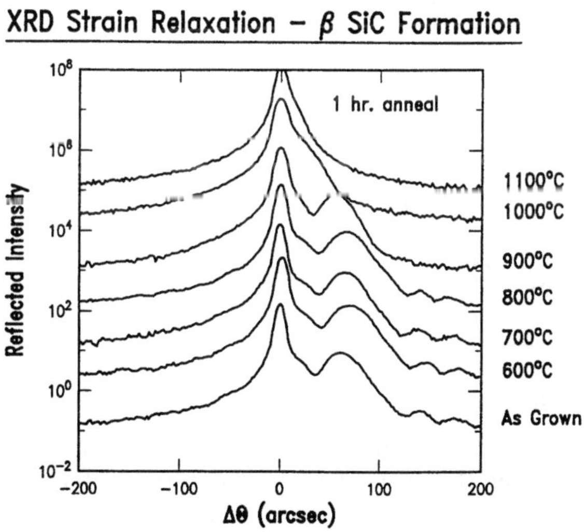

Fig. 6: X-ray rocking curves for a $Si(25\ nm)/Si_{1-y}C_y$ ($y = 0.003$, 5 nm) SLS as-grown and after annealing at different temperatures.

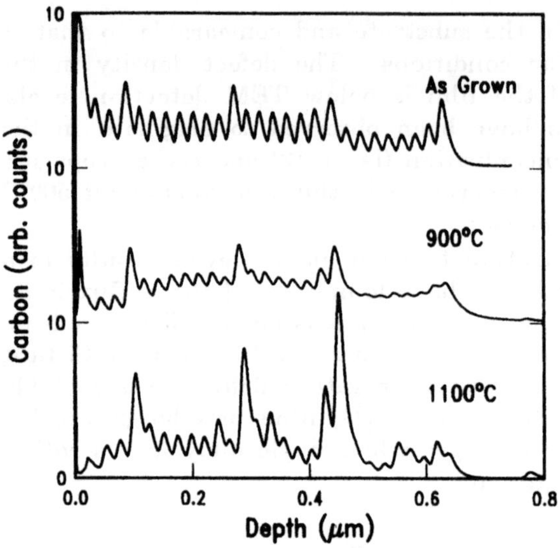

Fig. 7: SIMS profiles for C for the SLS of Fig. 6 showing breakdown by interdiffusion at 800°C and silicon carbide precipitation at 1100°C. The curves are offset for clarity.

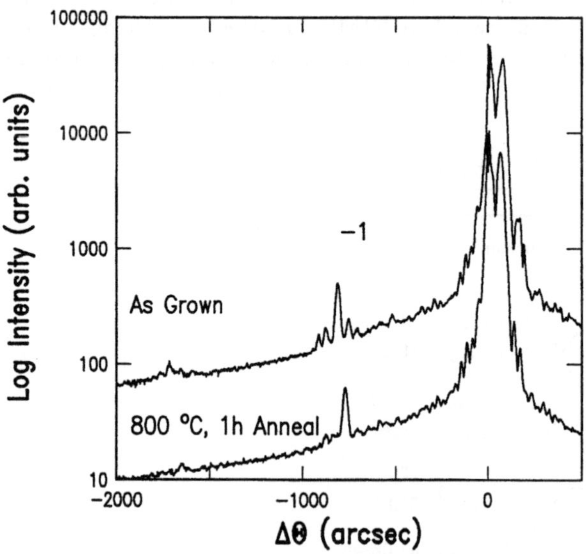

Fig. 8: High resolution X-Ray Diffraction, (a) as-grown and (b) after 800°C, 60 minute anneal showing the loss of the pendellosung fringes -1 order $Si_{1-y}C_y$ peak.

strain in the SLS. Higher resolution scans, shown in Fig. 8, over wider angle show well resolved higher order peaks with well defined pendellosung fringes for the as-grown SLS. This implies that the interfaces between the Si and $Si_{1-y}C_y$ layers are sharp and well defined, to a far better degree than can be ascertained by SIMS. Note also that we have intentionally put several marker layers of somewhat higher C concentration, the purpose of which is to study the effect of increase C concentration on thermal stability. First we consider the 900°C results. X-Ray diffraction shows that the average level of strain has not changed. However, with wider-angle high resolution scans, the fringes disappear. This is consistent with the loss of the sharply defined $Si_{1-y}C_y$ layers mainly due to interdiffusion. This conclusion is also borne out from the SIMS results which show a smearing of the profiles especially for the lower C levels. At 1100° C, not only do the higher order peaks disappear (not shown in Fig. 8), but the peak corresponding to the average strain in the layer moves towards the substrate peak. Since the SIMS profile shows that the carbon is segregating to the layers with high initial concentration, the loss of strain is most likely due to SiC precipitation, which represents the thermodynamically stable phase for this level of C in Si. Indeed we have observed such precipitates in TEM. We conclude that the SLS relaxes by interdiffusion at temperatures below 900°C. At temperatures above 1000°C, precipitation presumably of SiC, is the preferred mode of relaxation. Interestingly, misfit dislocation generation does not appear to be a preferred mode of relaxation.

IV. Conclusions

Our ability to grow high quality $Si_{1-y}C_y$ alloy and SLS layers adds a new dimension to Si-based bandgap engineering with the possibilit for integration of a potential wide bandgap semiconductor with Si technology. When combined with the SiGe system, several versatile and novel band configurations can be made possible. The key features of our growth technique are the use of elemental beams, the control of the C/Si flux ratio, the total Si flux and the growth temperature. Pseudomorphic $Si_{1-y}C_y$ layers with y upto 0.05 have been grown under the optimal conditions. If the C flux is allowed to increase without corresponding increase in the Si flux, or if too low a temperature is employed, twinned or even amorphous growth can result. Also, at high temperatures there is a tendency of the films to island and form SiC precipitates. Our layers are stable to about 800°C. Thereafter, breakdown occurs via two modes: interdiffusion at lower temperatures (≤900°C) and silicon carbide precipitation at higher temperatures (≥1000°C). The carbide precipitation process is nucleated at regions of pre-existing high C content and grow at the expense of the dissolved C.

REFERENCES

[1] S.S. Iyer, G.L. Patton, J.M.C. Stork, B.S. Meyerson, and D.L. Harame, IEEE Trans. Electron. Dev., **36**, 2043-2064, (1989).

[2] G.L. Patton, D.L. Harame, J.M.C. Stork, B.S. Meyerson, G.J. Scilla, and E. Ganin, IEEE Electon Device Lett., **11**, 171-3, (1990).

[3] S.L. Wright, H. Kroemer, and M. Inada, J. Appl. Phys., **55**, 2916-2927, Apr (1984).

[4] Y.H. Kwark and R.M. Swanson, Solid-State Electronics, **30**, 1121-1125, Nov (1987).

[5] T. Matsushita, H. Hayashi, N. Oh-Uchi, and H. Yamoto, Jap. J. Appl. Phys. (suppl.), **20**, 75-81, (1980).

[6] E. Yablonovitch and T. Gmitter, IEEE Electron Device Lett., **EDL-6**, 597-599, Nov (1985).

[7] M. Takahashi, M. Tabe, and Y. Sakakibara, IEEE Electron Device Lett., **EDL-8**, 475-476, Oct (1987).

[8] T. Sugii, T. Ito, Y. Furumura, M. Doki, F. Mieno, and M. Maeda, IEEE Electron Device Lett., **9**, 87-89, Feb (1988).

[9] K. Sasaki, T. Fukazawa, and S. Furukawa, in IEDM Tech. Dig., **1987**, 186-189, Dec (1987).

[10] J. Symons, M. Ghannam, A. Neugroschel, J. Nijs, and R. Mertens, Solid-State Electronics, **30**, 1143-1145, Nov (1987).

[11] H. Fujioka, T. Deguchi, K. Takasaki, and T. Takada, in IEDM Tech. Dig., **1987**, 574-577, Dec (1988).

[12] J.B. Posthill, R.A. Rudder, S.V. Hattangady, G.G. Fountain, and R.J. Markunas, Appl. Phys. Lett., **56**, 734-736, February (1990).

[13] J.B. Posthill, R.A. Rudder, S.V. Hattangady, G.G. Fountain, T.P. Humphreys, R.J. Nemanich, N.R. Parikh, and R.J. Markunas, Symp. Mat. Res. Soc., **198**, 497-502, (1990).

[14] T. Sugii, T. Aoyama, and T. Ito, J. Electrochem. Soc., **137**, 989-992, March (1990).

[15] H.C. Abbink, R.M. Broudy, and G.P. McCarthy, J. Appl. Phys., **39**, 4673-4681, (1968).

[16] S.S. Iyer, K. Eberl, M.S. Goorsky, F.K. Le Goues, and B.A. Ek, Proc. Mat. res. Symp., **220**, (1991).

[17] J.M. Matthews and A.E. Blakeslee, J. Crystal Growth, **27**, 118-125, (1974).

Investigation of the Growth of 3C-SiC and 6H-SiC Films on Low-Tilt-Angle Vicinal (0001) 6H-SiC Wafers

J.A. Powell[1], *D.J. Larkin*[1], *J.B. Petit*[2], *and J.H. Edgar*[3]

[1]NASA Lewis Research Center, Cleveland, OH 44135, USA
[2]Sverdrup Technology, Inc., Lewis Research Center Group, Brook Park, OH 44142, USA
[3]Kansas State University, Manhattan, KS 66506, USA

Abstract. Advances have been made in the understanding of the mechanisms that control polytype formation during the chemical vapor deposition of epitaxial SiC films on vicinal (0001) 6H-SiC wafers. Processes based on this new understanding have been developed. For example, 3C-SiC and 6H-SiC films were grown on separate 1-mm-square mesas on the same 6H-SiC substrate. Most of the 3C films were free of double positioning boundaries (DPBs), and had a reduced density of stacking faults. An important aspect of these results was an in situ pregrowth HCl/H_2 etch which eliminated unwanted 3C-SiC nucleation.

1. Introduction

Chemical vapor deposition (CVD) is a key process in the fabrication of semiconductor devices. Until recently, SiC CVD processes typically required temperatures above 1550°C [1-3]. This growth was carried out on the (0001) SiC basal plane, either the Si face or C face. In 1973, it was found that 6H-SiC could be grown on 6H-SiC in the temperature range 1320-1390°C when the growth direction was perpendicular to the [0001] direction [4]. The significance of this "90°-off-axis" was not appreciated until much later. Recently, (1) 3C-SiC films were achieved on the basal planes of 6H-SiC in the range 1350-1550°C [5-10], and (2) 6H-SiC films were achieved on "off-axis" 6H-SiC substrates that were oriented several degrees off the basal plane [6-8,11,12]. All of the above growth results were achieved on Acheson or Lely crystals. High quality 6H-SiC films were also grown on "off-axis" 6H-SiC wafers that were produced from sublimation-grown boules [13]. Very recently, regions of both 3C and 6H were grown on the same 6H substrate [14]. This paper reviews growth models that have been proposed to explain the above growth results, and describes further work directed toward achieving the controlled growth of high quality 3C and 6H films on 6H substrates.

2. CVD Growth Models

To explain the success of the 6H CVD at lower temperatures using "off-axis" 6H-SiC substrates, a model was suggested [7,8] whereby the density of atomic-scale steps on a vicinal (0001) 6H substrate determines the polytype of the grown film. Refer to fig. 1(a), a schematic cross-sectional view of a vicinal (0001) SiC crystal. The step density increases and the width of terraces between steps decreases as the tilt angle of the growth surface increases. According to the model, at tilt angles greater than

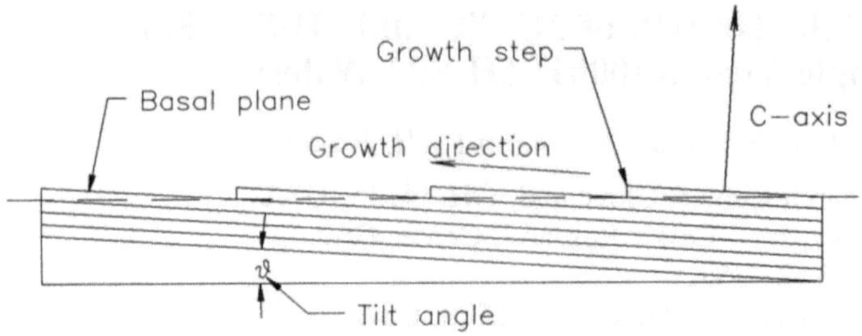

(a) Substrate showing surface growth steps.

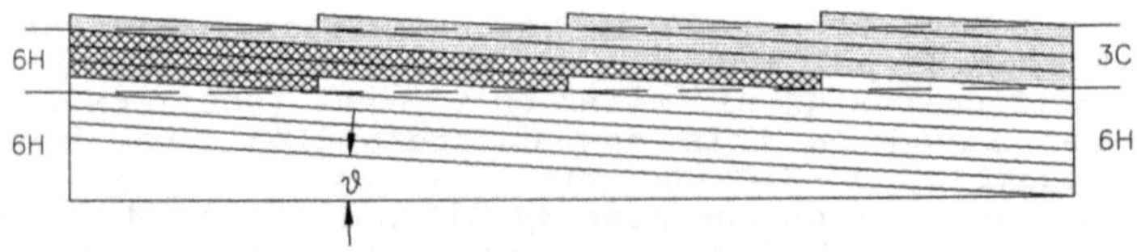

(b) Substrate with 3C and 6H film growth.

Fig. 1. Schematic cross section of a vicinal (0001) 6H-SiC substrate.

1.5°, arriving Si and C atoms can easily migrate to steps where lateral growth occurs parallel to the basal (0001) plane. In this case, grown films assume the ABCACB... stacking sequence of the 6H substrate. However, at smaller tilt angles (i.e. < 1.5°), terraces are larger and migration of Si and C atoms to the steps is less likely. Instead, nucleation of 3C occurs on the terraces. Depending on which terrace nucleation takes place, either the ABC... or the ACB... stacking sequence of the 3C polytype can occur. Coalescence of islands of both sequences create the double positioning boundaries (DPBs) that are commonly seen in 3C films grown on 6H substrates. Also according to the model, 6H films are achieved on the as-grown faces of Lely 6H crystals (very low tilt angles) at 1800°C because surface migration is more active and the Si and C atoms reach the growth steps.

Recent results at NASA Lewis [14] demonstrate that the above model [7,8] is not sufficient to explain the control of polytype in the CVD growth on low-tilt-angle vicinal (0001) 6H-SiC substrates. In these results [14], 6H films were grown on vicinal (0001) 6H substrates with tilt angles as small as 0.1°. In addition, 3C growth was intentionally produced on the low-tilt-angle substrates by introducing dislocations in the substrates with a diamond scribe. The 3C growth proceeded laterally from the point of intentional nucleation in a direction perpendicular to the growth steps. An essential part of this new result was a pregrowth etch that eliminated all sites of unwanted 3C nucleation. A new growth model [14] that explains the growth of SiC polytypes on vicinal (0001) SiC substrates is as follows. At moderate CVD growth temperatures (1300 to 1500°C), Si and C atoms can migrate to surface steps where lateral growth occurs even at tilt angles as small as 0.1°. In this case, the film will be the same polytype as the substrate. If there are defect sites on the surface (e.g. dislocations, contamination, etc.), then 3C will nucleate at these sites because 3C is the more stable polytype at these lower temperatures.

A practical application of the new growth model would be the growth of DPB-free 3C-SiC on vicinal (0001) 6H-SiC substrates. This would be very useful since 3C substrates have not been produced and there are no existing methods for producing low-defect 3C films on other substrates. In a proposed method [14] for growing 3C on 6H, the vicinal (0001) 6H substrate surface is divided into device-size (e.g. 1-mm-square) mesas. After a proper pregrowth etch to remove unwanted sites of 3C nucleation, a 3C "seed" is intentionally produced at a suitable site on each mesa. Then, a 3C film is grown laterally across the mesa from this 3C "seed". The proposed lateral growth is shown schematically in fig. 1(b). Note that during growth, both 3C and 6H grow laterally from steps until the 3C completely covers the mesa. This paper will present recent results which demonstrate the simultaneous growth of both 3C and 6H films on different areas of the same 6H substrate.

3. Experimental Procedures

The 6H-SiC wafers used in this work were produced by Cree Research Inc. [15] from sublimation-grown boules. The wafers were 25.4 mm in diameter and approximately 250 μm thick with one side polished. Tilt angles, which ranged from 0.1° to 0.6°, were measured to an accuracy of better than ±0.1° (for the smaller tilt angles) by an x-ray/laser technique (unpublished). The tilt directions were random and were measured to an accuracy of better than ±5°. The experiments reported herein used only Si face substrates. Because of limited availability, each wafer was cut into nine rectangular substrates, 6 x 5.5 mm^2. The growth surface of some substrates was further divided by cutting grooves in the surface with a dicing saw to form mesas that were 1 mm square. Before being used, substrates were sequentially cleaned with organic solvents, hot H_2SO_4, scrubbed with liquid detergent, rinsed with 18 MΩ-cm water, and then blown dry with nitrogen.

Pregrowth etching of the substrates and growth of the SiC films were carried out in a horizontal water-cooled CVD chamber. The H_2 carrier gas flow was maintained at 3 l/min, at one atm, in the 50 mm I.D. chamber. The substrates were heated by a rf-heated SiC-coated graphite susceptor, 80 mm long x 30 mm wide x 10 mm thick. Prior to growth, the substrates were subjected to a pregrowth HCl etch (3% in H_2) in order to eliminate sites of 3C nucleation. The temperature and time of an effective etch were determined to be about 1350°C and 25 min, respectively. Film growth was carried out at 1450±25°C with SiH_4 (300 ppm) and C_3H_8 (150 ppm) as the sources of Si and C, respectively. Typically, the growth rate was about 4 μm/hr.

The grown films were characterized in several ways. The polytype and distribution of DPBs and stacking faults in the films were mapped with optical microscopy after the films were decorated by a thermal oxidation process [16]. Based on differences in oxidation rates, the following conditions yielded "color maps" of the 3C and 6H regions: a <u>dry</u> oxidation of the Si face for 5 h at 1150°C yielded brown and light tan interference colors for 3C and 6H, respectively, and a <u>wet</u> oxidation of the C face for 0.5 h at 1150°C yielded dark blue and light blue for 3C and 6H, respectively. The DPBs in 3C films grown on the Si face were blue in color, contrasting sharply with the brown of the 3C film. In addition, the oxidation procedure also decorated stacking faults intersecting the film surface in 3C films.

4. Results and Discussion

The temperature and time of the pregrowth HCl etch were important parameters in controlling the polytype of the resulting film grown on vicinal (0001) 6H-SiC substrates. The presence of grooves in the surface also had a significant effect on the growth. Figure 2 illustrates some typical results. At low etch temperatures [fig. 2(a)], the resulting film was mostly 3C with a high density of DPBs (the many lines seen in the figure). When the etch temperature and time were increased as indicated for fig 2(b), the resulting film was mostly 6H. At higher temperatures [fig. 2(c)], the resulting film was also mostly 3C, but the density of DPBs was lower than in films etched at the lowest temperatures. The presence of grooves on the substrate [fig. 2(d)] illustrates that there is a significant edge effect in the etching process. Substrates of figs. 2(c) and 2(d) were processed at the same time but the presence of the grooves [fig. 2(d)] resulted in a film that was mostly 6H. This tendency toward more 6H film with grooved substrates was obtained for all etching conditions, except at the lowest temperatures. The mechanism of

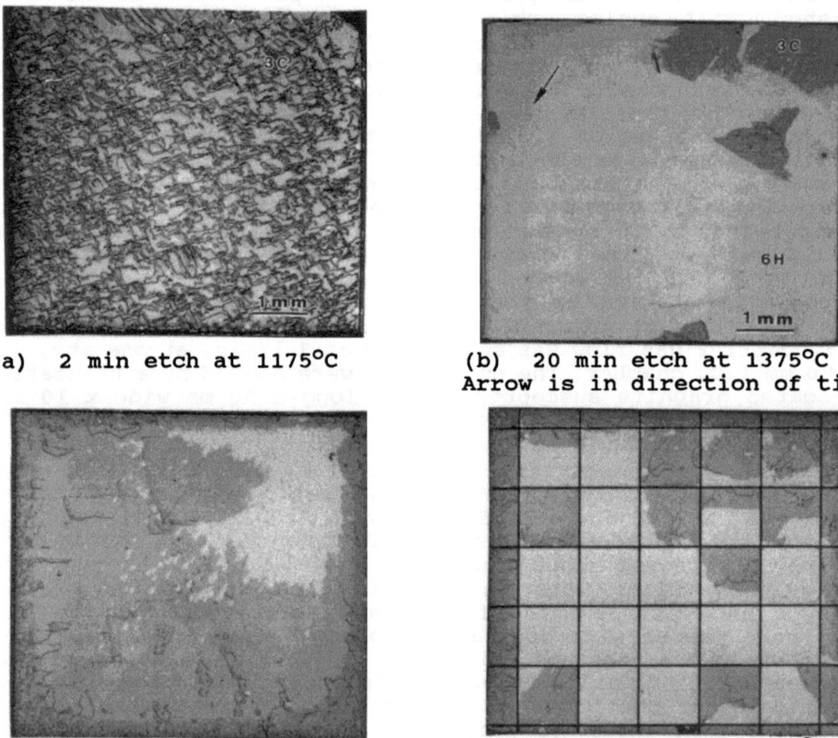

(a) 2 min etch at 1175°C

(b) 20 min etch at 1375°C
Arrow is in direction of tilt

(c) 10 min etch 1450°C, without grooves

(d) 10 min etch at 1450°C, with grooves

Fig. 2. Effect of time and temperature of the pregrowth HCl/H_2 etch on the polytype of SiC films grown on Si-face 0.2°-tilt angle vicinal (0001) 6H-SiC substrates. The film has been oxidized to "map" the polytypes: darker regions are 3C, lighter regions are 6H.

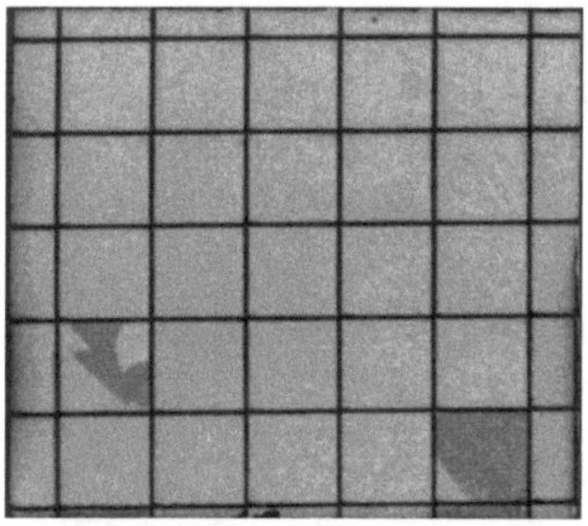

(a) 0.5° tilt angle, without intentional contamination

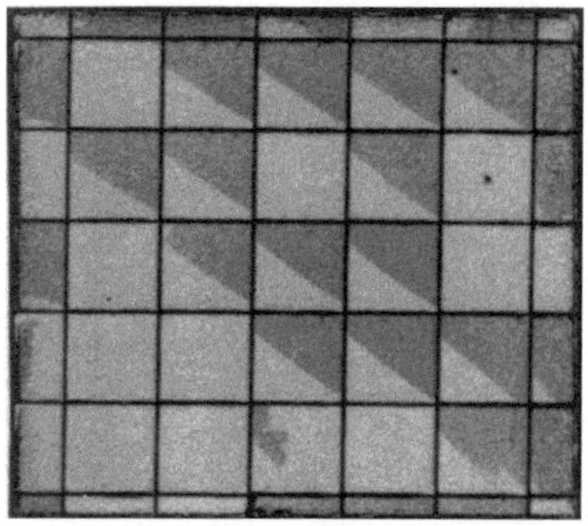

(b) 0.5° tilt angle, with intentional contamination

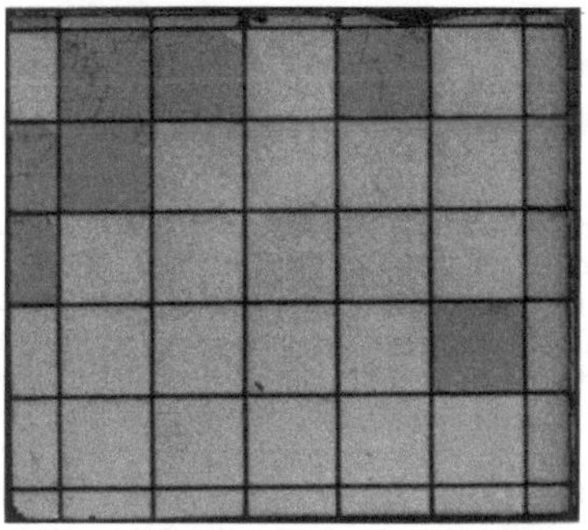

(c) 0.2° tilt angle, without intentional contamination

(d) 0.2° tilt angle, with intentional contamination

Fig. 3. Effect of tilt angle on the nucleation and lateral growth of 3C-SiC films on Si-face vicinal (0001) 6H-SiC substrates. Total film thickness: 8 μm. 1-mm-square mesas.

this edge effect is not understood at this time, but the presence of the grooves in the substrate was beneficial in controlling the nucleation of 3C or 6H.

In attempts to get a better understanding of the 3C nucleation process, the following experiments were carried out. Grooved substrates, with 0.2° and 0.5° tilt angles, were subjected to the "optimum" HCl pregrowth etch (25 min, 1350°C). In some experiments, SiC films were then grown immediately on the substrates with no further treatment. See figs. 3(a) and 3(c). As reported previously [14], there is a greater tendency for 3C to nucleate at lower tilt angles. For example, in fig. 3(a) (0.5°), two mesas are 3C, and in fig. 3(c) (0.2°), five mesas are 3C. Most 3C mesas are free of DPBs. This result is typical. In other experiments, the substrates were removed from the growth chamber for further treatment and examination (e.g. indenting the surface with a diamond scribe, examination with an optical microscope, etc.). Although indenting the surface with a diamond scribe did induce 3C nucleation, the resulting 3C films did contain some DPBs. We do not think that this is a viable approach to produce the lateral growth of 3C films.

The process of removing the substrates from the chamber after the pregrowth etch, exposing the substrates to a non-clean-room environment, and then returning the substrates to the chamber was

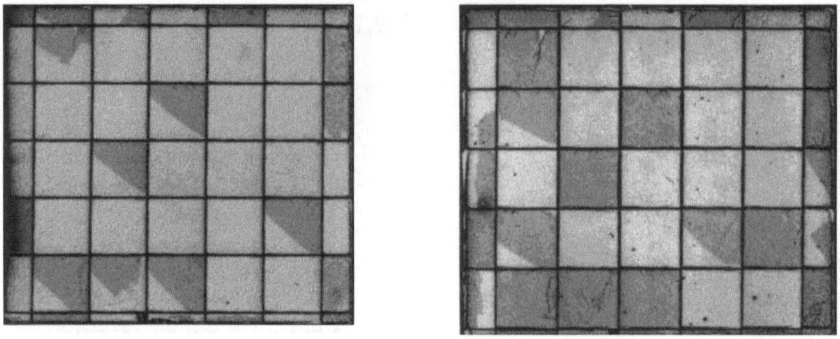

(a) Initial Growth (8 μm)   (b) Same substrate after additional growth (16 μm total)

Fig. 4. Demonstration of stopping and restarting 3C and 6H film growth on a Si-face, 0.5°-tilt-angle, vicinal (0001) 6H-SiC substrate. 1-mm-square mesas.

sufficient to increase the nucleation of 3C. The result of this extra handling and processing is shown in figs. 3(b) and 3(d). In each case, more than half of the mesas were 3C. In the substrate in fig. 3(b), 3C nucleation took place in the upper right hand corner of some mesas and then grew laterally from this location. Because of the surface tilt of this vicinal sample, the direction perpendicular to the growth steps was from the upper right to the lower left (i.e. the direction of lateral growth). The upper right was the location of the highest atomic plane on the mesa; the lateral growth of growth steps from this location eventually covers the whole mesa. The extent of lateral growth from this corner was 0.74 mm. This agrees fairly well with the value of lateral growth calculated by dividing the estimated film thickness (0.008 mm) by the tangent of the tilt angle (0.5°) (equal to 0.92 mm). In the case of the 0.2° substrate in fig. 3(d), the lateral growth was sufficient to cover the whole mesa after a two-hour growth. In all four samples of fig. 3, most of the 3C films are free of DPBs. The stacking fault density is also significantly reduced from that observed in 3C films with DPBs grown on 6H substrates.

The above growth experiments demonstrate that DPB-free 3C films can be grown on 1-mm-square mesas on vicinal (0001) 6H substrates. In order to grow 3C device structures using this approach, one must be able to stop and restart this CVD process. Figure 4 illustrates the results of such an experiment. After the growth shown in fig. 4(a), the sample was removed from the chamber, oxidized to reveal the 3C films, treated with HF to remove the oxide, and then returned to the chamber for another growth cycle. In seven of the squares, which were partially overgrown with 3C after the initial growth, the lateral growth continued and these squares were completely overgrown with 3C after the additional growth cycle. In three additional squares, 3C did nucleate in the second growth cycle. The second growth cycle did not introduce additional DPBs in the 3C films. This experiment did demonstrate that 3C and 6H film growth can be stopped and restarted without introducing additional DPBs in the 3C film.

The reduction of DPBs in most of the grown 3C films was accomplished by reducing <u>unwanted</u> 3C nucleation and forcing wanted 3C nucleation at desirable locations. The pregrowth etch presumably eliminates defects (i.e. 3C nucleation sites) caused by cutting and polishing the wafer. On each mesa, the location of the topmost atomic plane (the upper right position in figs. 3 and 4) was the site of 3C nucleation. Thus, reducing the number of 3C nucleation sites on each mesa reduced the probablility of DPB formation. In experiments so far, we have not correlated tilt direction with crystallographic direction. However, we have not observed a change from 6H growth to 3C growth during growth runs of up to 4 hours, as was observed by Ueda et al.[12] in growth on 6H substrates with tilt angles along certain crystallographic directions.

5. Conclusions

Factors that control the polytype formation of CVD-grown SiC films on vicinal (0001) SiC substrates include the tilt angle of the substrate and the defects (e.g. dislocations, contamination, etc.) on the growth surface. At small tilt angles, defects are the dominant factor. At growth temperatures of about 1450°C, defects are responsible for 3C nucleation. An effective pregrowth treatment for removing sites of unwanted 3C nucleation is HCl/H_2 etch at 1350°C for 25 min. Using a process of intentional 3C nucleation, films of 3C-SiC and 6H-SiC can be grown on different mesas on the same low-tilt-angle vicinal (0001) SiC substrate. Work does remain in getting better control of the 3C nucleation process.

References

1. R. B. Campbell and T. L. Chu, J. Electrochem. Soc. **113**, 825 (1966).
2. W. v. Muench and I. Pfaffeneder, Thin Solid Films **31**, 39 (1976).
3. S. Nishino, H. Matsunami, and T. Tanaka, J. Crystal Growth, **45**, 144 (1978).
4. J. A. Powell and H. A. Will, J. Appl. Phys. **44**, 177 (1973).
5. H. S. Kong, J. T. Glass, and R. F. Davis, Appl. Phys. Lett. **49**, 1074 (1986).
6. H. S. Kong, H. J. Kim, J. A. Edmond, J. W. Palmour, J Ryu, C. H. Carter, Jr., and R. F. Davis, in Materials Research Society Symposium Proceedings, edited by T. Aselage, D. Emin, and C. Wood (Materials Research Society, Pittsburgh, PA, 1987), Vol. 97, p. 233.
7. N. Kuroda, K. Shibahara, W. Yoo, S. Nishino, and H. Matsunami, Extended Abstracts 19th Conf. on Solid State Devices and Materials, Tokyo, 1987, pp. 227.
8. H. Matsunami, K. Shibahara, N. Kuroda, and S. Nishino, in Amorphous and Crystalline Silicon Carbide, ed. by G. L. Harris and C. Y.-W. Yang, Springer Proceedings in Physics, Vol. 34 (Springer-Verlag Berlin, Heidelberg, 1989), pp. 34-39.
9. H. S. Kong, J. T. Glass, and R. F. Davis, J. Mater. Res. **4**, 204 (1989).
10. J. A. Powell, D. J. Larkin, L. G. Matus, W. J. Choyke, J. L. Bradshaw, L. Henderson, M. Yoganathan, J. Yang, and P. Pirouz, Appl. Phys. Lett. **56**, 1353 (1990).

11. H. S. Kong, J. T. Glass, and R. F. Davis, J. Appl. Phys. **64**, 2672 (1988).
12. T. Ueda, H. Nishino, and H. Matsunami, J. Cryst. Growth **104**, 695 (1990).
13. J. A. Powell, D. J. Larkin, L. G. Matus, W. J. Choyke, J. L. Bradshaw, L. Henderson, M. Yoganathan, J. Yang, and P. Pirouz, Appl. Phys. Lett. **56**, 1442 (1990).
14. J. A. Powell, J. B. Petit, J. H. Edgar, I. G. Jenkins, L. G. Matus, J. W. Yang, P. Pirouz, W. J. Choyke, L. Clemen, and M. Yoganathan, Appl. Phys. Lett. **59**, 333 (1991).
15. Cree Research, Inc., Durham, NC 27713, USA
16. J. A. Powell, J. B. Petit, J. H. Edgar, I. G. Jenkins, L. G. Matus, W. J. Choyke, L. Clemen, M. Yoganathan, J. W. Yang, and P. Pirouz, Appl. Phys. Lett. **59**, 183 (1991).

Low Temperature Homoepitaxial Growth of 6H-SiC by VPE Method

T. Kimoto, H. Nishino[a], *A. Yamashita*[b], *W.S. Yoo, and H. Matsunami*

Department of Electrical Engineering, Kyoto University,
Yoshidahonmachi, Sakyo, Kyoto 606-01, Japan
Present address: [a]Fujitsu Corp., Kanagawa, Japan
[b]NEC Corp., Ibaragi, Japan

Abstract. Crystal growth of SiC on off-oriented 6H-SiC{0001} and 6H-SiC{01$\bar{1}$4} substrates was carried out at low temperatures of 1000-1500℃. Homoepitaxial growth of 6H-SiC is achieved at a temperature as low as 1200℃ governed by step-flow growth on off-oriented {0001} faces and at 1100℃ on (0$\bar{1}$1$\bar{4}$)C faces. The activation energy of growth rate shows a very small value of 3.0kcal/mole. This can be quantitatively analyzed on the basis of a stagnant layer model in which crystal growth is controlled by diffusion of reactants in a stagnant layer.

1. Introduction

6H-SiC is an attractive wide-gap semiconductor material for blue-light emitting devices, high-temperature and high-power devices. As for vapor phase epitaxial(VPE) growth of 6H-SiC, the authors' group has developed step-controlled epitaxy, by which single crystalline 6H-SiC with a very smooth surface can be grown on off-oriented 6H-SiC{0001} substrates at 1500℃[1-4]. The epitaxial growth at such a low temperature can be attributed to lateral growth from atomic steps, which leads to inheritance of the stacking order of substrates. TEM analyses, low-temperature photoluminescence and electrical measurements by other groups have also shown the excellent quality of grown layers[5-8].

Although the temperature for the homoepitaxial growth can be drastically reduced from 1800℃ to 1500℃ by step-controlled epitaxy, it is still high for device application. High-temperature process brings about impurity contamination from a growth system, redistribution of dopants and thermally induced damages. Therefore, further reduction of growth temperature is strongly needed. However, the growth of 6H-SiC at lower temperatures has never been reported.

Editors: C.Y. Yang · M.M. Rahman · G.L. Harris

In this paper, crystal growth of SiC by VPE growth at low temperatures is described. The effects of growth temperature and substrate polarity on the growth rate are investigated in detail. Based on these results, the growth mechanism such as kinetics of adsorbed species will be discussed.

2. Experiments

Crystal growth was carried out by an atmospheric pressure VPE method in a horizontal reaction tube using a SiH_4-C_3H_8-H_2 system. The growth temperature was varied in the range of 1000-1500℃. The flow rates of SiH_4, C_3H_8, H_2 were 0.15-0.30sccm, 0.10-0.20sccm and 3.0slm, respectively. The mean gas velocity on the susceptor is 6.0cm/s.

6H-SiC{0001} faces with 0-10° off-orientation toward [11$\bar{2}$0] and 6H-SiC{01$\bar{1}$4} were used as substrates. Off-oriented {0001} substrates were prepared by angle-lapping of a 6H-SiC basal plane grown by an Acheson's method. 6H-SiC{01$\bar{1}$4} substrates were obtained by a sublimation method. The 6H-SiC{01$\bar{1}$4} seeds were prepared on CVD(chemical vapor deposition)-grown 3C-SiC{001} films(~10μm thick) utilizing polytype conversion. The

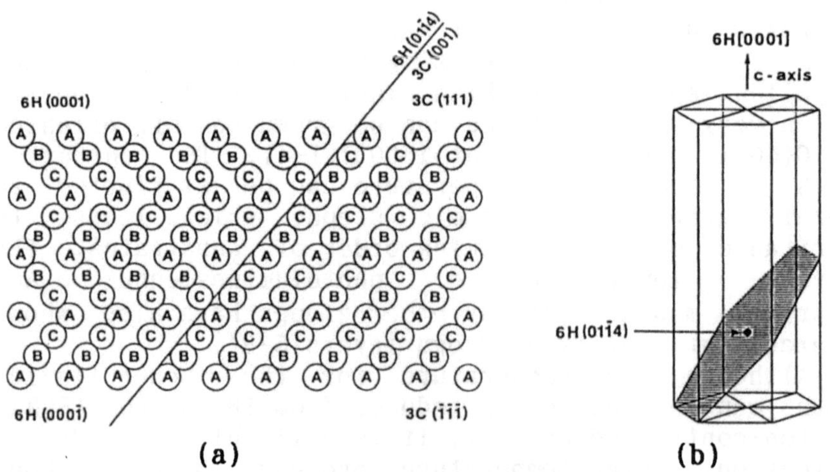

Figure 1 (a) The atomic arrangements of 6H-SiC(01$\bar{1}$4) and 3C-SiC(001) interface region. Each sphere represents a Si-C pair. (b) The relationship between 6H-SiC(01$\bar{1}$4) and the hexagonal unit cell.

details of polytype conversion and the ingot growth by sublimation are described elsewhere[9-11]. The atomic arrangements of 6H-SiC(01$\bar{1}$4) and 3C-SiC(001) interface region are schematically shown in Fig.1(a). The relationship between 6H-SiC(01$\bar{1}$4) and the hexagonal unit cell is illustrated in Fig.1(b). As seen in this figure, 6H-SiC{01$\bar{1}$4} has an angle of 54.74° to the basal plane of 6H-SiC{0001}. A 6H-SiC{01$\bar{1}$4} substrate has an advantage of its larger diameter because the original seed is a 3C-SiC(001) layer grown on Si(001).

Off-oriented {0001} substrates are either (0001)Si or(000$\bar{1}$)C faces. In this study, both faces were used to investigate polarity effect of substrates. For 6H-SiC{01$\bar{1}$4} faces, both (01$\bar{1}$4)Si and (0$\bar{1}$1$\bar{4}$)C faces were also employed.

3. Low Temperature Epitaxial Growth

As growth temperature decreases, the range of growth condition under which smooth surface is obtained becomes narrower. We investigated the effects of SiH_4 and C_3H_8 flow rate on the crystallinity of the grown layers and determined the optimum condition. For low temperature epitaxial growth, lower SiH_4 flow rate(corresponds to lower supersaturation) and lower Si/C ratio are preferable. Here, the former is effective to suppress two- or three-dimensional nucleation, and the latter compensates for lower decomposition ratio of C_3H_8 at low temperatures.

Figure 2 shows Nomarski micrographs and RHEED patterns of layers grown on off-oriented (000$\bar{1}$)C faces at 1200 and 1100℃, respectively. The flow rates of SiH_4 and C_3H_8 are 0.15 and 0.14sccm, and off-angle of substrates is 6° . The quite similar results are obtained on (0001)Si faces. At the temperature as low as 1200℃, homoepitaxial growth of 6H-SiC can be achieved. This indicates that the lateral growth from atomic steps occurs even at 1200℃ and the grown layers inherit the stacking order of substrates. As shown in Fig.2(b), twin crystalline 3C-SiC is grown at 1100℃. This is attributed to the suppressed surface diffusion of adsorbed species and two-dimensional nucleation on terraces. Since a nucleation rate on terraces can be restrained by lowering supersaturation, further reduction of SiH_4 flow rate might lead to homoepitaxial growth at lower temperatures.

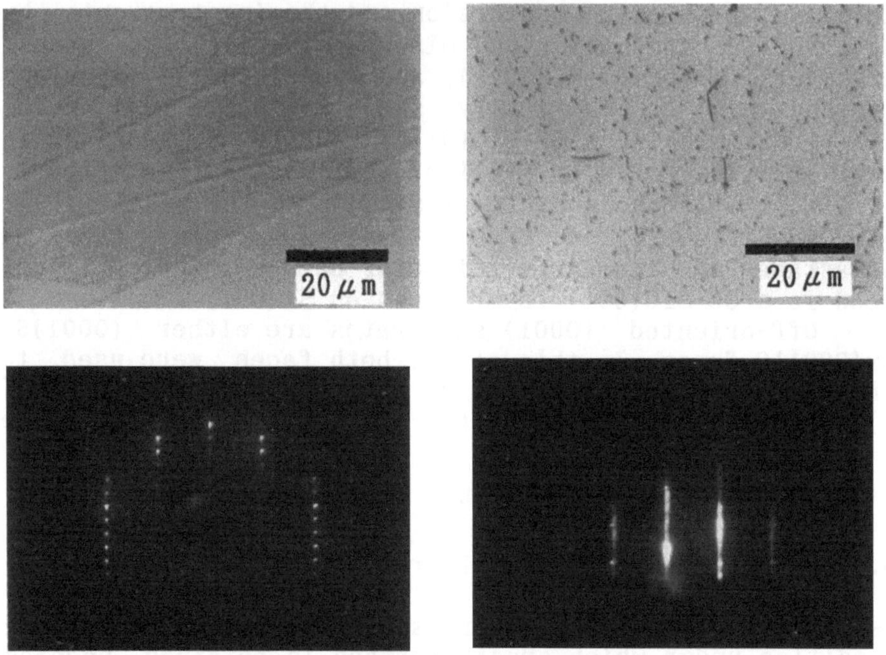

(a) (b)

Figure 2 Surface morphology and RHEED patterns of layers grown (a) at 1200℃ and (b) at 1100℃ on off-oriented 6H-SiC(000$\bar{1}$) faces. Off-orientation is 6° -off toward [11$\bar{2}$0].

For the VPE growth on 6H-SiC{01$\bar{1}$4}, 6H-SiC(0$\bar{1}$1$\bar{4}$)C faces were mainly used as substrates since a featureless surface was obtained only on 6H-SiC(0$\bar{1}$1$\bar{4}$) at 1500℃[12]. Nomarski micrographs and RHEED patterns of the layers grown on 6H-SiC(0$\bar{1}$1$\bar{4}$) at 1100 and 1000℃ are shown in Fig.3. The flow rates of SiH_4 and C_3H_8 are 0.15 and 0.20sccm, respectively. Single crystalline 6H-SiC(0$\bar{1}$1$\bar{4}$) can be grown at 1100℃. The increase of kink density on a surface by the inclination of 54.74° from {0001} faces brings about the homoepitaxy at lower temperatures than the growth on vicinal {0001} faces. At 1000℃, only polycrystalline Si is grown probably due to insufficient decomposition of C_3H_8. By using a carbon source, which decomposes at lower temperatures, further reduction of the homoepitaxial growth temperature is expected.

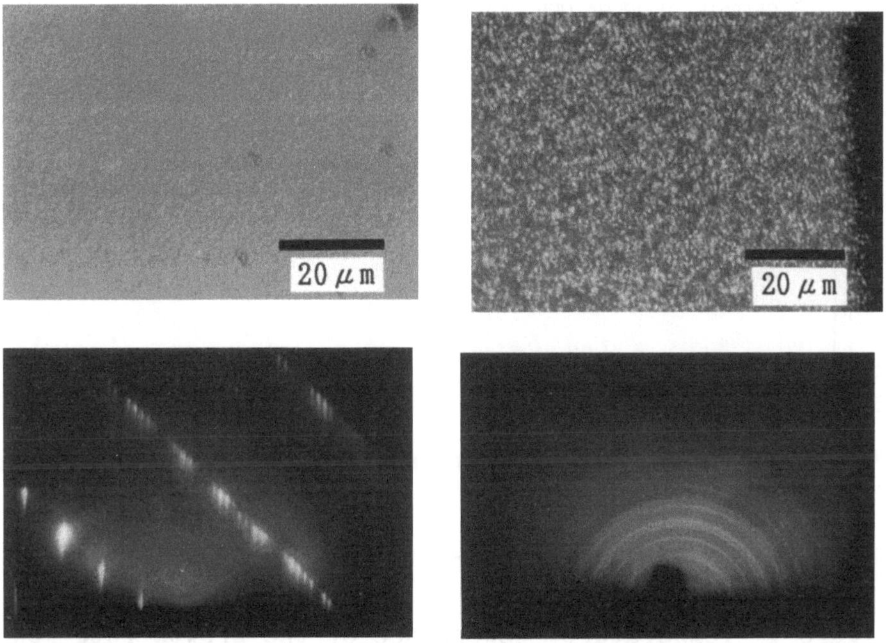

(a) (b)

Figure 3 Surface morphology and RHEED patterns of layers grown (a) at 1100℃ and (b) at 1000℃ on 6H-SiC($0\bar{1}1\bar{4}$) faces.

4. Growth Rate and Activation Energy

We have discussed the growth mechanism in step-controlled epitaxy from a viewpoint of dimensional relation between migration length and terrace widths[1,3,4]. Since the homoepitaxial growth at lower temperatures could be achieved, more detailed mechanism is studied from the temperature dependence.

Figure 4 shows the temperature dependence of growth rate on 6° off-oriented {0001} faces in the range of 1200℃ and 1500℃, in which epitaxial growth of 6H-SiC does occur. The flow rates of SiH_4 and C_3H_8 are 0.15 and 0.10-0.14sccm, respectively. As growth temperature decreases, the growth rate decreases as in the figure, but the activation energy shows a very small value of 3.0kcal/mole. For VPE growth on well-oriented {0001} faces, activation energies of 20 or 22kcal/mole were reported[13,14]. So, significant reduction of activation energy is obtained in step-controlled epitaxy. Moreover,

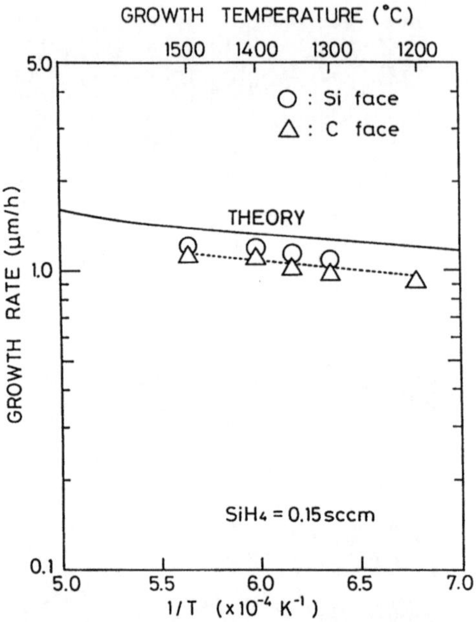

Figure 4 Temperature dependence of the growth rate on off-oriented 6H-SiC (0001)Si(○) and (000$\bar{1}$)C faces(△). The calculated result based on a stagnant layer model is shown by a solid curve. The flow rates of SiH_4 and C_3H_8 are 0.15 and 0.10-0.14sccm, respectively.

there is little difference between the growth rates on Si faces and on C faces in a wide range of growth temperature. This is notable because the growth on well-oriented C faces yields much higher growth rate than that on Si faces[13-16].

Here, the small activation energy obtained in step-controlled epitaxy and the growth mechanism are discussed. For VPE growth of Si in a horizontal reactor, Eversteyn has developed a stagnant layer model based on gas flow pattern experiments and explained the growth rates quantitatively[17]. In this case, diffusion of reactants from main gas stream to a crystal surface controls the growth. Since the growth rate of SiC is determined by the supply of Si species[18], the proposed model for Si growth will be applicable to the present system with some modification. The growth rate at the sample position of x_0(distance from the front edge of a susceptor : 3.0cm) can be described by the following expression[17] :

$$R(x_0) = \frac{hD_0T_sP_0}{n_0kT_0^2\delta(x_0)} \exp\left(-\frac{D_0T_s\alpha}{T_0V_0b}\right), \qquad (1)$$

where D_0 is the diffusion coefficient of SiH_4 in H_2 at 300K(0.20cm^2/s)[19], P_0 the partial pressure of SiH_4 at the inlet of a reactor(1.00x10^{-4}atm) calculated by the ratio of SiH_4 and the total flow rate, and V_0 mean gas velocity in non-heated condition(6.0cm/sec) given by the total gas flow rate divided by the free cross section of the tube at a susceptor. T_s is the substrate temperature, T_0 300K, b the free height of the reactor(1.7cm), h the spacing of the 6H-SiC{0001} face(2.52Å), n_0 the density of atomic sites on the surface(1.21x10^{15}cm^{-2}). $\delta(x_0)$ is the stagnant layer thickness at the point of x_0, and α is given by the following equation[17] :

$$\alpha = \int_0^{x_0} \frac{dx}{\delta(x)} = \int_0^{x_0} \sqrt{\frac{V_0\rho}{\mu x}}\, dx , \qquad (2)$$

where ρ and μ are gas density(8.13x10^{-5}g/cm^3) and gas viscosity(8.7x10^{-5}P)[19]. Although it is difficult to know the exact thickness of a stagnant layer in a complex system, $\delta(x)$ is estimated by the above simple relation. Using the above values for V_0, ρ and μ, $\delta(x_0)$ and α are calculated to be 0.73cm and 0.616, respectively.

We calculated the temperature dependence of growth rate based on eq.(1) neglecting complex flow dynamics caused by heating. The calculated result is shown by a solid curve in Fig.4. The absolute value of the growth rate shows good agreement between experimental and theoretical results. Small discrepancy in growth rate will be ascribed to underestimation of $\delta(x_0)$ or other factors due to a complex gas flow. An error in temperature measurement or contribution of desorption reaction from a growing surface might be also the origin for the discrepancy.

The predicted growth rate increases gradually with temperature increase. This is owing to, mainly, the enhancement of diffusion in a stagnant layer. Although eq.(1) is not an activation-type formula, the calculated curve in the range of 1200-1500℃ yields an apparent activation energy of 3.0kcal/mole by the method of least squares. This is the same value as the experimental one. Therefore, the growth would be limited by mass transport(diffusion of reactants in a stagnant layer) in step-controlled epitaxy. This is consistent with little difference in the growth rate between on Si and C faces, because no polarity dependence should be observed in mass transport-type growth.

As described previously, SiC growth on well-oriented substrates proceeds through two-dimensional nucleation on terraces, which needs a large activation energy. The growth may be controlled by a surface reaction, such as adsorption of Si species[13]. On off-oriented substrates, the lateral growth from atomic steps is induced, and fast surface reactions without nucleation may be realized. This will cause the growth rate to be limited by diffusion of reactants in a stagnant layer.

5. Conclusion

Crystal growth of SiC on off-oriented 6H-SiC{0001} and 6H-SiC{$01\bar{1}4$} substrates was carried out at low temperatures of 1000-1500℃. Homoepitaxial growth of 6H-SiC is achieved at a temperature as low as 1200℃ due to step-flow growth on off-oriented {0001} faces and at 1100℃ on ($0\bar{1}1\bar{4}$)C faces. The activation energy of growth rate shows a very small value of 3.0kcal/mole. This can be quantitatively analyzed on the basis of a stagnant layer model in which crystal growth is controlled not by surface reactions but by diffusion of reactants in a stagnant layer. Step-flow growth, which does not need two-dimensional nucleation, will bring about fast surface reactions and be controlled by mass transport. This is consistent with little difference in the growth rate between on off-oriented Si and C faces.

Acknowledgment

This work was partially supported by a Grant-in-Aid for Scientific Research from the Ministry of Education, Science and Culture of Japan.

References

[1] N.Kuroda, K.Shibahara, W.S.Yoo, S.Nishino and H.Matsunami : Extended Abstracts of the 19th Conf. on Solid State Devices and Materials, Tokyo(1987), p227.

[2] K.Shibahara, N.Kuroda, S.Nishino and H.Matsunami : Jpn. J. Appl. Phys., **26**(1987),L1815.

[3] T.Ueda, H.Nishino and H.Matsunami : J. Cryst. Growth, **104**(1990),695.

[4] H.Matsunami, T.Ueda and H.Nishino : Mat. Res. Soc. Symp. Proc., vol.162(1990),397.

[5] H.S.Kong, J.T.Glass and R.F.Davis : J. Appl. Phys., **64**(1988),2672.

[6] J.A.Powell, D.J.Larkin, L.G.Matus, W.J.Choyke, J.L.Bradshaw, L.Henderson, M.Yoganathan, J.Yang and P.Pirouz : Appl. Phys. Lett., **56**(1990),1442.

[7] T.Tachibana, H.S.Kong, Y.C.Wang and R.F.Davis : J. Appl. Phys., **67**(1990),6375.

[8] Y.C.Wang, R.F.Davis and J.A.Edmond : J. Electron. Mat., **20**(1991),289.

[9] W.S.Yoo, S.Nishino and H.Matsunami : J. Cryst. Growth, **99**(1990),278.

[10] W.S.Yoo, A.Yamashita, T.Kimoto and H.Matsunami : J. Cryst. Growth, (1991), in press.

[11] W.S.Yoo and H.Matsunami : J. Appl. Phys., in press.

[12] A.Yamashita, W.S.Yoo, T.Kimoto and H.Matsunami : to be submitted.

[13] B.Wessels, H.C.Gatos and A.F.Witt : **Silicon Carbide 1973**, edited by R.C.Marshall, J.W.Faust,Jr. and C.E.Ryen(University of South Carolina Press, 1974), p25.

[14] V.J.Jennings, A.Sommer and H.C.Chang : J. Electrochem. Soc., **113**(1966),728.

[15] W.v.Muench and I.Pfaffeneder : Thin Solid Films, **31**(1976),39.

[16] S.Nishino, H.Matsunami and T.Tanaka : J. Cryst. Growth, **45**(1978),144.

[17] F.C.Eversteyn, P.J.W.Severin, C.H.J.v.d.Brekel and H.L.Peek : J. Electrochem. Soc., **117**(1970),925.

[18] T.Kimoto, H.Nishino, T.Ueda, A.Yamashita, W.S.Yoo and H.Matsunami : to be submitted.

[19] **Handbook of Chemistry and Physics**, edited by R.C.Weast(CRC Press, 1975).

Heteroepitaxial Growth of 3C-SiC by LPCVD with Alternate Gas Supply

H. Nagasawa and Y. Yamaguchi

Materials Research Laboratory, Hoya Corporation,
3-3-1, Musashino, Akishima-shi, Tokyo 196, Japan

Abstract. A heteroepitaxial growth of 3C-SiC on Si substrates has been realized at 1000 °C by a hot-wall type LPCVD with an alternate supply of C_2H_2 and SiH_2Cl_2. The crystallinity of the grown SiC layer is influenced by orientation of Si substrates. A single crystalline 3C-SiC has been grown on Si(111) face, while a polycrystalline 3C-SiC on Si (001) face. A growth rate of SiC with alternate gas supply is almost 3 ML/cycle and independent of the duration of SiH_2Cl_2 supply from 3 s to 12 s. A simultaneous supply of SiH_2Cl_2 and C_2H_2 never produces the SiC film but SiC whiskers. This difference is supposed to be brought about through different mechanisms of SiC growth between the alternate gas supply and the simultaneous one. The growth mechanism of the SiC film is discussed using QMS measurement results and attributed to adsorbed $SiCl_2$ molecules at hollow bridge sites.

1. Introduction

Beta silicon carbide (3C-SiC) is a semiconductor for high temperature devices, which exhibits several outstanding properties such as high thermal stability, high electron mobility (about 1000 $cm^2/V{\cdot}s$), wide band gap (2.2eV at 300K) and high saturated drift velocity[1,2]. It is necessary for fabrication of SiC devices to develop a growth technology of high quality 3C-SiC over large area. Although the large area heteroepitaxy of 3C-SiC on Si substrates has been recently studied intensively[3,4], the required growth temperature around 1350 °C is too high to be suitable for the device material. The growth temperature of 3C-SiC lower than 1000 °C is desirable, where inter-diffusion is expected to hardly occur.

Molecular beam epitaxy (MBE) is one of the prominent methods for the low temperature growth of SiC, and the successful results in homoepitaxy of high quality 3C-SiC at 1000 °C have been reported [5]. Productivity of MBE is, however, far inferior to that of CVD.

We have attempted to grow the heteroepitaxial 3C-SiC on Si substrates at 1000 °C with a low pressure CVD (LPCVD). To realize the atomic level controlled surface reaction by LPCVD, SiC growth has been carried out by an alternate supply of the source gases. The heteroepitaxial single crystalline 3C-SiC film has been obtained on Si (111) face, and its crystallinity and surface morphology have been investigated. The growth rate of the SiC film has been nearly constant and mechanisms of SiC growth will be discussed.

Springer Proceedings in Physics, Vol. 71 Amorphous and Crystalline Silicon Carbide IV
Editors: C.Y. Yang · M.M. Rahman · G.L. Harris

2. Experimental procedure

A quartz reaction tube of the LPCVD was uniformly heated by a surrounding heater. The base pressure of the reaction tube evacuated by a mechanical booster pump was 1×10^{-3} Torr. C_2H_2 and SiH_2Cl_2 were used as C and Si source gases, respectively, and H_2 as a carrier gas. Each gas flow rate was precisely controlled by mass flow controller. To control the pressure in the reaction tube, a butterfly valve between the reaction tube and the mechanical booster pump was employed.

Two types of single crystal Si wafers with just-cut (111) face or just-cut (001) face were used as substrates. All wafers were treated by $NH_3:H_2O_2:H_2O$ and $HCl:H_2O_2:H_2O$, before they were loaded into the reaction tube.

A SiC growth procedure shown in fig.1 consists of two processes; carbonization and growth processes. The carbonization process was carried out first to convert the Si surface into a thin epitaxial SiC layer with C_2H_2, called carbonized layer[6], and is a key to obtain superior epitaxial layer on Si substrates. H_2 and C_2H_2 began to be supplied into the reaction tube at the beginning of the heating-up process. The flow rates of H_2 and C_2H_2 were kept at 300cm^3/min. and 10cm^3/min., respectively. The obtained partial pressure of C_2H_2 was 6.4mTorr, and the total pressure was kept at 0.2Torr. A mirror like carbonized layer of 80Å in thickness has been obtained by this carbonization process[7].

After carbonization of 90 min at 1020°C, the reaction tube was cooled down to 1000°C with continuous flow of H_2 and C_2H_2. Then the SiH_2Cl_2 at flow rate of 2 cm^3/min. and the C_2H_2 at that of 10 cm^3/min. were alternately introduced into the reaction tube. Throughout the whole process of the SiC growth, H_2 carrier gas was kept to flow at the flow rate of 300 cm^3/min. and the total pressure at 35 mTorr. The duration of SiH_2Cl_2 flow (t_s) was varied from 1 to 12 s, while that of C_2H_2 and interval between gas introductions were fixed at 5 s. The number of cycles of gas supply was changed from 300 to 2300. After the SiC growth, the reaction tube was cooled down to room temperature without any flow of gases.

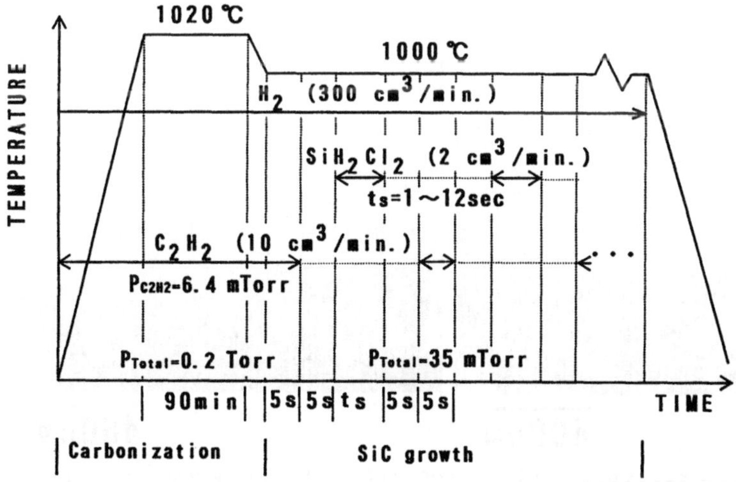

Fig.1. SiC growth procedure

A simultaneous supply of the SiH_2Cl_2, C_2H_2, and H_2 at the flow rate of 2 cm^3/min.,10 cm^3/min. and 300 cm^3/min., respectively, has been also carried out during 300 min. in order to investigate the effect of gas supplying method on the film growth.

3.Results and discussion

Stoichiometric SiC films have been grown on both Si(001) face and Si(111) face at 1000 °C by the alternate gas supply with t_s of 5 s. Fig.2. shows the cross-sectional SEM micrographs of the grown SiC films of 200 nm in thickness. The composition ratios (C/Si) of both films are shown to be equally 1.0 by XPS. Few voids have been observed at along the Si-SiC interface.

A heteroepitaxial 3C-SiC film has been obtained on Si (111) face. The TEM image in fig.3a and TED patterns in figs.3b and 3c of the grown SiC film on the Si (111) face exhibit the lattice image and strong spots ,respectively, all of which indicate a monocrystalline structure. On the other hand, the TEM image of SiC grown on the Si (001) face shows existence of some twins in the SiC layer, as shown in fig.4a. Many twins are oriented toward [111] directions of Si substrate, which are also suggested from TED pattern of the interface (fig.4b). Such orientation of twins results in disorder with progression of SiC growth, as shown in the TED pattern near the surface (fig.4c).

Two SiC films grown on (001) face and (111) face have clearly different morphologies each other, as shown in fig.5, which is supposed to result from difference in crystallinity of the grown SiC films. A much smoother surface (Ra=80Å) has been obtained for the SiC film grown on (111) face, while the SiC film grown on (001) face shows rougher surface morphology (Ra=200Å) due to many (111) facets formed on it.

Fig.6. shows that the SiC thickness is proportional to the number of cycles of gases supply, independent of orientation of substrates. The growth rate of SiC, defined as the grown SiC thickness per cycle, is about 7Å . The growth rate of 7Å /cycle corresponds to 3 mono-layer (ML) of SiC in case of

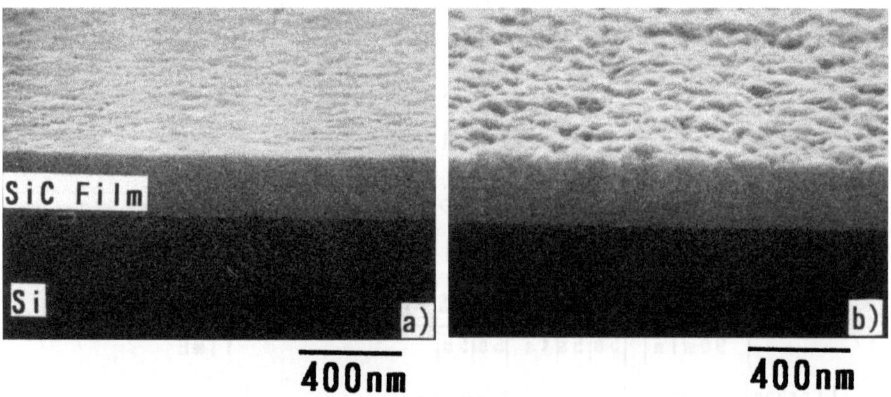

Fig.2. Cross-sectional SEM micrographs of the grown SiC films: (a) on (111) face; (b) on (001) face.

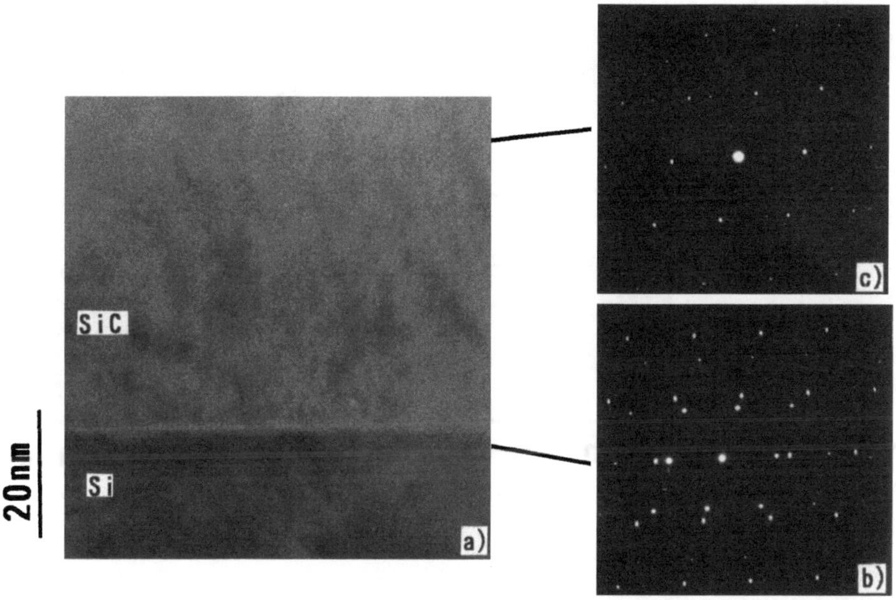

Fig.3. The TEM image and TED patterns of the grown SiC film on the Si (111) face:(a) cross-sectional TEM image; (b) TED pattern of the interface; (c) TED pattern near the surface.

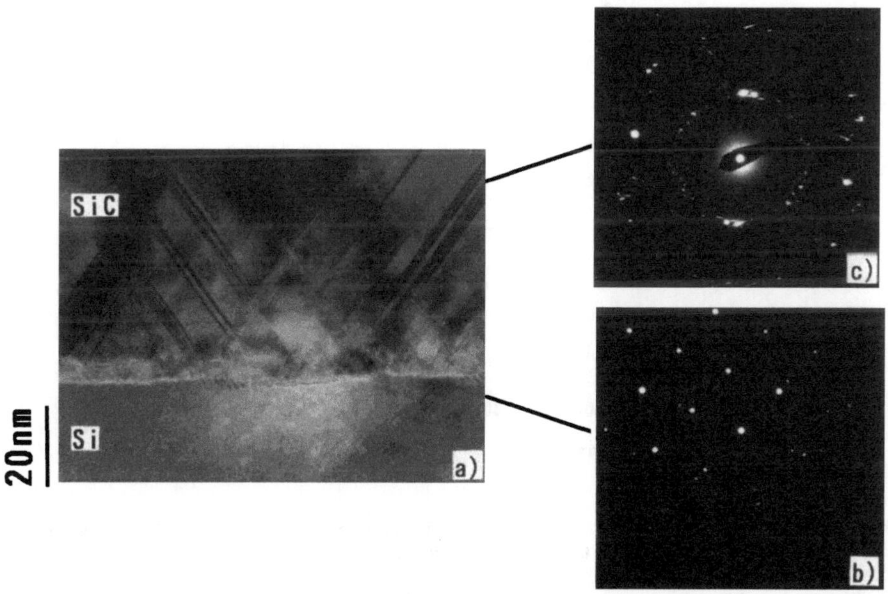

Fig.4. The TEM image and TED patterns of the grown SiC film on the Si (001) face:(a) cross-sectional TEM image; (b) TED pattern of the interface; (c) TED pattern near the surface.

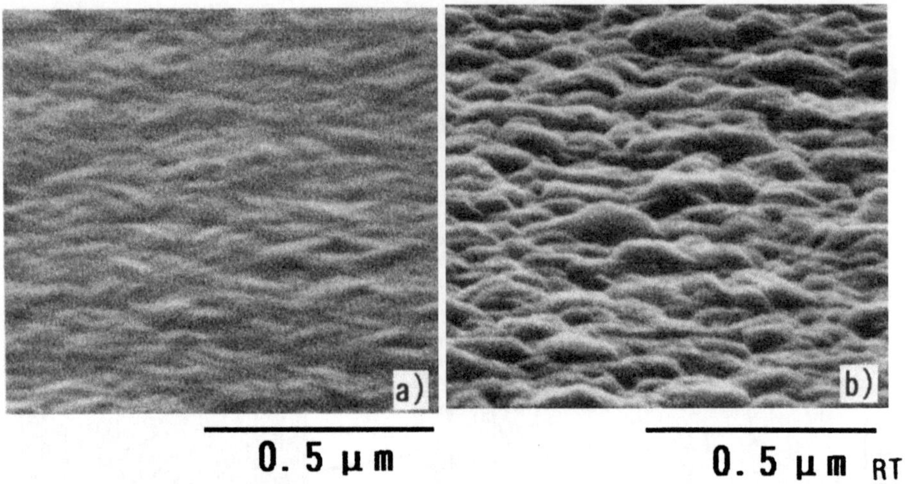

Fig.5. SEM micrographs of SiC surface: (a) on (111) face; (b) on (001) face.

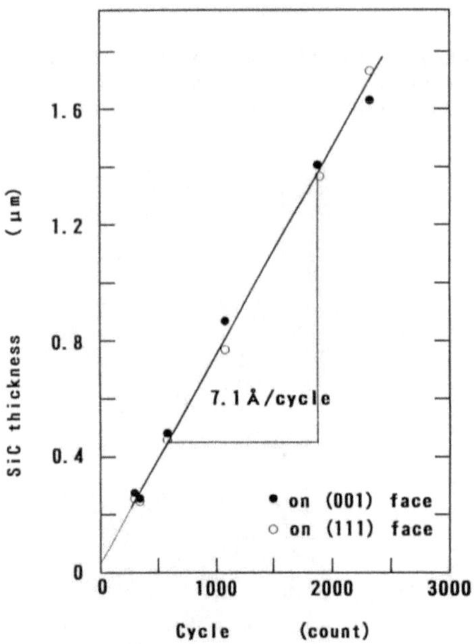

Fig.6. SiC thickness as a function of cycle count.

the epitaxially grown monocrystalline SiC on Si (111) face, and has been thought to occur by an atomic level controlled surface reaction [5].

The simultaneous supply of SiH_2Cl_2 and C_2H_2 has not formed such SiC films but many whiskers, as shown in fig.7. The difference is supposed to be brought about through different mechanisms of SiC growth between the alternate gas supply and the simultaneous one.

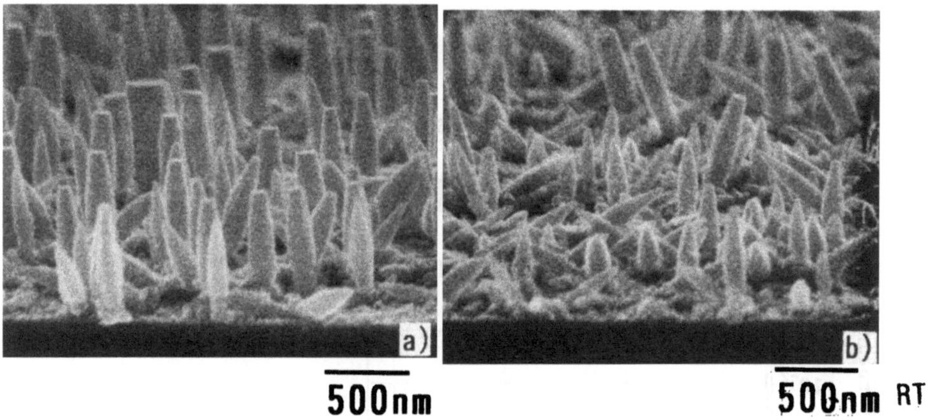

Fig.7. SEM micrographs of whiskers on Si substrates: (a) on (111) face; (b) on (001) face.

A quadrupole mass spectrometer (QMS) has been used to elucidate SiC growth mechanisms. The results of QMS measurement during the flow of SiH_2Cl_2 (2 cm^3/min.) at 1000°C, in addition to the continuous flow of H_2 (300 cm^3/min.), is shown in fig.8a. The SiH_2Cl_2 molecules are reported to be decomposed into $SiCl_2$,SiCl,HCl, and Cl molecules [8]. High partial pressure of HCl suggests that most of $SiCl_2$ molecules detected are formed by thermal decomposition at 1000°C, not by self-cracking through QMS. Cl and SiCl species is probably produced by the secondary cracking of HCl and $SiCl_2$ in QMS.

It has been reported that the adsorbed $SiCl_2$ molecules are ionized at hollow bridge sites of Si substrates, and change into $SiCl_2^-(^2B_1)$ state which has loose bonds between Si and Cl [9-11]. The $SiCl_2^-(^2B_1)$ state may be supposed to be also realized in this study by use of the alternate gas supply [8-11].

The partial pressures of HCl do not change when QMS measurement has been carried out within 1 min after stopping SiH_2Cl_2 supply, while that of $SiCl_2$ shows a large decrease, as shown in fig.8b. This indicates that the origin of residual HCl molecules can not be accounted for the decomposition of $SiCl_2$ in gas phase but that of the adsorbed $SiCl_2$ on substrates since most of the existing species in gas phase in the reaction tube is evacuated within 1s. The adsorbed $SiCl_2$ molecules may be slowly decomposed by H_2 and cause the Si growth.

Fig.8c shows an immediate change in QMS spectrum on introducing C_2H_2 at 5 s after stopping the SiH_2Cl_2 supply. The partial pressure of HCl decreases and that of $SiCl_2$ disappears. The partial pressures of HCl is then faded away from QMS spectrum at 5 s after stopping the C_2H_2 supply, as shown in fig.8d. The above results seem to suggest that the decomposition of adsorbed $SiCl_2$ is finished within 5 s after starting C_2H_2 supply, and additional SiC layers are immediately formed.

The amount of the adsorbed $SiCl_2$ seems to spontaneously determine the constant growth rate of SiC. The growth rate of both SiC films on (001) face and (111) face were 7-9Å /cycle, independent of t_s of longer than 3 s, as shown in fig.9. The

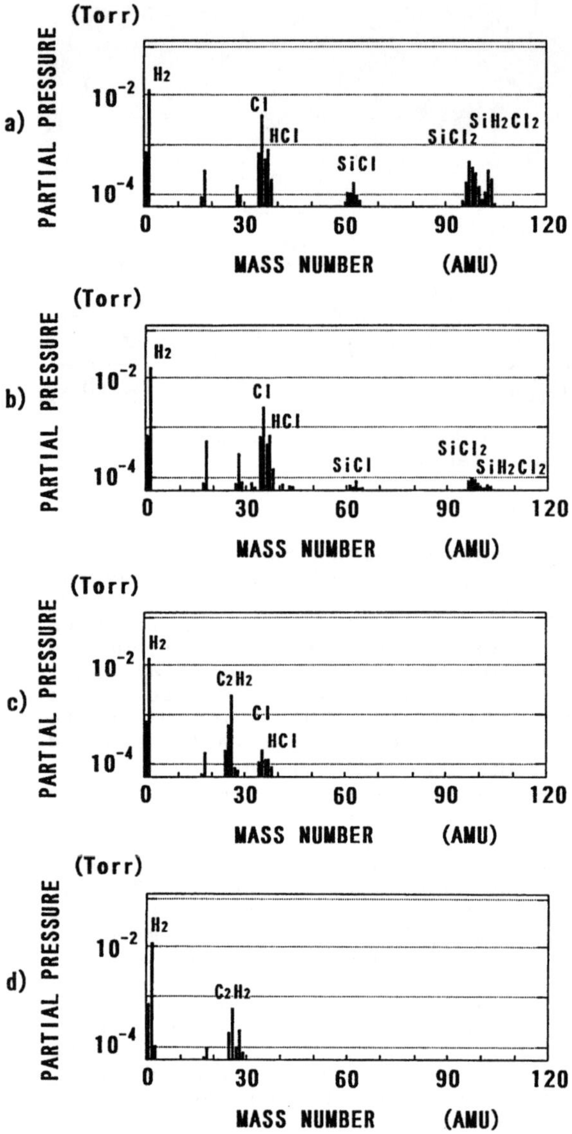

Fig.8. Quadrupole mass spectrometer (QMS) measurements in reaction tube: (a) simultaneous supply of SiH_2Cl_2 (2 cm^3/min.) and H_2 (300 cm^3/min.);(b) within 1 min. after stopping SiH_2Cl_2 supply; (c) C_2H_2 (10 cm^3/min.) supply at 5 s after stopping SiH_2Cl_2 supply; (d) 5 s after stopping the C_2H_2 supply.

amount of adsorbed $SiCl_2$ which incorporated in the SiC growth may be saturated with t_s of longer than 3 s.

As shown in Fig.10, the simultaneous supply of SiH_2Cl_2 and C_2H_2 has brought about an enormous increase in the partial pressure of $SiCl_2$, while no significant change is observed in

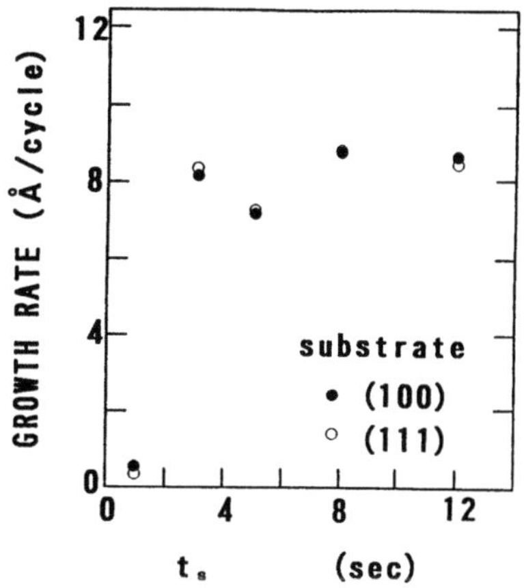

Fig.9. SiC growth rate versus duration of SiH_2Cl_2 supply (t_s).

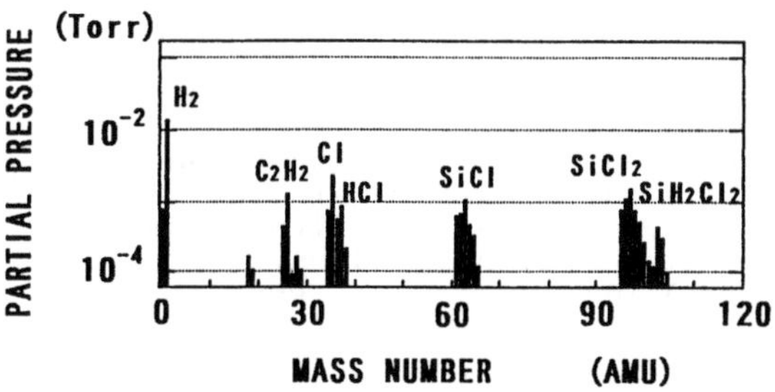

Fig.10. QMS measurement during simultaneous supply of SiH_2Cl_2 (2 cm^3/min.), C_2H_2 (10 cm^3/min.) and H_2 (300 cm^3/min.).

that of SiH_2Cl_2. The C_2H_2 molecules might be predominantly adsorbed on substrates and hinder $SiCl_2$ molecules from being adsorbed on substrates. The atomic level controlled surface reaction by the adsorbed $SiCl_2$ would not be realized. It is not understood yet what mechanisms would cause the formation of whiskers with the simultaneous supply of C_2H_2 and SiH_2Cl_2.

4. Conclusion

A single crystalline 3C-SiC has been grown on Si(111) face at 1000 °C by a hot-wall type LPCVD with an alternate supply of C_2H_2 and SiH_2Cl_2. On the other hand, a polycrystalline 3C-SiC has been grown on Si(001) face. A growth rate of SiC with the alternate gas supply is 7 - 9 Å/cycle, independent of duration of SiH_2Cl_2 supply of longer than 3 s. SiH_2Cl_2 molecules are supposed to be decomposed into $SiCl_2$ at 1000 °C and the decomposed $SiCl_2$ molecules might be adsorbed on hollow bridge sites of the substrates. The $SiCl_2$ molecules change into an ionized state of $SiCl_2^-$($^2B_1$) that has loose bonds between Si and Cl, and then the ionized $SiCl_2$ may be decomposed on introducing C_2H_2, giving rise to SiC growth with the constant rate of 7 Å/cycle.

A simultaneous supply of SiH_2Cl_2, C_2H_2, and H_2 have never produced the SiC films but many whiskers, resulting from the predominant adsorption of C_2H_2 molecules on the substrate.

References

1. G.L.Harris and C.Y.-W.Yang, eds., Amorphous and Crystalline Silicon Carbide (Springer, Berlin, Heidelberg, 1989), Springer Proceedings in Physics, **34**
2. M.M.Rahmann, C.Y.-W.Yang, and G.L.Harris, eds., Amorphous and Crystalline Silicon Carbide II (Springer, Berlin, Heidelberg, 1989), Springer Proceedings in Physics, **43**
3. S.Nishino, J.A.Powell, and H.A.Will, Appl . Phys. Lett. **42** (5), 1, March (1983)
4. Y.Furumura, M.Doki, F.Mieno, T.Eshita, T.Suzuki, and M.Maeda, J. Electrochm, Soc. **135**, No.5, 1255 (1988)
5. T.Fuyuki, N.Nakayama, T.Yoshinobu, H.Shiomi and H.Matsunami, J. Crystal Growth **95** 461 (1989)
6. C.J.Mogab and H.J.Leamy, J. Appl. Phys. **45** 1075 (1974)
7. H.Nagasawa and Y.Yamaguchi,Proceedings of the 7th International Conference on Vapour Growth and Epitaxy (Nagoya) (in press)
8. S.Motoyama,N.Morikawa,and S.Kaneda,J.Crys.Grow.,**100**,615 (1990)
9. Y.Ohshita et al., Extended Abstracts (The 47th Autumn Meeting,1986); The Japan Society of Applied Physics,p.718
10. A.Ishtani et al., Extended Abstracts (The 47th Autumn Meeting,1986); The Japan Society of Applied Physics,p.718
11. A.Ishitani, T.Takada, and Y.Ohshita, J.Appl.Phys. **63** (2),15 January (1988)

Mechanisms in the Low Pressure Growth of SiC-on-Si by RTCVD

A.J. Steckl and J.P. Li

Nanoelectronics Laboratory, Department of Electrical & Computer Engineering, University of Cincinnati, Cincinnati, OH 45221-0030, USA

Abstract. ß-SiC thin films have been grown on (100) Si by RTCVD at reduced pressures (LP-RTCVD) as low as 1 Torr. The growth process involved the carbonization of the Si surface by reaction with propane. The effect of propane partial pressure (fraction) on film morphology and growth rate at 5 Torr is similar to that observed at 760 Torr, but with a shift to higher propane fraction. This shift can be related to a lower density of C atoms on the growth surface at low pressure. Carbonization growth at 5 Torr using 100 sccm of C_3H_8 (5% in H_2) and 0.9 lpm H_2 resulted in a growth rate ranging from ~3 Å/sec at 1050°C to ~18 Å/sec at 1300°C, yielding an activation energy for the process of 1.5 ± 0.25 eV. A dramatic improvement in film thickness uniformity was observed at low pressure, resulting in a standard deviation of less than 5% over an entire 7.5 cm diameter wafer.

1. Introduction

Silicon carbide is a promising semiconductor exhibiting many desirable materials properties: wide energy band-gap, high melting point, high electric field breakdown, high thermal conductivity and physical and chemical stability. These properties are very attractive for SiC homostructure device applications requiring operation under extreme conditions (such as high voltage and/or current, high temperature, high radiation environment, high frequency) and for heterostructure applications, such as the SiC-Si heterojunction bipolar transistor which requires a large band-gap semiconductor. For device fabrication, the epitaxial growth of ß-SiC-on-Si has been accomplished by several techniques, including conventional chemical vapor deposition (CVD) [1-3], CVD with alternate supply of Si- and C-bearing gases [4] and molecular beam epitaxy [5-7]. We have previously reported [8] the epitaxial growth of ß-SiC films on (100) Si by rapid thermal chemical vapor deposition (RTCVD) using C_3H_8, SiH_4 and H_2 at atmospheric pressure (AP). RTCVD is a growth technique well suited for the growth of thin SiC films on Si due to its inherently limited exposure to high temperature obtained by rapidly ramping up and down the substrate temperature. This rapid temperature control is usually provided by W-halogen lamps with near-IR photoemission. AP-RTCVD growth of SiC-on-Si (100) using C_3H_8 and SiH_4 has been shown [8] to result in monocrystalline (100) SiC thin films by x-ray diffraction and transmission electron diffraction.

In the case of AP-RTCVD, the growth process is mass-transport-limited [9] at temperatures greater than 1200°C. Therefore, the film thickness uniformity is strongly affected by the gas flow pattern inside the reaction chamber. The gas flow pattern is a strong function of the geometry of the chamber and of the gas injection system and, therefore, cannot be significantly altered in an existing growth system. However, in the case of low pressure growth, because of the increased diffusivity of gas molecules through the boundary layer, the process becomes surface-reaction-rate-limited [10]. In this case, the thickness uniformity is a function of the temperature uniformity across the wafer which is relatively easy to control and improve. Therefore, films grown at low pressure are expected to exhibit superior thickness uniformity. Fujiwara et al. [11] have reported that conventional LP-CVD growth of SiC films on Si at 1.5 - 100 Torr showed improved film uniformity compared to AP-CVD growth. In this study we report on the epitaxial growth of ß-SiC-on-Si by LP-RTCVD at 5 Torr using C_3H_8 carbonization of the Si substrate.

2. Experimental Procedure

The growth system used in this study is a modified RTCVD system from AET/ADDAX Co. It consists of an oval quartz growth chamber and two banks of tungsten-halogen lamps on each side. The wafer is supported by three quartz pins of a quartz tray which slides into the chamber. The IR absorption by the Si wafer heats up the wafer rapidly to the desired temperature and a feed-back loop stabilizes the process. There are eight gas lines equipped with MFC's. The entire system operation is computer-controlled. A more detailed description of the system and its operation can be found in [8]. The gas precursor used in this study is propane (5%) diluted in H_2. Subsequent mention of C_3H_8 flow rate refers to the flow rate of the actual C_3H_8/H_2 mixture. In addition, hydrogen is also used as the carrier gas. The system pressure was measured by a capacitance manometer and controlled by a throttle valve with a pressure controller. The substrates were standard (100) n-type Si wafers with 4-6 ohm-cm resistivity. Samples were dipped in diluted HF and rinsed in DI H_2O just prior to being loaded into the chamber. The process diagram consists of two steps: in-situ cleaning in HCl/H_2 at 1200°C followed by carbonization in the presence of propane. For preliminary comparisons, we have used the same growth parameters for LP-RTCVD as those obtained for optimized AP-RTCVD growth: growth temperature of 1300°C, temperature ramp rate of 25-50°C/sec, growth time of 60-120 sec.

3. Results and Discussion

The advantage of low pressure growth in obtaining improved film thickness uniformity is graphically illustrated in Fig. 1, where thickness profiles of SiC films grown at atmospheric and reduced (5 Torr) pressures are shown. The profiles were obtained by ellipsometrically measuring the SiC film thickness across a 7.5 cm wafer at different locations. For atmospheric growth, strong non-uniformity is observed in Fig. 1(*i*) across the surface of the wafer, with a band of considerably thicker film being formed in the vicinity of the wafer edge. This effect is especially pronounced at the leading edge of the wafer. Only the film grown on the central portion of the substrate is at all uniform. We believe that this film thickness pattern is a replica of the gas flow pattern over the wafer. On the other hand, the profiles obtained from a SiC film grown by LP-RTCVD at 5 Torr, as shown in Fig. 1(*ii*), exhibit quite a uniform level of thickness at all locations along the wafer. No thick bands or other nonuniformities were observed in the wafer edge region. The average film thickness and standard deviation were calculated over the entire 3 inch wafer resulting in values of 1602 and 46 Å, respectively. This represents a standard deviation for LP-RTCVD growth of only 2.87% of the average thickness,

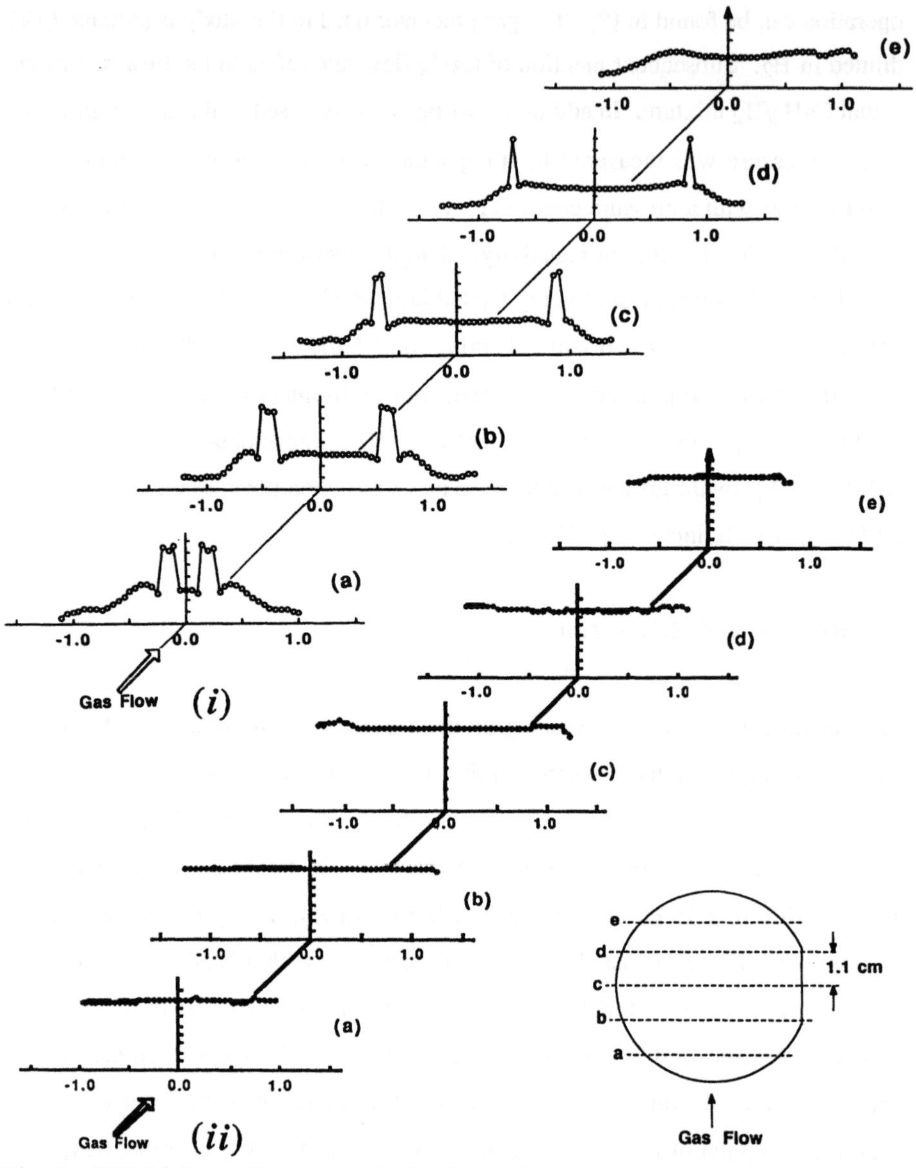

Fig. 1 SiC thickness line scans obtained by ellipsometry at different locations on the wafer: (*i*) atmospheric pressure; (*ii*) low pressure.

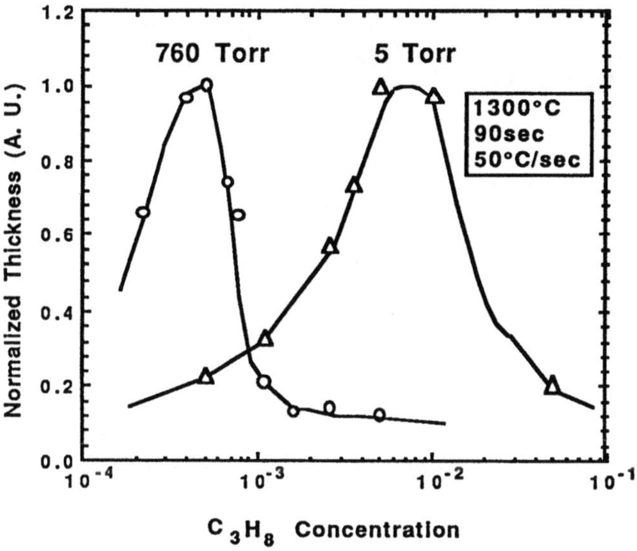

Fig. 2 SiC thickness vs. C_3H_8 fraction in the total gas flow.

and is the result of uniform growth over the entire specimen, including the edge region.

At atmospheric pressure, we have found that the C_3H_8 fraction in the gas stream (i.e. the ratio of C_3H_8 to the total gas flow) was extremely important in determining the nucleation process and the structural properties of the resulting SiC film. The SiC film thickness is shown in Fig. 2 as a function of C_3H_8 fraction at atmospheric and reduced pressures. At either pressure, a transition C_3H_8 fraction for maximum thickness is observed. For AP-RTCVD the transition occurs at a propane fraction of $5x10^{-4}$, while for LP-RTCVD at 5 Torr the value of the transition fraction has increased to $7x10^{-3}$. This characteristic pattern is reproducible, with some variation in the actual value of the transition propane fraction.

The morphology of films grown by LP-RTCVD at 5 Torr with selected propane fractions is shown in Fig. 3. At low C_3H_8 fraction ($5x10^{-4}$), dendritic growth occurs as shown in Fig. 3a. At the transition propane fraction ($5x10^{-3}$) a uniform and continuous film is observed in Fig. 3b, along with presence of voids in the underlying Si substrate. At the higher propane fraction of $5x10^{-2}$, the void size

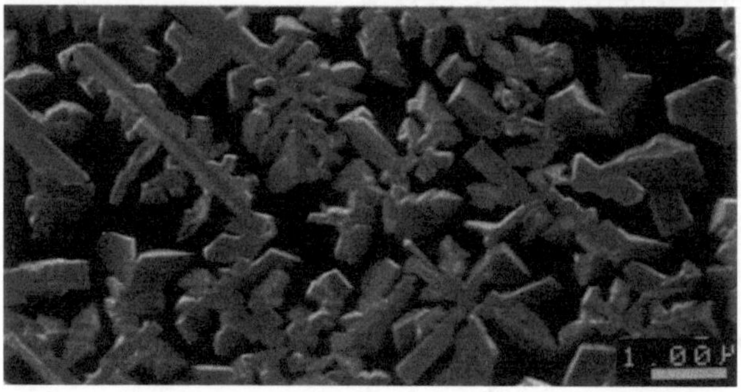

(a) 9 sccm C_3H_8
0.9 lpm H_2

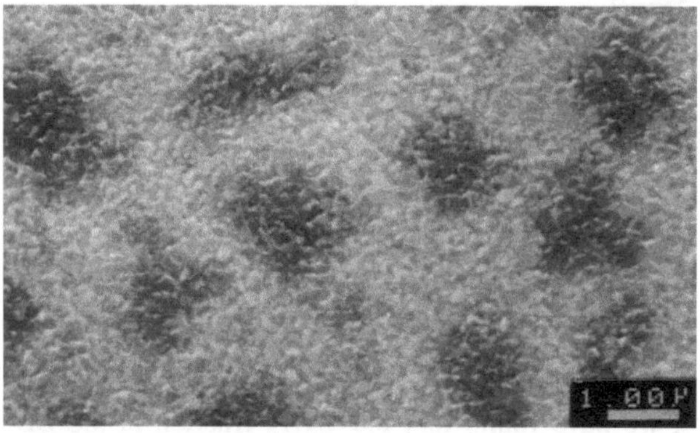

(b) 100 sccm C_3H_8
0.9 lpm H_2

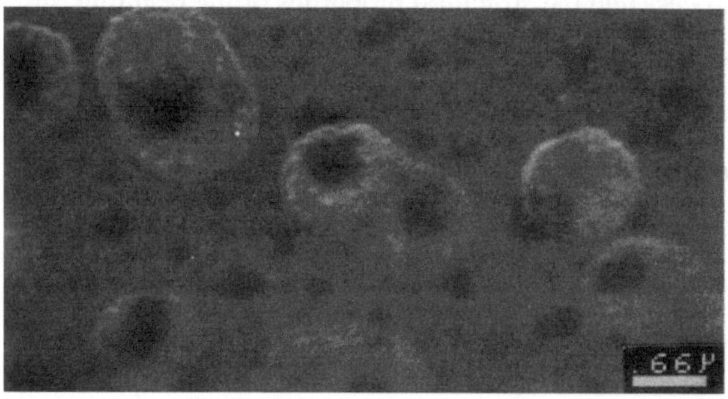

(c) 100 sccm C_3H_8

Fig. 3 SiC surface morphology for films grown at 5 Torr with different C_3H_8 fractions (corresponding to the values show in Fig. 2).

decreases and some toroidal structures appear on the surface. The origin and nature of these structures are not known at present. Their characterization is in progress. With the exception of the new features observed at high propane fraction, the progression in film morphology with propane fraction observed on films grown at 5 Torr is very similar to that obtained [8] with films grown at 760 Torr. We can, therefore, invoke the same nucleation site density argument to explain the change in film thickness and morphology with the propane flow rate at low pressure. In brief, at low propane flow rates dendritic growth can occur because the few nucleation sites present are widely separated and do not interact. On the other hand, at high propane flow a continuous film is grown rapidly due to the high density of nucleation sites, thus sealing the surface diffusion path for Si atoms from the substrate. This, in turn, greatly reduces the growth rate. At some intermediate flow rate, a maximum film thickness will occur. Now when the pressure is drastically reduced, from 760 Torr to 5 Torr, the propane fraction which will provide a density of C atoms on the growth surface comparable to that at atmospheric pressure has to be significantly higher to compensate for the lower gas density in the chamber. This, in essence, explains the low pressure shift of the film thickness vs. propane fraction curve toward higher propane values.

The structure of SiC films grown at low pressure was obtained by X-ray diffraction. The diffraction spectrum of a 2200 Å SiC film grown at 5 Torr with a propane fraction of 10^{-2} is shown in Fig. 4. The growth conditions were: 1300°C for 90 s, using a temperature ramp rate of 50°C/s with 100 sccm of C_3H_8 and 0.4 lpm of H_2. The spectrum is very similar to that of comparable films obtained [8] by AP-RTCVD carbonization. The main SiC peaks are present at $2\theta = 41.44°$ corresponding to the (200) reflection and at 89.98° for the (400) reflection, indicating the (100) cubic structure of the film.

The effect of temperature on SiC growth rate at low pressure (5 Torr) is shown in Fig. 5 for growth with and without H_2 carrier gas, corresponding to propane fractions of $5x10^{-3}$ and $5x10^{-2}$, respectively. Otherwise common conditions were used: 90 second reaction time at temperature, ramp rate of 50°C/s and 100 sccm of C_3H_8. A constant reaction time of 90 sec was used to produce films sufficiently

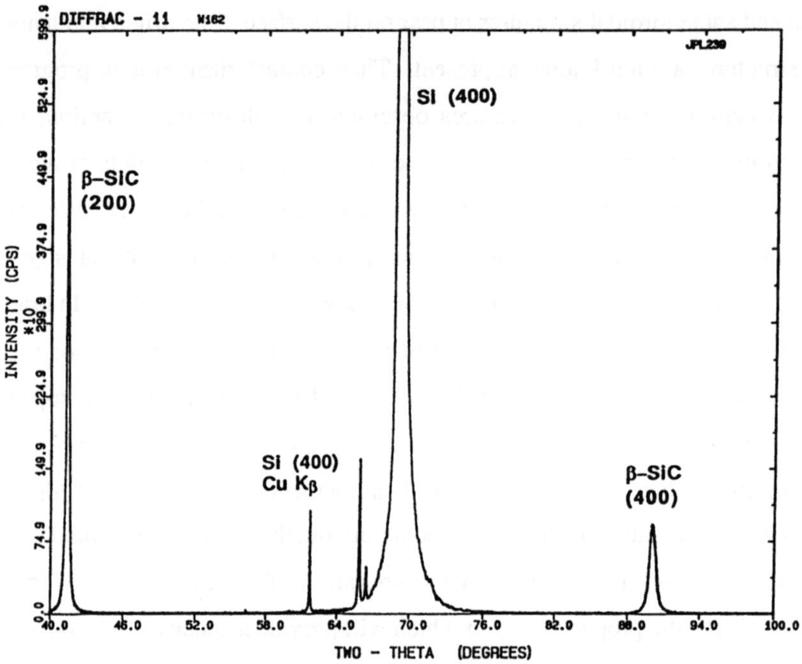

Fig. 4 X-ray diffraction spectrum of a SiC film on Si grown at 5 Torr: 1300°C, 90 sec, 100 sccm C_3H_8 and 0.4 lpm H_2.

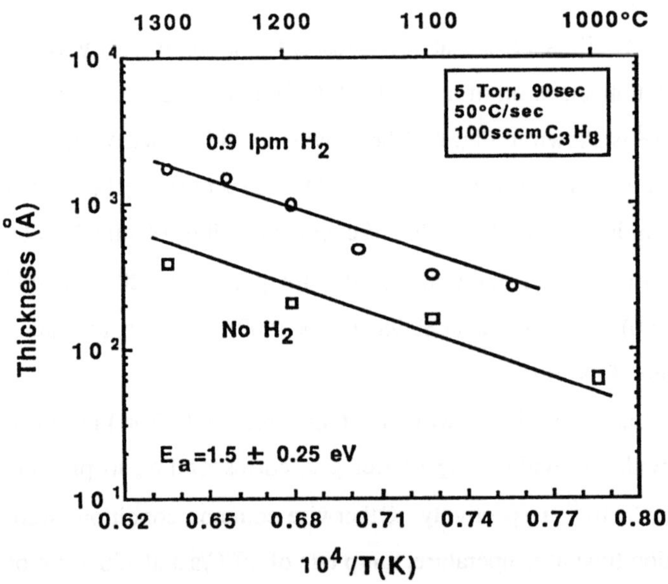

Fig. 5 SiC thickness vs. reciprocal temperature.

thick and uniform in all cases to be able to measure the thickness with reasonable accuracy. As expected from the effect of propane fraction on growth rate discussed above (see Fig. 2), the actual thickness values are affected by the presence, or absence, of hydrogen, over the temperature range explored (1000-1300°C). However, the slopes of the film thickness vs. inverse temperature are essentially the same, resulting in a calculated activation energy of 1.5 ± 0.25 eV. This indicates that the rate-limiting growth mechanism for these two cases with widely different propane fraction is the same. Growth at atmospheric pressure with the corresponding value of transition propane fraction (7-9 sccm) and all other conditions the same as for reduced pressure growth results [9] in two growth regimes: (a) for T < 1200°C, the activation energy is high, 3.1 ± 1.25eV, indicating reaction-rate-limited growth; (b) for T > 1200°C, the activation energy is rather low, 0.7 ± 0.2eV, indicating transport-rate-limited growth. Unfortunately, at this time the comparison of activation energies between atmospheric pressure and reduced pressure does not clearly provide an explanation for the mechanisms of low pressure growth rate.

4. Summary & Acknowledgements

Low pressure RTCVD growth of SiC using C_3H_8 carbonization has been shown to yield thin films with uniformity superior to that obtainable at atmospheric pressure. A summary of parameters and characteristics for SiC growth by atmospheric and low pressure (5 Torr) RTCVD is contained in Table I. The propane fraction, which determines the partial pressure of C_3H_8 on the growth surface, appears to be the determining parameter for the nucleation and subsequent growth of the film. Based on this mechanism, the shift in propane transition value with decreased pressure can be satisfactorily explained.

Table I. Comparison of atmospheric pressure and low pressure RTCVD of SiC

Parameters and Characteristics	Atmospheric Pressure	Reduced Pressure
Pressure	760 Torr	5 Torr
Flow Rates for Optimum Film Morphology and Thickness		
- Propane (5% in H_2)	9 sccm	100 sccm
- Hydrogen Carrier Gas	0.9 lpm	0.9 lpm
Propane Fraction of Total Gas Flow	$\sim 5 \times 10^{-4}$	$\sim 1 \times 10^{-2}$
Gas Velocity	~ 2 cm/sec	~ 300 cm/sec
Growth Temperature	1300°C	1300°C
Growth-Limiting Mechanism	Diffusion-Control	Reaction-Control
Thickness Uniformity	Strong Flow Pattern Effect	~ 3%
Uniformity-Limiting Mechanism	Fluid Flow	Temperature Profile

We acknowledge the support of this work by the Edison Materials Technology Center. Assistance in ellipsometry and other thickness measurements by G. Debrabander, P. H. Yih and J. T. Boyd is greatly appreciated.

References

[1] S. Nishino, J. A. Powell, and H. A. Will, "Production of large-area single-crystal wafers of cubic SiC for semiconductor devices", Appl. Phys. Lett., **42**, 460 (1983).

[2] P. Liaw and R. F. Davis, "Epitaxial growth and characterization of ß-SiC thin films", J. Electrochem. Soc., **132**(3), 642 (Mar. 1985).

[3] S. Nishino, H. Suhara, H. Ono and H. Matsunami, "Epitaxial growth and electric characteristics of cubic SiC on silicon", J. Appl. Phys., **61**(10), 4889, (May 1987).

[4] H. Nagasawa and Y. Yamaguchi, "Heteroepitaxial growth of 3C-SiC by LPCVD with alternative gas supply", Proc. of 4th SiC Conference (ICACSC'91), Oct.91, Santa Clara, CA.

[5] S. Kaneda, Y. Sakamoto, C. Nishi, M. Kanaya, and S. Hannai, "The growth of single crystal of 3C-SiC on the Si substrate by the MBE method using multi-electron beam heating", Jap. J. Appl. Phys., **25**(9), 1307 (Sept. 1986).
[6] T. Sugii, T. Aoyama, and T. Ito, "Low-temperature growth of β-SiC on Si by gas-source MBE", J. Electrochem. Soc., **137**(3), 9890 (1990).
[7] S. Motoyama, N. Morikawa, M. Nasu, and S. Kaneda, "Carbonization process for low-temperature growth of 3C-SiC by the gas-source MBE method", J. Appl. Phys., **68**(1), 101 (1990).
[8] A. J. Steckl and J. P. Li, "Epitaxial growth of β-SiC on Si by RTCVD with C_3H_8 and SiH_4", IEEE Trans. Elec. Dev., **39**(1), 64 (Jan. 1992).
[9] A. J. Steckl and J. P. Li, "RTCVD growth of nanometer-thin SiC-on-Si", Thin Solid Films Journal, (1992).
[10] W. Kern and G. L. Schnable, "Low pressure chemical vapor deposition for VLSI processing", IEEE Trans. Elec. Dev., **26**, 647 (1979).
[11] Y. Fujiwara, E. Sakuma, S. Misawa, K. Endo, and S. Yoshida, "Epitaxial growth of 3C-SiC on Si by low-pressure chemical vapor deposition", Appl. Phys. Lett., **49**(7), 388, (1986).

Effects of CH_3Cl Gas on Heteroepitaxial Growth of β-SiC on Si(111) by Chemical Vapor Deposition

K. Ikoma[1], *M. Yamanaka*[1], *H. Yamaguchi*[2], *and Y. Shichi*[2]

[1]Scientific Research Laboratory, Nissan Motor Co., Ltd., 1, Natsushima-cho, Yokosuka 237, Japan
[2]Nissan ARC, Ltd., 1, Natsushima-cho, Yokosuka 237, Japan

Abstract. ß-SiC epitaxial films were grown on Si(111) substrate by two-step chemical vapor deposition (CVD) using CH_3Cl as a C-source gas. Carbonized buffer layers were investigated by means of structural analyses. The buffer layers were formed with large voids underneath. This fact suggests that the Si atoms diffused out to the surface through the boundaries and defects in the buffer layer. Cl active species, which were produced by the decomposition of CH_3Cl, enhanced the supply of the Si atoms in the carbonization process. Furthermore, Cl active species enhanced two-dimensional growth of SiC nuclei on Si(111).

1. Introduction

ß-SiC has been reported to be grown epitaxially on Si(100) substrate by CVD method using a hydrocarbon-SiH_4-H_2 gas system [1,2]. The electrical and luminescent properties of ß-SiC(100) films have been studied [3,4]. However, the electrical properties of the heteroepitaxial ß-SiC(111) films have not been investigated well in comparison to ß-SiC(100) because of microcracks and warp [5].

We reported ß-SiC films were epitaxially grown on Si(111) substrate by two-step CVD using CH_3Cl gas as a C-source gas [6]. ß-SiC(111) epitaxial films with smooth surfaces (about 5nm in arithmetic average roughness) were obtained, and self-supported films could be handled without cracks and warp after the substrates removed by chemical etching. The carrier concentration and Hall mobility of the undoped *n*-type ß-SiC(111) film of 6μm thickness were $2\times10^{16}cm^{-3}$ and $150cm^2/Vs$, respectively.

In this paper, we will report the formation of the buffer layer on Si(111) using CH_3Cl-H_2 as well as C_3H_8-HCl-H_2 and C_3H_8-CH_3Cl-H_2, and discuss the carbonization mechanism and the effects of Cl active species in the carbonization process.

2. Experimental Procedure

ß-SiC films were grown on Si substrates by CVD using a cold-wall type reactor. Prior to the growth, the surface layers of Si substrates were etched by HCl-H_2 at 1200°C, and carbonized by CH_3Cl-H_2. The substrate temperature was raised to 1300°C at the rate of 500K/min and held at 1300°C for various carbonization times (t_c) in the carbonization process. t_c was varied from 0 to

Springer Proceedings in Physics, Vol. 71 Amorphous and Crystalline Silicon Carbide IV
Editors: C.Y. Yang · M.M. Rahman · G.L. Harris

15min. Thereafter, a SiC layer was grown on the buffer layer using CH_3Cl-SiH_4-H_2 at 1350°C. C_3H_8 was used instead of CH_3Cl for comparison in the carbonization and SiC growth process. Details of the epitaxial growth have been described elsewhere [6].

The structures of the buffer layers and the ß-SiC films were evaluated by scanning electron microscopy (SEM), reflection high energy electron diffraction (RHEED), and X-ray diffraction (XRD).

3. Results and Discussion

Figure 1 shows the SEM image of the buffer layer formed on Si(111) under the condition of carbonization using CH_3Cl-H_2 at 1300°C for 2min. The SEM image shows the mosaic structure of grains about 100nm in diameter. The buffer layer shows streaky RHEED pattern with spots originated from the ß-SiC mosaic crystals. The carbonization process on Si(111) using CH_3Cl-H_2 is characterized as follows in comparison to C_3H_8-H_2:

(i) Rapid growth of the buffer layer.
(ii) Formation of large voids under the buffer layer.
(iii) Enhancement of two-dimensional growth of the ß-SiC nuclei on the (111) plane.

The buffer layer of 100nm thickness was thicker than that of about 20nm formed using C_3H_8-H_2 at 1300°C for 2min. Figure 2 shows the thickness of the buffer layer formed using CH_3Cl-H_2 as a function of t_c. The thickness of the buffer layer did not increase in proportion to t_c, and was saturated to be around 150nm.

The RHEED patterns of the buffer layers for various t_c are shown in Fig.3. The buffer layer with t_c = 0min showed spot RHEED pattern. The streaky RHEED pattern was observed for the buffer layer with t_c = 15min. These results indicated that the crystallinity of the buffer layer was improved and the surface became smoother with increasing t_c. Figure 4 shows the

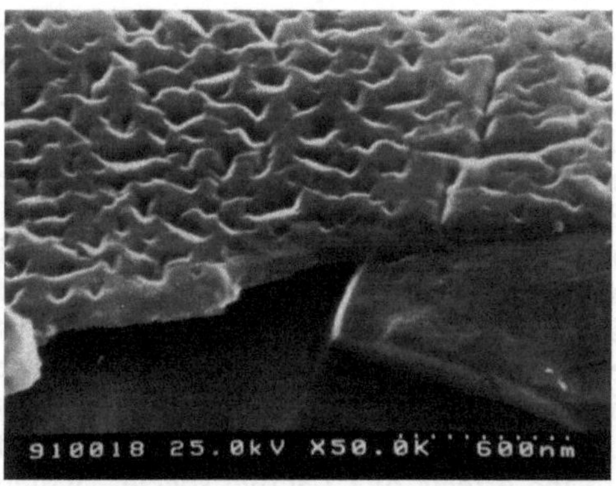

Fig.1. SEM image of the buffer layer formed using CH_3Cl-H_2 at 1300°C for 2min.

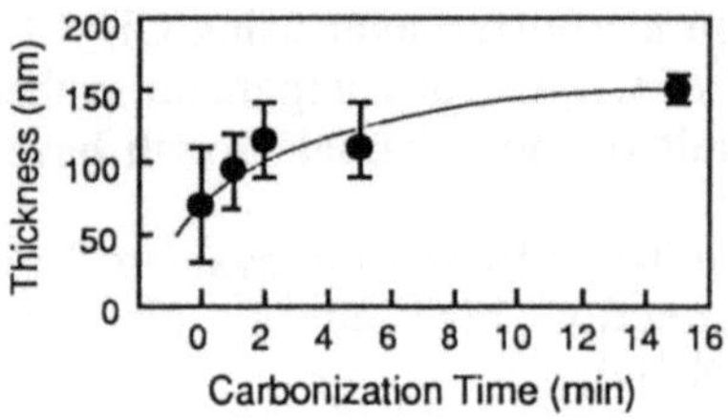

Fig.2. Thickness of buffer layers formed using CH_3Cl-H_2 as a function of carbonization time.

Fig.3. RHEED patterns of buffer layers formed using CH_3Cl-H_2 at 1300°C for (a) 0min, (b) 2min, and (c) 15min.

Fig.4. Optical microscopic image of a buffer layer formed using CH_3Cl-H_2.

optical microscopic image of the buffer layer with t_c = 2min. The white triangles indicate the voids under the buffer layer. The voids were 0.1~1µm in depth as shown in Fig.1. This suggests the buffer layer does not come in contact with the substrate in more than half of the area. The stress in the ß-SiC film may be relieved due to the large area of voids.

We discuss the formation mechanism of the buffer layer on the basis of the above results. The formation of a ß-SiC buffer layer requires the reaction of C species supplied from carbonization gas with Si species from Si substrate. Si atoms are considered to diffuse out to the surface through boundaries and defects in the buffer layer. The voids under the buffer layer were formed as the result of Si out-diffusion. Since the density of boundary and defect in the buffer layer decreases with increasing t_c, the thickness of the buffer layer is saturated due to Si diffusion limited. The rapid growth of the buffer layer and the formation of the voids are due to the effects of Cl active species, which enhance the out-diffusion of Si atoms.

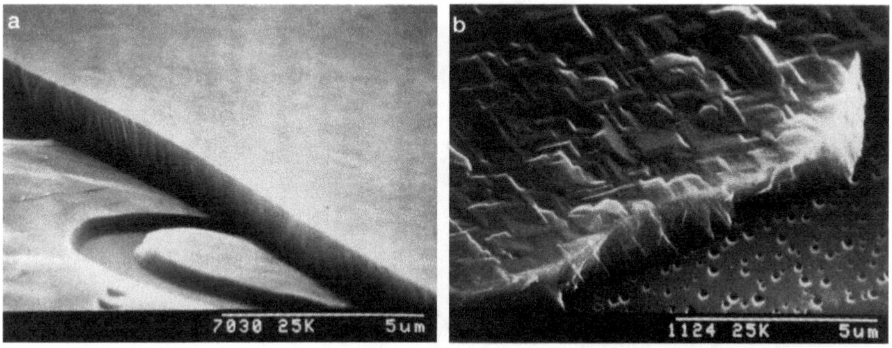

Fig.5. SEM images of ß-SiC layers grown using CH_3Cl-SiH_4-H_2 on (a) Si(111) and (b) Si(100) substrates. Buffer layers were formed using CH_3Cl-H_2.

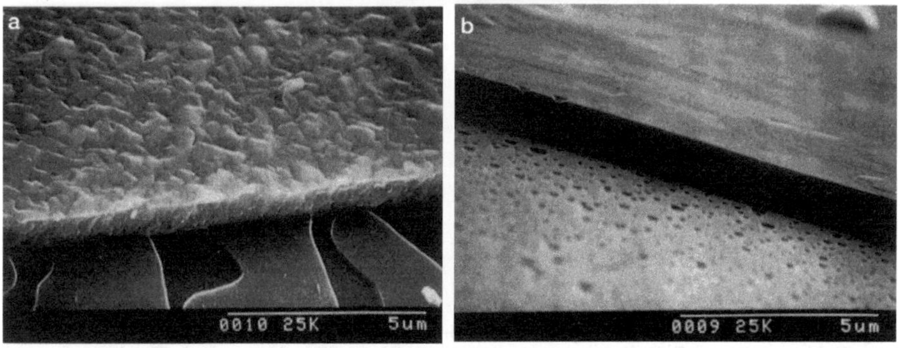

Fig.6. SEM images of ß-SiC layers grown using C_3H_8-SiH_4-H_2 on (a) Si(111) and (b) Si(100) substrates. Buffer layers were formed using CH_3Cl-H_2.

Next, we discuss the two-dimensional growth of ß-SiC on Si(111) substrate. The surface of the buffer layer, indicated in Fig.3, became smoother with increasing t_C. The difference of SiC growth mechanism between using CH_3Cl and C_3H_8 as a C-source gas is expected to become clearer in the SiC growth process. Figures 5 and 6 show SEM images of ß-SiC films grown using CH_3Cl-SiH_4-H_2 and C_3H_8-SiH_4-H_2, respectively, on the same buffer layers formed using CH_3Cl-H_2. On Si(111), the surface of the film grown using CH_3Cl was smoother. On the other hand, films with smooth surface tended to be grown on Si(100) using C_3H_8. Consequently, CH_3Cl as a C-source gas in the growth process enhanced two-dimensional growth of the ß-SiC on the Si(111). As the growth rate in the <111> direction is lower than the other directions, ß-SiC(111) films are considered to be grown two-dimensionally. The sticking probability of Si atoms on the (111) plane of ß-SiC may be suppressed by Cl active species.

The crystallinity of the buffer layer is expected to depend on the ratio of the supplied amount of C and Cl active species. Figure 7 shows the RHEED pattern and the SEM image of the buffer layer formed using C_3H_8-HCl-H_2 for

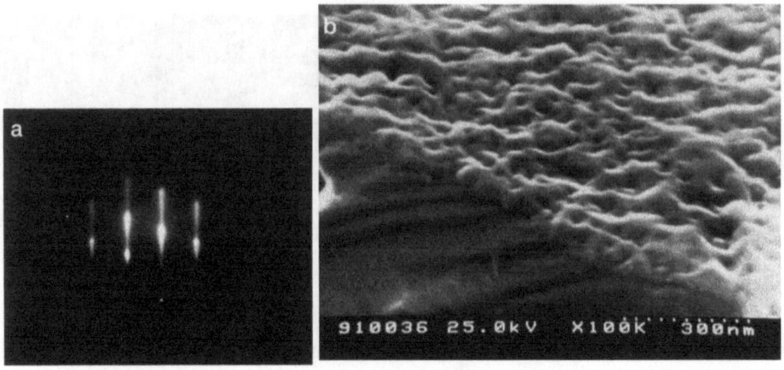

Fig.7. (a) RHEED pattern and (b) SEM image of the buffer layer formed using C_3H_8-HCl-H_2 for 30min.

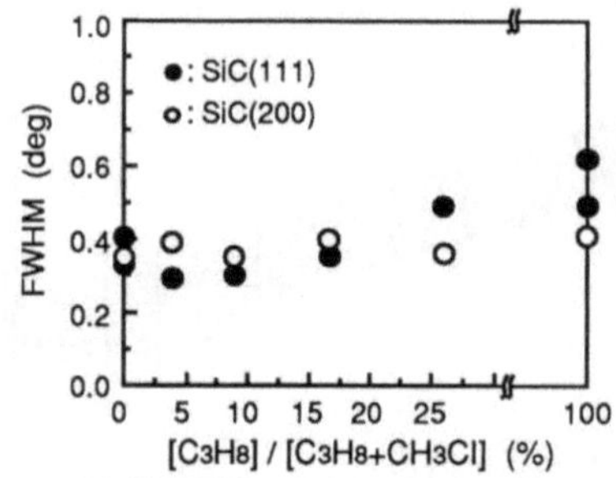

Fig.8. FWHM of X-ray rocking curves of ß-SiC on Si(111) and Si(100) as a function of C_3H_8 content. ß-SiC films were grown using CH_3Cl-SiH_4-H_2 on the buffer layer formed using C_3H_8-CH_3Cl-H_2.

30min. The flow rates of 1%C_3H_8 diluted with H_2 and HCl were regulated to be 175 and 12.5sccm, respectively. The buffer layer showed streaky RHEED pattern, and the full width of half maximum (FWHM) of SiC(111) diffraction of X-ray rocking curve of the SiC layer grown using CH_3Cl-SiH_4-H_2 was as small as 0.36°. However, the clear voids were not observed, and the surface of the SiC layer was rough. Therefore, Cl active species could not be supplied enough by HCl. Then, we investigated the buffer layer formed using C_3H_8-CH_3Cl-H_2 on the basis of the evaluation of the SiC growth layer on it. Figure 8 shows FWHM of the rocking curves of the SiC growth layers as a function of the C_3H_8 content ([C_3H_8]/[C_3H_8+CH_3Cl]) in the carbonization gas. The flow rate of 1%CH_3Cl diluted with H_2 was kept at 500sccm, and the C_3H_8 content was changed by adding the 1%C_3H_8. The FWHM of Si(111) diffraction became wider and the surface became rougher with increasing C_3H_8 content.

Figure 9 shows the SEM image of the surface of the SiC growth layer on the buffer layer formed using only C_3H_8. The surface of the SiC layer was rough with pits. The Cl active species have effects on the supply of Si atoms

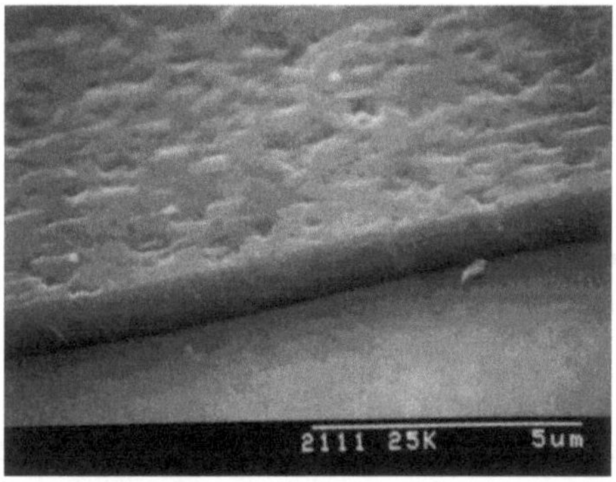

Fig.9. SEM image of ß-SiC layer grown using CH_3Cl-SiH_4-H_2 on the buffer layer formed using C_3H_8-H_2.

and the etching of Si atoms on the ß-SiC growth plane in the carbonization process. Although we cannot point out which effect is the most important to obtain good crystallinity of buffer layer, these results suggest that the deficiency of Cl active species causes nucleation of SiC with random distribution of crystal axes on Si(111) in the carbonization process.

4. Conclusion

We investigated the buffer layer on Si(111) substrate formed using a CH_3Cl gas. In the carbonization process, Si atoms were diffused out to the surface through the boundaries and defects in the buffer layer. Cl active species are considered to enhance the supply of Si atoms and to suppress the growth rate in the <111> direction. Cl active species are necessary for the formation of a good buffer layer on Si(111) substrate, and CH_3Cl could supply Cl active species effectively. It is difficult to form good buffer layers by HCl used as a source gas of the Cl active species in the carbonization process.

References

1. S.Nishino, J.A.Powell, and H.A.Will, Appl. Phys. Lett., **42**, 460 (1983).
2. P.Liaw and R.F.Davis, J. Electrochem. Soc., **132**, 642 (1985).
3. M.Yamanaka, H.Daimon, E.Sakuma, S.Misawa, and S.Yoshida, J. Appl. Phys., **61**, 599 (1987).
4. H.Okumura, M.Shinohara, E.Muneyama, H.Daimon, M.Yamanaka, E.Sakuma, S.Misawa, K.Endo, and S.Yoshida, Jpn. J. Appl. Phys., **27**, L116 (1988).
5. A.Suzuki, K.Furukawa, Y.Higashigaki, S.Harada, S.Nakajima, and T.Inoguchi, J. Cryst. Growth, **70**, 287 (1984).
6. K.Ikoma, M.Yamanaka, H.Yamaguchi, and Y.Shichi, J. Electrochem. Soc., **138**, 3028 (1991).

Growth Simulation of SiC Polytypes and Application to DPB-Free 3C-SiC on Alpha-SiC Substrates

W.S. Yoo and H. Matsunami

Department of Electrical Engineering, Kyoto University,
Yoshidahonmachi, Sakyo, Kyoto 606-01, Japan

An electrostatic model in which electrostatic Coulomb force is considered as the interatomic force is proposed to explain polytype formation and crystal growth of SiC. The Coulomb potential for atoms at two different sites on the surface of substrates, which induced by atomic arrays in substrates, was calculated by changing the interaction length. The dependence of surface potential on the stacking sequence of atoms in substrates was investigated. The growth simulation was tried taking energetic balance into consideration. The growth of 3C–SiC with double positioning boundaries (DPBs) on well–oriented 6H–SiC substrates can be explained. Possibility of DPB–free 3C–SiC growth on a 15R–SiC substrate is predicted. By simultaneous chemical vapor deposition growth on well–oriented 6H–SiC(0001) and 15R–SiC(0001), the prediction was experimentally verified.

1 Introduction

Polytypic phenomenon was first discovered in SiC by Baumhauer in 1912. [1] Various explanations have been proposed by several researchers, from time to time, to account for the polytypic phenomeon. Attempts to correlate the content of impurities and the temperature of crystallization with the polytype have been made very early, but clear correlation was not found. Frank's dislocation model, [2] Ramsdell and Kohn's polymer model, [3] Schneer's transformation model [4] and Jagodzinski's disorder model [5] have been proposed. The origin of the models can be classified into

Springer Proceedings in Physics, Vol. 71 Amorphous and Crystalline Silicon Carbide IV
Editors: C.Y. Yang · M.M. Rahman · G.L. Harris

two different concepts of cooperational phenomenon among equilibrium phases [4,5] and growth phenomenon on parent crystals. [2,3] These models seem fairy satisfactory when they are applied to the more common short–period polytypes such as 4H–, 6H– and 15R–SiC. However, the models originated from the cooperational–phenomenon concept can account for the polytypic phenomenon only in limited polytypes which show phase transformation. [4,5] All the models cannot account for the occurrence of short–period polytypes such as 2H– and 3C–SiC. There is as yet no complete model to account for polytype formation and polytype growth.

In this paper, a new electrostatic model in which electrostatic interaction is considered as the interatomic force is proposed to account for various phenomena related to the polytype of SiC.

2 Electrostatic model

Consider the bond structure of SiC and the difference between polytypes. In all polytypes, any Si or C atom has the same first and second coordination. Every atom has four first–nearest neighbors of the different kind and twelve second–nearest neighbors of the same kind. In a hexagonal close-packed system, if every other layer is in the different orientation from the preceding layer (Fig. 1(a)), hexagonal structure of 2H–type (wurtzite) is obtained. If every other layer is in the same orientation as the preceding layer (Fig. 1(b)), cubic structure of 3C–type (zinc blende) arises. The difference arises only in the third and fourth coordination of atoms between 2H–type and 3C–type.

In this study, it is assumed that the interatomic force for bonding comes from only electrostatic interaction. The outermost electrons of Si and C are treated as negative point charges. The atomic cores of Si and C are considered to have positive +4 point charges. The eight outermost electrons which contribute to the bonding are assumed to be slightly shifted to the C side because

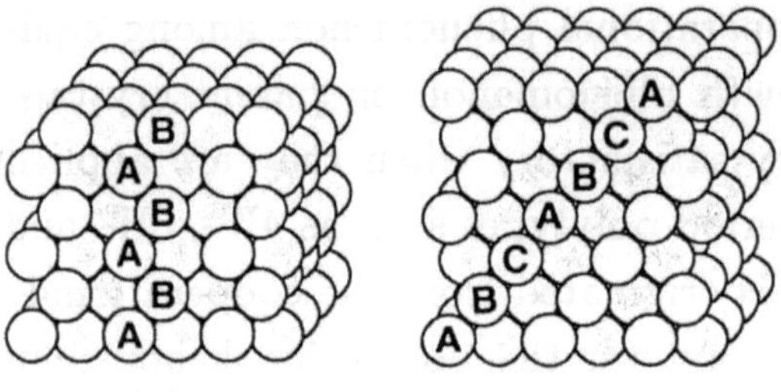

(a) (b)

Fig. 1

Schematic diagram of (a) wurtzite 2H– and (b) zinc blende 3C–structures. Individual sphere represents a pair of Si and C atoms.

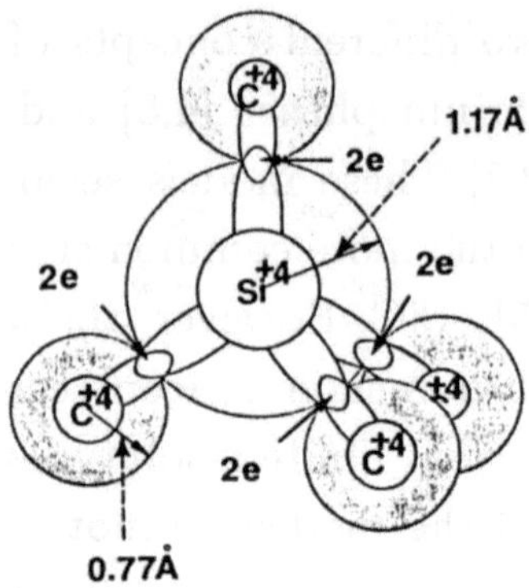

Fig. 2

Schematic diagram of tetrahedral bonds of SiC.

of the difference in characteristic covalent radii of Si (r_{Si} = 1.17 Å) and C atoms (r_{C} = 0.77 Å). The four electron pairs (2e) are placed at the position determined by the radius ratio of Si and C atoms along the tetrahedral bonds between Si^{+4} and C^{+4} (Fig. 2). The Coulomb potential is in inverse proportion to the distance r between the charges. The total electrostatic potential energy can be written by

$$E_{total} = E_{ce}+E_{cc}+E_{ee} = -\frac{1}{2}\sum_{i=1}^{n}\sum_{k=1}^{m}\frac{Z_i q^2}{r_{ik}}+\sum_{i>j}^{n}\frac{Z_i Z_j q^2}{r_{ij}}+\sum_{k>l}^{m}\frac{q^2}{r_{kl}}, \quad (1)$$

where E_{ce} is core–electron interaction energy, E_{cc} core–core interaction energy and E_{ee} electron–electron interaction energy, q electronic charge, Z_i positive charge of i-th atomic core, Z_j positive charge of j-th atomic core, respectively. The details of calculation will be presented elsewhere.

3 Simulation of polytype formation

Polytypism is defined, in general, as the ability of a substance to crystallize into various different structures owing to the existence of two possible sites on close–packed surfaces. The two sites are regarded as equivalent in geometrical sense. The atom arrangement in substrates may influence the surface potential, and the difference in the surface potential would exist between the sites. Polytype formation occurs during crystal growth, and it would be determined by the difference between the surface potential energy and the kinetic energy of an impinging atom.

For calculating the surface potential, the electrostatic interaction among the point charges of Si^{+4}, C^{+4} and electron pairs within a finite interaction length (radius) was considered. In the calculation, the interaction length r_{int} was changed. The following formula was used as a boundary condition.

$$\frac{4}{3} \times L_B < r_{int} < (n+1) \times \frac{4}{3} \times L_B \ , \qquad (2)$$

where L_B is the bond length of Si–C (1.89 Å) and n integer, respectively. The $(4/3) \times L_B$ is equal to the height of Si–C unit layer (2.52 Å) along the c–axis.

3.1 Growth on 6H–SiC

The Coulomb potentials at two different kinds of sites on the surface were calculated. One kind of sites takes the same orientation as the preceding layer, and the other kind of sites takes the different orientation from the preceding layer. Only those atoms within a certain interaction length were taken into account to calculate the potential. The interaction length r_{int} was varied from 1 to 20 of the Si–C double layer ($n = 1$ to 20 in eq. (2)).

The calculation of surface potential for a 6H–SiC substrate and growth simulation on it were carried out. As shown in Fig. 3,

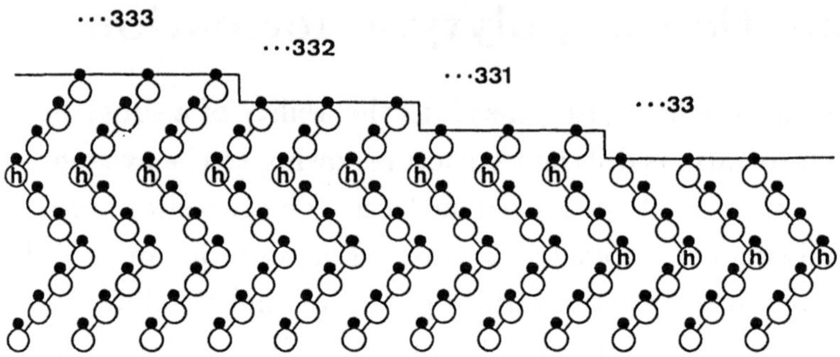

Fig. 3 Three kinds of terraces on the surface of 6H–SiC(0001).

6H-SiC has a zigzag stacking sequence and there are three kinds of terraces which preceded by 1 (...331: ...*hkkhkkh*), 2 (...332: ...*hkkhkkhk*), and 3 layers (...333: ...*hkkhkkhkk*) above a hexagonal–site layer (denoted by h : turning point) in the substrate. We consider two sites of ad–atom on each terrace ; non–6H site and 6H site. The surface potentials for two sites were calculated using eq.(1) where only atoms within an interaction length (r_{int}) were taken into consideration. The difference ($\Delta E = E_{\text{non-6H}} - E_{\text{6H}}$) between non–6H site and 6H site on the three terraces for various interaction length is shown in Fig. 4, where 6H site means the site by which the layer keeps 6H-SiC sequence, and non-6H site means the site by which the layer cannot keep 6H-SiC sequence. If the difference is positive, an ad–atom occupies the 6H–site. The difference in surface potential between the two sites becomes significant in the range of $n = 3 \sim 10$. The difference becomes small with the distance away from the hexagonal–site layer denoted by h, and it becomes a negative value on the (...333) terraces. As shown in Fig.4, on the (...331) and (...332) terraces, atoms prefer 6H site, whereas on the (...333) terraces, atoms prefer non-6H site. Assuming that atoms with lower kinetic energy occupy potential depressions, the above results predict the growth of 3C–SiC on 6H–SiC substrates. 3C–SiC on 6H–SiC may be accompanied with 180° twinning around the c–axis, because 6H–SiC has a zigzag stacking sequence as seen in Fig. 3.

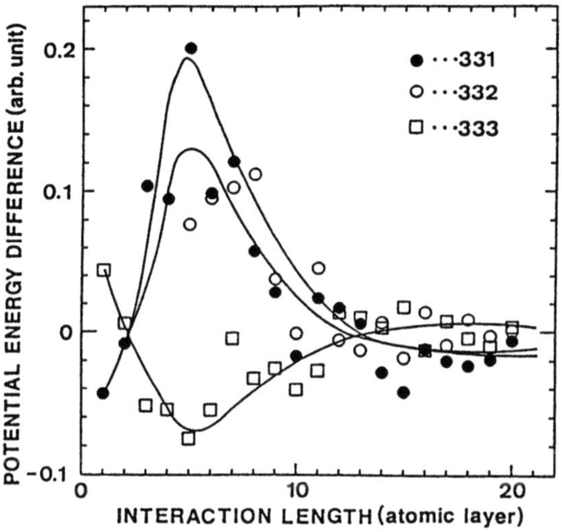

Fig. 4 Difference in potential energy between sites of 6H and non–6H on three kinds of 6H–SiC terraces.

Growth procedure is schematically shown in Fig. 5. Three initial terraces exist in 6H-SiC. Consider an ad–atom on the (...331) terrace. The potential energy difference takes a positive value with a large magnitude. An ad–atom may occupy a 6H–site, and a new (...332) terrace which is same as the second initial terrace appears. Similarly, an ad–atom on the second initial terrace occupies the 6H–site, and the third initial (...333) terrace appears. An ad–atom on the third initial terrace favors a non–6H site(the potential energy difference takes negative values). By repetition of growth on non–6H sites, 3C–SiC with double positioning twins grows on 6H–SiC{0001} substrates. However, the absolute difference of potential energy is the smallest on the (...333) terrace as shown in Fig.4. If an appropriate kinetic energy is given, the ad–atom may overcome the potential difference most easily and occupy a 6H site on the (...333) terrace. Then, the surface becomes (...3331: ...*hkkhkkhkkh*) terrace which is identically same as the first initial (...331) terrace. Therefore, growth of 3C–SiC on 6H–SiC is expected at low temperatures and 6H–SiC on 6H–SiC at high temperatures.

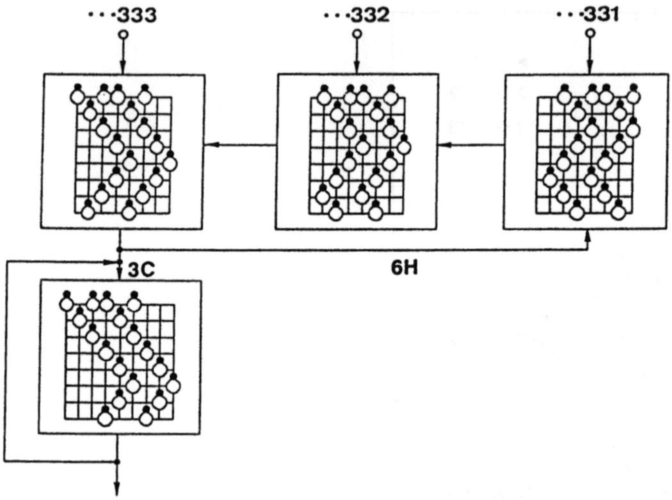

Fig. 5 Schematic illustration for growth sequence of 3C–SiC and 6H–SiC on 6H–SiC(0001) substrate.

Polytype of a layer grown on 6H–SiC{0001} basal planes changes from cubic 3C– to hexagonal 6H–type with increasing substrate temperature. [6–9] The critical temperature for 6H–SiC growth was determined to be about 1700°C. [8,9] The 3C–SiC layer grown on 6H–SiC{0001} substrates has special type of boundaries caused by 180° twinnings (*i.e.* double positioning boundary : DPBs). [10–13]

Several other experimental results on the growth of SiC on 6H–SiC can be well explained by this simulation. [14]

3.2 Possibility of DPB–free 3C–SiC growth

The stacking sequence of substrates influences the surface potential. Different results will be expected on substrates with different stacking sequences. Assuming that the n value is between 3~10, ad–atoms prefer cubic sites which has the same orientaton as the preceding layer. Since 6H–SiC has three types of terraces on its surface, the area ratio of ABC... stacking to ACB... stacking is 1:1 in the 3C–SiC grown layer.

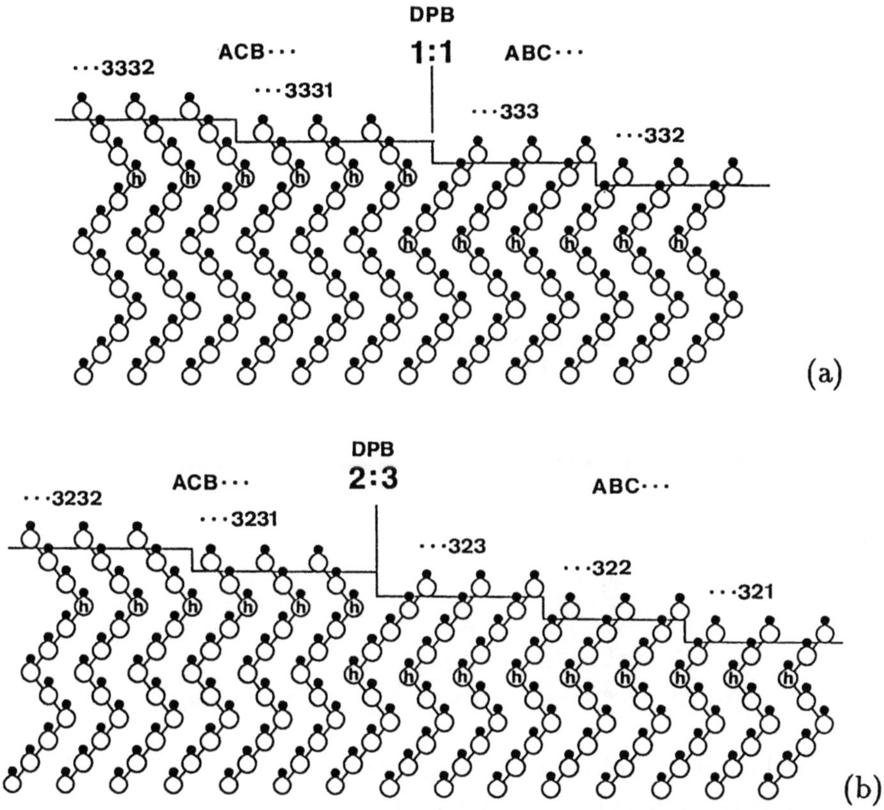

Fig. 6 Illustrations of twin crystalline 3C–SiC(111) containing DPB grown on (a) 6H–SiC (0001) and (b) 15R–SiC(0001).

On the other hand, 15R–SiC has a stacking sequence of ABCACBCABACABCB... $((32)_3 : (hkkhk)_3)$ and five kinds of terraces on its surface. It is expected by the same analogy that three of them lead to ABC... stacking in the 3C–SiC grown layer, and two of them bring about ACB... stacking. Therefore, the area ratio of ABC... stacking to ACB... stacking becomes 3:2 at the initial stage. Although DPBs still remain in 3C–SiC layers grown on 15R–SiC, the balance of the area ratio (1:1) in the 3C–SiC layer grown on 6H–SiC substrates is broken, and the area ratio becomes 3:2 (Fig. 6(b)). The area imbalance in the 3C–SiC layer grown on 15R–SiC may be enhanced during the growth.

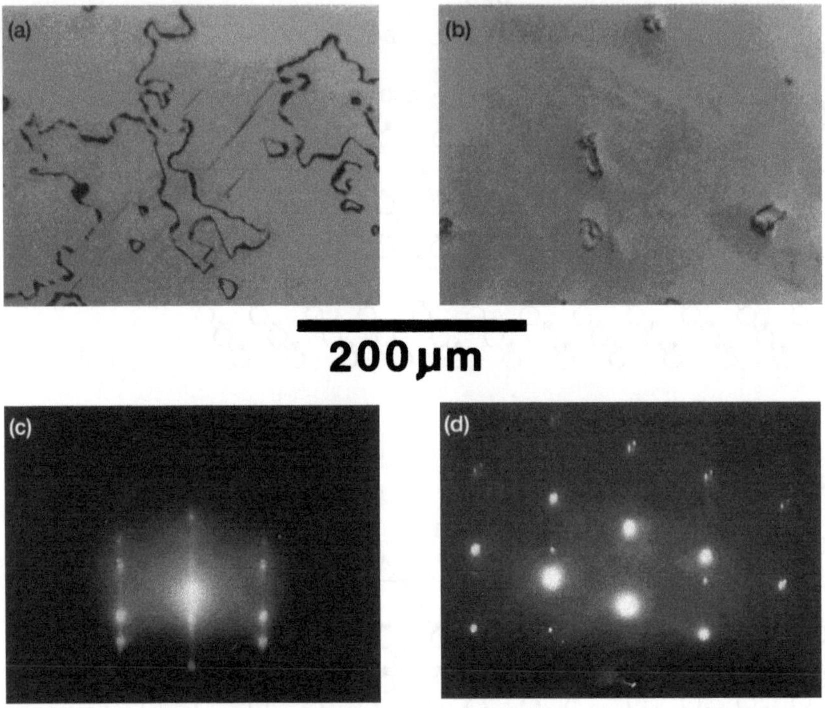

Fig. 7 Surface morphology of layers grown on (a) 6H–SiC(0001) and (b) 15R–SiC(0001), and RHEED patterns of layers grown on (c) 6H–SiC(0001) and (d) 15R–SiC(0001).

To ascertain this prediction, CVD growth on well-oriented 6H–SiC(0001) and 15R–SiC(0001) was simultaneously carried out at 1500°C. SiH_4(0.30sccm), C_3H_8 (0.20sccm) and H_2(3.0slm) were used as source gases and a carrier gas. The growth pressure and growth time were 1atm and 2h. Much different morphology was obtained as shown in Figs. 7(a) and (b). These results are typical over the entire wafer. On 6H–SiC, mosaic pattern caused by DPBs is observed (Fig.7(a)). However, small mosaic pattern is observed in the layer grown on 15R–SiC, and some of the boundaries do not come out clear (Fig.7(b)). While the area ratio seems to be 1:1 in the layer grown on 6H–SiC, a nearly single domain is grown on 15R–SiC.

Polytype of the both grown layers were examined by RHEED pattern observation. The observed RHEED patterns from the layers grown on 6H–SiC and 15R–SiC are shown in Figs. 7(c) and (d), respectively. The area irradiated with an electron beam in RHEED analysis is estimated to be 0.1x10mm. Thus, the area ratio of two domains appears as the diffraction intensity. The intensity of diffraction spots is much different each other. The layer grown on 6H–SiC gives equal intensity for the spots corresponding to the ABC... domain and ACB... domain. This supports the area ratio in the 3C–SiC layer grown on 6H–SiC is approximately 1:1. From the layer grown on 15R–SiC, much stronger spots corresponding to the ABC... domain than those corresponding to the ACB... domain(the area ratio of two domains is about 20:1) are observed. Considering the intensity ratio of the ABC... spots to ACB... spots, the layer grown on 15R–SiC is almost single domain 3C–SiC(111) (ABC... stacking). Thus, the experimental results agree well with the prediction.

4 Summary

A new electrostatic model for polytype formation and crystal growth was proposed. The Coulomb potential was calculated by changing the interaction length. The dependence of surface potential on the stacking sequence of atoms in substrates was investigated.

Growth simulation was tried using calculated results on the surface potential. The growth of 3C–SiC with DPBs on well–oriented 6H–SiC substrates can be explained. The possibility of DPB–free 3C–SiC growth on 15R–SiC was predicted by using the electrostatic model. By simultaneous CVD growth on well–oriented 6H–SiC(0001) and 15R–SiC(0001) at 1500°C, the prediction was verified.

Several experimental results on growth and polytype formation were explained using the model, and some growth features on different polytype substrates were predicted. The model proposed in this paper is not only very useful for understanding polytypism inherent in SiC but also for making predictions of resulting polytypes on different polytype substrates.

References

[1] *Polymorphism and Polytypism in Crystals* A.R.Verma and P.Krishna, (John Wiley & Sons, Inc. New York, 1966), Chapter 4.

[2] F.C.Frank, Phil. Mag. **42** (1951) 1014.

[3] L.S.Ramsdell and J.A.Kohn, Acta Cryst. **5** (1952) 215.

[4] C.J.Schneer, Acta. Cryst. **8** (1955) 279.

[5] H.Jagodzinski, Acta Cryst. **7** (1954) 300.

[6] V.J.Jennings, A.Sommer and H.Chang, J. Electrochem. Soc. **113** (1966) 728.

[7] W.v.Muench and I.Pfaffeneder, Thin Solid Films **31** (1976) 39.

[8] S.Nishino, H.Matsunami and T.Tanaka, J. Crystal Growth **45** (1978) 144.

[9] S.Yoshida, E.Sakuma, H.Okumura, S.Misawa and K.Endo, J. Appl. Phys. **62** (1987) 303.

[10] N.Kuroda, K.Shibahara, W.S.Yoo, S.Nishino and H.Matsunami, *Extended Abstracts 19th Int. Conf. on Solid State Devices and Materials, Tokyo, 1987* (Business Center for Academic Societies Japan, Tokyo, 1987) p.249.

[11] K.Shibahara, N.Kuroda, S.Nishino and H.Matsunami, Jpn. J. Appl. Phys. **26** (1987) L1815.

[12] T.Ueda, H.Nishino and H.Matsunami, J. Crystal Growth **104** (1990) 695.

[13] H.S.Kong, J.T.Glass and R.F.Davis, J. Appl. Phys. **64** (1988) 2672.

[14] W.S.Yoo and H.Matsunami, to be submitted.

Epitaxial Growth of Cubic SiC Using Various Alkyl-Silicon Compounds by Chemical Vapor Deposition

K. Takahashi[1], S. Nishino[1], J. Saraie[1], and K. Harada[2]

[1]Department of Electronics and Information Science, Faculty of Engineering and Design, Kyoto Institute of Technology, Matsugasaki, Sakyo-ku, Kyoto 606, Japan

[2]Electronic Materials Development, Toagosei Chemical Industry Co., Ltd., Funami-cho, Minato-ku, Nagoya 455, Japan

Cubic SiC was grown on Si(100) and (111) substrates by atmospheric pressure chemical vapor deposition using hexamethyldisilane (HMDS) or tetramethylsilane (TMS) as a source gas. The growth rate of SiC was independent of the growth temperature above 1050 °C in both HMDS/H_2 and TMS/H_2 system. Single crystalline 3C-SiC(111) was grown at 1100 °C and 1150 °C for HMDS/H_2 and TMS/H_2 system, respectively. In this case, the growth rate was as high as 700 Å/min. Crack lines were observed in the 5 μm thick epilayer, even though the growth temperature was as low as 1100 °C.

1. Introduction

Cubic silicon carbide (3C-SiC) is eminently suitable semiconductor material for high temperature, high power, and high frequency electronic devices, because of wide band gap (2.2 eV), high electron mobility (1000 cm^2/Vs[1]), high saturated electron drift velocity (calculated as 2.7×10^7 cm/s at 2×10^5 V/cm[2]), and thermal stability. 3C-SiC is also considered as a candidate material for an emitter of silicon hetero-junction bipolar transistors (HBTs). So much efforts have been devoted to produce single crystalline 3C-SiC on Si substrate at low temperature. However growth temperature to obtain epitaxial layer reproducibly by conventional atmospheric pressure chemical vapor deposition (CVD) method was quite high being around 1300 °C[3].

From the point of view of low temperature epitaxy, one source material containing both silicon and carbon atoms (alkyl-silicon compound) is suitable compared with two source materials[4-6]. Because this kind of compounds contain Si-C bonds in original molecule, it can be expected to bring crystal growth of SiC with ease even at low temperature. There are many reports on the use of alkyl-silicon compounds to obtain fine particles and ceramics of SiC, however few reports are published in the semiconductor field. Avigal et al. reported the crystal growth of 3C-SiC using various alkyl-silicon compounds, such as tetramethylsilane (TMS), diethylsilane (DES), and tripropylsilane (TPS)[7]. They obtained single crystalline 3C-SiC up to 2000 Å thickness on a Si(111) substrate and discussed the composition of obtained films. Powell et al. reported low-temperature epitaxy of 6H-SiC on 6H-SiC($11\bar{2}0$) substrate using methyltrichlorosilane (MTS)[8]. They only showed the effect of crystal orientation of 6H-SiC substrate and did not touch the characteristics of source gas. We previously reported the epitaxial growth of 3C-SiC on a Si(111) substrate using MTS[9]. However the advantages of the use of alkyl-silicon compounds compared with two source gas system have not been discussed. In this report, we will describe the crystal growth of 3C-SiC

Springer Proceedings in Physics, Vol. 71 Amorphous and Crystalline Silicon Carbide IV
Editors: C.Y. Yang · M.M. Rahman · G.L. Harris

on Si substrate using hexamethyldisilane ($Si_2(CH_3)_6$: HMDS) or tetramethylsilane ($Si(CH_3)_4$: TMS) as a source gas by atmospheric pressure CVD. The effects of substrate temperature on growth rate and crystallinity of SiC will be discussed. Using these alkyl-silicon compounds as a source gas, we will demonstrate the low-temperature epitaxy with relatively high growth rate.

2. Experiment

3C-SiC films were grown on Si(100) substrate with the <100> inclined 2° toward <011> and Si(111) substrate with the <111> inclined 4° toward $\langle 2\bar{1}\bar{1} \rangle$ by atmospheric pressure CVD using HMDS/H_2 and TMS/H_2 gas system. The source gases of HMDS (4N:research grade) and TMS (4N:research grade) used in this study were produced by Toagosei Chemical Industry co. The flow rate of HMDS and TMS was controlled by H_2 flow rate for bubbling. The temperature of the bubbler was held at 20 °C and -30 °C for HMDS and TMS, respectively. The horizontal reaction tube with inner diameter of 30 mm was used except for the experiment discussed in section 3.1. The susceptor was made of graphite coated with SiC and heated by 15 kW RF generator with a frequency of 300 kHz. The substrate temperature was estimated by measuring the temperature inside a 15 mm deep hole in the down-stream end of the susceptor with an optical pyrometer through the chamber end window. The difference between the measured and actual temperature was determined by melting a Si substrate under actual growth conditions.

To study the low-temperature epitaxy in various conditions, two different temperature programs were used. One is crystal growth using a carbonized buffer layer formation method[3,4]. This temperature program consisted of two steps; carbonization of a Si substrate by raising a substrate temperature from room temperature to 1250 °C under C_2H_2 and H_2 ambient, and subsequent CVD growth. By carbonizing a Si substrate, thin SiC epilayer (< 200 Å) was formed on a Si substrate. The other is crystal growth without buffer layer formation. In both temperature programs, gas phase etching of a Si substrate using HCl and H_2 mixture was employed prior to CVD growth. The substrate temperature was varied in the range of 900 - 1350 °C and the flow rate of HMDS and TMS was changed from 0.2 to 1.5 sccm. H_2 flow rate was kept at 3.0 slm.

The growth rate was measured by monitoring intensities of Ar laser (wavelength:488 nm) reflected from the surface of the growing layer. During crystal growth, the oscillation of reflection intensity took place by the interference effect of transparent SiC film. This method made the *in-situ* monitoring of the growth rate possible.

The surface morphology of the grown layer was observed by optical microscope (Nomarski interference contrast). To study the crystallinity of the grown layers, the high energy electron diffraction (RHEED) and X-ray diffraction (XRD) were employed.

3. Results and Discussions

3.1 Effect of Temperature on Growth Rate

Temperature dependence of the growth rate was studied in the range of 900 to 1350 °C. Only this experiment was carried out using a reaction tube with inner diameter of 50 mm. Fig.1 shows the Arrhenius plots of the growth rates and gas flow rates examined. This figure shows that the growth rates are saturated above substrate temperature of 1050 °C in both HMDS/H_2 and TMS/H_2 system. In this temperature region, the growth rate is limited by diffusion of chemical species through stagnant layer above the substrate. This saturation was not observed in $Si_2H_6/C_2H_2/H_2$ system as previously reported[4,10]. Below 1050 °C, the growth rate is exponentially decreased and controlled by certain chemical reaction. The

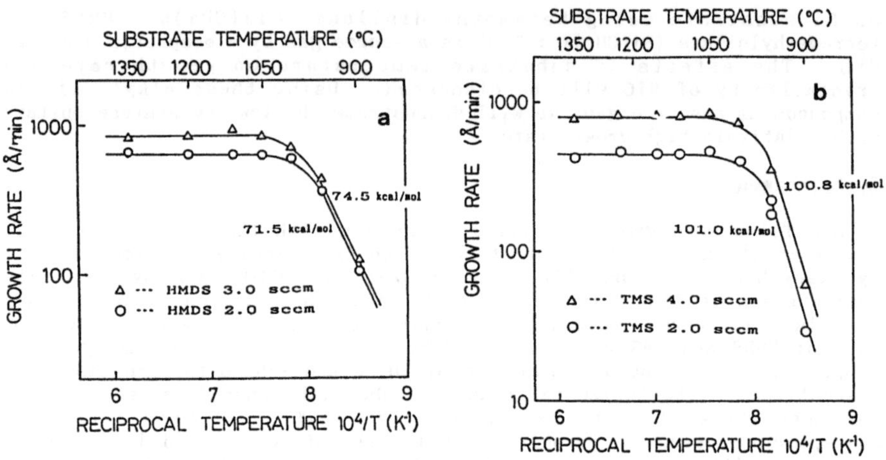

Fig.1 Influence of growth temperature on the growth rate of 3C-SiC films prepared by (a)HMDS/H_2 and (b)TMS/H_2 system. Only this experiment was carried out using a reaction tube with inner diameter of 50 mm. H_2 flow rate was kept at 4.0 slm.

activation energies (E_a) obtained in the exponential decrease of this curve is estimated as 71.5-74.5 kcal/mol and 100.8-101.0 kcal/mol for HMDS/H_2 system and TMS/H_2 system, respectively. These values are considered to be related with the rate limiting decomposition process of the source materials. The dissociation energies of $(CH_3)_3Si-Si(CH_3)_3$ bond in HMDS and $CH_3-Si(CH_3)_3$ bond in TMS is 80.2 kcal/mol[11] and 89.0 kcal/mol[12], respectively. Taking an experimental error into consideration, these dissociation energies almost correspond to E_a of SiC deposition. After first order dissociation of HMDS or TMS, the chemical species decomposed from source gas, which is thought to be radical molecule of $(CH_3)_3Si^*$, would show the same reaction in both cases of HMDS and TMS, because the growth rates show the same temperature dependence in both case.

3.2 Effect of Temperature on Crystallinity

Temperature dependence of the crystallinity was examined for HMDS/H_2 system with HMDS flow rate of 0.6 sccm, which was the optimum flow rates at substrate temperature of 1100 °C as shown in the next section. The Si(111) was used as a substrate. Crystallinity of 7000 Å thick layer was characterized by RHEED measurement. Grain growth and polycrystalline SiC was obtained at a substrate temperature of 1350 - 1300 °C without buffer layer. Using carbonized buffer layer formation before CVD growth, mirror-like single crystals were obtained at these temperature regions. The buffer layer suppressed the 3-dimensional nucleation and yielded quasi 2-dimensional mirror-like thin films even at high growth temperature. At a substrate temperature of 1200 - 1100 °C, single crystalline SiC containing twin crystals was obtained with and without buffer layer as shown in Fig.2(a)(b). At these low temperature regions, 2-dimensional nucleations occurred and mirror-like thin films were obtained even without buffer layer formation. Highly oriented SiC crystal was grown at 1050 °C and polycrystalline SiC was produced at 1000 °C as shown in Fig.2(c)(d). The epitaxial temperature of this system is as low as 1100 °C, which is about 150 °C lower than that of $Si_2H_6/C_2H_2/H_2$ system[4].

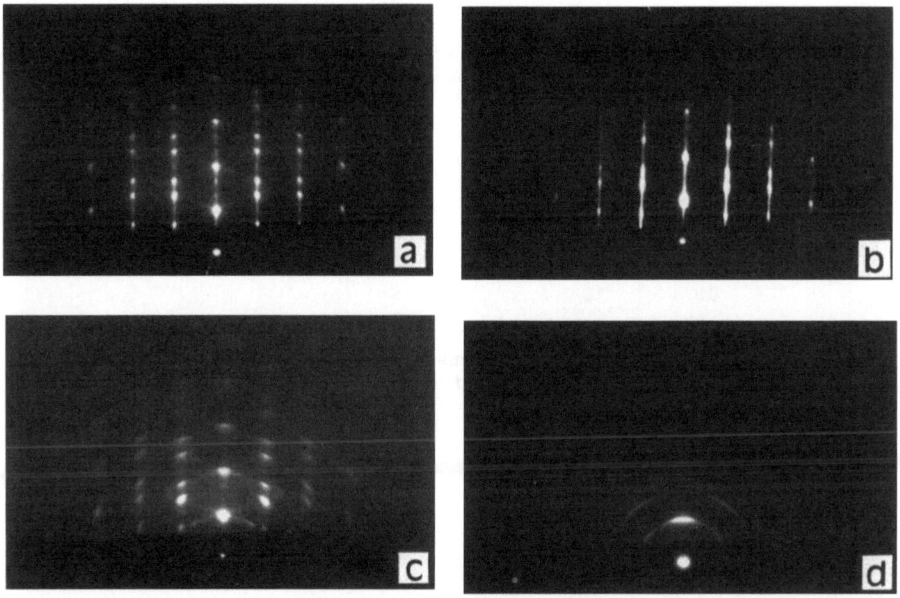

Fig.2 RHEED patterns from the layers on Si(111) grown at various substrate temperatures using HMDS/H_2 system. The growth temperature were (a)1200 °C, (b)1100 °C, (c)1050 °C, and (d)1000 °C. Carbonized buffer layer was not employed. The flow rate of HMDS was kept at 0.6 sccm.

Almost the same results were obtained in TMS/H_2 system. Single crystalline 3C-SiC films were grown on Si(111) at low temperature of 1150 °C with TMS flow rate of 0.5 sccm. This indicates that the decomposed species from source gases show the same reaction in both HMDS/H_2 and TMS/H_2 system.

We carried out the same experiment using Si(100) as a substrate. Without buffer layer formation, crystallinity of the layer grown on Si(100) was poorer than on Si(111) in both HMDS/H_2 and TMS/H_2 system. We could not obtain single crystal on Si(100) substrate without buffer layer formation at any substrate temperature. However, using buffer layer, single crystal was obtained at 1350 - 1100 °C. In contrast to the crystal growth on Si(111), buffer layer was necessary to the crystal growth on Si(100).

In the temperature region which single crystal could be grown, the E_a calculated from the Arrhenius plots of the growth rates has been reported for various gas system; for example, 20-25 kcal/mol, 15 kcal/mol, and 3.6 kcal/mol for $SiCl_4/C_3H_8/H_2$[13], $SiH_4/C_3H_8/H_2$[14], and $Si_2H_6/C_2H_2/H_2$[10] system, respectively. In case of HMDS/H_2 and TMS/H_2 system, E_a is almost 0 kcal/mole as described in section 3.1. These values are interpreted as the apparent potential barrier of the chemical reaction required for epitaxy. Therefore, the source gas system which bring smaller E_a is considered as more suitable gas system for low temperature epitaxy.

3.3 Low Temperature Growth without Buffer Layer

At substrate temperature of 1100 °C, we examined the effects of the flow rate of HMDS on the growth rate and crystallinity. The growth rates and RHEED patterns of the layer grown at various flow rates of HMDS were shown

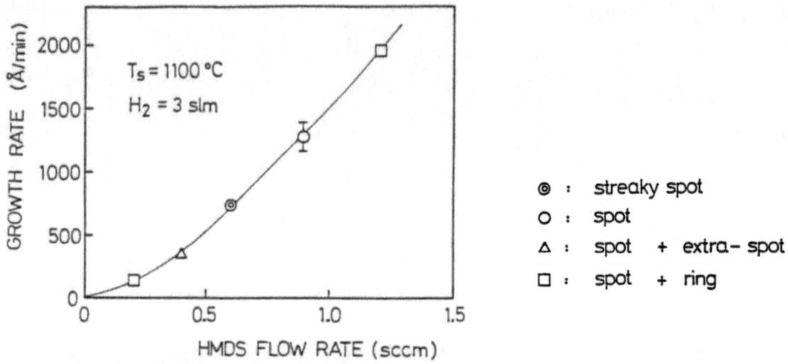

Fig.3 Growth rates and RHEED patterns of the layer grown at 1100 °C using $HMDS/H_2$ system. Si(111) was used as a substrate and buffer layer formation was not employed.

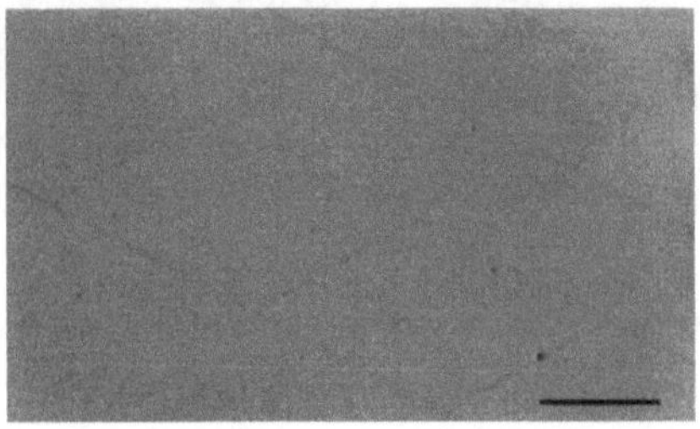

Fig.4 Nomarski microphotograph of the layer with 5 μm thickness prepared at 1100 °C. Si(111) was used as a substrate. Scale bar shows 200 μm.

in Fig.3. The thickness of the layers examined was approximately 7000 Å, and Si(111) was used as a substrate. Buffer layer was not employed. The growth rate was linearly increased by increasing the flow rate of HMDS. The crystallinity of the grown layer was very sensitive to the flow rate of HMDS, because 1100 °C was the lowest temperature for single crystal growth. We found that crystallinity was degraded if HMDS flow rate was too much or too low. At HMDS flow rate of 0.6 sccm, streaky spots pattern, which indicated single crystalline SiC with smooth surface, was obtained as shown in Fig.3 and 2(b). In this case, the growth rate was as high as 700 Å/min. The use of HMDS as a source gas has the advantage of growing SiC layer with desired thickness in short time even at low temperature. When Si(100) was used as a substrate, polycrystal was grown at any flow rate of HMDS without employing buffer layer.

If SiC layers were used for hetero-emitter of Si-HBTs, thin layers with thickness of less than a few thousand angstrom was enough. However the another interest arose to us whether we can obtain thick SiC layer in low temperature or not. The SiC layer with 5 μm thickness was grown on Si(111) at 1100 °C with the optimum growth conditions. The RHEED pattern from the thick grown layer shows strong spot patterns, which indicates single crystalline SiC was grown at the top surface of the layer. The XRD measurement revealed only 3C-SiC(111) plane was grown and no other planes of SiC and no Si phase were included. The full width at half-maximum of the rocking curve of SiC(111) peak was 0.80-0.85 degrees. The color of this layer after removing Si substrate was brownish yellow, and transparent. In case of 3C-SiC(111)/Si(111) grown at higher temperature than 1300 °C, it is reported that crack lines are observed in the SiC epilayer. They come from the tensile stress in the epilayer caused by large difference of the coefficient of thermal expansion between 3C-SiC and Si[15]. So, to decrease the growth temperature for epitaxy has been expected to solve the problem of cracking. The Nomarski microphotographs of 5 μm thick sample grown at 1100 °C was shown in Fig.4. The crack lines along the <110> direction were observed. Although we decreased the epitaxial temperature down to 1100 °C, we could not avoid the appearance of these crack lines. However, the existence of the crack lines with crystallographical symmetry is the another evidence for epitaxial growth of thick SiC layer even at low growth temperature.

4. Summary

3C-SiC was grown on Si substrates by atmospheric pressure CVD method using HMDS or TMS as a source gas. The growth rates of SiC was independent of the growth temperature above 1050 °C in both HMDS/H_2 and TMS/H_2 system. Below 1050 °C, the activation energy of SiC deposition was 71.5-74.5 kcal/mol and 100.8-101.0 kcal/mol for HMDS/H_2 and TMS/H_2 system, respectively. Single crystalline 3C-SiC(111) was grown at 1100 °C and 1150 °C for HMDS/H_2 and TMS/H_2 system, respectively. The crystallinity of the grown layer strongly depends on the orientation of the substrate, when buffer layer was employed. The Si(111) substrate was suitable for low temperature epitaxy comparing with the Si(100) substrate. Even though growth temperature was as low as 1100 °C, crack lines were observed in the 5 μm thick epilayer grown on Si(111).

References

[1] W.E.Nelson,F.A.Halden and A.Rosengreen: J.Appl.Phys.**37**(1966)333.
[2] D.K.Ferry: Phys.Rev.**B12**(1975)2361.
[3] S.Nishino,J.A.Powell and H.A.Will: Appl.Phys.Lett.**42**(1983)460.
[4] S.Nishino and J.Saraie: Springer Proceedings in Physics,**43**(1989)8.
[5] H.Matsunami,K.Shibahara,N.Kuroda,W.Yoo,and S.Nishino: Springer Proceedings in Physics,**34**(1989)
[6] H.S.Kong,J.T.Glass,and R.F.Davis: Appl.Phys.Lett.**49**(1986)1074.
[7] Y.Avigal,M.Schieber,and R.Levin: J.Crystal Growth**24/25**(1974)188.
[8] J.A.Powell and H.A.Will: J.Appl.Phys.,**44**(1973)5177.
[9] S.Nishino and J.Saraie: Springer Proceedings in Physics,**34**(1989)45.
[10] K.Takahashi,S.Nishino,and J.Saraie:ICVGE-7(J.Crystal Growth in press)
[11] By Iain M.T.Davidson and Anthony V.Howard:J.Chem.Soc.,Faraday Trans.1 **71**(1975)69.
[12] Robin Walsh: Acc.Chem.Res.**14**(1981)246.
[13] H.Matsunami,S.Nishino,and T.Tanaka: J.Crystal Growth**45**(1978)138.
[14] S.Nishino,Y.Hazuki,H.Matsunami,and T.Tanaka: J.Electrochem.Soc. **127**(1980)2674.
[15] H.P.Liaw and R.F.Davis: J.Electrochem.Soc.**131**(1984)3014.

Growth and Characterization of β-SiC Films Grown on Si by Gas-Source Molecular Beam Epitaxy

L.B. Rowland, S. Tanaka, R.S. Kern, and R.F. Davis

North Carolina State University, Department of Materials Science and Engineering, P.O. Box 7907, Raleigh, NC 27695-7907, USA

Abstract. Films of β-SiC have been grown on 4° off-axis Si (100) substrates from 1198 to 1398 K by gas-source MBE using Si_2H_6 and C_2H_4. Monocrystalline films were obtained at temperatures as low as 1248 K, as confirmed by electron diffraction. The latter films were specularly reflective. However, SEM revealed growth pits extending to the SiC surface as well as preferential growth on the pit edges. Cross-sectional TEM analysis confirmed the epitaxial relationship between the film and the Si substrate. Misfit dislocations and microtwins were also observed.

1. Introduction

At present, economics and availability make Si and a two-step chemical vapor deposition (CVD) the most commonly used substrate and process route for the growth of SiC films. However, mismatches in lattice parameters and coefficients of thermal expansion of 20% and 8%, respectively, exist between the two materials. The CVD process involves the reaction of the Si substrate surface with a C-containing gaseous species to form a "converted" layer of β-SiC [1] prior to subsequent deposition of the β-SiC film using both C- and Si-containing precursors.

However, the high growth temperatures and level of control of growth parameters and doping species obtained with CVD have sparked interest in the technique of gas-source molecular beam epitaxy (GSMBE) for growth of SiC [2-4]. The latter technique allows for precise control of growth parameters and minimization of sample contamination during deposition and was used in the studies described below. Experiments were conducted to determine the feasibility of SiC growth directly on Si (100) and to offer a comparison to films to be grown on converted layers in the near future. Effects of the C_2H_4/Si_2H_6 ratio and growth temperature on film morphology and quality were examined.

2. Experimental Procedures

Growth experiments were conducted using Si substrates primarily cut 4° off (100) toward the [011] direction. The use of off-axis substrates eliminates inversion domain boundaries (IDB's) [5,6]. The chemically cleaned (5 min H_2SO_4 (60°C), 1

Springer Proceedings in Physics, Vol. 71 **Amorphous and Crystalline Silicon Carbide IV**
Editors: C.Y. Yang · M.M. Rahman · G.L. Harris

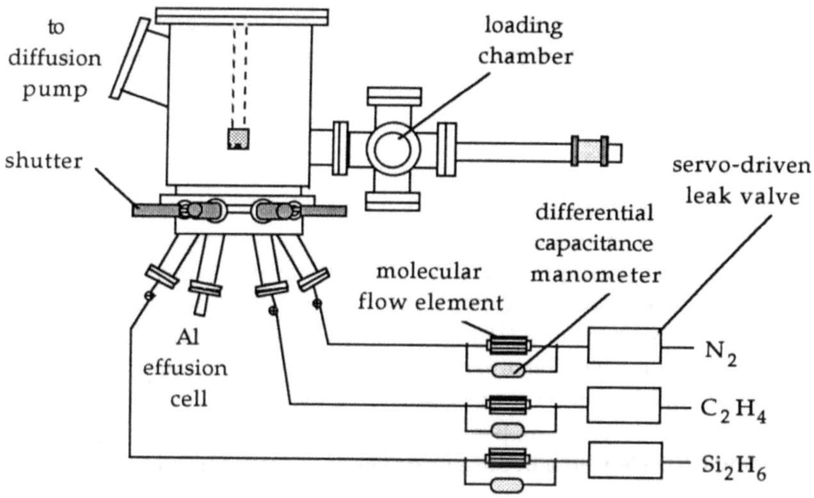

Figure 1. Schematic of molecular beam epitaxy apparatus

min DI rinse, 5 min 1:1 H_2O_2/NH_4OH (60°C), 1 min DI rinse, 5 min BOE, 1 min DI rinse) substrates were introduced into the growth chamber shown in Figure 1 (6×10^{-10} torr base pressure) through a load lock (5×10^{-8} torr base pressure) and resistively heated for 180 s at the growth temperature to desorb the native oxide layer. The heating assembly consisted of a tungsten wire coil surrounded by a SiC-coated graphite shell. Layers of W and Mo heat shielding were placed on the inside of this shell to reflect heat onto the backside of the sample. This heater is capable of long-term operation at constant temperatures (±3°C) to 1600 K, as measured with an infrared thermometer.

The introduction of ethylene and disilane (both 99.99% purity) into the growth chamber were controlled by a differential pressure transducer which measured the pressure drop across a flow element of essentially constant conductance [7,8]. The pressure drop across the element, ΔP, and the conductance of the element, C, were then converted to flow rate, Q, using the relation $Q = C\Delta P$. A servo-driven leak valve was used to stabilize the flow at the desired level via a feedback loop. In this manner the flow rate of a gaseous species (and consequently species flux) were known and controlled at flow rates substantially less than 1 sccm.

Growth conditions used in this study were as follows: temperature, 1298-1498 K; pressure, $4\text{-}6 \times 10^{-5}$ torr; Si_2H_6 flow rate, 0.40-2.0 sccm; C_2H_4 flow rate, 2.0 sccm; and time, 90 min. The resulting films were 30-50 nm in thickness. The chemical composition and depth profile of each sample were obtained using a scanning Auger microprobe. Samples which showed a 1:1 Si to C ratio from the Auger results were examined using an optical microscope and a field-emission scanning electron microscope (SEM) to observe surface morphology. Reflection high-energy electron diffraction (RHEED) was used after growth was completed to determine the crystalline quality of the SiC surface. Cross-sectional transmission electron microscopy (XTEM) also was employed to analyze representative β-SiC films. XTEM sample preparation procedures were similar to those used in Ref. 9.

3. Results and Discussion

Auger spectra from the surface of each sample showed the presence of an oxide from atmospheric exposure between growth and analysis. However, the underlying films in most samples were stoichiometric SiC. Figure 2 shows the Auger depth profile adjusted for differences in Auger yield of Si and C from a sample grown directly on off-axis (100) Si at 1298 K using 2.0 sccm C_2H_4 and 0.4 sccm Si_2H_6 for 90 min. The results indicate essentially a 1:1 SiC ratio. The oxygen profile within the film is representative of the background level for this Auger system. A monocrystalline β-SiC film previously grown in our laboratory on (100) Si using CVD showed a depth profile nearly identical to that found in the SiC film shown in the figure. Figure 3(a) shows the RHEED pattern ([110] beam azimuth) of the same sample used to obtain the Auger profile of Fig. 2. The RHEED pattern of a sample grown for 90 min on off-axis (100) Si at 1248 K with 2.0 sccm C_2H_4 and 1.0 sccm Si_2H_6 is shown in Fig 3(b). Both samples were monocrystalline β-SiC and were of similar crystalline quality. Twin spots were detected in the RHEED micrographs.

Auger results indicated that only certain C_2H_4:Si_2H_6 flow ratio and temperature combinations resulted in growth of single phase SiC. For example, Si-rich films were obtained for 2.0 sccm C_2H_4 and 0.4 sccm Si_2H_6 at 1248 K and at 1348 K. However, at 1298 K, the film grown using these flow rates was stoichiometric SiC. Films grown using 2.0 sccm C_2H_4 and 2.0 sccm Si_2H_6 were

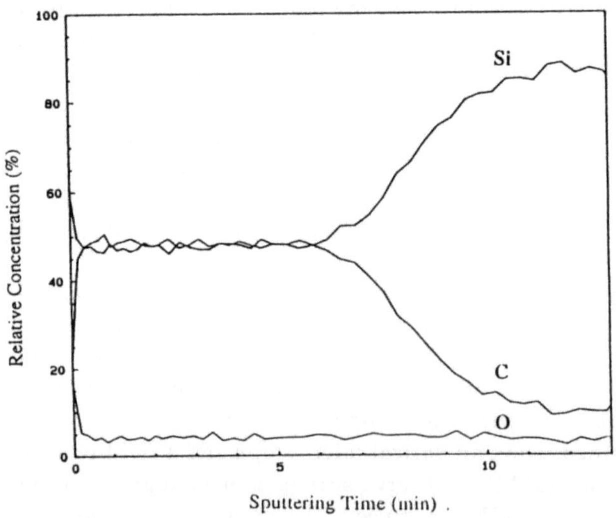

Figure 2. Auger depth profile of sample grown at 1075°C

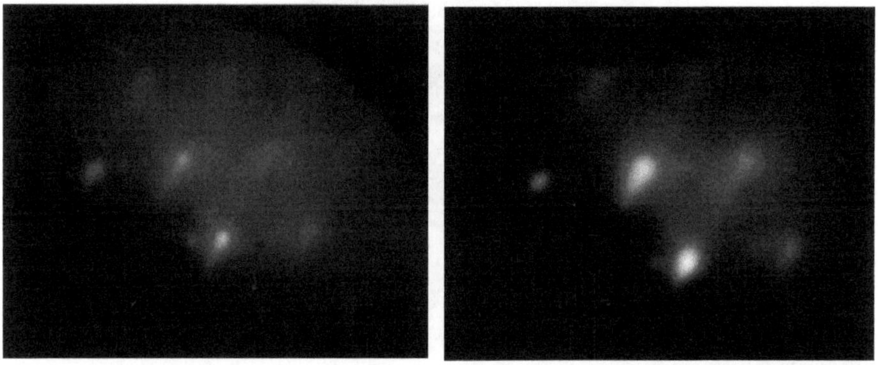

a) 1075°C, C/Si: 5:1 b) 975°C, C/Si: 2:1

Figure 3. RHEED of β-SiC films on Si (100) substrates

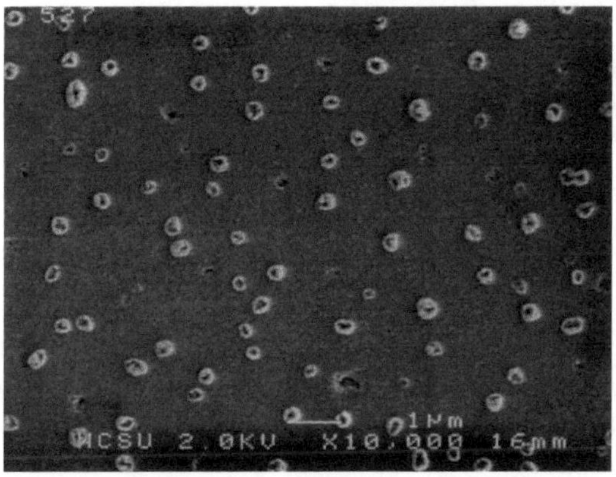

Figure 4. SEM micrograph of β-SiC grown at 1075°C showing preferential growth on pitted regions

also Si-rich at 1348 K, but were stoichometric at 1298 and 1248 K. Films grown using 2.0 sccm C_2H_4 and 1.0 sccm Si_2H_6 were nominally 1:1 SiC over the entire temperature range investigated. The reasons for these variations are being studied.

Films which were monocrystalline SiC appeared smooth to the naked eye and had a yellowish tinge. Figure 4 is an SEM micrograph of the surface of a sample grown at 1298 K using 2.0 sccm C_2H_4 and 0.4 sccm Si_2H_6. At this resolution, the surface continued to appear fairly smooth except for the presence of pitted regions on the SiC surface. Many of these pits, which are believed to be pyramidal and bounded by {111} planes [10], have preferential growth on their edges. Similar examination of other samples with differing gas source ratios and temperatures provided evidence that growth on the pit edges eventually converges above the center of the pit, forming a hillock above each pit [11].

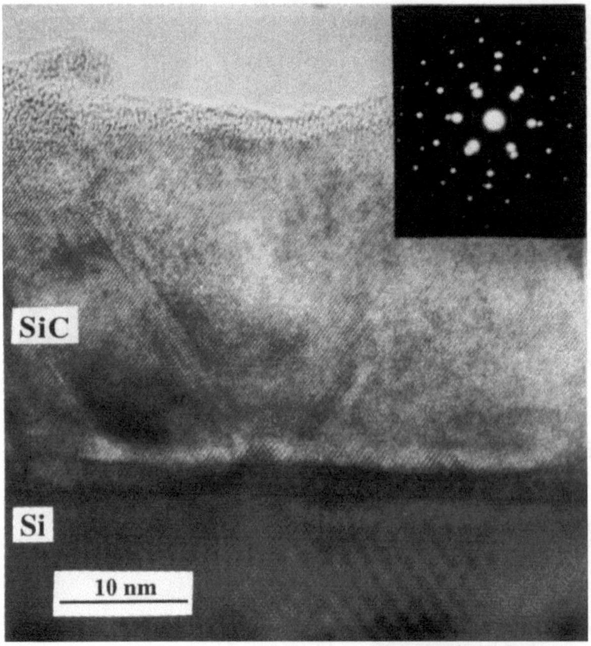

Figure 5: Cross-sectional TEM micrograph of SiC/Si sample grown at 1075°C with inset of selected area diffraction pattern of the SiC film and Si (100) substrate

Transmission electron microscopy (TEM) was used to evaluate the structural quality of the films. Figure 5 shows a high-resolution cross-sectional micrograph and corresponding (110) electron diffraction patterns of both the β-SiC film grown at 1298 K for 90 min and the Si substrate. Both the SiC/Si interface and the SiC surface were fairly rough. The region at the substrate/film interface appears to be amorphous in the micrograph; this is believed to be caused by damage from the ion milling used as part of the TEM sample preparation procedure [1]. A high concentration of defects, namely [111] microtwins and misfit dislocations, are readily visible in the SiC film in the figure. An epitaxial relationship between the SiC film and the Si substrate can clearly be seen in the figure.

Growth thicknesses were estimated from Auger depth profiles and correlated with thicknesses obtained for samples examined using XTEM. The average values were 30-50 nm for 90 min of growth. Experiments designed to increase the SiC growth rate by changing source flow rates or using more reactive, carbon-bearing species are currently in progress.

4. Conclusions

Monocrystalline β-SiC films were grown by GSMBE on off-axis Si (100) substrates using Si_2H_6 and C_2H_4 at temperatures as low as 1248 K. This temperature is much lower than that typically used for SiC CVD growth. Films appeared smooth and specular to the naked eye, but {111} pyramidal growth pits were present on the film surface. Preferential growth on the pit edges under certain growth conditions was also observed. Cross-sectional TEM analysis of selected SiC films showed the presence of [111] microtwins and visually confirmed the epitaxial relationship between the film and the Si substrate.

Acknowledgements

The authors wish to thank Z. Sitar and M. J. Paisley for the RHEED analysis and S. H. Rogers for the Auger analysis. This research was supported by the Office of Naval Research (Grant No. N00014-88-K-0341/P00002).

References

1. H. J. Kim, R. F. Davis, X. B. Cox, and R. W. Linton, J. Electrochem. Soc. **134**, 2269 (1987)
2. S. Motoyama and S. Kaneda, Appl. Phys. Lett. **54**, 242 (1989).
3. T. Fuyuki, M. Nakayama, T. Yoshinobu, H. Shiomi, and H. Matsunami, J. Cryst. Growth **95**, 461 (1989).
4. T. Sugii, T. Aoyama, and T. Ito, J. Electrochem. Soc. **137**, 989 (1990).
5. K. Shibahara, S. Nishino, and H. Matsunami, Appl Phys. Lett **50**, 1888 (1987).
6. J. A. Powell, L. G. Matus, M. A. Kuczmarski, C. M. Chorey, T. T. Cheng, and P. Pirouz, Appl. Phys. Lett. **51**, 823 (1987).
7. R. A. Kiesling, J. J. Sullivan, and D. J. Santeler, J. Vac. Sci. Technol. **15**, 771 (1978).
8. H. Ishikawa, H. Ando, K. Kondo, A. Sandhu, E. Miyauchi, T. Fujii, and S. Hiyamizu, J. Vac. Sci. Technol. A **8**, 805 (1990).
9. C. H. Carter, Jr., R. F. Davis, and S. R. Nutt, J. Mater. Res. **1**, 811 (1986).
10. A. Addamiano and J. A. Sprague, Appl. Phys. Lett. **44**, 525 (1983).
11. L. B. Rowland et al., to be published.

Atomic Layer Control of β-SiC(001) Surface

S. Hara[1], Y. Aoyagi[1], M. Kawai[1], S. Misawa[2], E. Sakuma[2], and S. Yoshida[2]

[1]The Institute of Physical and Chemical Research (RIKEN), 2-1 Hirosawa, Wako, Saitana 351-01, Japan
[2]Electrotechnical Laboratory, 1-1-4 Umezono, Tukuba-shi, Ibaraki 305, Japan

Abstract. Atomic layer control of cubic SiC(001) was investigated using gas reactions on the surface. Self-limiting growth at about one atomic monolayer at 1050°C is observed both for Si_2H_6 exposures onto a carbon terminated surface and for C_2H_2 exposures onto a silicon covered surface. Further, about one monolayer of a silicon-covered surface is etched by exposing to oxygen at 1050°C. The resultant etched surface is carbon-covered surface.

1. Introduction

Atomic layer epitaxy (ALE) of silicon carbide provides an opportunity for multi-hetero epitaxy such as Si/SiC/Si/SiC..., 6H/3C/6H/3C..., etc. which are new bandgap-controllable materials. There is only one report on precise-controlled epitaxy by T.Fuyuki *et al.*[1] They demonstrated homoepitaxial growth of β-SiC at 1000°C by alternative supply of Si_2H_6 and C_2H_2 under a partial pressure of 5×10^{-5} Torr. In this experiment, they observed an increase in growth rate per one cycle with increasing Si_2H_6 flux. In other words, a self-limiting growth was not found under this condition.

Another interest in a gas supply is atomic layer etching, which is also necessary to achieve complete atomic layer control. So far, although an attempt to etch one GaAs monolayer has been made by T.Meguro *et al.*[2], accurate one monolayer etching has not been achieved. They used thermal etching effect of an etching precursor adsorbed at room temperature. After raising the substrate temperature, they observed the removal of only half of the surface topmost atoms.

To deposit or etch one monolayer, knowledge of surface elemental compositions and surface atomic configurations is required. We have investigated the β-SiC surface elemental composition and the surface reconstruction elsewhere[3]. In that report, correlations between reconstruction phases of low energy electron diffraction (LEED) and surface composition were investigated. (2×1) and c(4×2) phases have a full silicon topmost layer, whereas a c(2×2) phase has a full carbon topmost layer. (3×2) and (5×2) phases have additional dimer rows on top of a silicon terminated crystal.

In this work, we investigate one monolayer homoepitaxy of β-SiC(001) crystal by alternative exposures of C_2H_2 and Si_2H_6, and one monolayer etching (desorption) of the surface by O_2 exposure.

Springer Proceedings in Physics, Vol. 71 **Amorphous and Crystalline Silicon Carbide IV**
Editors: C.Y. Yang · M.M. Rahman · G.L. Harris

2. Experimental

β–SiC(001) samples were grown on a Si(001) wafer by chemical vapor deposition with the buffer layer technique[4,5] that relaxes the lattice mismatch between Si(001) and β–SiC(001). The details of such growth have been described elsewhere[6]. The thickness of the heteroepitaxial β–SiC layer is about 2μm.

These samples were degreased by trichloroethylene, followed by a HF rinse to remove the native oxide. After cleaning, the samples were annealed at 1150°C for 10 minutes by direct current heating in a UHV chamber. After this treatment, a sharp c(2×2) LEED pattern was always observed at room temperature. In the previous work[3], the c(2×2) pattern was found to have a carbon terminated surface without any contamination and oxygen.

A silicon atomic layer was grown on the carbon terminated sample surface at 1050°C by Si_2H_6 exposure whose partial pressure was 1×10^{-6} Torr. After every 20 seconds' exposure, the Si_2H_6 supply was stopped, and the sample was cooled down to room temperature to observe the surface reconstruction with LEED and the surface composition with Auger electron spectroscopy (AES), and subsequently it was raised up to 1050°C, followed by the gas supply again.

A carbon atomic layer was grown at 1050°C onto the silicon covered sample by C_2H_2 exposure whose partial pressure was 1×10^{-6} Torr. The surface characterization procedure was the same as that for Si_2H_6.

Oxygen gas was introduced in a manner similar to the two growth experiments to etch one atomic layer of silicon–covered surface. The sample temperature during etching was 1050°C, and the oxygen partial pressure plays 1×10^{-7} Torr.

3. Results and Discussion

3.1. Silicon self–limiting growth

Fig.1 shows surface composition versus Si_2H_6 exposure time on a β–SiC(001)–c(2×2) carbon–terminated surface. The vertical axis indicates the AES Si_{LVV}/C_{KLL} peak–to–peak ratio estimated from raw data without using a relative sensitivity correction. The changes in the phases and the ratios from the c(2×2) to the (3×2) correspond to the surface composition change from a carbon–terminated to a silicon–covered surface. We believe that this change occurred by silicon epitaxial growth on the carbon terminated surface.

The saturation of the composition in the (3×2) region is seen in the figure. The LEED pattern had no change in the saturation region. Evidently these phenomena are attributed to the silicon self–limiting growth on the carbon–terminated surface. If extra silicon deposition continued, the composition ratio should increase because of the depletion of the inner carbon intensity.

The c(2×2) and the next (1×1) have the same composition ratio, suggesting that Si_2H_6 makes a role in releasing the c(2×2) structure from dimerization before the start of Si_2H_6 deposition. The reason for the different atomic configurations with the same Si/C composition may be ascribed to a change in surface dangling bond

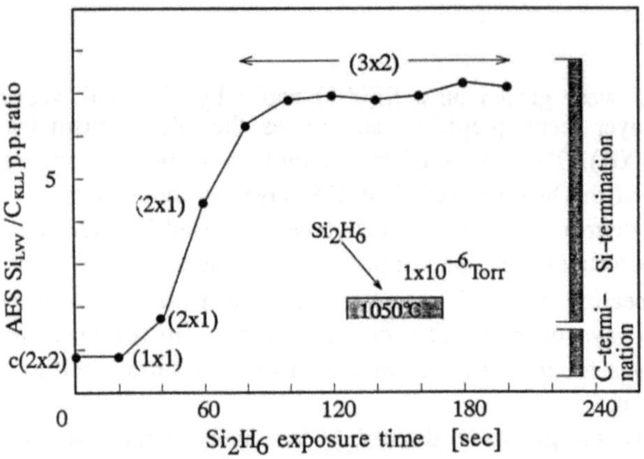

Fig.1 The transition of elemental composition from carbon to silicon termination by exposing Si_2H_6 at 1050°C. The vertical bars on the right side are just eye guides.

termination by hydrogen that can terminate each dangling bond generated by the collapse of the c(2×2) dimers. On the Si(001) surface, the phase transition from (2×1) to (1×1) by exposing to atomic hydrogen was observed[7], and the increase in surface hydrogen in the transition was also reported[8].

The pattern changes from the initial carbon terminated c(2×2) to the final silicon-covered (3×2) through the (1×1) and the (2×1) are consistent with our previous work on phase changes at a high temperature using medium ion energy scattering spectroscopy (MEIS) except that a (5×2) phase between the (3×2) and the (2×1) was observed in the measurements[3]. Since the (5×2) phase requires a long-range correlation among additional dimer rows that exist at every 5th normal dimer row, skipping the (5×2) phase is probable.

3.2. Carbon self-limiting growth

Fig.2 shows surface composition versus C_2H_2 exposure time on a β-SiC(001)-(3×2) silicon covered surface. The initial silicon-covered (3×2) phase changes into a carbon-terminated c(2×2) phase after 80 seconds' exposure. The AES Si_{LVV}/C_{KLL} peak-to-peak ratio decreases with the surface phase transition, leading to a constant composition at the c(2×2) region. We believe that about 1ML carbon deposits epitaxially on the silicon-covered (3×2) surface during the transition, followed by termination of the growth. Evidently the stopping is attributed to the carbon self-limiting growth on the silicon-covered surface similar to the silicon self-limiting growth discussed in section 3.1.

The straight line without data dispersion in the c(2×2) region suggests that the c(2×2) is somewhat a stabler structure than the (3×2) structure shown in Fig.1. We observed that the c(2×2) phase continues up to 1200 seconds' exposure, followed by a new (1×1) phase. The c(2×2) region is so much longer in duration than the transition region from the (3×2) to the c(2×2) that one can distinguish the c(2×2)

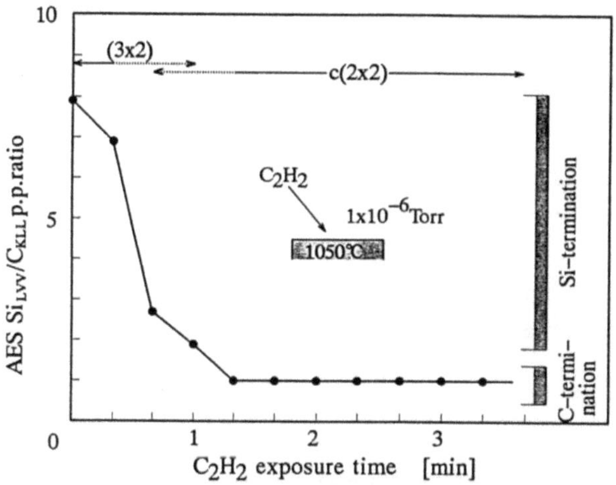

Fig.2 The transition of elemental composition from silicon–covered to carbon–terminated surface by exposing C_2H_2 at 1050°C. The dotted lines indicate a mixed region of (3×2) and c(2×2). The vertical bars on the right side are just eye guides.

phase from the (1×1), much more conveniently for ALE. The (1×1) phase that appears after the c(2×2) has been reported by Kaplan[9] and by Parrill *et al*[10]. From their electron energy loss spectroscopy measurements, they concluded that this (1×1) phase has the graphite structure.

The (3×2) and the c(2×2) phases are superimposed in the region from 40 to 60 seconds. In our additional dimer row model[3], the (3×2) structure has straight *silicon* dimer rows along the <110> directions whereas the c(2×2) has *carbon* dimer rows slanted along <100> directions. This suggests that the two phases co–exist at different locations with each other.

We did not observe any different phases between the (3×2) and the c(2×2) phases. All the spots in the transitional (2×1) and (1×1) LEED patterns observed in the experiments of the Si_2H_6 exposure can be overlapped on the higher ordered (3×2) spots, leading to hidden phases. Since a part of the (3×2) region endures after the c(2×2) structure appears, the confirmation of the transitional phases is impossible as far as macroscopic LEED is used. The transitional (5×2) phase is also difficult to observe as we discussed earlier, although it has different spots as compared to the (3×2) phase.

3.3. Atomic layer etching by oxygen

We used O_2 as an etching precursor without using a halogen gas that often causes damage to a UHV chamber because of its strong etching effect. Although it is well known that oxygen often oxidizes a surface, the etching effect of oxygen is not well known. J.J.Lander and J.Morrison first reported the etching effect of oxygen for Si(001) and (111) surfaces[11]. They found that in the regime of high substrate temperatures and low O_2 pressures etching occurs, whereas in the regime of low

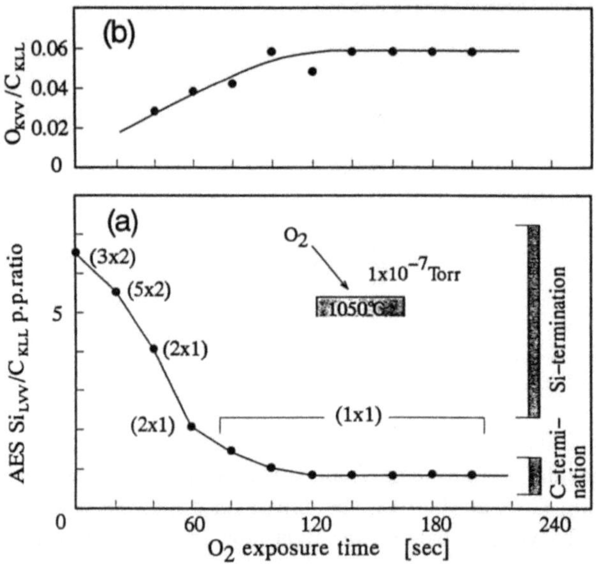

Fig.3 The transition of elemental composition from the silicon-covered (3×2) to the carbon-covered (1×1) surface by exposing to O_2 at 1050°C. (a) AES Si_{LVV}/C_{KLL} peak-to-peak ratio. (b) AES O_{KVV}/C_{KLL} p.p.ratio. From figure (b), it is found that the surface oxygen amount is considerably smaller than the carbon amount.

temperature and high O_2 pressure oxidation occurs. In the former regime, a surface silicon atom desorbs by forming a SiO molecule.

Fig.3(a) shows surface composition versus O_2 exposure time on the (3×2) silicon-covered surface. The initial silicon-covered (3×2) phase changes into a carbon-covered (1×1) phase after 80 seconds exposure. The AES Si_{LVV}/C_{KLL} peak to peak ratio decreases with the surface phase transition, leading to a constant ratio at the (1×1) region. We believe that about 1ML silicon desorbs during the transition, followed by termination of the desorption when the next carbon layer appears.

Fig.3(b) shows that the oxygen on the surface is saturated at a considerably smaller amount of 0.06 AES O_{KVV}/C_{KLL} peak-to-peak ratio. Since the usual relative sensitivity of oxygen KVV transition is several times larger than the carbon KLL transition, the actual O/C ratio on the carbon surface is less than 0.06.
This low oxygen concentration on the surface indicates that oxygen at 1050°C under 1×10^{-7} Torr on the β-SiC(001) surface is in the desorption mode, not in the oxidation mode.

The absence of the c(2×2) phase as shown in the Fig.3(a) may be due to surface oxygen atoms that can disturb the surface periodicity as well as the oxygen on silicon surfaces. This disturbance does not penetrate to deeper layers because we can always observe a typical (3×2) surface without oxygen by exposing to Si_2H_6 on the (1×1) surface again. Note the effect of a silicon flux to remove oxygen on the β-SiC(001) surface.

Extremely low etching effect of oxygen from the carbon surface may be attributed to a larger heat of formation of the product CO(−110.57kJ/mol) than that of SiO(−100.4kJ/mol).

4. Conclusions

We have found that the β−SiC(001) surface has self-limiting mechanisms of growth for both the silicon surface and the carbon surface. Silicon growth by Si_2H_6 exposure at 1050°C under 1×10^{-6} Torr onto the c(2×2) carbon-terminated surface stops at about 1ML growth, which shows the (3×2) silicon-covered surface. Whereas carbon growth by C_2H_2 exposure onto the (3×2) surface under the same condition as the Si_2H_6 exposure stops at about 1ML growth, which shows the c(2×2) carbon surface.

Further, we have found that oxygen etches the (3×2) topmost silicon layer under an exposure condition of 1×10^{-7} Torr at 1050°C. The resultant carbon surface shows no change in the surface elemental composition and the (1x1) LEED pattern by exposing to extra O_2.

Acknowledgment

This work was partially supported by a Grant-in Aid for Scientific Research of the Ministry of Education, Science and Culture, Japan.

References

1 T.Fuyuki, M.Nakayama, T.Yoshinobu, H.Shiomi, and H.Matsunami, J.Cryst.Growth, **95**,461 (1989).

2 T.Meguro, M.Ishii, H.Kodama, M.Hanagaki, T.Hara, Y.Yamamoto, and Y.Aoyagi, Extended Abstracts of the 22nd(1990 International) Conference on Solid State Devices and Materials, Sendai, 893 (1990).

3 S.Hara, W.F.J.Slijkerman, J.F.van der Veen, I.Ohdomari, S.Misawa, E.Sakumka, and S.Yoshida, Surf.Sci. **231**,L196 (1990).

4 S.Nishino, J.A.Powell, and H.A.Will, Appl.Phys.Lett. **42**,460 (1983).

5 A.Suzuki, K.Furukawa, Y.Higashigaki, S.Harada, S.Nakajima, and T.Inouguchi, J.Cryst.Growth, **70**,287 (1984).

6 S.Yoshida, K.Sasaki, E.Sakumka, S.Misawa, and S.Gonda, Appl.Phys.Lett. **46**,766 (1985).

7 J.J.Boland, Phys.Rev.Lett. **65**,3325 (1990).

8 K.Oura, J.Yamane, K.Umezawa, M.Naitoh, F.Shoji, and T.Hanawa, Phys.Rev. **B41**,1200 (1990).

9 R.Kaplan, Surf.Sci. **215**,111 (1989).

10 T.M.Parrill and Y.W.Chung, Surf.Sci. **243**,96 (1991).

11 J.J.Lander and J.Morrison, J.Appl.Phys. **33**,2089, (1962).

Growth and Characterization of 6H-SiC Bulk Crystals by the Sublimation Method

K. Koga, Y. Fujikawa, Y. Ueda, and T. Yamaguchi

Semiconductor Research Center, Sanyo Electric Co., Ltd.,
1-18-13 Hashiridani, Hirakata, Osaka 573, Japan

Observations regarding a number of defects in 6H-SiC single crystals grown by the sublimation method and the dependence of these defects on source materials and pressure are reported in this paper. A SiC source with unintentional impurities resulted in the creation of a number of striated defects during the initial stage of crystal growth. Three kinds of etch pits (EP-1, EP-2 and EP-3) were observed. The incidence of small-size EP-3 etch pits was found to be effectively reduced by increasing the processing pressure. 6H-SiC single crystals with a lower defect density were fabricated using pure source material at 2300°C: Etch pit densities were $<1\times10^3$, 5.2×10^3 and 2.2×10^4 cm^{-2} for EP-1, EP-2 and EP-3, respectively.

1. Introduction

Since controlled fabrication of single-crystalline ingots of 6H-SiC using the sublimation method was begun [1-4], several types of reliable semiconductor devices have been produced utilizing the ingots as substrates, [5-10]. These include high temperature devices, a blue light emitting diode (LED) and a full-color LED lamp. From the standpoint of device capabilities and mass-productivity, single crystals of high quality and homogeneity are desirable. Defects in single crystals should therefore be reduced as much as possible.

Inherent defects in ingots prepared by the sublimation method using seed crystals, however, have not been thoroughly investigated [2]. Although a few papers did report etch pits and dislocations in single crystals prepared by the Acheson process and conventional Lely methods [11-13]. In this paper, we describe the effect of certain defects on LED characteristics in substrates prepared by the sublimation method. We also report on the growth of 6H-SiC single crystals using pure source material and the dependence of etch-pit densities on reaction chamber pressure.

2. Effect of certain defects on LED characteristics

SiC blue LED's were fabricated using the usual processes as reported in reference [14]. Single crystalline ingots prepared by the sublimation method were used as substrates for the blue LED's. In the fabrication processes for blue LED's, certain striated defects in the substrates were found in a few of the LED's. A cross-sectional optical observation of one LED chip with defects and the surface morphology of the chip are shown in Fig. 1. Three lines, corresponding to pore-like defects, were observed by optical transmission microscope. These striated defects were almost perpendicular to the substrate with an off-oriented {0001} face and reached the top surface of the substrate. This resulted in the formation of pits in the epitaxial layer, as shown in Fig. 1 (b). Figure 2 shows the I-V characteristics of the blue LED with defects shown in Fig. 1, as well as those of a defect-free one. In the forward-bias part of the I-V curve, leakage current of the LED with defects was a remarkable 0-2 V. The break-down voltage of the defective LED was substantially lower than that of a normal LED due to leakage currents in the reverse bias part of the I-V curve.

Most LED's having striated defects and pits under the electrode showed leakage currents and low break-down voltages. The above findings are explained as follows. Small pores, corresponding to striated defects, existed in the surface of the substrate. In the epitaxial growth process, epitaxial layers might not have grown on parts of the pores, and pits on the surface resulted. Therefore, a p-n junction might not be formed on those parts. In the next process,

Springer Proceedings in Physics, Vol. 71 Amorphous and Crystalline Silicon Carbide IV
Editors: C.Y. Yang · M.M. Rahman · G.L. Harris

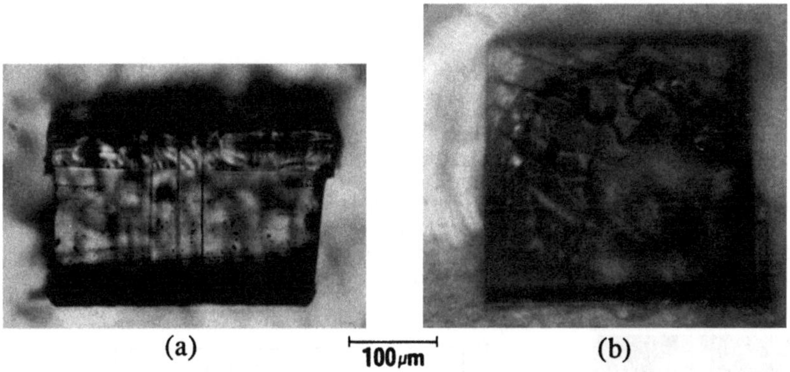

Fig. 1 Micrographs of the LED chip with defects: (a) cross-sectional view, three lines in the substrate were observed due to the transparency of the substrate. (b) plan view. Surface morphology was shown through the substrate. The round pattern was the p-type electrode.

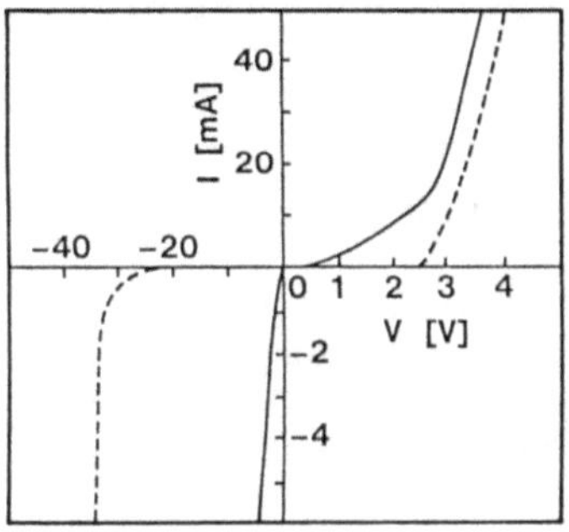

Fig. 2 I-V characteristics of the blue LED with the defects and a defect-free one. The solid line denotes the LED with defects and the dotted line denotes the defect-free one.

metallization for the p-type electrode, Al/Ti used for the electrode might have penetrated into the pits and reached the substrate. Leakage current was thus able to flow through the part of the pits filled with Al/Ti.

3. Experiment

In order to improve the quality of single crystals and study defect dependency, crystal-growth experiments were performed under the following conditions. Single crystals of 6H-SiC were grown on 6H-SiC substrates. Seed-crystals with a (0001) Si-face made by the sublimation method in an Ar ambient were used. Graphite crucibles, where the seed and the polycrystalline SiC powder source materials were placed, were heated using the R. F. induction method. The growth temperature, as monitored at the bottom of the crucible, was varied from 2300°C to 2400°C. The temperature gradient between the seed and the source material was 10-20°C/cm. The pressure in the reaction chamber was varied from 2 Torr to 10 Torr. Two kinds of SiC powders were used as source materials: industrial-grade powder made by the Acheson process and a pure powder created by the carbothermal reduction. The two were quite different in their levels of unintentional impurities: 1200 ppm for the industrial grade powder and <45 ppm for the pure powder.

Grown crystals were characterized by cross-sectional optical observation and etching evaluation. The etching was carried out at 400°C and 500°C for 1-30 minutes in molten KOH.

4. Results and Discussion

4-1. Relationship between etch pits and striated defects

A typical etched surface for the grown crystals is shown in Fig. 3(a). Three kinds of etch pits, all having the same hexagonal shape but different diameters, were found in the surface. The dependence of pit-diameter on etching time is shown in Fig. 3(b). This result shows that the three pits have different etching rates, ~1, ~0.6 and ~0.1 μm/h, for EP-1 (large), EP-2 (medium) and EP-3 (small), respectively. Through repeated lapping and etching in the c-axis direction, these etch pits were found to be continuous and identified as dislocations.

Striated defects like pores, observed in a cross-sectional view by optical transmission microscope, were found to correspond exactly to EP-1s in the surface with the (0001) Si-face.

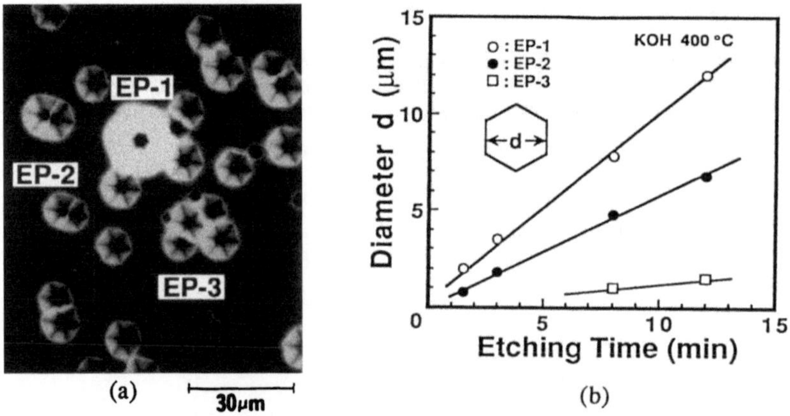

Fig. 3 (a) Micrograph of the etched surface with (0001) Si-face. The etching was carried out using molten KOH at 500°C. Three types of etch pits with different diameters were observed. (b) dependence of pit diameter: d on etching time.

4-2. Effect of unintentional impurities in the source

Two types of SiC powders were used as sources for the crystal growth. Figure 4 shows cross-sectional views of crystals grown near the interface between the seed and grown layers. A dark layer and striated defects were observed in the initial stage of crystal growth

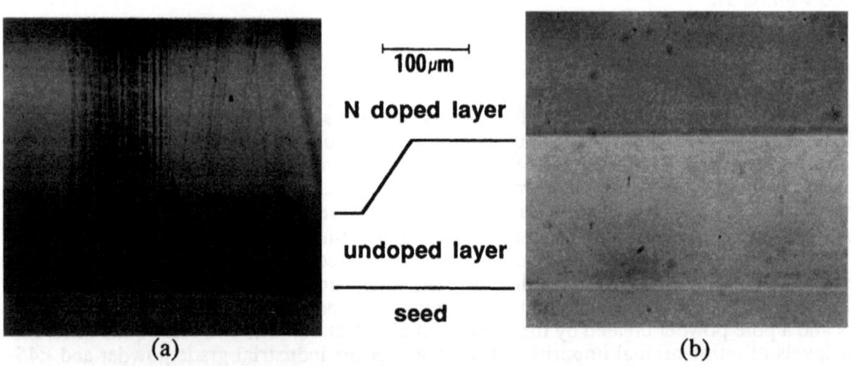

Fig. 4 Cross-sectional optical microscopic views of grown crystals near the interface between the seed and grown layers, when (a) industrial-grade SiC powder and (b) pure SiC powder were used as source materials.

when industrial-grade SiC powder was used. Some of the striated defects extended as far as the crystal surface. On the other hand, no dark layer and only rare striated defects were seen in the case of pure SiC powder. As mentioned in Sec. 4-1, EP-1 etch pits in the surface corresponded to striated defect. The number of EP-1s could be reduced to lower than 10^3 cm^{-2} by using pure SiC powder as the source material. Dislocations which correspond to EP-2 and EP-3 type defects could not be observed by optical transmission microscope. It was thought that differences in impurity levels between the two types of SiC powders would divide the initial stage of growth into two patterns: A dark layer would be generated by the high level impurities found in industrial-grade SiC powder. Excess impurities might act as a disturbance in well-controlled epitaxy, and result in the generation of striated deffects related to EP-1s.

4-3. Dependence of EPD's on reaction chamber pressure

Grown crystals were examined by etch pit density (EPD). We first observed dislocation behavior during crystal growth. Figure 5 shows the dependence of EPD's on the distance from the seed in the c-axis direction. EPD's of EP-2 and EP-3 type were slightly reduced with increased distance. The quality of the grown crystal was also noted to improve with a longer growth time. Figure 6 shows the dependence of EPD's of types EP-2 and EP-3 on the pressure in the chamber. EPD's of all samples were measured at a point 800 μm from the seed.

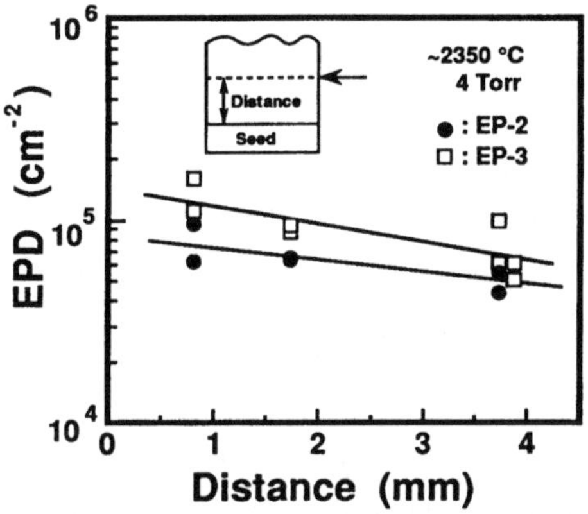

Fig. 5 Relation between EPD's and the distance from the seed in the c-axis direction.

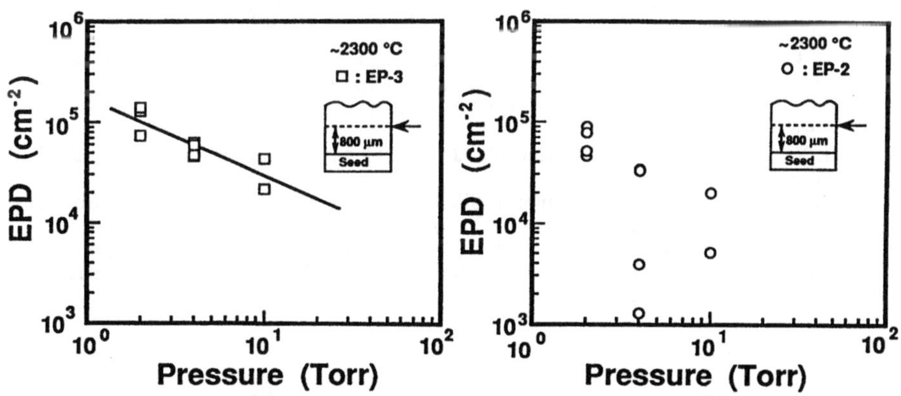

(a) (b)

Fig. 6 Dependence of EPD's on pressure in the chamber: (a) EP-3 (small-size etch pit, (b) EP-2 (medium-size etch pit). Growth temperature was 2300°C.

The growth temperature was fixed at 2300°C. The growth rates were 740, 400 and 170 μm/h for 2, 4 and 10 Torr, respectively.

EPD's of type EP-3 increased as the chamber pressure was decreased. This result is explained as follows. The pressure in the chamber can control the gas-transportation process from source to seed. Transportation of gases, sublimated from SiC sources, increases with a decrease in pressure. Therefore, with decreased pressure, supplied gases increase near the crystal growth region. Excess gases generate defects in the crystal. The density of EP-3 defects was 2.2×10^4 cm^{-2} at 2300°C and a pressure of 10 Torr. Clear dependence of EP-2s on pressure was not observed. EP-2s were often found at the center of spiral patterns on the surface. Therefore EP-2s may be caused by screw dislocations. We believe the density of EP-2s to be partially dependent on conditions at the initial stages of crystal growth. The density of EP-2s was 5.2×10^{-3} cm^{-2} at 2300°C and a pressure of 10 Torr.

5. Conclusion

It was confirmed that the leakage current in blue LED's was caused by striated defects corresponding to EP-1s. 6H-SiC single crystals with low defect density were grown by the sublimation method using pure source material. Densities of EP-1, EP-2 and EP-3 were $<1\times10^3$, 5.2×10^3 and 2.2×10^4 cm^{-2} at conditions of 2300°C and 10 Torr. Small-size etch pits (EP-3s) were found to be dependent on the pressure in the reaction chamber and their density was reduced when the pressure was increased.

References

1. Yu. M. Tairov and V. F. Tsvetkov: J. Cryst. Growth 52 (1981) 146.
2. G. Ziegler, P. Lanig, D. Theis and C. Weyrich: IEEE Trans. Electron Devices, ED-30 (1983) 277.
3. K. Koga, T. Nakata and T. Niina: Extended Abstr. 17th Conf. Solid State Devices and Materials (Tokyo 1985) 249.
4. D. L. Barrett, R. G. Seidensticker, W. Gaida, R. H. Hopkins and W. J. Choyke: J. Cryst. Growth 109 (1991) 17.
5. L. Hoffmann, G. Ziegler, D. Theis and C. Weyrich: J. Appl. Phys. 53 (1982) 6962.
6. H. S. Kong, J. T. Glass and R. F. Davis: J. Appl. Phys. 64 (5) (1988) 2672.
7. L. A. Tang, J. A. Edmond, J. W. Palmour and C. H. Carter, Jr.: in Extended Abstracts of the 1989 Fall meeting of the Electrochemical Society (1989) 705.
8. Y. Matsushita, T. Nakata, Y. Ueda, T. Uetani, Y. Fujikawa, K. Koga and T. Niina: Proc. Japan Display '89 (Kyoto, 1989) 696.
9. Y. Matsushita, T. Nakata, T. Uetani, T. Yamaguchi and T. Niina: Jpn. J. Appl. Phys. 29 (1990) L343.
10. K. Koga, T. Nakata, Y. Ueda , Y. Matsushita and Y. Fujikawa: in Extended Abstracts of the 1989 Fall Meeting of the Electrochemical Society (1989) 689.
11. M. J. Jennings, In Silicon Carbide 1968, H. K. Henisch and R. Roy: Eds., New York, Pergamon Press (1969) S199.
12. J. W. Faust, Jr.: in Silicon Carbide, J. R. O'Connor and J.Smiltens, Eds. London, Pergamon Press (1960) 403.
13. S. Amelinckx and G. Strumane: in Silicon Carbide, J. R. O'Connor and J. Smiltens, Eds., London, Pergamon Press (1960) 162.
14. T. Nakata, K. Koga, Y. Matsushita, Y. Ueda and T. Niina: in Amorphous and Crystalline Silicon Carbide II, M. M. Rahman, C. Y. W. Yang and G. L. Harris, Eds., Springer-Verlag (1989) 20.

Liquid Phase Epitaxy of SiC-AlN Solid Solutions

V.A. Dmitriev, L.B. Elfimov, I.Yu. Lin'kov, Ya.V. Morozenko, I.P. Nikitina, V.E. Chelnokov, A.E. Cherenkov, and M.A. Chernov

A.F. Ioffe Physical-Technical Institute, Russian Academy of Sciences, Politechnicheskaya 26, 194021 St. Petersburg, Russia

Abstract: In this paper we described $(SiC)_{1-x}(AlN)_x$ layers and pn structures grown by container free liquid phase epitaxy. Auger measurements have indicated that the concentration of the AlN component in the epitaxial layers range from 2 to 10 percent, depending on the growth conditions. Element's distribution across the layer was uniform. X-ray difraction data indicated that layers are monocrystalline. Intense cathodoluminescence from the layers was recorded at 85 K. The short-wave wing of the luminescence spectrum extending up to 3.6 eV in energy. It was found that the conductivity type of solid solution depends on growth conditions. Using this dependence, we fabricated $(SiC)_{1-x}(AlN)_x$ solid solution pn junction, for the first time. Also hetero pn junction between n-type 6H-SiC substrate and p type solid solution has been made. Both pn junctions have effective electroluminescence at room temperature.

1. Introduction

According to [1], the pseudobinary system $(SiC)_{1-x}(AlN)x$ consists of a continuous series of solid solutions. The system energy gap varies with composition between ~3eV and ~6eV; in addition, the excess of x over 0.7 causes the transition from indirect to direct band structure [2]. Areas of potential application for this solid solutions include semiconductor emitters in the UV region, high-temperature electronics, and widegap emitters for SiC devices [3,4]. Practical way to fabricate $(SiC)_{1-x}(AlN)_x$ solid solution device structures is a hetero-epitaxial growth. The possibility of $(SiC)_{1-x}(AlN)_x$ hetero-epitaxial growth on SiC substrates by sublimation has been demonstrated in [3,5], where the growth temperature was ~2000°C. It was mentioned that solid solutions of both types of conductivity can be achieved by sublimation technique [3]. Attempts to make pn junction in this solid solution system have not, to our knowledge, been reported. This paper reports a liquid phase epitaxial growth of $(SiC)_{1-x}(AlN)_x$ layers and pn junctions.

2. Epitaxial growth and layers characteristics

The growth technique is based on the method of container free liquid phase epitaxy described previously in [6]. The substrates used were (0001) oriented 6H-SiC single crystals grown by the Lely technique. Growth temperature ranged from 1400°C to 1500°C. The epitaxial layers produced were from 1 to 40 μm in thickness.

Auger-spectroscopic analysis was carried out on C, Si, Al, N, and O (the KLL) spectra. The coordinate resolution was 10μm, with energy resolution of 0.75%. Quantitative analysis was done as described in [7] using the relative sensitivity factors given in [8]. Spacial inhomogeneity of elemental spread over the layer was estimated based on Auger measurements done on oblique (~89°) sections.

All of the initial elements (Si,C,Al and N) were found to be present in the layers. It was also found that Auger maxima for Si, Al, and N were shifted relative to their energy positions in original elements. A comparison of Auger spectra for Al in $(SiC)_{1-x}$-$(AlN)_x$

epitaxial layers with similar spectra for (1) metallic aluminium, (2) epitaxial layers of pure AlN, and (3) Al_2O_3 samples has allowed a conclusion of the Al - N bond as the most probable. It is clear correlation in percentage, between Al and N. The energy positions of Si and C peaks were found to be the same as those in monocrystalline SiC. Auger measurement have indicated that the concentration of the AlN component in the epitaxial layers range from 2 to 10 percent, depending on the growth conditions. Element's distribution across the layer was uniform.

The layer crystallographic patterns were determined with a dispersionless double-crystal X-ray spectrometer. (00012) CuKα reflection pendulum curves showed two maxima, one from the 6H-SiC substrate and one from the epitaxial layer. The calculated lattice mismatch parameter $\Delta c/c$ was found to range from -7.8×10^{-5} to -1.1×10^{-3} for different layers. The presence of a narrow peak originating from the epitaxial layer (its half-width varied for different layers from a few angular seconds to few dozens angular seconds) was taken as evidence of the monocrystalline structure. The fact that the $\Delta c/c$ value is negative is an indication that the layer's lattice is compressed relative to the lattice of the 6H-SiC substrate.

Cathodoluminescence measurements employed beams of 15 keV electrons, and were done on layer oblique sections. Intense luminescence from the layers was recorded at 85 K in the form of a band having maximum in the blue region. The position and halfwidth of the band are very nearly the same as those of the band produced by N-Al donor-acceptor recombination in 6H-SiC. The band itself, however, has a pronounced distinctive feature in the form of a short-wave wing extending up to 3.6eV in energy. Since the percentage of AlN in the layers grown is within 10%, the energy gap in $(SiC)_{1-x}(AlN)_x$ must, according to [2], be approximatly the same as for SiC. This accounts for the coincidence in position between $(SiC)_{1-x}(AlN)_x$ luminescence line and the 6H-SiC line given by N-Al donor-acceptor recombination (Fig. 1).

It was found that the conductivity type of solid solution depends on growth conditions (Fig. 2). If the melt contained a large concentration of aluminum, p type layer has been grown with a high nitrogen pressure n type solid solution has been grown. Using this dependence, we fabricated $(SiC)_{1-x}(AlN)_x$ solid solution pn junction, for the first time. Also hetero pn junction between n type 6H-SiC substrate and p type solid solution has been made. Both pn junctions have effective electroluminescence at room temperature (Fig.3). Electroluminescence spectra were measured from the diodes structures made

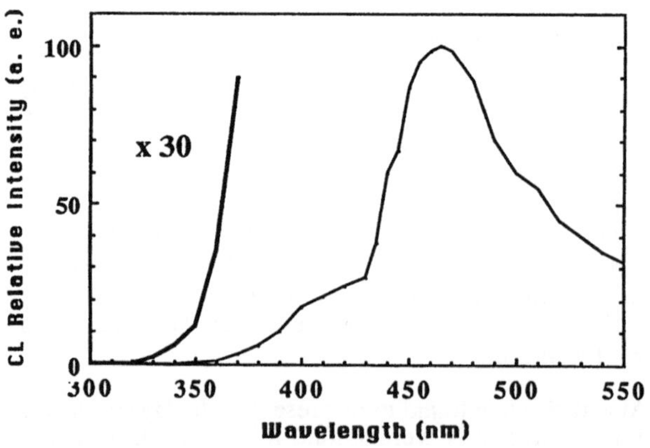

Figure 1. (SiC)i-x(AlN)x solid solution cathodoluminecence spectrum (85k).

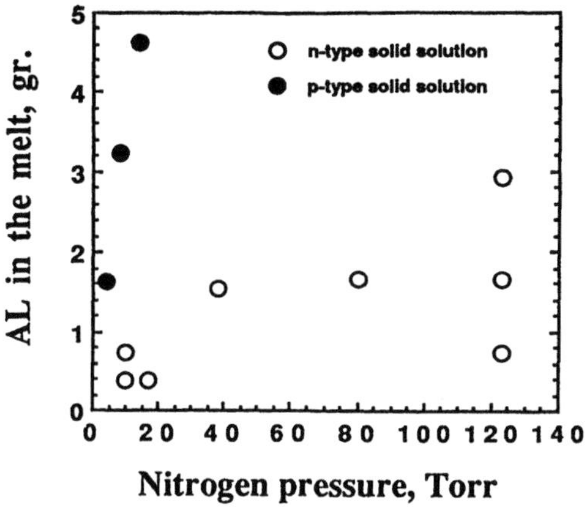

Figure 2. Dependence of solid solution conductivity type on growth conditions.

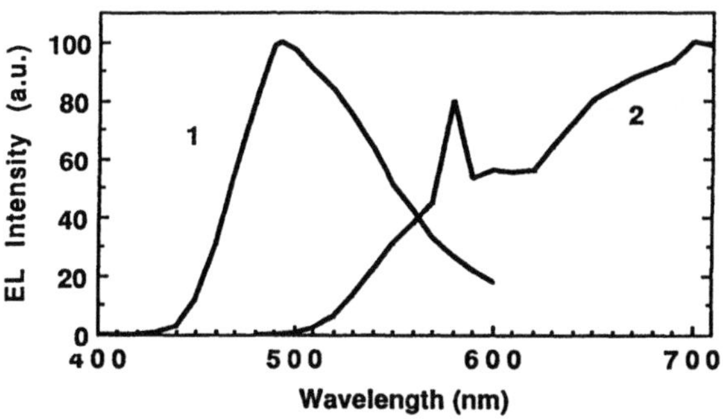

Figure 3. Electroluminescence spectrum of (1) n 6H-SiC - p $(SiC)_{1-x}$ $(AlN)_x$ heterojunction and (2) n $(SiC)_{1-x}(AlN)_x$ -p $(SiC)_{1-x}(AlN)_x$ junction. (I=20 mA, 300K).

using a mesa geometry (s ~ $3x10^{-4}cm^{-2}$). We believe that blue electroluminescene from the pn junction between n 6H-SiC and p solid solution layer is devoted to recombination in the silicon carbide. For red electroluminescence from solid solution pn junction two preliminary explanations may be suggested: (1) solid solution layer near the pn junction has cubic polytype structure which has smaller value of energy gap then hexagonal material, (2) solid solution layer near the pn junction has hexagonal polytype structure but red luminescence is due to recombination via deep levels. Characterization of the pn junctions is presently underway.

3. Conclusion

$(SiC)_{1-x}(AlN)_x$ epitaxial layers have been grown from Si-C-Al-N melt by liquid phase epitaxy. AlN concentration in the epitaxial layers ranged from 2 to 10 mol.%. The layers are monocrystalline in structure and have effective cathodoluminescence in the blue region of the spectrum. First, $(SiC)_{1-x}(AlN)_x$ pn junction have been made as well as pn heterojunction between n type 6H-SiC substrate and p type solid solution.

The authors are thankful to I.D. Kovalenko for technical assistance and to N.V. Aibabina for processing of samples.

References

1. Zangvil A., Ruh R. Journal Amer. Ceram. Soc., v.17, No 10, p.884 (1988)

2. Nurmagomedov Sh. A., et. al. Fiz. Tekh.Poluprovodn., v. 23, No 1, p.162 (1989) [in Russian]

3. Safaraliev G.K. in Wide-gap Semiconductors, Makhachkala, 1988, p.34 [in Russian]

4. Dmitriev V.A. SiC-Based Solid Solutions: Technology and Properties. Proceedings of the 3rd ICACSC

5. Rutz R.F. United States Patent, No 4382837, May 10 (1983)

6. Dmitriev V.A. et. al. Sov. Tech. Phys. Lett., v.11, No 2, p.98 (1985)

7. Hall P.M., Morabito J.M. Surf. Sci., v. 83, p.391 (1979)

8. Davis L.E. et. al. Handbook of Auger Electron Spectroscopy. 2nd ed.

Comparison of Dilutely Doped p-Type 6H-SiC from a Variety of Sources

L.L. Clemen[1], *W.J. Choyke*[1], *R.P. Devaty*[1], *J.A. Powell*[2], *and H.-S. Kong*[3]

[1]University of Pittsburgh, Pittsburgh, PA 15260, USA
[2]NASA Lewis Research Center, Cleveland, OH 44135, USA
[3]Cree Research, Inc., 2810 Meridian Parkway, Durham, NC 27713, USA

Abstract. We have observed a closely spaced series of very sharp and shallow lines in the low temperature photoluminescence spectrum of lightly doped p-type epitaxial single crystal films of 3C and 6H SiC from a variety of sources. We use a model based on group theory to argue that the origin of these lines is the recombination of an exciton in an acceptor four particle complex.

1. Introduction

The recombination of an exciton in a nitrogen donor four-particle complex (4D) produces well-known features in the low temperature photoluminescence spectra of SiC that have been observed in many polytypes, including 3C, 6H, 2H, 4H, 15R, 21R, and 33R [1]. Donor-acceptor pair (DAP) spectra have been observed in 3C, 4H, 6H, and 15R SiC [2-6], although a detailed assignment of shell numbers has only been performed for 3C SiC [2,6]. However, luminescence associated with an acceptor four-particle complex (4A) has not been reported for any SiC polytype.

We have observed very sharp and shallow luminescence lines at low temperatures in lightly doped p-type epitaxial single crystal films of 6H SiC fabricated in different CVD reactors. Bogdanov and Gubanov [7] have reported lines in the same spectral region in platelets of 6H SiC and argued that they are associated with the recombination of many-exciton-impurity complexes. Haberstroh [8] has found similar spectral features in p-type SiC boules produced at the Siemens Research Laboratories. We have also observed these new lines in epitaxially grown 3C SiC films lightly doped with Al and having a low nitrogen background.

In this paper, we report very high resolution low temperature photoluminescence spectra of the new lines in 3C and 6H SiC samples from a variety of sources. We also present preliminary results on Zeeman spectroscopy. It is argued that the new lines are associated with the recombination of an exciton in a 4A complex. For 3C SiC, we make use of a model based on group theory to explain complex luminescence spectra associated with recombination of 4A complexes in GaP [9]. We construct a similar model for 6H SiC. Although we are not yet able to assign specific transitions for our data, the model predicts a multiplicity of closely spaced emission lines for a 4A complex.

Springer Proceedings in Physics, Vol. 71 Amorphous and Crystalline Silicon Carbide IV
Editors: C.Y. Yang · M.M. Rahman · G.L. Harris

2. Experiment

Very high resolution low temperature (2K) photoluminescence is performed using a Bayard-Atomic 2 meter spectrometer with an 1800 l/mm grating, blazed at 5000 Å to yield an inverse linear dispersion of about 3 Å/mm. The spectra are recorded on Kodak 103aF photographic plates and digitized using a Jarrell-Ash densitometer. Zeeman spectra are obtained using a Janis dewar with a 7 T split coil superconducting magnet. Other spectra are taken on a 3/4 meter Spex monochromator with a GaAs/CsO photomultiplier (PMT) thermoelectrically cooled to −15°C. The output is fed into a Stanford Research Associates lock-in amplifier or, for weak signals, a photon counting system. An Omnichrome 8 mW He-Cd laser is the excitation source. The penetration depth of the 3250 Å laser radiation in 3C-SiC is about 1.5 μm [10].

Thin single crystal epitaxial films of both 3C and 6H SiC were grown by chemical vapor deposition (CVD) at the NASA Lewis Research Center on 6H SiC substrates. A trace of trimethyl aluminum (TMA) was added during growth to obtain p-type material. Details regarding the growth of these films have been published [11].

P-type single crystal epitaxial films were also grown on 6H SiC substrates by CVD at Cree Research. Aluminum concentrations in the range 1.7-$4.0 \times 10^{16}/cm^3$ were obtained by the introduction of TMA during the growth process.

3. Results

Figure 1 shows luminescence spectra for a lightly Al-doped p-type 3C SiC film grown on a 6H SiC substrate at the NASA Lewis Research Center. The no-phonon line of the nitrogen 4D complex and its phonon replicas are labeled, as are lines associated with close nitrogen-aluminum pairs. The new feature can be resolved into two lines (possibly three) for the best sample, as shown in the inset. Phonon replicas of this center are not clearly seen, probably due to the close proximity of stronger nitrogen 4D phonon replicas and the DAP lines. We have also observed this new feature in a 3C CVD film produced by CS GmbH, Munich, Germany [12].

Figure 2 shows two spectra for a lightly Al-doped 3 μm thick 6H SiC film grown by Cree Research on a 6H SiC substrate. The new features appear between the P_0 and the R_0, S_0 lines [13] of the nitrogen 4D complex. The very sharp lines show splittings as small as 0.1 meV. The fact that the strongest of these lines is twice as intense as P_0 immediately rules out an interpretation as multiple bound excitons (MBE). The fact that the new lines are only observed in p-doped material also rules out MBE.

Figure 3 shows a spectrum for a lightly Al-doped 4 μm thick 6H SiC film grown on the silicon face of a 6H SiC substrate at the NASA Lewis Research Center. This spectrum confirms the reproducibility of the new lines and the

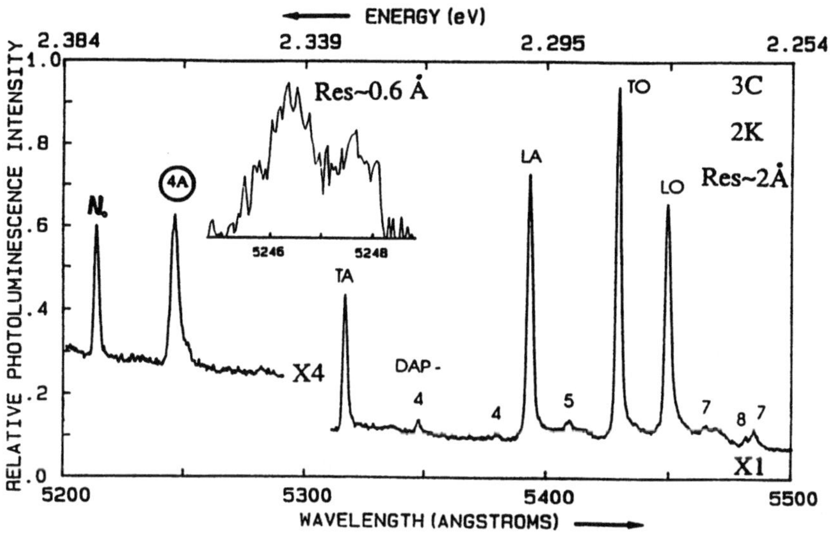

Figure 1. Photoluminescence spectrum of sample NASA 1324-5. The sample consists of an 8 μm thick layer of 3C-SiC grown with TMA on a 4 μm undoped layer of 3C-SiC, which was grown on the carbon face of a 6H-SiC wafer oriented 0.2° off the (0001) plane. The inset shows part of a high resolution spectrum taken photographically. The new (4A) lines are indicated, as are the nitrogen (4D) no phonon line and its phonon replicas and the shell numbers for nitrogen-aluminum close pair lines.

existence of weaker lines not shown on the high resolution scan. A high resolution spectrum (not shown) reveals some broadening in the fine structure, indicating that the sample may be slightly strained. The same features were observed for a film grown on the carbon face of a 6H SiC substrate, but the lines were considerably weaker than the P_0 line.

Recently, we have examined a thicker 8 μm film grown by NASA with a lower nitrogen background. The new lines are as sharp as those observed in the Cree sample discussed above, but without the very fine splitting shown in Figure 2. We do not believe that the extra fine structure is due to stress, because planar and axial stresses do not lead to additional splittings according to group theory. Figure 4 shows spectra taken with this sample in a 6 T magnetic field. The observed luminescence is predominantly polarized with $\vec{E} \perp \hat{c}$. Figure 5 shows a fan diagram for the two strongest lines at zero field (at 3.0034 and 3.0024 eV, labeled l_1 and l_2 in the figures) of the new multiplet and the R_0 and S_0 lines. The g-values for R_0 and S_0 are consistent with published results [14] on the nitrogen 4N complex to within experimental uncertainty. The new lines show comparable g-values. A detailed assignment of magnetic field induced splittings requires further experimental work at high resolution.

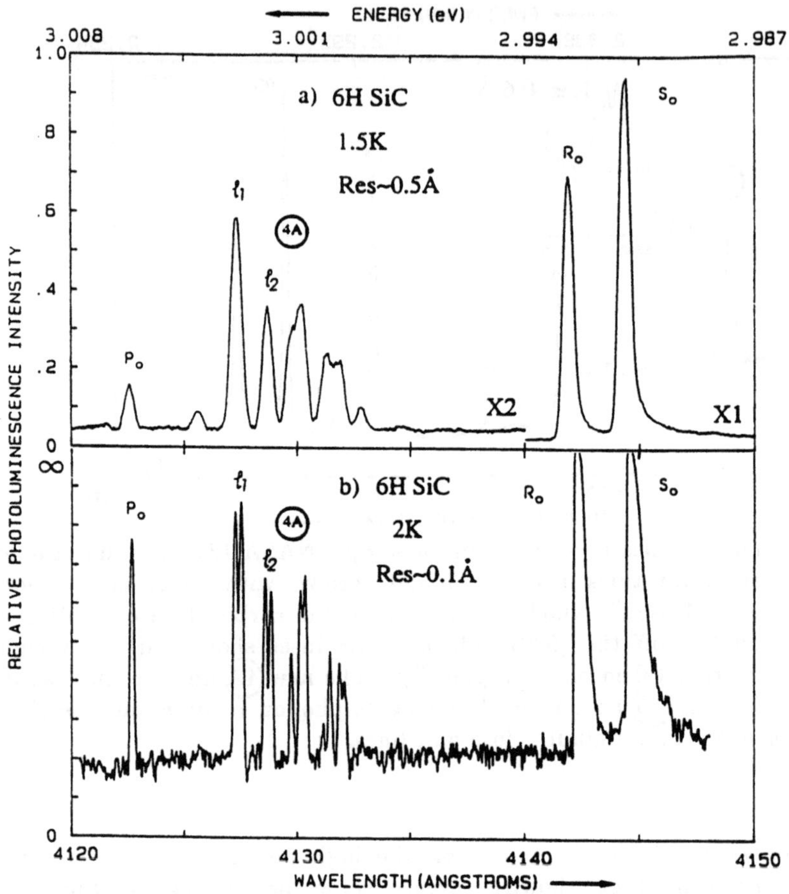

Figure 2. Photoluminescence spectra of sample Cree B0270-8-np, which is a 3 μm thick lightly Al-doped 6H-SiC film grown on a 6H-SiC substrate. The nitrogen (4D) complex no phonon lines P_0, R_0, and S_0 and the two strongest lines from the new (4A) complex l_1 and l_2 are labeled. a) Spectrum obtained using the PMT/photon counting system. b) High resolution spectrum taken photographically with a one hour exposure.

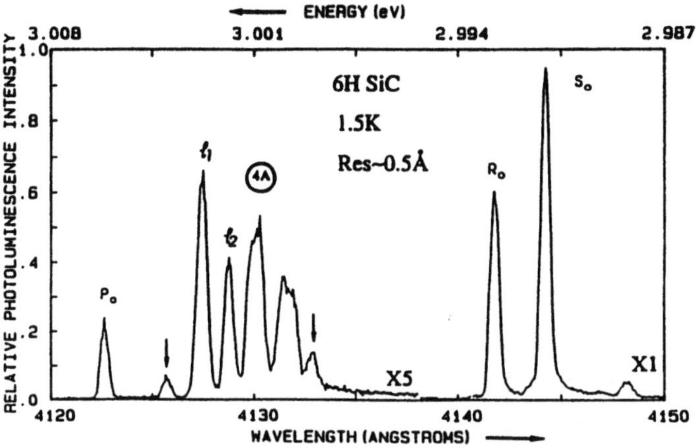

Figure 3. Photoluminescence spectrum taken with the PMT/photon counting system of NASA 1203-6, which is a 4 μm thick 6H-SiC film grown on the silicon face of a 6H-SiC wafer, oriented 3.5° off the (0001) plane. Spectral features are labeled as in Figure 2. The arrows point to lower intensity peaks which are too dim to be observed using the high resolution photographic system.

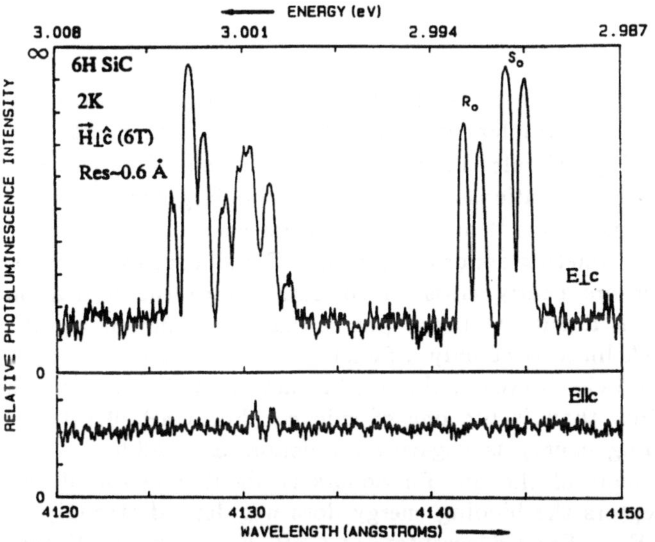

Figure 4. Photoluminescence spectra of sample NASA 1324-1, which is an 8 μm thick Al-doped 6H-SiC film grown on a 6H-SiC substrate oriented 3° off the (0001) plane. The spectra were taken with the sample in a 6 T magnetic field oriented perpendicular to the symmetry axis of the crystal. The spectra, which were taken photographically with a two hour exposure for $\vec{E} \perp \hat{c}$ and a 2.5 hour exposure for $\vec{E} \parallel \hat{c}$, show a strong polarization dependence. The $\vec{E} \parallel \hat{c}$ spectrum at H = 6 T is identical to the zero field $\vec{E} \parallel \hat{c}$ spectrum and shows no field induced splittings.

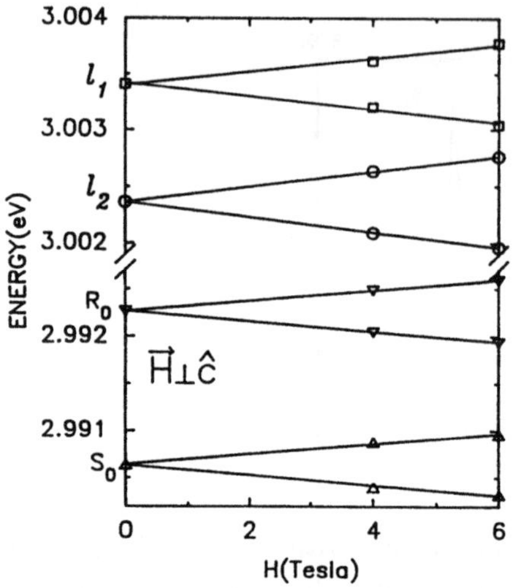

Figure 5. Fan diagram for sample NASA 1324-1, an Al-doped 6H-SiC film described in Figure 4. The experimental uncertainty in energy is specified by the size of the symbols.

4. Discussion

We identify the new spectral features with the recombination of an exciton in an acceptor four-particle complex, involving Al_{Si}. This statement is in apparent conflict with Haynes' rule [15] as applied to silicon. Haynes' rule states that there is a linear relationship between the binding energy of the complex ("exciton binding energy") and the ionization energy for a shallow impurity of a given type, whether donor or acceptor. For silicon, the lines for both donors and acceptors nearly cross the origin. Since the ionization energy for Al acceptors is about five times greater than for N donors in SiC, we would not expect 4A lines to be only a factor of two deeper than the 4N line, as observed in 3C SiC. However, the band structure of GaP is a much closer analog to 3C SiC than is the one of silicon. For GaP [9,16], the intercept at zero binding energy is negative for donors and +3.5 meV for acceptors. Also, the slope of the line for donors is about a factor of two larger. Thus, for acceptors the binding energy does not depend strongly on the ionization energy E_A. For Cd acceptors in GaP, $E_{BX} = 8.4$ meV even though $E_A = 96.5$ meV [9]. With this perspective, it is not surprising that we observe a shallow 4A complex in 3C SiC.

Dean *et al.* [9] made use of group theory to argue that a large number of closely spaced lines are expected for the recombination of shallow 4A complexes in GaP. We can apply this model directly to 3C SiC. Figure 6 illustrates the scheme which we now describe. The Lampert [17] four particle complex consists of an ion and its associated hole and the electron and hole associated with the exciton. The binding energy arises from the exchange

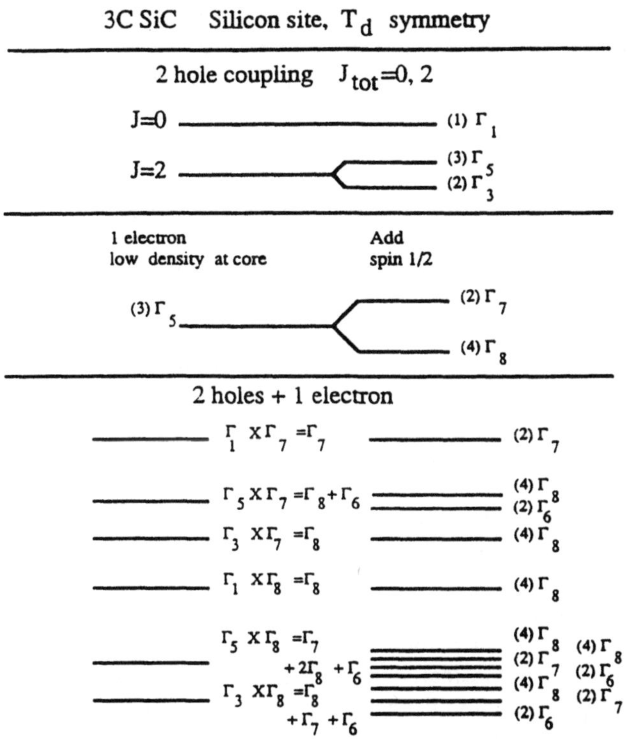

Figure 6. Schematic representation of the symmetry model for an acceptor four particle complex at a silicon site in 3C-SiC. This model was originally applied to 4A complexes at Ga sites in GaP by Dean *et al.* [9]. The diagrams show how two holes and an electron are combined to obtain a multiplet of closely spaced initial states of the complex. The degeneracies of levels are shown in parentheses. The ordering of the levels is arbitrary. The final states, not shown, are those of a neutral acceptor in its ground state; the single hole belongs to Γ_8 in T_d.

interaction between the indistinguishable holes [9,15,17]. The topmost valence band in 3C SiC is four-fold degenerate at the Γ-point with symmetry Γ_8 and $j = 3/2$. When two holes in the valence band interact via j-j coupling, the J=0 and J=2 multiplets are allowed by the Pauli exclusion principle. In the T_d symmetry of the Si site, the singlet J=0 state belongs to Γ_1 and the J=2 multiplet splits into a Γ_3 doublet and a Γ_5 triplet. The conduction band minima in 3C SiC are at the X-point [18]. There are three inequivalent minima. Since the carbon site is electron attractive in SiC, we can use an argument similar to Morgan's [19] to assert that an electron participating in a four-particle complex at a silicon site is repelled by the core. Therefore, a spatially antisymmetric wavefunction is appropriate and the conduction band miniumum belongs to X_3. The three-fold degenerate conduction band states made up of superpositions of states associated with the three inequivalent conduction band minima belong to Γ_5 in T_d. When the spin of the electron is included, there are six levels: a four-fold degenerate Γ_8 multiplet and a two-fold degenerate Γ_7 multiplet. When the two holes and the electron are combined, 12 distinct energy levels (ignoring accidental degeneracies) are predicted for the ground state multiplet of the 4A complex. The degeneracies are indicated in Figure 6. We have neglected j-j coupling between the electron and holes since the holes are localized near the ion, whereas the electron is more distant. The ground state multiplet of a neutral acceptor, which specifies the final states of the system after recombination, is four-fold degenerate and transforms as Γ_8 in T_d. As discussed by Dean *et al.* [9], a large number of transitions are predicted for Zeeman spectroscopy. Dean *et al.* [9] did not assign transitions in their work in GaP, and we are not yet able to do so for 3C SiC. However, this model is a plausible explanation for the complexity of the observed spectra. In 3C SiC, only two lines have been observed thus far. This result is analogous to the observations for the Cd acceptor in GaP, the deepest acceptor studied [9], which is reasonable since the binding energy that we observe, $E_{BX} = 24$ meV, is even larger.

We propose a similar model to explain the multiplicity of transitions associated with the recombination of a four particle acceptor complex in 6H SiC. In this case, illustrated schematically in Figure 7, the topmost valence band, belonging to Γ_9 of C_{6v}, is two-fold degenerate. When we combine two Γ_9 holes, we obtain four nondegenerate states: $\Gamma_9 \times \Gamma_9 = \Gamma_1 + \Gamma_2 + \Gamma_3 + \Gamma_4$. Only the Γ_1 state is compatible with the Pauli exclusion principle according to tabulated coupling coefficients [20]. Colwell and Klein [21] showed that the electron pockets are located along the $M-L$ lines in the Brillouin zone. Lyle Patrick [22] used the Kohn-Luttinger interference effect to argue that the conduction band minima are at points U along the $M-L$ lines, rather than at the points M or L. Thus, there are six nonequivalent electron minima, three above and three below the plane of hexagonal symmetry containing the Γ point. Following Patrick [22], we break up the electron wavefunctions into planar and axial parts. An electron bound at a silicon site is repelled by the acceptor core. Thus, if we place the origin at the silicon site substitutionally occupied by the acceptor, we must choose the planar part of the wavefunction to be asymmetric with respect to inversion so that the electron density at the origin is zero. Next, we must consider the symmetry of the impurity site, which is C_{3v}. If the wavefunction of the complex is sufficiently large, it may be appropriate to use the full C_{6v} symmetry of the bulk crystal, as discussed by Patrick [22]. We assume that C_{3v} symmetry is appropriate. The sixfold degeneracy of the conduction band minima reduces to $2\Gamma_1 + 2\Gamma_3$. When electron spin is included, there are twelve levels, with symmetries $4\Gamma_4 + 2\Gamma_5 + 2\Gamma_6$. The singlet levels belonging to Γ_5 and Γ_6 are degenerate by time reversal symmetry, so we write the combination as Γ_{56}. When the singlet hole state with Γ_1 symmetry in C_{3v} is combined with the electron states to obtain the symmetries of the states of

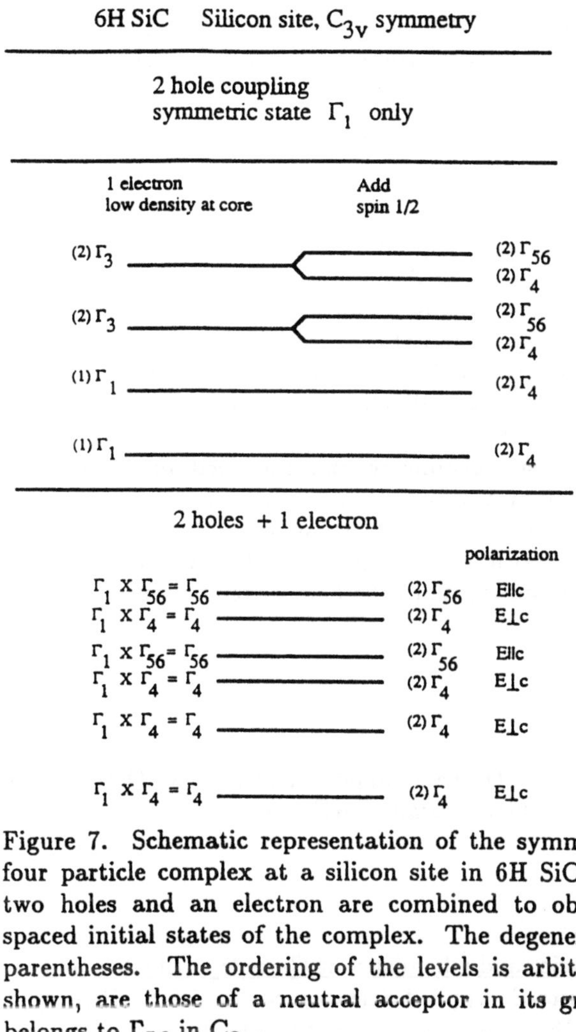

Figure 7. Schematic representation of the symmetry model for an acceptor four particle complex at a silicon site in 6H SiC. The diagrams show how two holes and an electron are combined to obtain a multiplet of closely spaced initial states of the complex. The degeneracies of levels are shown in parentheses. The ordering of the levels is arbitrary. The final states, not shown, are those of a neutral acceptor in its ground state; the single hole belongs to Γ_{56} in C_{3v}.

the 4A complex, we obtain six doublets with symmetries $4\Gamma_4 + 2\Gamma_{56}$ for the initial states. The final state is a neutral acceptor in its ground state, with symmetry Γ_{56}. The transitions allowed by symmetry with the electric field vector of the emitted radiation parallel or perpendicular to the symmetry axis are shown in Figure 7. There are three inequivalent sites for the acceptor in 6H SiC, so the maximum number of distinct energies for initial states is 18. Two of the sites have k (cubic) symmetry; the remaining site is h (hexagonal) [23]. It is possible that the levels associated with the distinct k sites are sufficiently closely spaced that they are not resolved. For neutral nitrogen 4D complexes, the R_0 and S_0 lines associated with the cubic sites are closely spaced with respect to the energy of the P_0 line [13]. The splittings of levels predicted by group theory when a magnetic field is applied are easily worked out.

5. Conclusion

Although we are not able to assign specific transitions at this time, it is clear that a model of the type described here is able to explain the multiplicity of observed transitions. Thus, we believe that the observed lines are indeed associated with the recombination of an acceptor four particle Lampert complex, presumably associated with aluminum impurities. Further work, including Zeeman spectroscopy, will allow us to test the model more rigorously. Possible changes to the model include the possiblility of C_{6v} symmetry as the appropriate site symmetry and the need to include the nearby valence band (about 10 meV away) that is split off due to the spin-orbit interaction. The spin-orbit splitting is smaller than the exciton localization energy associated with this complex, so it is conceivable that additional valence band states contribute to the hole wavefunction(s).

Acknowledgments

L. L. C., W. J. C., and R. P. D. were supported in part by ONR Grant N00014-89-J-1517 and NASA Contract Grant NAG 3-603.

References

1. W. J. Choyke, The Physics and Chemistry of Carbides, Nitrides, and Borides, NATO ASI Series E: Applied Sciences, ed. R. Freer (Kluwer, Dordrecht, 1990), Vol. 185, p. 863, and references therein.
2. W. J. Choyke and L. Patrick, Phys. Rev. B2, 4959 (1970).
3. I. S. Gorban', V. A. Gubanov, and V. M. Efimov, Fiz. Tverd. Tela (Leningrad) 14, 2324 (1972) [Sov. Phys. Solid State 14, 2010 (1973)].
4. S. H. Hagen, A. W. C. Van Kemenade, and J. A. W. Van der Does de Bye, J. Lumin. 8, 18 (1973).
5. A. Suzuki, H. Matsunami, and T. Tanaka, J. Phys. Chem. Solids 38, 693 (1977).
6. S. Yamada and H. Kuwabara, SiC – 1973, eds. R. C. Marshall, J. W. Faust, Jr., and C. E. Ryan (Univ. of South Carolina Press, 1974), pp. 305-312.
7. S. V. Bogdanov and V. A. Gubanov, Fiz. Tekh. Poluprovodn. 22, 728 (1988) [Sov. Phys. Semicond. 22, 453 (1988)].
8. Ch. Haberstroh, private communication.

9. P. J. Dean, R. A. Faulkner, S. Kimura, and M. Ilegems, Phys. Rev. B4, 1926 (1971).
10. W. J. Choyke, Z.C. Feng, and J. A. Powell, J. Appl. Phys. 64, 3163 (1988).
11. J. A. Powell, J. B. Petit, J. H. Edgar, I. G. Jenkens, L. G. Matus, J. W. Yang, P. Pirouz, W. J. Choyke, L. Clemen, and M. Yoganathan, Appl. Phys. Lett. 59, 333 (1991).
12. L. Muelhoff, private communication.
13. W. J. Choyke and L. Patrick, Phys. Rev. 127, 1868 (1962).
14. P. J. Dean and R. L. Hartman, Phys. Rev. B5, 4911 (1972).
15. J. R. Haynes, Phys. Rev. Lett. 4, 361 (1960).
16. P. J. Dean, R. A. Faulkner, and S. Kimura, Phys. Rev. B2, 4062 (1970).
17. M. A. Lampert, Phys. Rev. Lett. 1, 450 (1958).
18. L. A. Hemstreet and C. Y. Fong, in Silicon Carbide, eds. R. C. Marshall, J. W. Faust, Jr., and C. E. Ryan (University of South Carolina Press, Columbia, SC, 1973), p. 284.
19. T. N. Morgan, Phys. Rev. Lett. 21, 819 (1968).
20. G. F. Koster, J. O. Dimmock, R. G. Wheeler, and H. Statz, Properties of the Thirty-Two Point Groups (MIT Press, Cambridge, MA, 1963).
21. P. J. Colwell and M. V. Klein, Phys. Rev. B6, 498 (1972).
22. L. Patrick, Phys. Rev. B5, 2198 (1972).
23. A. R. Verma, Crystal Growth and Dislocations (Buttersworth Scientific Publishers, Ltd., London, 1953), Chap. 7.

Growth of Cubic Silicon Carbide on Silicon Using the C_2HCl_3-SiH_4-H_2 System

I.M. Baranov[1], *V.A. Dmitriev*[1], *I.P. Nikitina*[1], *and T.S. Kondrateva*[2]

[1]A.F. Ioffe Physical-Technical Institute, Russian Academy of Sciences, Politechnicheskaya 26, 194021 St. Petersburg, Russia
[2]NPO "Svetlana", 194021 St. Petersburg, Russia

1. Introduction

Cubic silicon carbide is a widegap semiconductor whose chemical and physical properties make it applicable for fabrication of electronic devices operating in extreme environments: high temperature, radiation etc. In recent years considerable progress has been achieved in fabrication of semiconductor devices in 3C-SiC. However, the problem of producing good-quality and low-cost substrates of large enough dimensions is still unsolved. CVD-growth of large-area 3C-SiC on other materials, e.g. silicon[1-5], has become extensively used. With a 20% lattice mismatch between the 3C-SiC layer and the Si substrate the growth of good-quality epitaxial layers presents considerable technological difficulties.

In most studies on the growth of 3C-SiC layers on Si substrates, a CVD technique is employed at temperatures in excess of 1300°C. We are aware of only two works [4,5] in which lowering of temperature in the CVD growth of single crystal 3C-SiC on Si has been achieved. The 3C-SiC layer of 0.4 µm in thickness has been grown on Si by CVD at T=1100°C [4]. No buffer layer in between substrate and epilayer was formed. By introducing HCL into the growth cell, it was found possible to lower the growth temperature of 3C-SiC to 1150°C. The grown layers had a thickness of 0.5 µm.[5]

In the present study, single crystal 3C-SiC layers with thicknesses of 0.5 - 11 µm on Si substrates were grown at a temperature 1100°C using a new gas mixture, SiH_4- C_2HCl_3-H_2. The growth was carried out under atmospheric pressure and without forming a buffer layer between Si and SiC.

2. Experimental procedure

Epitaxial layers of silicon carbide were grown using a CVD technique with monosilane SiH_4 and 3-chloroethylene C_2HCL_3, as Si- and C-sources, respectively. The growth process was carried out in an installation with an RF heating utilized in industrial production of silicon layers by CVD [6]. 24 substrates with 3 inch diameter were placed on a graphite pedestal coated with silicon carbide. As substrates, silicon wafers were used with their surfaces tilted by 4° off from the (111) plane. In some experiments Si (100) substrates were also used. The graphite pedestal was heated to 1100-1200°C and substrates were annealed for 10-20 min in a flow of hydrogen. HCl was then flown through the reactor to remove a substrate surface layer. After the flow of HCl was stopped, the required substrate temperature (in the range 1000-1250°C) established and the reagent gases introduced into the reactor. In our study the total flow rate of reagent gases through the reactor (**F**) were varied in such a way that a ratio of C_2HCl_3 and SiH_4 flows remained constant.

Springer Proceedings in Physics, Vol. 71 **Amorphous and Crystalline Silicon Carbide IV**
Editors: C.Y. Yang · M.M. Rahman · G.L. Harris

3. Limiting stage of 3C-SiC epitaxial growth

Depending on conditions, the epitaxy proceeds under domination of one of the two deposition regimes, either surface kinetic or mass-transfer. One way to find which type is the growth-limiting process is by the investigation of the epitaxial growth rate (Vg) as a function of deposition conditions. A simple way of determining Vg is to measure the thickness of a layer grown in a given time. We performed growth runs of various durations with fixed growth parameters fixed.Thicknesses of the grown layers were determined using ball-lapping. It was found that for the growth procedure used the growth rate does not change during the epitaxial process. From the dependences of the growth rate on flow rate **F** (Fig.1) a conclusion can be drawn that the limiting stage in the described growth process is the mass transfer. To a certain value of the gas flow rate F the growth rate rises linearly, corresponding to mass transfer of type I [7] when the growth rate is determined by a delivery of reagents to the deposition zone. With further increase in the flow of reagents the dependence deviates from linearity but the growth rate continues to rise, i.e. the limiting stage here is mass transfer of type II when the growth is dependent on reagent diffusion at the substrate-gas boundary.

One more argument in favour of the conclusion that mass transfer is the limiting stage is provided by a weak dependence of the growth rate on the substrate temperature (Fig.2). Also, in mass-transfer limited processes the growth rate does not depend on the substrate orientation. Our experiments with simultaneous deposition of 3C-SiC on both Si (100) with simultaneous deposition of 3C-SiC on both Si (111) and Si (100) substrates revealed no difference in the growth rates. This is one more evidence of the mass-transfer limiting stage.

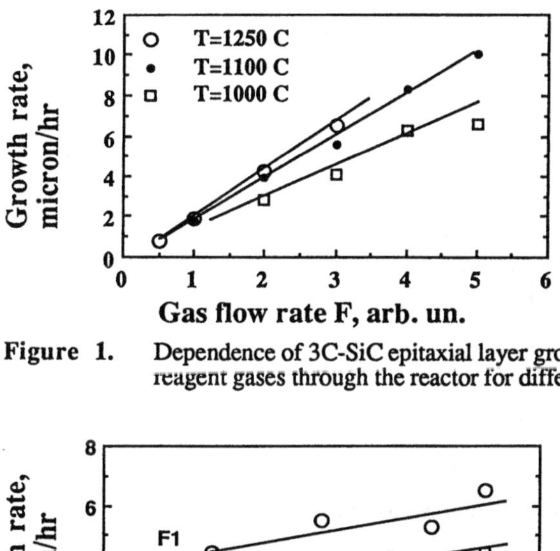

Figure 1. Dependence of 3C-SiC epitaxial layer growth rate on the total flow rate of reagent gases through the reactor for different substrate temperature.

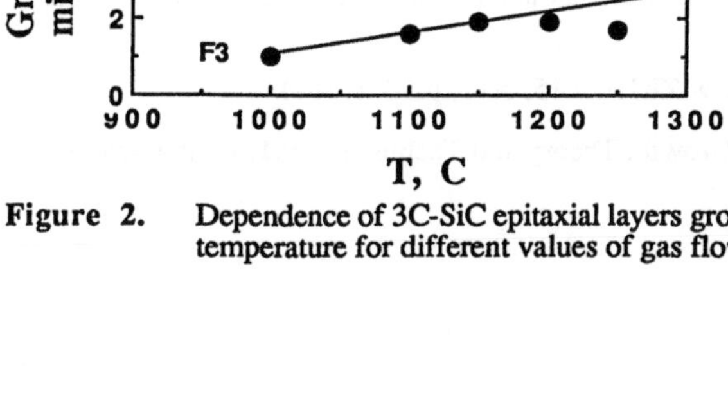

Figure 2. Dependence of 3C-SiC epitaxial layers growth rate on the substrate temperature for different values of gas flow rate **F**: $F_1 > F_2 > F_3$.

4. Crystalline structure of the 3C-SiC layers

The growth layers had, as a rule, a mirrorlike surface. The crystalline structure of these layers was investigated by X-ray diffraction. It was found that crystal structure of 3C-SiC depends both on temperature and the flow rate of reagents (Fig.3). Growth of monocrystal layer is possible only at a substrate temperature of 1100°C and on the (111) 4° off substrate. At other growth temperatures the layers were polycrystalline. Half-widths of rocking curves for 3C-SiC layers grown at different gas flow rates (T=1100°C) varied from 1.3° to 3.4° with increasing gas flow rate. The X-ray analysis also gave evidence that the single crystal 3C-SiC does not grow on the (111) 4° off Si substrate even at 1100°C if the flow of reagents is too low (Fig.3).

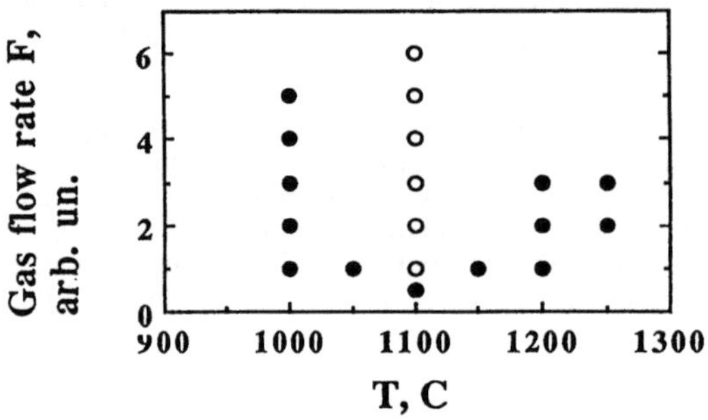

Figure 3. Dependence of the 3C-SiC crystalline structure on the growth conditions: closed circles indicate polycrystal structure, open circles indicate single crystal structure.

5. Conclusion

CVD technique with the use of a mixture of gases SiH_4, C_2HCL_3, and H_2 allows to grow single crystal 3C-SiC layers on silicon substrates at ~1100°C; The growth-limiting stage in the regimes studied is a mass transfer process.

References

1. Nishino S. et al. J. Electrochem. Soc., v.127, N12, p.2674 (1980).
2. Nishino S. et al Appl. Phys. Lett., v.42, N5, p.460 (1983).
3. Powell I.A. et al. J. Electrochem. Soc., v.134, N6, p.1558 (1987).
4. Furumura V. et al. J. Electrochem. Soc. v.135, N5 p.1255 (1988).
5. Nishino S. et al. Amorphous and Crystalline Silicon Carbide, 2 Int. Conf., Santa Clara, 1988.
6. Baranov I.M. et al. Pis'ma w ZhTF, v.15, N12, p.50, (1989).
7. Goodman C.H.L. Crystal Growth: Theory and Techniques, v.1, London-New York, 1974.

AC Plasma-Assisted Chemical Vapor Deposition of Cubic Silicon Carbide on Silicon Substrate

H. Shimizu, K. Naito, and S. Ishio

Department of Technology, Aichi University of Education, Hirosawa, Igaya, Kariya, Aichi 448, Japan

Abstract. Cubic SiC films were deposited on Si substrates by AC plasma-assisted chemical vapor deposition (CVD) using SiH_4 and CH_4 as source materials at a substrate temperature of 750 °C–1150°C. When the ratio of SiH_4 to CH_4 (Si/C) increases, the structures of films changed from amorphous to polycrystalline loaded with excess Si at a low substrate temperature less than 1000°C, reflecting a Si radical steeply increases and a H radical is saturated. Excess Si in the SiC films disappeared by applying a substrate bias, suggesting H radicals excited by electrons extract excess Si from the films. The plasma assists the dissociation of source materials and the cleaning effects of the growing surfaces by reactive species. The substrate bias enhances the surface reactions and crystallization.

1. Introduction

Cubic SiC/Si heteroepitaxial film formation, to fabricate electronic and optical devices was difficult for a long time because of large lattice mismach between SiC and Si. Since techniques for cubic SiC/Si heteroepitaxial formation by CVD method were established, many devices characteristics have been reported [1-3]. These formation techniques, however, require high growth temperature around 1300°C. From the viewpoint of device fabrications, low temperature growth is desired. Recently new approach for low temperature growth has been perfomed by various techniques, such as CVD method using reactive gas additions [4], plasma-assisted CVD [5], and reactive ion beam deposition [6]. These approach suggested that cleaning effects of growing surfaces by reactive species are important for low temperature growth.

In relatively low pressure conditions, we propose AC plasma-assisted CVD technique for crystal growth of cubic SiC at a low temperature, from experience of a synthesis of diamond films at a low temperature substrate utilizing DC plasma CVD technique [7,8]. In this report, as a first attempt for low temperature growth, the correlations between microstructures of SiC films prepared by AC plasma-assisted CVD method and optical emission spectroscopic (OES) studies of the AC plasma were investigated.

Springer Proceedings in Physics, Vol. 71 Amorphous and Crystalline Silicon Carbide IV
Editors: C.Y. Yang · M.M. Rahman · G.L. Harris

2. Experimental

Cubic SiC films were deposited on Si(111) substrates by AC plasma-assisted CVD using SiH_4 and CH_4 as source materials. A schematic diagram of the AC plasma-assisted CVD system is shown in Fig. 1(a). A tungsten plate mounted on a copper cooled by water is used as electrodes for exciting AC plasma. A Si wafer (7×16mm) is mounted on the graphite susceptor coated with SiC. The substrate temperature was measured by an optical pyrometer. The plasma was

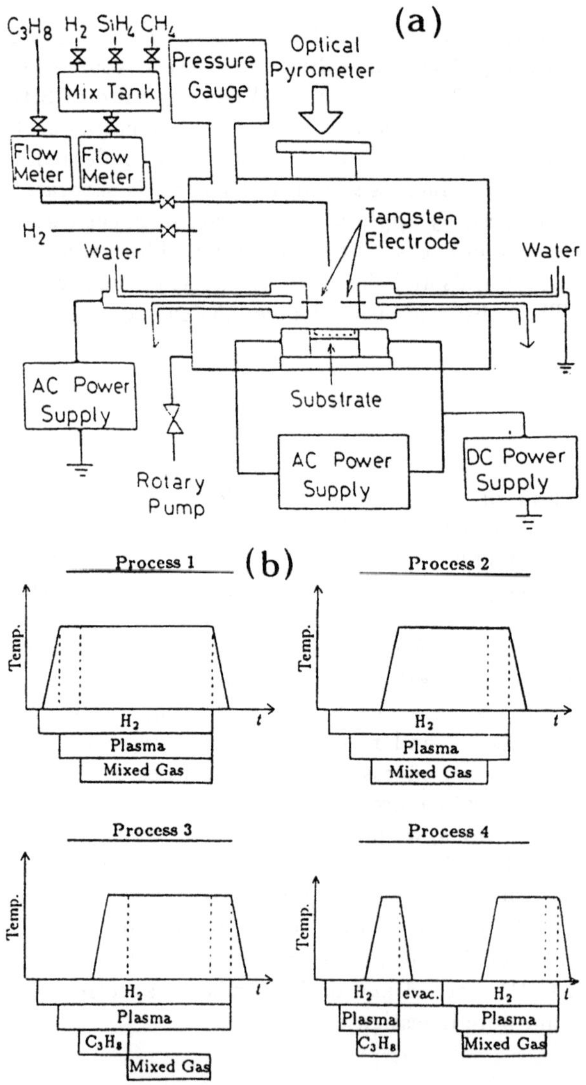

Fig. 1. A schematic diagram of the AC plasma-assisted CVD system (a), and flow charts of procedures for the depositions (b).

excited by AC power supply (60Hz), 10–12mm apart from the substrate. The source gases mixed by mixing tank were introduced through a nozzle (3mmϕ) above the substrate. These gases were directly decomposed through hydrogen plasma, in parallel to the substrate. Optical emission from the plasma was measured through the side viewing port with quartz window by multi-channel photodetector systems.

Crystal growth of cubic SiC on Si substrates was carried out by four types of growing processes as shown in Fig. 1(b). A carbonization step was introduced in process 3 and 4, but was not in process 1 and 2. In process 2 and 4, mixed gases were flowed before heating up the substrate, but in process 1 and 3 after heating up the substrate. After the chamber was pre-evacuated by rotary pump, a H_2 gas was introduced, the pressure was kept at 170 Torr, and growing processes as shown in Fig. 1(b) was followed. Crystal growth was carried out at the substrate temperature in the range of 750°C–1150°C. The total ratio of (SiH_4+CH_4) to (SiH_4+CH_4+H_2) and its flow rate were kept at 10 % and 20 sccm, respectively. The ratio of SiH_4 to CH_4 (Si/C) was changed from 0.1 to 1.0. A substrate bias was controlled by changing DC currents from 0 to 150 mA. AC plasma was excited by AC power supply of about 250 W (500V/0.5A). The deposition time was usually 10–15 min. The microstructures of the films were examined by reflection electron diffraction (RED), scanning electron microscopy (SEM), and fourier-transform infrared spectrometer (FT-IR). The thickness of the films was measured by ellipsometer.

3. Result and Discussion

The first objective in this investigation was to obtain the microstructure informations for the deposited films with varying the ratio of Si/C. To obtain the informations of plasma effects, the deposition must be carried out at low substrate temperature less than 1000°C. Fig. 2 shows RED patterns and SEM micrographs (**R&S**) of deposited films by growing process 1, when the ratio

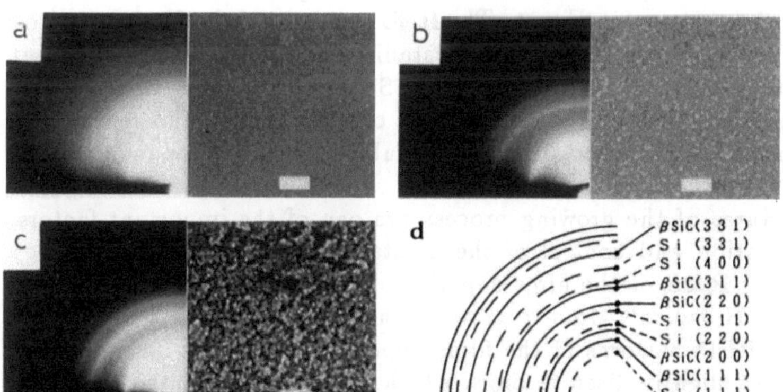

Fig. 2. **R&S** of the films deposited at substrate temperature of 770°C with various ratio of Si/C: (a) 0.1, (b) 0.3, (c) 0.5. Scale is 1 μm.

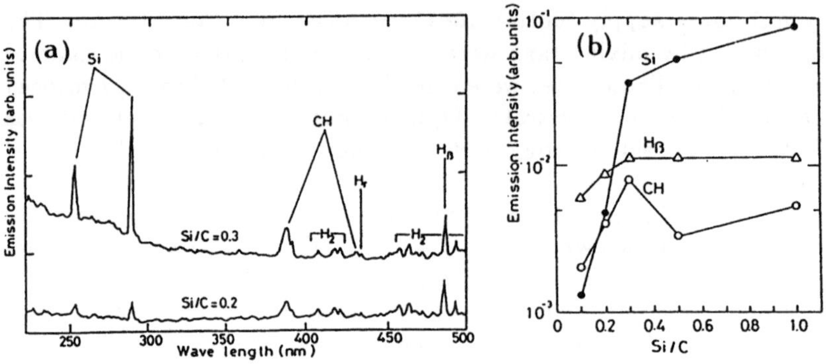

Fig. 3. Optical emission spectra of AC plasma (a), and the ratio of Si/C dependence of the emission intensity of Si, CH, and H_β (b).

of Si/C was changed from 0.1 to 0.5 at 770°C. When the ratio of Si/C is low, the RED patterns are broad though their d-values are in agreement with those of β-SiC from ASTM cards and the surfaces of films are smooth, as shown in Fig. 2(a). Upon increasing of the ratio of Si/C, the RED patterns become sharp, however contain rings in agreement with the Si and the surfaces of films become rough,as shown in Fig. 2(c). These results show that the structures of films become from amorphous to polycrystalline with increasing the ratio of Si/C. Typical optical emission spectra obtained from AC plasma are shown in Fig. 3(a). Fig. 3(b) illustrates the correlations with emission intensities of the Si radical (288nm), the CH radical (387nm), and H_β radical (486nm) and the ratio of Si/C. The emission intensities of the Si radical steeply increase at Si/C of about 0.3. Those of the CH and H_β radical increase up to Si/C of 0.3 and are saturated. Reflecting these results of optical emission spectra, the microstructures of films change as shown in Fig. 2. These results suggest that H radicals play the important role for the crystallization of cubic SiC. To confirm the effects of the plasma, **R&S** of the films deposited with and without plasma at 950°C are shown in Fig. 4. The RED patterns of the films deposited with plasma show that the film is polycrystalline β-SiC, while those without plasma show that the film is polycrystalline Si though Si/C is small at 0.2. These results suggest that the decompositions of the CH_4 and the generations of the H radicals are insufficient to grow the cubic SiC at the low temperature less than 1000°C without plasma.

The procedures of the growing process are one of the important factors, in the determination and control of the crystal quality, surface morphology and growth rate. **R&S** of the films deposited with applying a substrate bias as for increasing H radicals close to growing surfaces, in process 2, are shown in Fig. 5. The RED patterns of the films deposited at 1000°C, as shown in Fig. 5(a), present slightly discrete rings, which indicate an oriented structure in the film. At a substrate temperature of 1100°C, highly oriented film is obtained, as shown in Fig. 5(b). The surfaces of films change from smooth to rough with increasing the substrate temperature, as shown in Fig. 5. To

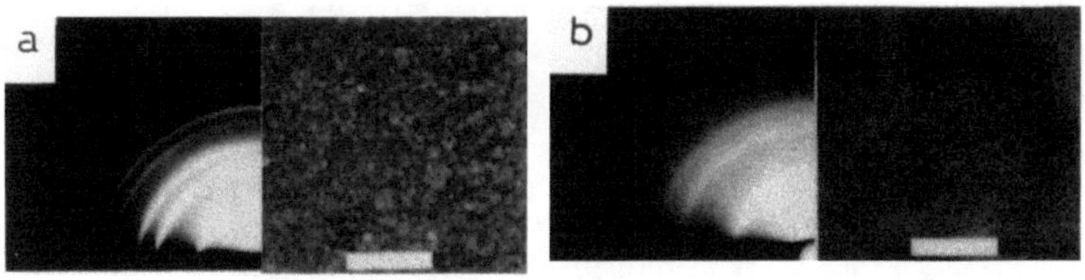

Fig. 4. **R&S** of the films deposited at 950°C without (a) and with plasma (b). The ratio of Si/C is 0.2. Scale is 1 μm.

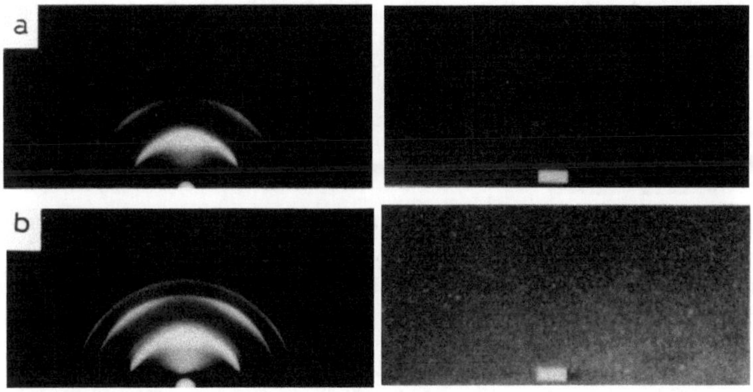

Fig. 5. **R&S** of the films deposited at 1000°C (a) and 1100°C (b) in process 2 with substrate bias. The ratio of Si/C is 0.3. Scale is 1 μm.

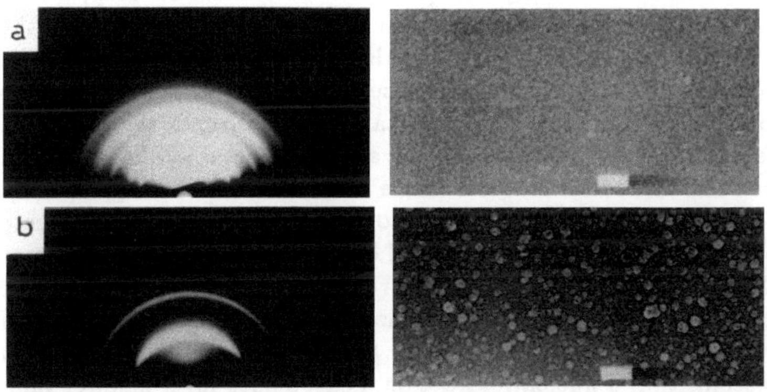

Fig. 6. **R&S** of the films deposited at 1100°C in process 2 without (a) and with substrate bias (b). The ratio of Si/C is 0.5. Scale is 1 μm.

confirm the effects of the substrate bias, **R&S** of the films deposited with and without the substrate bias under the higher Si/C (0.5) are shown in Fig. 6. The RED patterns of the films deposited with the substrate bias show that the film is polycrystalline β-SiC, while those without the substrate bias show that the film is polycrystalline Si included with β-SiC. The surfaces of the films

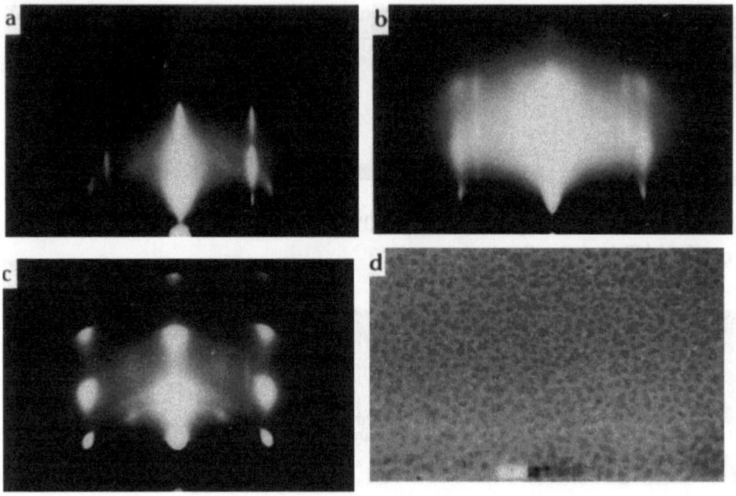

Fig. 7. RED patterns of the carbonized Si surfaces at various substrate temperatures: (a) 910°C, (b) 1000°C, (c) 1170°C. And SEM micrographs of the carbonized Si surfaces at 1170°C (d). Scale is 1 μm.

deposited without the substrate bias are relatively smooth, as shown in Fig. 6(a), because of a high nucleation rate of the Si in process 2. These results indicate that excess Si is extracted from the growing surfaces by H radicals, which are increased by the substrate bias.

Finally, to obtained better crystal quality, a carbonization step, which was carried out by flowing a diluet C_3H_8 with H_2 (5%) before heating up the substrate, was introduced. **R&S** of the carbonized Si surfaces are shown in Fig. 7. The RED patterns of the surfaces carbonized at 1170°C present faint rings and strong spots, which indicate a single crystalline structure, as shown in Fig. 7(c). The surfaces are rough and a triangular shape appears, as shown in Fig. 7(d). When the substrate temperature become lower, as shown in Fig. 7(a)(b), the RED patterns change from strong spot to diffused streak, which indicate the surfaces to be smooth. The films deposited on a carbonized layer, in process 4, show that the surfaces are rough and the structures are polycrystalline. The thickness of the films deposited on a carbonized layer was about one fifth of those on Si without a carbonized layer, under the same deposition time. From these results it seems that a nucleation rate on a carbonized layer is lower than that on Si.

4. Conclusions

Cubic SiC films were deposited on Si substrate by AC plasma -assisted CVD. Highly oriented cubic SiC was obtained at a substrate temperature of 1100°C. The investigations of the correlations with the microstructures of the deposited film and OES studies of the plasma suggest that H radicals play the important role for the crystallization of cubic SiC. The substrate bias enhances the surface reactions and crystallization.

Acknowledgment

The authors would like to thank Dr. S. Nishino at Kyoto Institute of Technology for his instructive advices.

References

1) S.Nishino, et al; Appl.Phys.Lett.**42**,460(1983).
2) S.Nishino, et al; J.Appl.Phys.**61**,4889(1987).
3) K.Shibahara et al; IEEE Electron Devices **EDL-7**,692(1986).
4) S.Nishino and J.Saraie;Springer Proceedings in Physics,**43**,8(1989).
5) S.Nishino and J.Saraie; Springer Proceedings in Physics,**43**,186(1989).
6) H.Yamada; J.Appl.Phys.**65**,2084(1989).
7) S.Nakao et al; J.Crystal Growth **99**,1215(1990).
8) S.Nakao et al; Jpn.J.Appl.Phys.**29**,1511(1990).

Part II

Characterization of Crystalline Silicon Carbide

Hall Effect and Infrared Absorption Measurements on Nitrogen Donors in 6H-SiC

W. Suttrop[1], *G. Pensl*[1], *W.J. Choyke*[2], *A. Dörnen*[3], *S. Leibenzeder*[4], *and R. Stein*[4]

[1]Institut für Angewandte Physik, Universität Erlangen-Nürnberg, Staudtstr. 7, W-8520 Erlangen, Fed. Rep. of Germany
[2]University of Pittsburgh, Pittsburgh, PA 15260, USA
[3]4. Physikalisches Institut, Universität Stuttgart, Pfaffenwaldring 57, W-7000 Stuttgart 80, Fed. Rep. of Germany
[4]Siemens AG, Research Laboratories, Scharowskystr., W-8520 Erlangen, Fed. Rep. of Germany

Abstract. Nitrogen-doped 6H-SiC samples are investigated by Hall and FTIR measurements. Four series of nitrogen-related absorption lines are observed in the spectral regions from 400 to 700 cm^{-1} and 1000 to 1250 cm^{-1}. The observed lines are assigned to electronic transitions of nitrogen donors residing at three inequivalent lattice sites (h,k_1,k_2). A valley-orbit splitting of the ground state of 12.6 meV is determined for nitrogen donors on hexagonal sites; the effective-mass approximation (EMA) is found to hold for the excited states.

1. Introduction

Isolated nitrogen atoms which reside at carbon lattice sites [1,2] in a silicon carbide (SiC) single crystal represent the prevailing shallow donor species; they govern the n-type conductivity of this semiconductor. The polytype 6H provides three inequivalent carbon lattice sites - one hexagonal (h) and two differing cubic (k_1, k_2) ones - that can be occupied by nitrogen atoms. Nitrogen impurities on inequivalent lattice sites are expected to show differing electronic properties.

Hall effect [3,4,5,6,7], luminescence [8], and IR absorption [9,10,11] measurements were conducted to determine ionization energies and binding energies of excited states of the nitrogen donors. O.V.Vakulenko and O.A.Guseva [11] put together the available data on the nitrogen donor and proposed an energy level scheme that is based on the effective-mass approximation (EMA) taking into account the anisotropy of the effective electron mass [12]. Their assignment of observed absorption lines to calculated transitions, however, is accompanied by energy shifts that are greater than the experimental uncertainty and by assignment of strong lines to parity forbidden transitions.

Recent progress in the growth of high-quality 6H-SiC single crystals [13] motivated us to revisit the structure of nitrogen-related energy levels. We use Hall effect and FTIR absorption on selected 6H-SiC samples with varying nitrogen concentration and compensation in order to obtain improved experimental data. A number of new absorption lines can be resolved in the spectral regions from 400 to 750 cm^{-1} and 1000 to 1250 cm^{-1}. In addition, we investigated thermalization effects (7K to 80K) and polarization dependences of nitrogen-related absorption lines.

As a result, we propose a new nitrogen-donor-model that is based on the EMA considering anisotropy of the effective electron mass [12,14], valley-orbit splitting of the donor ground state ($1s(A_1)$ - $1s(E)$) and three inequivalent lattice sites. This model is consistent with our refined experimental data.

2. Experimental

Nitrogen-doped 6H-SiC single crystals are grown from the vapor phase by the modified Lely method [13]. The investigated samples are cut either perpendicular or parallel to the c-axis. Square pieces of 5x5 mm^2 size are used for the Hall measurements (van der Pauw arrangement). Ohmic contacts are fabricated by evaporation of titanium and by a subsequent heat treatment at 900°C for 5 min. Hall measurements are conducted in a temperature range from 60K to 700K. For the IR absorption measurements, we use a Nicolet 740 FTIR spectrometer equipped with a KBr beam splitter and a DTGS detector. IR transmission spectra are taken with a resolution of 1 cm^{-1}.

3. Hall Effect Results

The free electron concentration $n=r_H/e \cdot R_H$ and the resistivity ρ versus reciprocal temperature of differing N-doped 6H-SiC samples are plotted in Fig.1a) and b). The Hall scattering factor r_H is assumed to be 1, because its magnitude and temperature dependence are not known in 6H-SiC. Donor concentrations $N_{N,i}$, ionization energies $\Delta E_{N,i}$, and compensation N_K are obtained from a fit of Eq. (7) in Ref. 15 to the experimental data in Fig. 1a). The effective density of states mass and the degeneracy factor g are taken equal to 0.9 m_0 and 2, respectively. The effective density of states in the conduction band is calculated assuming that 6 minima contribute to the conduction band. The Hall data evaluated by least-squares fits (fit curves are not inserted in Fig. 1a) are listed in Table 1. The resistivity (Fig.1b) at room temperature is given in column 7. It turns out that two donors $N_{N,1}$ and $N_{N,2}$ with differing ionization energies $\Delta E_{N,1}$ and $\Delta E_{N,2}$ are required to fit the measured

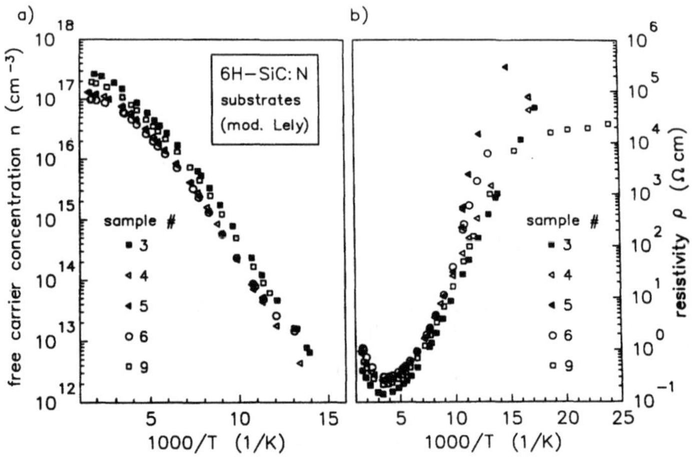

Fig.1a) Free electron concentration vs. reciprocal temperature as obtained from Hall measurements on various N-doped 6H-SiC samples; b) resistivity ρ vs. reciprocal temperature (for the same set of samples)

sample #	$\Delta E_{N,1}$ (meV)	$N_{N,1}$ (cm^{-3})	$\Delta E_{N,2}$ (meV)	$N_{N,2}$ (cm^{-3})	N_{comp} (cm^{-3})	ρ (20°C) (Ωcm)
1	-	-	-	-	-	> 7×10^7
2	80.0	1.2×10^{16}	120	1.3×10^{17}	9×10^{15}	0.4
3	85.5	1×10^{17}	125	2×10^{17}	2×10^{16}	0.2
4	85.5	5×10^{16}	125	1×10^{17}	1.8×10^{16}	0.3
5	85.5	4.3×10^{16}	125	1×10^{17}	1.6×10^{16}	0.3
6	85.5	2.7×10^{16}	125	8.6×10^{16}	1×10^{16}	0.33
7	83	1.2×10^{17}	125	6.7×10^{16}	5.1×10^{16}	0.24
8	85.5	1.8×10^{17}	125	1.3×10^{17}	5.9×10^{16}	0.32
9	85.5	8.6×10^{16}	125	1.4×10^{17}	2×10^{16}	0.21

Table 1. Donor concentrations and ionization energies obtained from a least-squares fit of Eq. (7) in Ref. 15 to the experimental data in Fig. 1a). The resistivity at 20°C is given in column 7.

free electron concentration over the entire temperature range. The ratio of the donor concentrations in all investigated samples is approx.

$$N_{N,1} : N_{N,2} = 1 : 2$$

This value gives credence to the assumption - which is supported by the IR absorption results reported in the following sections - that the shallow and deep donor species belong to nitrogen atoms residing at hexagonal (h) and cubic (k_1,k_2) lattice sites, respectively, whereas the two differing cubic sites cannot be resolved in energy by the Hall analysis.

The Hall mobility μ (not plotted) reaches a maximum value of approx. 1000 cm^2/Vs around T=110K and an average value of 320 cm^2/Vs at room temperature.

4. IR-Absorption Results

Fig. 2a) displays transmission spectra of 6H-SiC samples containing differing nitrogen concentrations. The spectra are recorded in the spectral region from 400 to 750 cm^{-1} with polarization E⊥c and at a sample temperature of 14K. The dips in the transmission (absorption lines) are marked by arrows and the corresponding wave numbers. Sample #1 has a very high resistivity (ρ > 7 $\cdot 10^7$ Ω cm at 20°C); it does not show any nitrogen-related absorption line. Samples #2, #3, and #4 show a series of absorption lines; these lines are

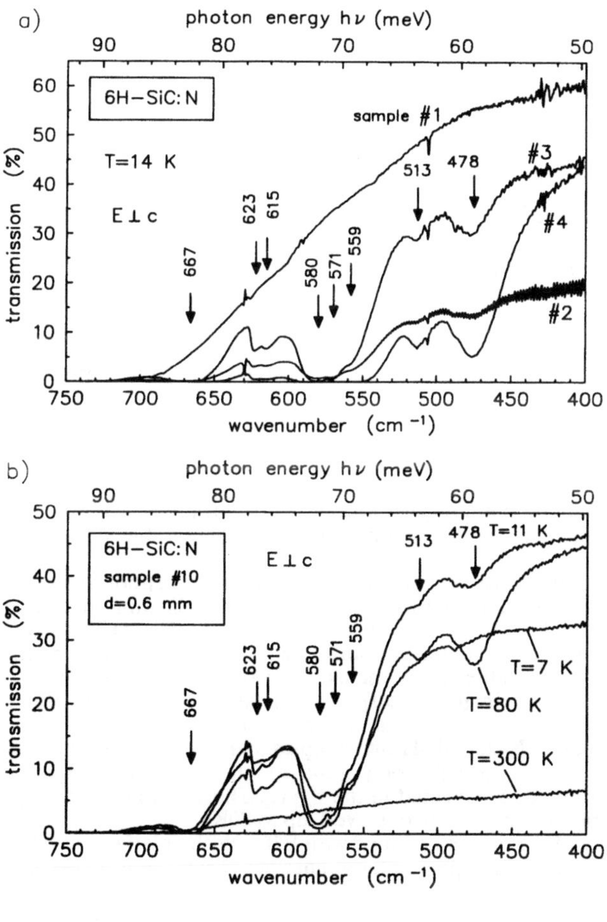

Fig.2a) IR transmission spectra (750 to 400 cm^{-1}) of 6H-SiC samples containing differing nitrogen concentrations (see Table 1 columns 3 and 5); b) IR transmission spectra (750 to 400 cm^{-1}) of sample #10 taken at various temperatures.

series	wavenumber (cm^{-1})	$h\nu$ (meV)	observed at $T = 7$ K	$T = 80$ K	polarization
I.	478	59.3	no	yes	$E \perp c$
(h)	513	63.6	no	yes	$E \perp c$
	559	69.3	weak	yes	$E \perp c$
II.	571	70.8	yes	yes	$E \perp c$
(h)	580	71.9	yes	yes	$E \perp c$
	615	76.2	yes	yes	$E \perp c$
	623	77.2	yes	yes	$E \perp c$
	633	78.5	yes	yes	$E \parallel c$
	667	82.7	yes	yes	$E \perp c$
III.	1050	130.1	yes	yes	$E \perp c$
(k_1)	1069	132.5	yes	yes	$E \perp c$
	1081	134.0	yes	yes	$E \parallel c$
	1123	139.2	yes	yes	$E \perp c$
IV.	1086	134.6	yes	yes	$E \perp c$
(k_2)	1106	137.1	yes	yes	$E \perp c$
	1119	138.7	yes	yes	$E \parallel c$
	1162	144.0	yes	yes	$E \perp c$

Table 2. Absorption lines observed at 7K and/or 80K in nitrogen-doped 6H-SiC samples.

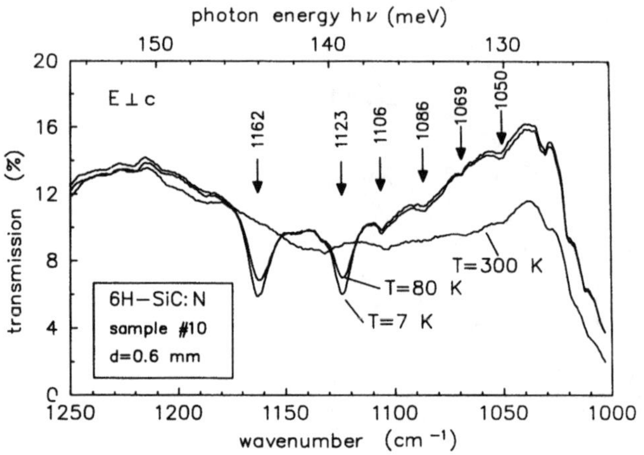

Fig.3) IR transmission spectra (1250 to 1000 cm^{-1}) of sample #10 taken at various temperatures.

stronger in samples #3 and #4 which contain higher nitrogen concentrations (see Table I). A detailed analysis reveals that the marked absorption lines are only observable in N-doped 6H-SiC samples.

Fig.2b) demonstrates the temperature dependence of N-related absorption lines. All marked lines are observed at $T \leq 80$ K and disappear at T = 300K indicating that these transitions belong to excitations of the N-donors which are completely ionized at T = 300K. Based on the energy position, the measured optical transitions can be summarized in four series and we have carried this out in Table 2. Thermalization is observed for line series I, II (see Table 2). Lines of series II have higher absorption constants at low temperatures (7K), lines of series I are only observed at elevated temperatures (80K).

Fig. 3 examines the transmission of N-doped 6H-SiC samples in the spectral region from 1000 to 1250 cm^{-1} with polarization $E \perp c$ as a function of the sample temperature. Again a series of absorption lines is observed as indicated in the figure. These lines disappear at 300K; however, no thermalization effects can be observed up to 80K.

Absorption lines observed with $E \perp c$ cannot be found with $E \parallel c$. With polarization $E \parallel c$, three new lines appear in the transmission spectra; these lines are also listed in Table 2.

5. Nitrogen-Donor-Model

Line series I to IV are attributed to h-, k_1-, and k_2-nitrogen donors (see Table 2) to be consistent with the energy level scheme given in Fig. 4. In the following, we present arguments that support the proposed nitrogen-donor-model.

The observed thermalization of line series I and II in the temperature range from 7 to 80K leads to the assumption that both series are caused by the same donor species (h). The transitions are assumed to start at different ground states (1s(A_1), 1s(E)) split by valley-orbit interaction and having the same final excited states. The population of the 1s(E) ground level increases at elevated temperatures resulting in stronger absorption lines of series I. The pairs 478/580 cm^{-1}, 513/615 cm^{-1}, and 559/667 cm^{-1} combined with lines of both series have approximately the same energy difference of 12.6 meV. This value corresponds to the valley-orbit splitting, 1s(A_1) - 1s(E), of the ground state. Our result is confirmed by Colwell and Klein [16] who observed the direct transition from the 1s(A_1) to the 1s(E) state by Raman scattering and obtained an energy splitting of 13.0 meV for the h-nitrogen-donor.

Line series III and IV do not show any thermalization; they are thought to belong to different donor species (k_1,k_2). Raman scattering experiments between split ground states (1s(A_1), 1s(E)) of k_1- and k_2-nitrogen donors provide a value of 60.3 and 62.6 meV, respectively, for their energy difference [16]. This energy spacing is too large to populate the 1s(E) level at 80K; thermalization can therefore not be observed for k_1- and k_2-nitrogen donors.

To identify the excited states of N-donors, we apply the EMA taking into account the anisotropy of the effective electron mass [12, 14]. According to the EMA, the order of excited states depends on the magnitude of the anisotropy factor $1-\gamma^{1/3}$, where $\gamma = m_\perp / m_\parallel$. For $1-\gamma^{1/3} < 0.52$, we obtain the order of excited states that is given in Fig.4. This order is in agreement with the relative intensities of measured absorption lines taking into account the fact that the line intensities decrease with increasing quantum numbers n and that p_0-states are weaker than the $p_\pm$-states. Excited EMA states are calculated on the basis of Ref. 16 (valid for $1-\gamma^{1/3} \leq 0.3$). All theoretical binding energies fit well to the experimental data (h-donor: energy difference $\leq$ 0.2 meV; k_1-, k_2-donor: energy difference $\leq$ 1.5meV) as demonstrated in Table 3.

The effective electron masses are determined by the fit of the EMA to the obsered absorption lines:

$$m_\perp = 0.24\, m_0, \quad m_\parallel = 0.34\, m_0$$

resulting in an anisotropy factor of $1-\gamma^{1/3} = 0.11$.

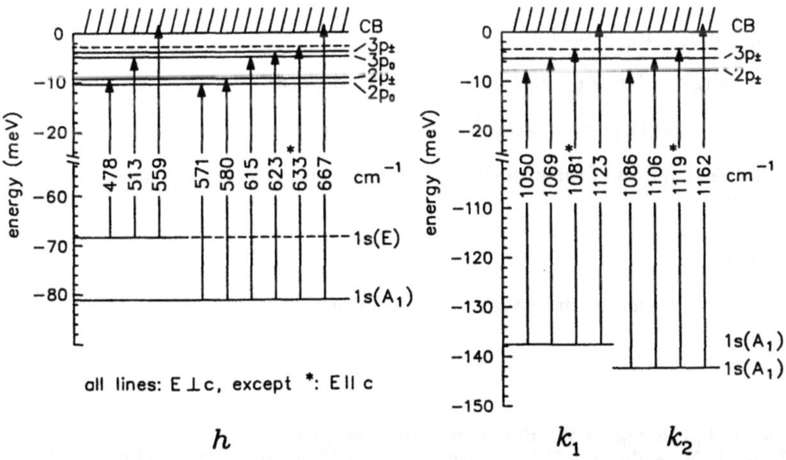

Fig.4) Energy level scheme of nitrogen donors residing at h-, k_1-, and k_2-lattice sites in a 6H-SiC single crystal.

donor site	final state	wavenumber (cm^{-1})	binding energy (meV) experiment	theory (EMA)
h	$2p_0$	571	−10.26	−10.36
	$2p_\pm$	580	−9.15	−9.06
	$3p_0$	615	−4.81	−4.6
	$3p_\pm$	623	−3.82	−4.02
	$4p_0$	n.o.		−2.59
	$4p_\pm$	n.o.		−2.26
	CB	667	+1.6	0
k_1	$2p_\pm$	1049	−7.6	−9.06
	$3p_\pm$	1069	−5.1	−4.02
	CB	1123	+1.6	0
k_2	$2p_\pm$	1086	-7.8	-9.06
	$3p_\pm$	1106	-5.3	-4.02
	CB	1162	+1.6	0

CB: conduction band n.o.: not observed

Table 3. Binding energies of excited nitrogen donor states (h-, k_1-, and k_2-lattice sites) obtained from IR absorption and from the effective-mass approximation calculated after Gerlach and Pollmann [14].

The binding energy of the $1s(A_1)$ ground state (ionization energy) of the various N-donors is not correctly predicted by the EMA. The ionization energies are therefore taken from the strong absorption lines at 667, 1123, and 1162 cm^{-1} which are attributed to transitions from the $1s(A_1)$ ground state to the conduction band of 6H-SiC. The observed values are:

$\Delta E(h) = 81.0$ meV, $\Delta E(k_1) = 137.6$ meV., $\Delta E(k_2) = 142{,}4$ meV.

6. Summary

N-doped 6H-SiC samples are investigated by Hall effect and FTIR absortion. Four series of absorption lines are observed that are attributed to N-donors residing at inequivalent lattice sites (series I, II at h-site, series III at k_1-site, series IV at k_2-site). A valley-orbit splitting of the ground state of 12.6 meV is determined for the h-site. The effective mass approximation (EMA) is found to hold for the observed $2p_0$, $2p\pm$, $3p_0$, and $3p\pm$ excited states. The effective electron mass of 6H-SiC is determind by the fit of the EMA to the IR data; it results in:

$m_\perp = 0.24\ m_0$ and $m_\parallel = 0.34\ m_0$

The ionization energies of h-, k_1-, and k_2-nitrogen-donors are:

$\Delta E(h) = 81.0$ meV, $\Delta E(k_1) = 137.6$ meV, $\Delta E(k_2) = 142{,}4$ meV.

A more detailed description of the experiments and the theoretical analysis will appear in future publications.

Acknowledgements

This research was supported by the BMFT under contract No. 03M2708B9

References

1. Woodbury, H.H., Ludwig, G.W.: Phys.Rev. 124 (1961) 1083
2. Kalabukhova, E.N., Kabiding, N.N., Lukin, S.N.: Sov.Phys.Solid State 29 (1987) 1461
3. Rosengreen, A.: Mat.Res.Bull. 4 (1968) 355
4. Hagen, S.H., Kapteyns, C.J.: Philips Res.Repts. 25 (1970) 1

5. Alekseenko, M.V., Zabrodskii, A.G., Timofeev, M.P.: Sov.Phys.Semicond. 21 (1987) 494
6. Zhang, H.,Pensl, G.,.Glasow, P., Leibenzeder, S.: The Electrochemical Society, Extended Abstracts Vol. 89-2, p.714 (1989), Hollywood, Florida
7. Tachibana, T., Kong, H.S., Wang, Y.C., Davis, R.F.: J.Appl.Phys. 67 (1990) 6375
8. Choyke, W.J.,.Patrick, L.: Phys.Rev. 127 (1962) 1868
9. Dubrovskii, G.B., Radovanona, E.I.: Phys.Status Solidi (b) 48 (1971) 875
10. Vakulenko, O.V., Govorova, O.A., Sergeev, O.T.: Sov.Phys.Semicond. 6 (1972) 829
11. Vakulenko, O.V., Guseva, O.A.: Sov.Phys.Semicond. 15 (1981) 886
12. Faulkner, R.A.: Phys.Rev. 184 (1969) 713
13. Ziegler, G., Lanig, P., Theis, D., Weyrich, C.: IEEE Trans. Electron. Dev. ED-30 (1983) 277
14. Gerlach, B., Pollmann, J.: Phys. Status Solidi (b) 67 (1975) 93
15. Schaub, R., Pensl, G., Schulz, M., Holm, C.: Appl.Phys A 34 (1984) 215
16. Colwell, P.J., Klein, M.V.: Phys.Rev. B6 (1972) 498

Nitrogen Impurities in 3C-SiC Epilayers

I. Nashiyama[1], *S. Misawa*[1], *H. Okumura*[1], *S. Yoshida*[1], *and Y. Hirabayashi*[2]

[1]Electrotechnical Laboratory,
1-1-4 Umezono, Tsukuba, Ibaraki 305, Japan

[2]Industrial Research Institute of Kanagawa Prefecture,
3173, Showa-machi, Kanazawa-ku, Yokohama 236, Japan

Abstract. Undoped 3C-SiC epilayers with residual carrier densities as low as 1x 10^{15} cm^{-3} were obtained by chemical vapor deposition using pure SiH_4 and C_3H_8 source gasses instead of diluted source gasses. Hall-effect, ESR, and SIMS measurements were carried out for these 3C-SiC epilayers. The epilayers having carrier densities below $1x10^{15}cm^{-3}$ showed no significant 3-line ESR signal associated with nitrogen impurities. The reduction in residual carrier density is explained in terms of the nitrogen impurity concentration in 3C-SiC epilayers. Lattice site location of nitrogen atoms in 3C-SiC was determined by the $^{15}N(p,\alpha_0)^{12}C$ nuclear reaction analysis combined with channeling.

1. Introduction

Undoped 3C-SiC epilayers grown on Si substrates by chemical vapor deposition (CVD) show n-type conduction with residual carrier densities above $1x10^{16}cm^{-3}$[1]. Epilayers with less carrier density are necessary for electronic device fabrication. Origin of the residual carriers has been studied by several authors using Hall-effect, electron spin resonance (ESR), and photoluminescence (PL) measurements[2-5]. Since nitrogen-related 3-line structures are observed by ESR, nitrogen impurities are considered to be one of the sources of residual carriers.

We found that the 3-line structures can be separated into a nitrogen-related hyperfine structure(hfs) and a narrow single-line components having the same g-value[6]. Quantitative evaluation of their spin densities was made for 3C-SiC epilayers having various carrier densities, and correlations between carrier and spin densities were studied[7]. It was found that the densities of the ESR center corresponding to the hfs component are about an order of magnitude lower than the residual carrier densities and that the major source of the residual carriers is related to the narrow single-line component. Origin of this narrow single-line component is not clear so far. Possible explanations are 1) exited states induced by valley-orbit splitting of the nitrogen donor level[8], 2) exchange interaction of nitrogen donor electrons[9], and 3) unknown defect centers which have no relation with nitrogen impurities. Considering that the g-value and the temperature dependence of the narrow single-line component are almost the same as those of the hfs component, 1) or 2) seems more probable.

Since the nitrogen impurity is considered to be unintentionally doped in 3C-SiC epilayers from diluted source gasses used for CVD, pure source gasses of SiH_4 and C_3H_8 are applied in the present work instead of diluted source gasses. We have studied the influence of the gas dilution on the residual carrier density, the nitrogen concentration, and the ESR spin density in 3C-SiC epilayers. In order to make clear that impurity nitrogen atoms are active as donors in 3C-SiC epilayers, we measured atomic location of nitrogen atoms in the 3C-SiC lattice by nuclear reaction analysis(NRA) combined with a channeling technique.

Springer Proceedings in Physics, Vol. 71 **Amorphous and Crystalline Silicon Carbide IV**
Editors: C.Y. Yang · M.M. Rahman · G.L. Harris

2. Experiments and Results

2.1 Growth and Electronic Properties of 3C-SiC Epilayers

In the previous works, 3C-SiC epilayers were grown epitaxially on Si(100) substrates at 1400°C using a horizontal CVD system with H_2-diluted SiH_4(1%) and C_3H_8(1%) reaction gasses, and with H_2 carrier gas refined with a Pd-alloy purifier. Details of the epitaxial growth procedure have been described elsewhere [10]. Using the diluted source gasses, we obtained high mobility epilayers with carrier density as low as $1x10^{16}$ cm^{-3}. In the present work, we apply pure 5-nine glade SiH_4 and pure 6-nine glade C_3H_8 instead of the diluted source gasses. The 7-nine grade H_2 carrier gas is also used for several epilayers. Residual carrier densities and Hall mobilities at room temperature are listed in Table I for typical 3C-SiC epilayers grown by using diluted and pure source gasses. The epilayers are grouped into the following three categories depending on the differences of the CVD source gasses.

(a) 1%SiH_4, 1%C_3H_8, and H_2 refined with a Pd-alloy purifier
(b) pure SiH_4, pure C_3H_8, and H_2 refined with a Pd-alloy purifier
(c) pure SiH_4, pure C_3H_8, and pure H_2

For type(b) source gasses, several epilayers were grown with different gas flow rates and different C/Si atomic ratios.

Carrier density and electron mobility of these epilayers vary depending on the crystal-growth conditions as shown in Table I, suggesting that adjustment of the growth parameters is crucial for the development of high quality 3C-SiC epilayers. It should be noticed that some of the epilayers grown using type(b) source gasses show carrier densities about $1x10^{15}cm^{-3}$, which are an order of magnitude lower than those of the epilayers grown using type(a) source gasses. However, the epilayers with low carrier densities show rather low mobilities.

Table I. 3C-SiC epilayers grown by CVD using different source gasses (a), (b) and (c) (see text). Mobilities and residual carrier densities at room temperature are listed together with ESR spin densities at 30K and nitrogen impurity concentra-tions obtained by SIMS measurements.

Sample	gas	Mobility $cm^2/(V.s)$	carrier $x10^{15}cm^{-3}$	ESR cm^{-3}	SIMS cm^{-3}
vcs221-43	(a)	637	$1.6x10^{16}$	$2.9x10^{16}$	$4.6x10^{17}$
vcs181-42	(a)	729	$2.4x10^{16}$	$1.5x10^{16}$	$7.9x10^{17}$
IS85-3B	(b)	388	$2.9x10^{15}$	$7.1x10^{15}$	$1.5x10^{16}$
IS86-4B	(b)	326	$8.0x10^{14}$	$6.0x10^{15}$	
IS87-4B	(b)	365	$2.7x10^{15}$	$2.0x10^{16}$	
IS90-3C	(c)	440	$1.6x10^{16}$	$2.2x10^{16}$	$2.4x10^{16}$
IS92-4D	(b)	744	$1.8x10^{16}$		

2.2 Spin Density of Nitrogen-related ESR Signal

Intensities of the 3-line ESR signals associated with nitrogen atoms were measured at 30K for various 3C-SiC epilayers grown with different CVD source gasses. Some of the typical 3-line signals are shown in Fig.1. It should be noticed in Fig.1 that sample IS86-4C with the carrier density below $1x10^{15}$ cm^{-3} shows quite low signal intensity which is almost at the back ground level. Overall spin densities of the 3-line signals were evaluated using standardized Mn^+ in MgO, and results are plotted against the carrier density in Fig.2. There exists an apparent correlation between the carrier and the spin densities, indicating that some of the residual carriers are closely related to the nitrogen impurity.

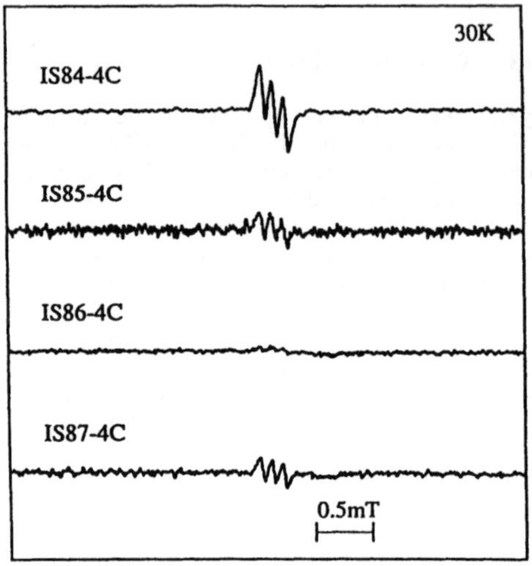

Fig.1. The 3-line structure signals of ESR spectra measured at 30K for low carrier-density 3C-SiC epilayers. Signal intensities are adjusted so as to be proportional to the spin densities.

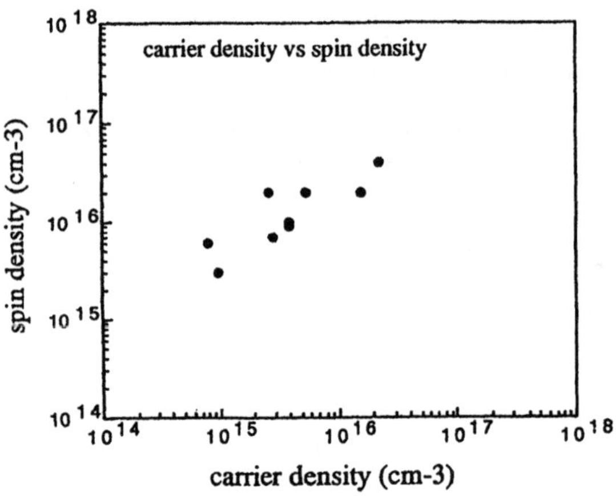

Fig.2. Overall spin densities at 30K corresponding to the 3-line ESR signals are plotted against the residual carrier density at 300K.

2.3 Nitrogen Impurity Concentration

Atomic densities of the nitrogen impurities in 3C-SiC epilayers were measured by high-resolution secondary ion mass spectroscopy (SIMS). Since its sensitivity to nitrogen ions is too low to measure nitrogen concentrations lower than ppm level, the intensity of $^{12}C^{14}N^-$ ions at M=26.0031 was measured instead of nitrogen ions, where mass resolution above 7000(m/Δm) was required to separate the peak from that of $^{12}C_2^-$ at M=26.0067. We used a nitrogen-implanted 3C-SiC epilayer as a standard sample for the quantitative evaluation of nitrogen concentration. Figure 3 shows a depth profile of the nitrogen concentration in the standard sample. The ordinate is calibrated using the total fluence of nitrogen implanted at 60keV and 160keV. The background level of the SIMS system was 1×10^{16} cm^{-3} ±10%. Nitrogen concentrations obtained by SIMS are listed in Table I, where the nitrogen concentration decreases by an order of magnitude by applying the pure source gasses.

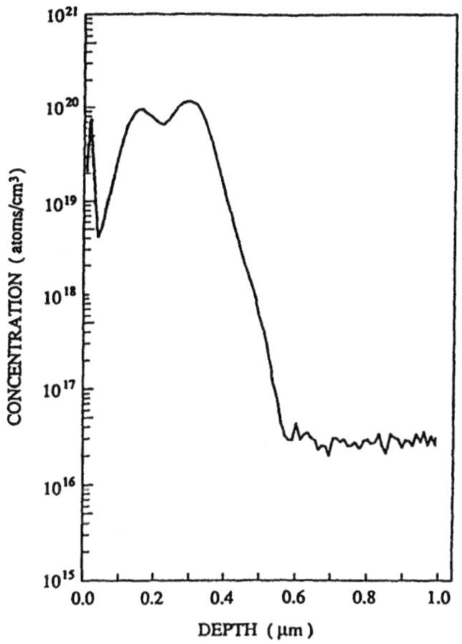

Fig.3. A depth profile of nitrogen impurity concentration in a 3C-SiC epilayer implanted with 60keV and 160keV nitrogen ions. The nitrogen concentration is calibrated by adjusting the total area of the depth profile to the implantation fluence of $2.5 \times 10^{15} cm^{-2}$.

2.4 Lattice Site Location of Nitrogen Atoms

In order to compare the residual carrier density with the nitrogen concentration measured by SIMS, we measured the lattice site location of nitrogen atoms in the 3C-SiC lattice using the ion channeling technique. Because of the very low atomic concentration of impurity nitrogen and its low mass number compared with the matrix, we used the $^{15}N(p,\alpha_0)^{12}C$ nuclear reaction analysis (NRA). A ^{15}N-doped 3C-SiC epilayer was grown for this purpose. The Q-value of this reaction is 4.964 MeV, and 4.0 MeV α-particles are emitted to the scattering angle 135° for 920 keV proton incidence. Emitted α-particles were detected by a silicon surface barrier (SSB) detector with 12.5 mm thick Kapton absorber, which allows to pass only the high energy α-particles and stops all the backscattered protons with extremely high intensity but low energy. The spectra of the α-particles are shown in Fig. 4 (a) for the random and the <111> axis aligned incidence of 920 keV proton beam,

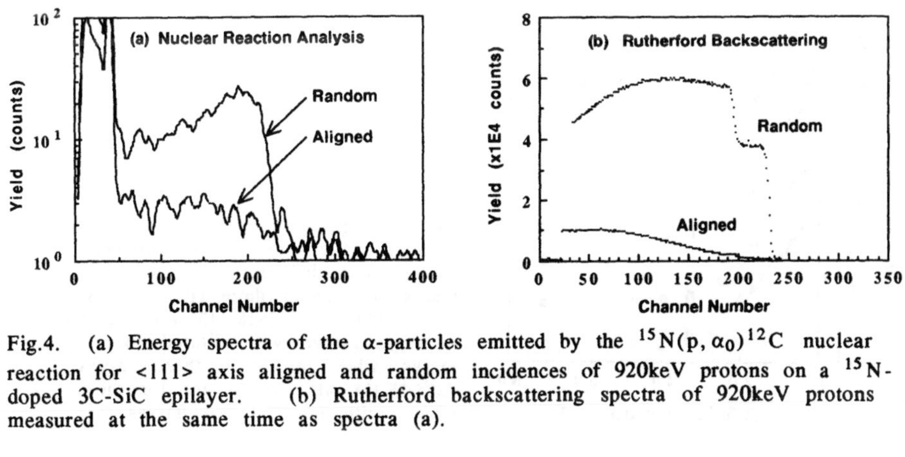

Fig.4. (a) Energy spectra of the α-particles emitted by the $^{15}N(p,\alpha_0)^{12}C$ nuclear reaction for <111> axis aligned and random incidences of 920keV protons on a ^{15}N-doped 3C-SiC epilayer. (b) Rutherford backscattering spectra of 920keV protons measured at the same time as spectra (a).

where the normalized minimum yield, χ_{min}, is calculated to be 2.4%. Figure 4 (b) shows the Rutherford backscattering (RBS) spectra measured at the same time as NRA spectra by a SSB placed at about 85cm away from the sample. The χ_{min} of 1.0% is obtained for the silicon sublattice. From the comparison of the NRA χ_{min} and the RBS χ_{min}, we conclude that almost all the ^{15}N atoms are substitutional. This implies that the nitrogen impurities are active as donors in 3C-SiC epilayers.

We evaluated ^{15}N atomic concentration to be 7.3×10^{17} atoms/cm^3, using a thin Si_3N_4 epilayer as a standard sample which contains 0.366% (natural abundance) ^{15}N atoms. Whereas 5.5×10^{17} atoms/cm^3 was obtained by SIMS measurement, being in fairly good agreement with the NRA result.

3. Discussions

In the previous section, we showed an order of magnitude reduction in residual carrier density in the 3C-SiC epilayers grown using the pure source gasses. Since 3C-SiC epilayers are compensated by unknown acceptors, this carrier density reduction does not necessarily mean the reduction in donor density. It is also probable that the reduction arises from the increase in acceptor density. Therefore, we discuss in this section whether the carrier reduction is due to the reduction in donor density or to the increase in acceptor density.

In Fig.5, carrier densities(n) are plotted as a function of the reciprocal of temperature($1/T$) for various 3C-SiC epilayers grown using different CVD source gasses. The upper three curves are for the epilayers grown with type(a) source gasses, whereas the lower three curves are for type(b) source gasses. The carrier density and their temperature dependence differ significantly from sample to sample depending on crystal growth conditions even for the epilayers grown with the same source gasses. Nevertheless, there exists an apparent reduction in carrier density for the epilayers grown with type(b) source gasses.

In the temperature region between 200K and 50K, the n-$1/T$ plots can be fitted by the theoretical curves calculated using the two-donors model[6]. The curve fitting analyses reveal that the reduction in residual carrier density arises mostly from the reduction in donor level density. For example, the donor level densities are 2×10^{16}cm^{-3} and 1.9×10^{14}cm^{-3} for vcs181-43 and IS86-4B epilayers respectively. This result, together with the ESR, SIMS and NRA results, indicates that the carrier reduction is mostly due to the reduction in the concentration of nitrogen impurity which is likely to be incorporated during the dilution process of CVD source gasses.

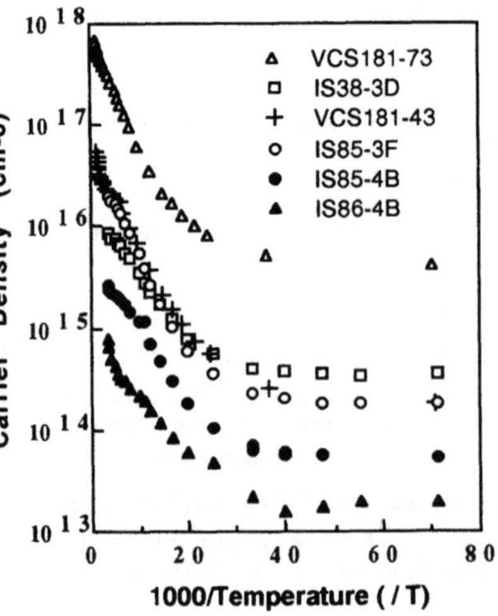

Fig.5. Temperature dependences of residual carriers in 3C-SiC epilayers grown using different CVD source gasses. The upper three curves (VCS 181-73, IS38-3D, VCS181-43) are for the epilayers grown with diluted source gasses, while the lower three curves (IS85-3F,IS85-4B,IS86-4B) are for those grown with pure source gasses.

Above 200 K, carrier density increases rapidly with temperature, suggesting that there exist other donor levels and that the residual carrier density at room temperature is strongly affected by these donors. This is one reason why residual carrier density at room temperature depends strongly on crystal growth conditions.

4. Summary

We obtained 3C-SiC epilayers with residual carrier densities about $1\times10^{15}cm^{-3}$ by applying pure CVD source gasses instead of diluted ones. Hall-effect, ESR, SIMS and NRA measurements were made on these epilayers. No significant 3-line ESR signal associated with nitrogen impurities was observed for the epilayers with carrier densities below $1\times10^{15}cm^{-3}$. The SIMS measurements showed that nitrogen concentration is about $1\times10^{16}cm^{-3}$ in the epilayers grown with the pure source gasses. By applying the $^{15}N(p,\alpha_0)^{12}C$ nuclear reaction analysis combined with channeling, we found that almost all the nitrogen atoms are at the substitutional sites of the 3C-SiC lattice and active as donors. From the temperature dependence of the carrier density, the reduction in carrier density is attributed mostly to the decrease in nitrogen donor concentration.

The authors would like to thank Dr. I. Sakata and Dr. H. Sekiguchi of Electrotechnical Laboratory for assisting ESR and NRA measurements.

References

1. M.Shinohara,M.Yamanaka,H.Daimon,E.Sakuma,H.Okumura,S.Misawa,K.Endo,and S.Yoshida: Jpn.J.Appl.Phys. 27(1988)L434.
2. A.Suzuki,A.Umemoto,M.Shireta,K.Furukawa,and S.Nakajima: Appl.Phys.Lett.49 (1986)450.
3. B.Segall,S.A.Alterovits,E.J.Haugland,and L.G.Matus: Appl.Phys.Lett. 49(1986)584.
4. B.Segall,S.A.Alterovits,E.J.Haugland,and L.G.Matus: Appl.Phys.Lett. 50(1987)1533.
5. J.A.Freitas Jr.,S.G.Bishop,P.E.R.Nordquist,Jr.,and M.L.Gipe: Appl.Phys.Lett. 52 (1988)1695.
6. I.Nashiyama,H.Okumura,E.Sakuma,S.Misawa,K.Endo,and S.Yoshida: in 3rd. Int. Conf. on Amorphous and Crystalline Silicon Carbide and other Group IV-IV Materials, (Washington, D.C., April,1990). (to be published)
7. H.Okumura,M.Shinohara,S.Kuroda,K.Endo,E.Sakuma,S.Misawa,and S.Yoshida Jpn.J.Appl.Phys. 27(1988)1712.
8. D.J.Lepine : Phys.Rev. B2(1970)2429.
9. G.Feher, R.C.Fletcher, and E.A.Gere: Phys.Rev. 100(1955)1784.
10. S.Yoshida,K.Sasaki,E.Sakuma,S.Misawa,and S.Gonda:Appl.Phys.Lett.46(1985)766.

Electron Nuclear Double Resonance Investigations of Nitrogen Donors in 6H and 4H-SiC

S. Greulich-Weber, R. Müller, and J.-M. Spaeth

University of Paderborn, FB 6 – Experimentalphysik,
Warburger Str. 100A, W-4790 Paderborn, Fed. Rep. of Germany

Nitrogen is the most important unavoidable impurity in silicon carbide. It was investigated intensively by many different methods. Nevertheless, the microscopic structures of nitrogen defects are not known in detail up to now. Early Electron Nuclear Double Resonance (ENDOR) measurements by Hardeman [1] could not give any information on the structures of those defects. Therefore we carried out ENDOR experiments on two SiC polytypes. We got detailed informations on the behavior of nitrogen on the three different lattice sites possible in 6H and those two sites in 4H-SiC. Different hyperfine (hf) interactions were obtained for nitrogen on all possible sites in both materials. Also ligand hf interactions with ^{29}Si could be resolved. From the size and magnitude of the N hf interactions it is clear that the different sites have markebly different electronic structures. The microscopic structures as determined from our ENDOR analysis will be discussed for the nitrogen defects in both polytypes.

[1] G.E.G. Hardeman, J. Phys. Chem. Solids, Pergamon Press 1963, Vol. 24, pp 1223-1231

Springer Proceedings in Physics, Vol. 71 **Amorphous and Crystalline Silicon Carbide IV**
Editors: C.Y. Yang · M.M. Rahman · G.L. Harris

Recrystallization and Electrical Properties of High-Temperature Implanted (N,Al) 6H-SiC Layers

W. Suttrop[1], *H. Zhang*[1,*], *M. Schadt*[1], *G. Pensl*[1], *K. Dohnke*[2], and *S. Leibenzeder*[2]

[1]Institut für Angewandte Pgysik, Universität Erlangen-Nürnberg, Staudtstr. 7, W-8520 Erlangen, Fed. Rep. of Germany

[2]Siemens AG, Research Laboratory, Scharowskystr., W-8520 Erlangen, Fed. Rep. of Germany

*Present address: Robert Bosch GmbH, Postfach 30 02 40, W-7000 Stuttgart 30, Fed. Rep. of Germany

Abstract. Ion implantation of aluminum and nitrogen in 6H-SiC is investigated using Rutherford backscattering in the channeling mode (RBS/C) and Hall effect for structural and electrical characterization of the implanted layers. RBS/C spectra reveal that in the case of room temperature implantations heavily damaged layers are formed approx. 0.2 μm below the SiC surface. The use of elevated sample temperatures (T=800°C to 1000°C) during the N or Al implantation leads to strongly reduced crystalline defect densities. Subsequent heat treatments by rapid isothermal annealing (RIA) at 1060°C result in a nearly perfect recrystallization of the implanted layers. While N can be electrically activated up to nearly 100% in the case of RT-implantations, Al is only activated in the high-temperature implanted samples. p^+n-diodes are fabricated by hot Al implantation; they show good I-V characteristics (RT: 22 A/cm^2 at 5V forward bias, $1\cdot 10^{-5}$ A/cm^2 at 10V reverse bias).

1. Introduction

Ion implantation of aluminum and nitrogen as the prevailing dopants for the semiconductor silicon carbide (SiC) is the most promising doping technique of a planar device technology since the low diffusivity of these elements in SiC does not allow diffusion at process temperatures of T ≤1500°C [1,2]. Ion implantation of Al and N has been primarily studied at room-temperature (see e.g. [3,4]). Amorphization by N- and Al-implantations has been reported to occur at ion fluences above $1\cdot 10^{15}$ cm^2 and $4.9\cdot 10^{14}$ cm^2, respectively [4]. In the present paper, it is demonstrated that high-temperature ion implantation (T_i=800°C or 1000°C) leads to strongly reduced crystalline defects compared to room-temperature implantation (see also [5]). In addition, electrical activation of aluminum as an acceptor is observed only for implantation temperatures above T_i=500°C. Al-implantation has been used in the past for the fabrication of p^+n-diodes [6,7]. It is demonstrated that p^+n-diodes which are Al-implanted at T_i=1000°C can be fabricated by using anneal temperatures as low as 1060°C. This low processing temperature enables the use of oxide masks for a planar device technology.

2. Experimental

The 6H-SiC crystals used for the current investigations are grown by a modified Lely method [8]. Specimens for backscattering experiments are polished and prepared by reactive ion etching (RIE) with O_2-gas in an rf plasma in order to remove the surface damage introduced by the polishing procedure. For electrical measurements, 6H-SiC epitaxial layers are grown from the liquid phase [9]. Nitrogen and aluminum were implanted either at room temperature (T=20..40°C) or at temperatures of 800°C and 1000°C. Post-implantation annealing is performed either by rapid isothermal annealing (RIA) at temperatures up to 1060°C or in a furnace with a SiC-coated graphite crucible at temperatures up to 1500°C.

Rutherford backscattering in the ion channeling mode (RBS/C) is performed to investigate the crystalline quality of implanted SiC layers. A 1.5 MeV or 2 MeV He^+beam (ion current: 30nA) supplied by a 2 MeV Van-de-Graaff accelerator is used for the experiments. The samples are mounted on a 3-axis goniometer in high-vacuum ($p=4\cdot10^{-7}$ mbar). Axial channels in [0001]-direction are used for the aligned spectra. Random spectra are recorded by rotating the specimen around an axis normal to its surface. The He^+ ion fluence and the scattering angle are 20μC and 170°C, respectively.

Hall effect experiments are conducted on nitrogen-implanted LPE layers. Nitrogen is implanted at room temperature by multiple ion implantation in order to achieve a nearly rectangular doping profile. The respective ion energies and fluences employed are: 40 keV, $5.5\cdot10^{13}$ cm^{-2}; 70 keV, $8.2\cdot10^{13}$ cm^{-2};and 110 keV $1\cdot10^{14}$ cm^{-2}.

3. Backscattering Results

RBS/C-spectra recorded of Al- and N-implanted SiC (6H) crystals are shown in Figs. 1 and 2. After implantation of Al (fluence: $2\cdot10^{15}$ cm^{-2}, ion energy: 280 keV) at room temperature, a heavily damaged surface-near region is generated. This is reflected in the RBS/C spectrum (labelled "as implanted" in Fig. 1 a)) by an

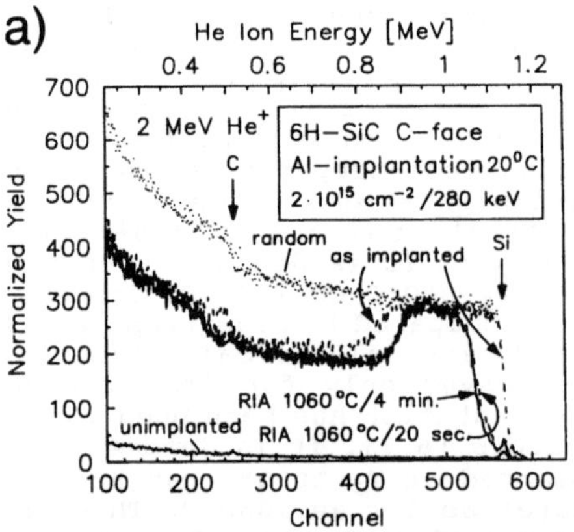

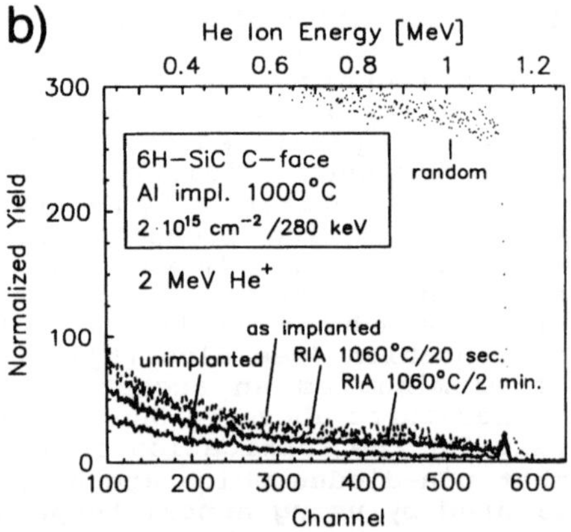

Fig. 1: Rutherford backscattering spectra ([0001] aligned) of Al-implanted 6H-SiC:
a) implantation at room temperature, b) implantation at T_i=1000°C

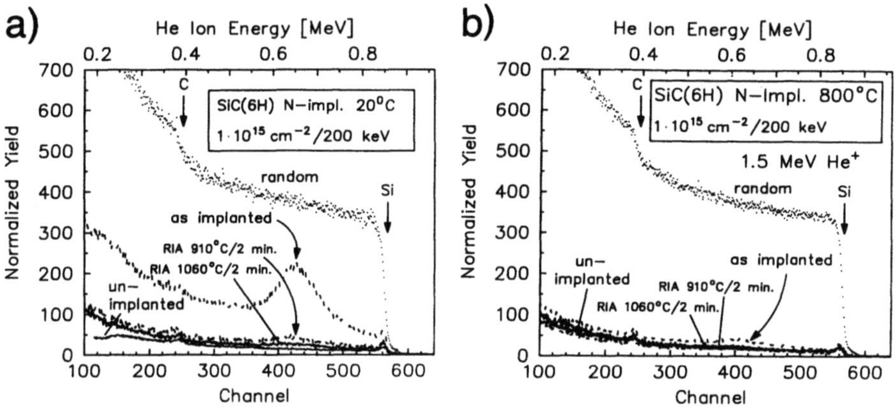

Fig. 2: Rutherford backscattering spectra ([0001] aligned) of N-implanted 6H-SiC:
a) implantation at room temperature, b) implantation at T_i=800°C.

increased backscattering yield, which approaches random yield for ions back-scattered in a depth of approx. 0.2 μm below the surface. Annealing by RIA at 1060°C for 20 sec results in partial recrystallization at the surface-near and bulk-near edges of the damaged layer. Additional RIA annealing at the same temperature for 2 min does not lead to significant changes in the RBS/C spectra. Based on these results, we suggest that the core of the damaged layer is amorphous.

Implantation of the same fluence of Al at a sample temperature of T_i =1000°C leads to a strongly reduced defect concentration as demonstrated by the RBS/C-spectra in Fig. 1 b). Stepwise annealing by RIA at 1060°C for 20 sec and for 2 min subsequently to the implantation yields a further reduction of the residual lattice defects. However, no complete removal of the damage is achieved for the particular annealing parameters employed.

Comparable results are obtained for nitrogen implantations Fig.2 a) shows RBS/C-spectra after room temperature implantation and subsequent annealing steps. The fluence and ion energy for this experiment were $1 \cdot 10^{15}$ cm^{-2} and 200 keV, respectively. A damaged surface-near layer is observed. However, in this case we conclude that no complete amorphization occurs, because a heat treatment by RIA at 910°C for 2 min reduces the implantation damage. The defect density is further reduced by RIA annealing at 1060°C for 2 min, but is not completely removed. This is demonstrated by the backscattering yield being higher than that of the unimplanted specimen. High-temperature implantation (T_i=800°C) of the same fluence of nitrogen results in a much lower defect density. The RBS/C spectra in Fig. 2 b) indicate that this damage can be removed by RIA annealing at 1060°C.

4. Hall-Effect Results

Figs. 3 a) and b) show the electron concentration and the electron mobility of N-implanted SiC layers of four samples annealed at different conditions. Ionization energy E_N, concentration N_N, compensation N_A, and the maximum electron mobility (T=200K) are

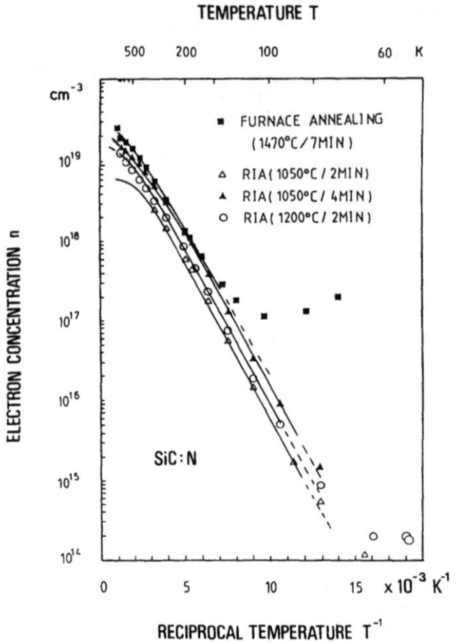

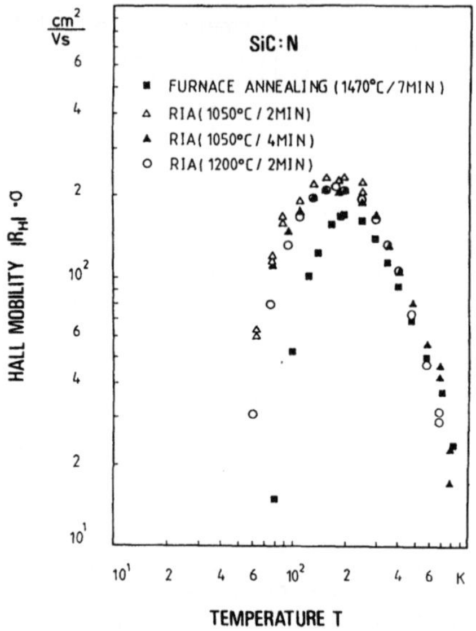

Fig. 3 a): Electron concentration as determined from Hall effect experiments for nitrogen-implanted 6H-SiC layers as a function of the reciprocal temperature.b) Electron Hall mobilitiy for nitrogen-implanted 6H-SiC layers as a function of temperature.

Table 1: Hall-effect results of investigated N-implanted 6H-SiC LPE layers. Annealing parameters, ionization energy, doping concentration, compensation, and maximum electron mobility are listed in columns 2 to 6.

SAMPLE No.	ANNEALING $^{o}C/min$	IONIZATION-ENERGY E_N meV	CONCENTRATION N_N cm^{-3}	COMPENSATION N_A cm^{-3}	MAX.MOBILITY (200K) cm^2/Vs
1	furnace 1470/ 7	70	$3{,}0 \times 10^{19}$	$4{,}0 \times 10^{18}$	180
2	RIA 1050/ 2	70	$9{,}8 \times 10^{18}$	$2{,}2 \times 10^{18}$	250
3	RIA 1050/ 4	70	$2{,}3 \times 10^{19}$	$2{,}4 \times 10^{18}$	230
4	RIA 1200/ 2	70	$1{,}4 \times 10^{19}$	$2{,}6 \times 10^{18}$	230

listed in Table 1. An electrical activity of the implanted nitrogen of nearly 100% is achieved by furnace annealing at 1470°C for 7 min. By RIA anneals at 1050°C for 4 min, the electrical activation is 77%. However, the RIA process leads to a higher mobility (μ_n = 230 cm^2/Vs) compared to the furnace annealing (μ_n = 180 cm^2/Vs).This observation can be attributed to the increased activation of the compensation by furnace annealing at 1470°C in comparison to RIA annealing at 1050°C.

5. Al-Implanted pn-Diodes

p^+n-diodes were fabricated by Al implantation at T_i=1000°C. A schematic view of the device is shown in Fig. 4. A 3-fold implantation with a maximal ion energy of 100keV and total fluence of $5.7 \cdot 10^{14}$ cm^{-2} is employed to achieve a uniformly doped p^+-region. Post-implantation annealing is performed either by RIA at 1060°C for 2 min or in a furnace at 1500°C for 2 min. Contacts are prepared by evaporation of aluminum and titanium on the p^+- and the n-side, respectively, followed by a heat treatment at 900°C for 2 min. The active diode area was defined by RIE (reactive ion etching) using the Al p^+-contacts as a mask. Fig. 5 shows I-V characteristics of two diodes at room temperature. A forward current density of 22A/cm^2 at 5V bias and a leakage current density of $1 \cdot 10^{-5}$ A/cm^2 at 10V reverse bias are obtained for RIA-annealed diodes (T_A=1060°C).

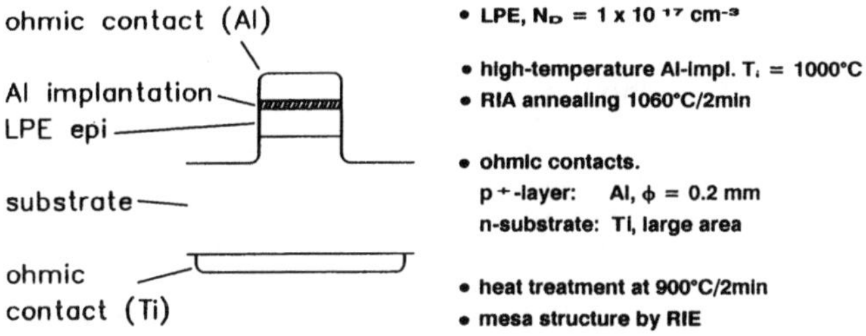

Fig. 4: Schematic view of an Al-implanted p^+n-diode.

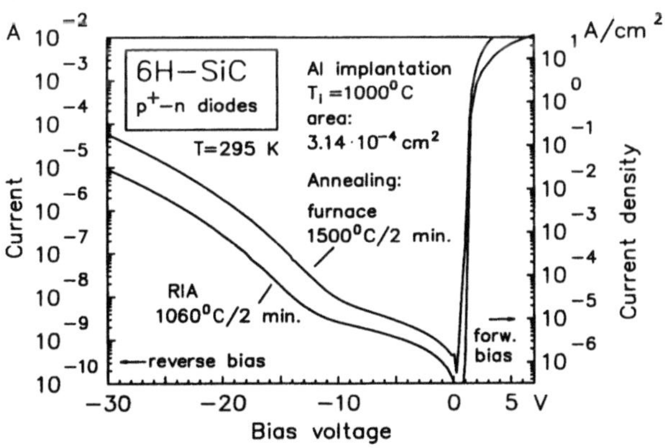

Fig.5: I-V characteristics of two selected Al-implanted p^+n-diodes annealed at 1060°C and at 1500°C, respectively.

6. Summary

It is demonstrated that implantations of nitrogen or aluminum in SiC samples at elevated temperatures (800°C or 1000 °C) lead to largely reduced crystal damage. Electrical activation of N as a donor can be achieved after room temperature implantation and subsequent annealing whereas electrical activity of Al as an acceptor is observed only after implantation at T_i=1000°C. With respect to the electrical properties of nitrogen-implanted layers, rapid isothermal annealing (RIA) at 1060°C is favorable. Furnace annealing at 1450°C activates the compensation to a higher degree. High-temperature Al-implanted p^+n-diodes show good I-V characteristics at room temperature.

Acknowledgement

This research work was supported by the BMFT under contract No. 03M2708B9.

References

1. Vodakov, Y.A., Mokhov, E.N. in: Silicon Carbide 1973, Ed. R.C.Marshall, J.W.Faust, Jr., E.E.Ryan, University of South Carolina Press, Columbia (1973), p.508
2. Tajima, Y., Kijima, K., Kingery, W.D.: J.Chem. Phys. 77 (1982) 2592
3. Marsh, O.J., in: Silicon Carbide 1973, Ed. R.C.Marshall, J.W.Faust, Jr., C.E.Ryan, University of South Carolina Press, Columbia (1973), p. 471
4. Wood, S., Spitznagel, J.A., Choyke, W.J., Bradshaw, J., Greggi, J., Jr.,. Doyle, N.J.: Mat. Res.Soc. Symp. Proc. 60 (1986) 459
5. Edmond, J.A., Withrow, S.P., Wadlin, W., Davis, R.F.: Mat.Res.Soc.Symp. Proc. 77 (1987) 193
6. Anikin, M.M., Lebedev, A.A., Popov, I.V., Sevast'yanov, V.E., Syrkin, A.L., Suvorov, A.L., Chelnokov, V.E., Shpynev, G.P.: Sov.Tech. Phys.Lett. 10 (1985) 444
7. Edmond, J.A., Das, K., Davis, R.F.: J.Appl.Phys. 63 (1988) 922
8. Ziegler, G., Lanig, P., Theis, D., Weyrich, D.: IEEE Trans. ED-30 (1983) 277
9. Hoffmann, L., Ziegler, G., Theis, D., Weyrich, D.: J.Appl.Phys. 53 (1982) 6962

EPR in the 2-mm Range and Optical Absorption of the Native Defect in 4H-SiC Epilayers

E.N. Kalabukhova[1], *S.N. Lukin*[1], *B.D. Shanina*[1], *Yu.A. Vodakov*[2], *A.A. Lepneva*[2], and *E.N. Mokhov*[2]

[1]Institute of Semiconductors, Ukrainian Academy of Sciences, Pr. Nauki 45, 252650 Kiev 28, Ukraine
[2]A.F. Ioffe Physical-Technical Institute, Russian Academy of Sciences, Politechnicheskaya 26, 194021 St. Petersburg, Russia

Abstract. EPR spectra at frequency $\nu = 140$ GHz and optical absorption spectra are studied for 4HSiC epilayers produced by means of vacuum sublimation "sandwich" method with nitrogen concentrations from 10^{16} to $10^{18}cm^{-3}$.

A new paramagnetic center ND-1 is discovered in the 4HSiC epilayers, which have deviations from the stoichiometry towards Si excess. Dynamics of the EPR spectrum of 4HSiC epilayers are studied in the temperature interval from T = 4.2K to 34K. Comparing theoretical to experimental data, the binding energy for the paramagnetic carrier captured by the ND-1 center is found to be E = 12.8 MeV.

A model of the ND-1 center is proposed, which interprets it as an antisite defect Si_C^- i.e., a silicon atom in the hexagonal position of a carbon site with captured electron.

1. Introduction

It is well known [1-3] that the luminescence properties of silicon carbide epilayers (EL), especially of those grown at low temperatures, strongly depend on the preparation technology. These peculiarities of SiC EL are usually ascribed to the existence of native point defects, in particular, stoichiometrical with excess Si or inequilibrium carbon vacancies (V_C).

The data of Refs.[4,5] show that such non-stoichiometric defects exist in the EL of cubic modification of SiC, grown on silicon substrates. These defects are shown to be able to lessen the electron mobility [2], to favour the increase of photosensitivity in the infrared range. Non-stoichiometric effects have been also discovered in the EPR studies. However, the nature of these centers is still unknown.

In this paper we study EPR and optical absorption spectra in the SiC epilayers of the 4H polytype, grown by the vacuum

Editors: C.Y. Yang · M.M. Rahman · G.L. Harris

sublimation technology. A new center of pure defect origin is discovered and assumed to be the antisite defect Si_C^-.

The 4HSiC epilayers were grown by means of the sublimation "sandwich" technology [6] in vacuum of 10^{-3} at temperatures 1700– 1900C. The grown samples exhibited n-type conductivity with the noncompensated donor concentration $(N_D-N_A) \propto 2\times10^{16}$–$5\times10^{17}cm^{-3}$. The doping with acceptor impurity N_A did not exceed $(1-2)\times 10^{16}cm^{-3}$.

2. Optical absorption spectra

The absorption spectra were obtained from the analysis of experimental transmission spectra with regard for multiple reflection. Measurements were carried out for T = 300K in the 0.4–1.2 μm wavelength range with the light polarization $E \perp c$.

The results for 4H and 6H SiC epilayers are shown in Fig.1 together with the absorption spectra of usual industrial Lely grown crystals having similar concentrations of noncompensated donors.

As follows from Fig.1, the SiC EL absorption spectra are appreciably different from the traditional spectra of Lely grown crystals.

1. The absorption band associated with the nitrogen impurity is weakly manifested in SiC EL.
2. For $\lambda>0.8$ μm, absorption in the nearer infrared range in SiC EL is weaker than in Lely grown crystals.

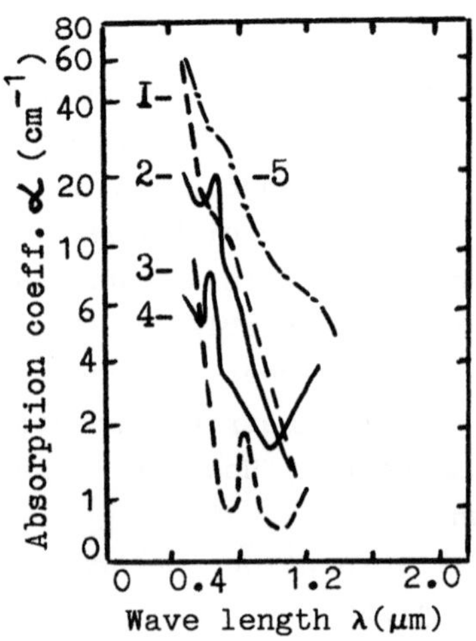

Fig.1. Absorption spectra for 4HSiC and 6HSiC crystals. $E \perp c$, T=300K, $(N_D-N_A)\propto (1-3)\times10^{17}cm^{-3}$. 1-6HSiC(EL); 2-4HSiC(EL); 3-6HSiC (Lely); 4-4HSiC (Lely); 5-4HSiC (Lely), irradiated by neutron beam with $\phi_n \propto 10^{18}cm^{-2}$.

3. Absorption in the visible range in SiC EL is stronger than in Lely grown crystals.

The above observations suggest that the properties of usual donor nitrogen atoms are weakly manifested in SiC EL. Instead, new absorption mechanisms arise which might be associated with the presence of native defects which are formed in the course of the epilayer grown. This assumption is confirmed by the similarity of SiC EL spectra to the absorption spectra of crystals irradiated by reactor neutrons which give rise to numerous defect states. Fig.1 (curve 5) shows the data for Lely grown 4HSiC, irradiated by a neutron beam from the reactor with $\Phi_H \propto 10^{18} cm^{-2}$.

Unfortunately, the absorption spectra are not structuralized and thus it is impossible to analyse the defect states formed under the growth of SiC epilayers. Therefore, approaches providing complete or partial solutions such problems are of great interest.

3. EPR spectra of 4H-SiC epilayers

In this paper we study the EPR spectrum of nitrogen atoms in the 4HSiC epilayers at the frequency 140 GHz. Fig.2 shows the

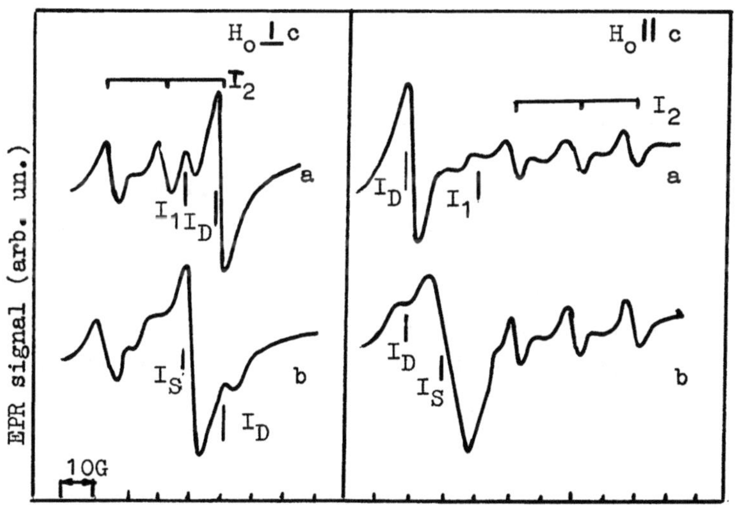

Fig.2. EPR spectrum of nitrogen and native defect I_D in 4HSiC epilayers. T = 4.2K, ν = 142 GHz. Nitrogen concentration (N_D-N_A): a - $5\times10^{16} cm^{-3}$; b - $5\times10^{17} cm^{-3}$.

EPR spectrum of 4HSiC EL with nitrogen concentrations $(N_D-N_A) \propto 5\times10^{16}cm^{-3}$ and $5\times10^{17}cm^{-3}$ for T = 4.2K, $H_o \parallel c$ and $H_o \perp c$. The epilayer EPR spectrum consists of a hyperfine structure (HFS) line triplet I_2 corresponding to the cubic "c" position of the nitrogen atom and a single line I_1 with anisotropic width which corresponds to the hexagonal ("h") position of the nitrogen atom. The parameters of these lines are the same as those of the nitrogen EPR spectrum in the Lely grown 4HSiC [7] (see Table 1). Along with the EPR spectrum of single paramagnetic nitrogen centers, the EPR spectrum of 4HSiC EL contains a single line I_D whose parameters are given in Table 1.

As nitrogen concentration increases to $5\times10^{17}cm^{-3}$, an additional wide intensive line I_S appears in the EPR spectrum of the EL, similarly to the case of Lely grown 4HSiC. The parameters of this line are given in Table 1.

Let us consider the characteristics of the I_D-line and compare the I_S-line behavior in 4HSiC epilayers and Lely grown 4HSiC.

1. The I_D-line has an axially symmetric g-factor and corresponds to the defect center with spin S = 1/2. The angle dependence of the g-tensor of the I_D-line is analogous to that of the "h" center, but with greater $g_{\parallel}$ (see Table 1).
2. The I_D-line has isotropic width ΔH = 2.5 G.
3. Estimates of the integral intensities ratio of the defect center to nitrogen atom for a series of samples show that defect center concentration is of the same order of magnitude as nitrogen concentration in the sample.
4. As follows from Table 1, the g-factors of I_S-lines for T= 4.2K are different. The g-factor for the Lely grown 4HSiC crystal corresponds to the center of gravity of the two EPR spectra of the nitrogen atom in "c" and "h" positions while the g-factor of the I_D-line in 4HSiC EL corresponds to the center of gravity of the EPR spectra of the defect center I_D and nitrogen in the "h" position I_1.
5. I_S-line intensity for both Lely grown crystals and 4HSiC EL is considerably greater than the EPR spectrum intensity for paramagnetic nitrogen atom.
6. The width of the I_S-line in the 4HSiC EL is anisotropic in contrast to that for Lely grown 4HSiC, see Table 1.

What is the reason for the observed difference between the EPR spectra of 4HSiC crystals, produced by Lely and vacuum

Table 1. Parameters of the nitrogen atom and native defect EPR spectra in the 4HSiC samples, grown by the Lely and by the epitaxial "sandwich" methods with nitrogen concentrations from 10^{16} to 5×10^{17} cm^{-3}

		$g_{\parallel}$				$g_{\perp}$				A, ΔH,G	A, ΔH,G
I_I	T, K	Lely (N_D-N_A) 5×10^{16}	Lely (N_D-N_A) 5×10^{17}	EL (N_D-N_A) 5×10^{16}	EL (N_D-N_A) 5×10^{17}	Lely (N_D-N_A) 5×10^{16}	Lely (N_D-N_A) 5×10^{17}	EL (N_D-N_A) 5×10^{16}	EL (N_D-N_A) 5×10^{17}	Lely	EL
I_1	4.2	2.0055 ±0.0002	–	2.0055 ±0.0002	–	2.0010 ±0.0002	–	2.0010 ±0.0002	–	$\Delta H_{\parallel}$ =9.8 $\Delta H_{\perp}$ =7.4	$\Delta H_{\parallel}$ =9.8 $\Delta H_{\perp}$ =7.4
	12	–	–	2.0055 ±0.0002	–	–	–	2.0010 ±0.0002	–	$\Delta H_{\parallel}$ =9.8 $\Delta H_{\perp}$ =7.4	$\Delta H_{\parallel}$ =9.8 $\Delta H_{\perp}$ =7.4
I_2	4.2	2.0043 ±0.0002	2.0043 ±0.0002	2.0043 ±0.0002	2.0043 ±0.0002	2.0013 ±0.0002	2.0013 ±0.0002	2.0013 ±0.0002	2.0013 ±0.0002	18.2 ±0.2	18.2 ±0.2
	12	–	2.0043 ±0.0002	2.0043 ±0.0002	2.0043 ±0.0002	–	2.0013 ±0.0002	2.0013 ±0.0002	2.0013 ±0.0002	18.2 ±0.2	18.2 ±0.2
I_D	4.2	–	–	2.0063 ±0.0002	2.0063*	–	–	2.0006 ±0.0002	2.0006*	–	$\Delta H_{\parallel}$= 2.5 $\Delta H_{\perp}$= 2.5
	12	–	–	2.0063 ±0.0002	2.0063*	–	–	2.0006 ±0.0002	2.0006*	–	$\Delta H_{\parallel}$= 2.5 $\Delta H_{\perp}$= 2.5
I_S	4.2	–	2.0051 ±0.0002	–	2.0059 ±0.0002	–	2.0011 ±0.0002	–	2.0009 ±0.0002	$\Delta H_{\parallel}$= 12 $\Delta H_{\perp}$= 12	$\Delta H_{\parallel}$= 12 $\Delta H_{\perp}$= 7.5
	12	–	2.0055 ±0.0002	–	2.0058 ±0.0002	–	2.0010 ±0.0002	–	2.0009 ±0.0002	$\Delta H_{\parallel}$= 12 $\Delta H_{\perp}$= 12	$\Delta H_{\parallel}$= 12 $\Delta H_{\perp}$= 7.5
	34	–	–	–	2.0055 ±0.0002	–	–	–	2.0009 ±0.0002	–	$\Delta H_{\parallel}$= 8 $\Delta H_{\perp}$= 3.5

* I_D – line g-factor cannot be determined because of overlap with I_s – line.

"sandwich" technologies? The epilayer samples are grown under the conditions of considerable silicon excess [1-3]. Hence, the additional intensive single I_D-line in the EPR spectrum of epilayer 4HSiC may be associated with the native defects existing in the epilayers due to the deviations from the stoichiometry.

To reveal the nature of such defect centers, we studied EPR spectra of EL 4HSiC with nitrogen concentrations from 10^{16} to $10^{18} cm^{-3}$ in the temperature range from 4.2K to 34K.

4. Study of the temperature dependence of the 4H SiC epilayers EPR spectrum

The intensity of the EPR I_D-line of the native defect in the sample with nitrogen concentration $(N_D-N_A) \propto 5\times10^{16} cm^{-3}$ strongly dependents on temperature. Fig.3 gives its temperature function

$$F_D(T)=\frac{I_D(T)}{I_C(T)}\, th(\frac{h\nu_C}{2kT}) \tag{1}$$

Here I_D, I_C are the integral intensities of EPR lines associated with the defect and nitrogen in the "c" position respectively, $I_D(T)/I_C(T)$ is the experimentally observed ratio of these (the width of the I_D-line is temperature independent), $th(h\nu_C/2kT)$ is the Curie-Weiss temperature function for $I_C(T)$, $F_D(T)$ is the temperature dependence of the I_D-line intensity. The curve $F_D(T)$ is plotted by experimental points in Fig.3; a good approximation of this curve is given by the function

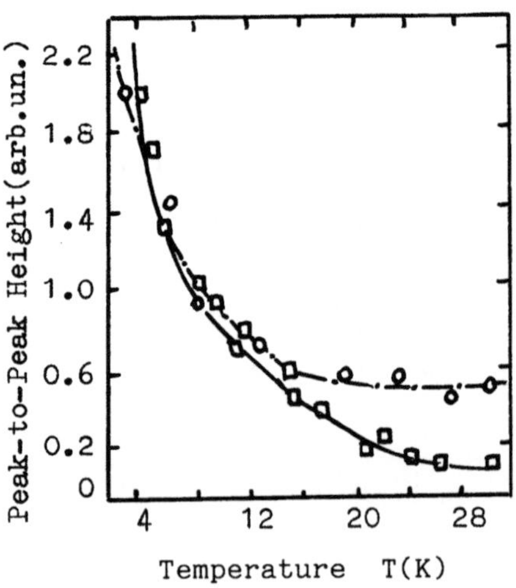

Fig.3. Temperature dependence of EPR line intensities for native defect I_D and exchange I_s-line in 4HSiC epilayers. "□"-$F_D(T)=I_D/I_C th(h\nu_c/2kT)$-experimental; "-" - $F_D(T)$ calculation by (2); "o"-$F_s(T)=I_s/I_c th(h\nu_c/2kT)(\Delta H_s)^2$-experimental "-.-"-$F_s(T)$ calculation by (6).

$$F_D(T) = \frac{8\ \mathrm{th}\ (h\nu_D/2kT)}{1 + 901\ \exp(-148.4/T)} \qquad (2)$$

The temperature dependence of this type is obtained by solving the population balance equations for a pair of spin levels (with $M_S=\pm1/2$) of a paramagnetic defect in the forbidden band taking into account thermal ionization of the defect, i.e.,

$$\begin{aligned} \dot{n}_1 &= -W_{21}n_1 + W_{21}n_2 - b(T)\ n_1 \\ \dot{n}_2 &= -W_{21}n_2 + W_{12}n_1 - b(T)\ n_2 \end{aligned} \qquad (3)$$

where $n_{1,2}$ are the spin sublevel populations for the native defect ground state.

Considering equilibrium state, we substitute in (3) $\dot{n}_1=0$, $\dot{n}_2=0$ and thus obtain

$$n_1 - n_2 = \frac{W_S\ \mathrm{th}(h\nu_D/2kT)}{W_S + b(T)} \qquad (4)$$

where $b(T)=b_o(-\varepsilon_i/kT)$ is the thermal decay rate, W_S is the spin-lattice relaxation rate for the levels 1,2 given by

$$W_{12} = W_S \frac{\exp(-h\nu_D/kT)}{1 + \exp(-h\nu_D/kT)}, \quad W_{21} = \frac{W_S}{1 + \exp\ (-h\nu_D/kT)}$$

ν_D is the difference of the defect center level energies. Comparing (4) and (2) we see that function (2) is governed by the temperature dependence ratio of the local center paramagnetic susceptibility $\mathrm{th}(h\nu_D/2kT)$ to the defect thermal ionization $b(T) \propto \exp(-\varepsilon_i/kT)$. Then the thermal ionization energy is ε_i =-148.4K = 12.8 MeV.

It is be noted that the decrease of the EPR I_D-line intensity with growing temperature is not accompanied by the line intensity increase in the nitrogen EPR spectrum. Carrier ionization does not cause their transition to the conductivity band followed by the capture into shallow nitrogen levels in the forbidden band. Therefore, one can assume that the above value of the defect thermal ionization energy is associated with the paramagnetic carrier defect binding energy rather than with the carrier ionization energy in the conductivity band.

Let us consider the behavior of the I_S-line as temperature varies in the interval from 4.2K to 34K. The dynamics of EPR

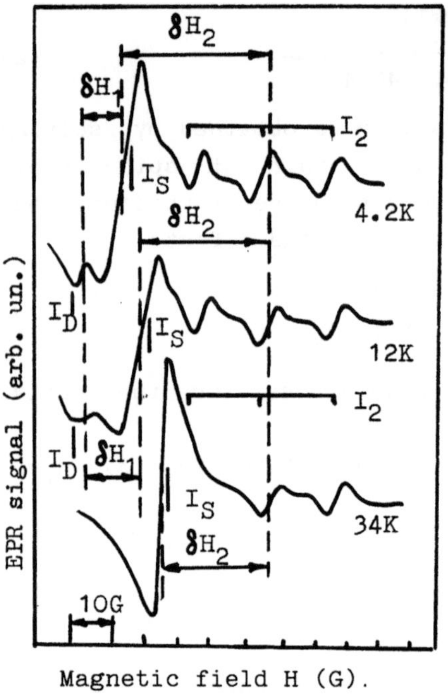

Fig.4. EPR spectra of nitrogen atom and the native defect in the 4HSiC EL at various temperatures. ν = 142 GHz. Nitrogen concentration $(N_D-N_A) \propto 5\times10^{17}cm^{-3}$.

spectra temperature dependences for 4HSiC epilayers with nitrogen concentration $(N_D-N_A)\propto 5\times10^{17}cm^{-3}$ and Lely grown 4HSiC turn out to be different.

Let us compare the temperature behavior of I_S for the Lely grown and epilayer samples of 4HSiC.

1. In both types of samples,the temperature variations lead to the shift of the I_S-line g-factor (Δg=0.0003). However, in the Lely grown samples of 4HSiC it attains the value of the g-factor of the EPR I_1-line of the nitrogen in the "h" position for 7K, while in the 4HSiC epilayers this value is attained at T =34K, for which the defect center is ionizied and the EPR spectrum of the I_D-line disappears. Fig.4 gives the nitrogen EPR spectrum for the 4HSiC EL with nitrogen concentration $(N_D-N_A)\propto 5\times10^{17}cm^{-3}$ for the magnetic field orientation $H_0 \parallel c$ and T = 4.2K, 12K, 34K. The values δH_1 and δH_2 are the distances between the resonance fields of the I_D- and I_S- lines, I_S and I_2 respectively.

2. The I_S-line intensity for the Lely grown 4HSiC samples is not changed with temperature variations; it shows strong temperature dependence for 4HSiC epilayers. Fig.3 given the temperature function of this line

$$F_s(T) = -\frac{I_s(T)}{I_c(T)}\,\mathrm{th}[\frac{h\nu_c}{2kT}]\,(\Delta H_s)^2 \qquad (5)$$

where ΔH_s is the "spike to spike" width of the I_s-line for $H_o \parallel c$. ΔH_s decreases from 12 G to 7.5 G as temperature varies from 4.2K to 34K.

As follows from Fig.3, $F_s(T)$ reproduces $F_D(T)$ in the temperature interval 4.2K to 12K. This means that for $T < 12K$ the I_s-line integral intensity is determined by the process of filling the defect center electron levels. For higher temperatures, the mechanism of the exchange line ionization is modified. As the center is ionized, the electrons, in contrast to the case of the samples with low nitrogen concentration, pass directly to "h"-nitrogen atoms, without entering the conductivity band. For T = 34K, when the defect center is completely ionized, the g-factor of the I_s-line is equal to the factor of the EPR spectrum of the nitrogen atom in the "h" position. The exchange narrowing of the I_s-line starts, its intensity being determined by the number of nitrogen paramagnetic centers in the 4HSiC epilayers.

Taking these processes into account, one obtains the following balance equation for the spin energy level populations for the electron contained in the nitrogen atom in the "h" position $N_{s\pm}$, conductivity electron $n_\pm$ and defect center electron $N_{D\pm}$:

$$\dot{N}_{s-} - \dot{N}_{s+} = -W_h(N_{s-}-N_{s+}) + W_r(n_- - n_+) + W_{Dh}(N_{D-}-N_{D+}) \qquad (6)$$

Here W_{Dh} is the probability of electron exchange hopping from the defect center to the "h"-nitrogen; W_r is the probability of conductivity electron capture by the nitrogen atom; W_h is the probability of electron escape from paramagnetic nitrogen atom in the "h" position.

Under stationary conditions, we have $\dot{N}_{s-} - \dot{N}_{s+} = 0$. Note that $I_s \propto (N_{s-}-N_{s+})$, $I_D \propto (N_{D-}-N_{D+})$, $(n_- - n_+) \propto (1 - I_D/I_{Do})$, where I_{Do} is the total number of paramagnetic defect centers at low temperatures. Then, making use of (6) and (2), we obtain

$$F_s(T) = F_D(T)[1 + \frac{W_r}{W_h}\,901\,\exp\,(-\frac{148.4}{T})]\,\frac{W_{Dh}}{W_h} \qquad (7)$$

Good agreement between the theoretical expression (7) and the experimental curve $F_s(T)$ of Fig.3 confirms the adequacy of the above assumption concerning the dynamics of centers charge exchange. Moreover, both W_{Dh} and W_h are temperature - independent, i.e., electron hopping between the defect center and nitrogen atom is due to the overlapping of electron wave functions rather than to the Mott mechanism.

Thus, the above analysis of the temperature dependence $F_S(T)$ proves the assumption that the I_S-line is produced by the paramagnetic electron that is weakly bound both to the "h"-nitrogen atom and to the defect center. This is also confirmed by the I_S-line width anisotropy ($\Delta H_S^{\parallel} = 12$ G, $\Delta H_S^{\perp} = 7$ G), analogous to the anisotropy of the "h"-nitrogen EPR line width. Anisotropy of the I_S-line g-factor ($\Delta g_{\parallel}$) is the arithmetical mean of the Δg of the I_D-line and the "h"-nitrogen I_1-line (see Table 1).

5. Results and discussions

We have obtained a wealth of experimental data concerning the properties of the native defect by studying the EPR spectra of the 4HSiC epilayers. Comparing the data obtained by the technology of 4HSiC epilayer growing, we can come to the definite conclusions concerning the nature of the defect under consideration.

1. The single EPR I_D-line of the native defect with isotropic width $\Delta H = 2.5$ G, observed in the 4HSiC epilayers corresponds to the one-electron center with spin $S = 1/2$.
2. The absence of the triplet structure in the defect center EPR spectrum suggests that the defect does not contain nitrogen atoms.
3. Values of the g-tensor components of the native defect (ND-1) and it's anisotropy indicate that the local symmetry of the site, in which the defect is located, is similar to that of the nitrogen in the hexagonal position.
4. Low paramagnetic carrier-defect binding energy E = 12.8 MeV, obtained from the temperature dependence of the I_D-line intensity shows that the electron is weakly bound in the defect and has long-range wave function.
5. Analysis of the I_S-line temperature dependence shows that the cross-section of electron capture by the ionized defect is greater than that of the electron capture by the paramagnetic "h"-nitrogen. Therefore, the range of action of the defect potential must be longer than that of the nitrogen atom Coulomb potential.

Thus, ND-1 is a shallow one-electron center containing no nitrogen atoms and having long-range wave function.

Indeed, the observed peculiar features of EPR and optical absorption spectra of the 4HSiC EL evidently are not associated with the usual paramagnetic nitrogen atom (N_C) which substitutes carbon in SiC lattices. Inasmuch as these peculiarities are not observed for the Lely grown 4HSiC samples with analogous impurity content, it is natural to assume then that it is produced by native point defects.

We have already mentioned that the 4HSiC epilayer was grown from the vapour phase with excess stoichiometric content of silicon. Earlier papers [8] studying analogous samples reported detection of defect centers with energy E = 0.6 eV which contained carbon vacancies and activated emissive recombination with the spectrum D_1.

With boron traces and native point defects produced by Si excess in the vapour phase being present, luminescent centers are formed, which evidently are associates of (B-V_C) with ionization energy E_a= 0.7 eV. We employed the EPR data to find the paramagnetic carrier defect binding energy. The data obtained do not allow one to make a definite conclusion as to whether the value ε_i= 12.8 MeV is associated with the defect level in the forbidden band.

The literature data does not report such a level. However, possibility of its existence is confirmed by the analysis of the SiC epilayer optical absorption spectra, given in Fig.1. The epilayers were shown to possess lower (as compared to the Lely grown samples) fraction of impurity absorption, caused by the optical transition into the first miniband. Since the values of (N_D-N_A) are nearly similar for all the samples, one can assume that the presence of centers shallower than nitrogen atoms is highly probable.

Moreover, donor states with ionization energy lower that of the nitrogen impurity, were detected by G.A. Lomakina in the study of electrical characteristics of SiC epilayers, especially 4HSiC. Concentration of such centers was not lower than the nitrogen concentration.

The ND-1 centers hardly can be produced by the monovacancies, since the temperature of epilayer growing is sufficiently high and cooling must induce these defects to annihilate or to form more complex associates. This is confirmed by the data on the fast quenching of high temperature samples, which results in the formation of vacancy pairs or other associates.

Different approaches in the study of annealing process for the SiC samples irradiated by high energy particles, also shows that the dominant effects for T = 500C, probably, for lower temperatures, are associates (Frenkel pairs, nitrogen-vacancy pairs). At the same time, the assumption that the ND-1 is a complex containing a vacancy or impurity nitrogen, is not confirmed experimentally.

EPR spectra of vacancy defects produced by radiation treatment or quenching are mainly observed for temperatures equal to or higher than 77K, when the EPR spectrum of nitrogen is absent. For 4HSiC EL, the spectrum of ND-1 is registered for 4.2K simultaneously with the nitrogen EPR spectrum; for T = 34K the defect is ionized and the spectrum disappears.

One may assume that the shallow center ND-1 is associated with the electron capture by the antisite defect Si_C^- in the hexagonal position of the carbon site. An evidence that favours this assumption is that the ND-1 spectrum is observed only for 4HSiC epilayer samples grown with considerable silicon excess. The formation of the antisite defect is favoured by comparatively low creation energy. According to [9], E = 3.5 eV for the Si_C^- defect. The antisite defect Si_C may be responsible for the increase of the SiC lattice parameter with growing silicon excess. At last, certain reasons indicate that antisite defects exist in the SiC samples irradiated by ions or neutrons. Unfortunately, information concerning probable energy levels in SiC, associated with antisite defects is very poor. The only paper [10] which analyses native defects in β SiC, show that antisite defects do not introduce new levels in the forbidden band.

In Ref.[11],we studied the EPR spectra of 6HSiC epilayers grown in the vapour phase with excess Si content. The analysis of the EPR spectrum shows that the excess silicon is clustered in the intersites and the lines similar to I_D are not observed. One of the reasons for the differences between the properties of 6H and 4H SiC epilayers is higher stability of defect centers in 4HSiC.

In conclusion we note that the analysis of optical absorption spectra for epilayer 4HSiC shows that the ND-1 center is not the only defect existing in SiC epilayers. Stronger light absorption in the visible spectrum of SiC epilayers indicates that deep levels exist in the forbidden band. They exhibit a broad spectrum of ionization energies and with concentrations much higher than in the Lely grown crystals.

Acknowledgments

The authors are grateful to A.A.Maltsev and G.A.Lomakina for their assistance in the sample preparation and performance of the experiments.

References

1. Yu.A.Vodakov, N.Zhumaev,V.P.Zverev,G.A.Lomakina,E.N.Mokhov, V.G.Oding,V.V.Semenov,Yu.F.Simakhin : Phys.Techn.Polupr.**11** (1977), 373.
2. V.I.Levin, Yu.M.Tairov, V.F.Tsvetkov : Phys.Techn.Polupr.**18** (1984) 1194.
3. Yu.A.Vodakov, E.E.Goncharov, G.A.Lomakina, A.A.Maltsev, E.N.Mokhov, V.G.Oding, G.G.Ryabova, M.G.Ramm : Phys. Techn. Polupr. **21**(1987) 207.
4. H.Okumura, M.Shinohara, S.Kuroda, K.Endo,S.Misawa,E.Sakuma, S.Yoshida .: Jap.J.Appl. Phys. **27** (1988) 1712.
5. J.A.Freitas, S.G.Bishop, I.A.Edmond,I.Ryu,R.F.Davis:J.Appl. Phys. **61** (1987) 2011.
6. Yu.A.Vodakov, E.N.Mokhov, M.G.Ramm, A.D.Roenkoy : Krist und Teckn. **14** (1979) 729.
7. E.N.Kalabukhova, N.N.Kabdin, S.N.Lukin, E.N.Mokhov, B.D.Shanina:Phys.Tverd.Tela **32** (1990) 818.
8. Yu.A.Vodakov, G.A.Lomakina, E.N.Mokhov, V.G.Oding,M.G.Ramm, V.I.Sokolov : Phys.Tverd.Tela **20** (1986) 1433.
9. G.Pensl and R.Helbig: Festkorperprobleme 30 (1990) 133.
10. P.J.Lin-Chuang and J.Li:Mater. Sci.Forum. **10-12** (1986) 1247
11. E.N.Kalabukhova, S.N.Lukin, B.D.Shanina, E.N.Mokhov : Phys. Tverd. Tela **32** (1990) 789.

The Structure of the D-Center in Silicon Carbide – A Study with Electron Nuclear Double Resonance

S. Greulich-Weber, R. Müller, and J.-M. Spaeth

University of Paderborn, FB 6 – Experimentalphysik,
Warburger Str. 100A, W-4790 Paderborn, Fed. Rep. of Germany

Boron-doped 6H-SiC has been studied for many years intensively. With DLTS the isolated boron acceptor and a deep donor called D-center was observed ([1] - [5]). The D-center was assumed to be a boron-vacancy complex. Up to now neither the structure nor the chemical composition of the D-center is known. In order to determine the electronic and microscopic structure of this defect we investigated boron doped 6H-SiC with Electron Nuclear Double Resonance (ENDOR). To correlate our ENDOR measurements with the energy levels known from DLTS for boron-related defects Photo-EPR (Electron Paramagnetic Resonance) and Photo-ENDOR measurements were carried out. From those measurements it is clear that with ENDOR indeed the D-centre was investigated. Our ENDOR measurements show that boron is a component of the D-center. From symmetry and quadrupole interaction observed we conclude that the D-center is not an isolated boron defect, but perturbed by an adjacent defect. The microscopic structure of the D-center as determined from our ENDOR analysis will be discussed.

[1] W.Suttrop, G.Pensl, and P.Lanig, Appl. Phys. A **51**, 231-237 (1990)
[2] M.M.Anikin, A.A.Lebedev, A.L.Syrkin, A.V.Suvorov, Sov. Phys. Semicond. **19**, 69 (1985)
[3] S.H.Hagen, A.W.C. van Kemenade, Phys. Status Solidi A **33**, 97 (1976)
[4] A.Suzuki, H.Matsunami, T.Tanaka, J. Electrochem. Soc. **124**, 241 (1977)
[5] M.Ikeda, H.Matsunami, T.Tanaka, Phys. Rev. B**22**,2842 (1980)

Springer Proceedings in Physics, Vol. 71 **Amorphous and Crystalline Silicon Carbide IV**
Editors: C.Y. Yang · M.M. Rahman · G.L. Harris

Oxidation Studies for 6H-SiC

C.S. Patuwathavithane[1], *J.B. Crofton*[1], *J.R. Williams*[1], *C.C. Tin*[1], *Z.C. Feng*[1,*], *M.J. Bozack*[1], *P.A. Barnes*[1], *R. Ramesham*[2], and *C.D. Ellis*[2]

[1]Department of Physics, Auburn University, Auburn, AL 36849, USA
[2]Department of Electrical Engineering, Auburn University, Auburn, AL 36849, USA
*Permanent address: Department of Physics, Emory University Atlanta, GA 30322, USA

Abstract. Polytype 6H silicon carbide samples having polar faces terminated with either carbon (000$\bar{1}$) or silicon (0001) have been oxidized at 810°C for varying times using a microwave plasma source. Silicon dioxide layer thicknesses have been determined using Rutherford backscattering techniques, and the oxidized samples were further characterized using capacitance-voltage techniques and the methods of Raman spectroscopy. Results indicate that oxide layers of comparable quality to those produced using standard thermal oxidation methods can be obtained in less time at lower temperature.

1. Introduction

Standard techniques for growing silicon dioxide layers on silicon carbide include thermal oxidation using wet or dry molecular oxygen at temperatures in the range 1000-1200°C [1,2]. Different oxidation rates result in different oxide layer thicknesses which can be used to distinguish SiC polytypes (e.g. 3C or 6H) [2]. The electrical characteristics of the oxide layers are important from the standpoint of device fabrication where high quality layers produced through the most efficient methods are required. This last consideration has led us to investigate the problem of SiC oxidation using atomic oxygen - from either a thermal, ground-state source [3] or an oxygen plasma source [4]. Future work will result in a comparative study for both systems; however, the results reported herein concern the oxidation of 6H-SiC in a microwave plasma source. Oxidized samples have been characterized using capacitance-voltage (C-V) techniques together with the techniques of Rutherford backscattering spectroscopy (RBS) and Raman spectroscopy. Results are summarized and compared with similar results for oxide layers grown using standard thermal oxidation techniques.

2. Experimental Results

Bulk, 6H-Sic wafers with both C- and Si-terminated faces [(000$\bar{1}$) and (0001), respectively] purchased from CREE Research, Durham, NC were used for this work. The polished (both sides), one inch diameter wafers were oriented 3.5 ±

0.5° off the (0001) axis and were unintenionally doped n-type at approximately $10^{18}cm^{-3}$. The termination of one face on each wafer was specified. Individual samples of average area $0.6cm^2$ were cut from these wafers for the oxidation studies.

The amount of 6H material available for this work was limited, and as a result, individual samples were oxidized repeatedly. Cumulative exposure times and oxide layer thicknesses were recorded for each sample, and between successive exposures, each sample was subjected to a standard, six minute cleaning procedure consisting of rinses (in order) of chloroform, acetone, methanol and deionized water. Additionally, each sample was etched in a dilute HF solution for one minute and rinsed in dionized water prior to the sample's initial exposure. Metals were thermally evaporated onto the samples for the C-V measurements. Ohmic contacts were formed using Al (≈ 3000Å) deposited over the entire unoxidized side of a sample, and Au gate contacts (≈ 3000Å) were deposited in 0.5mm diameter dots on the oxidized face. The plasma oxidation system and the diagnostic facilities (RBS, C-V and Raman) have been described previously [5-7].

3. Results

Oxide layer thicknesses are shown in Fig. 1 for C- and Si-faced 6H samples exposed to the oxygen plasma. The layer thicknesses were determined from RBS spectra such as those shown in Fig. 2 using the simulation program RUMP [8]. The layer thicknesses for the Si-faced samples are less accurate (perhaps ± 30%) than the thicknesses for the C-faced samples (± 10%) because of the limitation on depth resolution characteristic of the RBS technique. The shorter exposure times are also more uncertain because of a time interval of approximately four minutes over which the samples reached the oxidation temperature of 810°C starting from ambient.

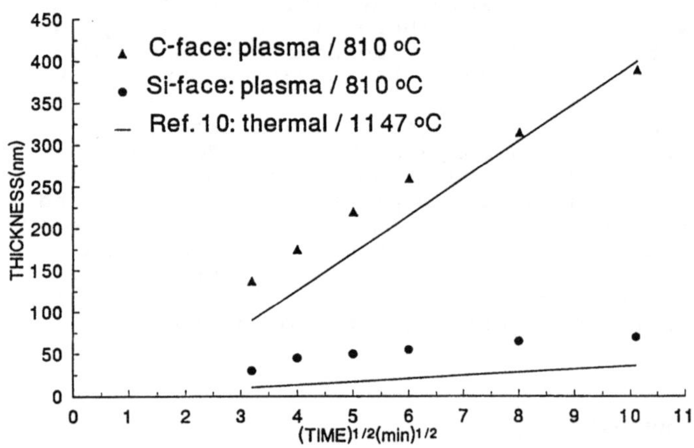

Figure 1. SiO_2 layer thicknesses as a function of exposure time for the microwave plasma oxidation of 6H-SiC. The straight lines represent fits to wet, thermal oxidation data reported in Ref. 10.

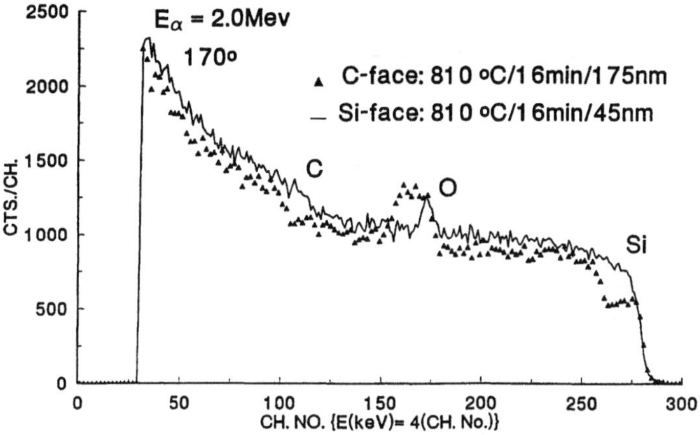

Figure 2. Typical RBS spectra for oxidized 6H-SiC samples.

Samples were exposed simulataneously - i.e., one C- and one Si-faced sample were exposed together at one time.

Layer thicknesses for both the C- and Si-faced samples deviate from the parabolic growth law first proposed by Deal and Grove [9]. However, other authors have reported that the oxidation of the different faces of 6H-SiC can be governed by different growth laws (e.g. parabolic and linear)[1]. In this work, the plasma oxidation rate for the C-faced samples is approximately 4-6 times that for the Si-faced samples. These rates are shown in Fig. 1 together with the results of fits to data reported previously for wet, thermal oxidations at 1147°C [10]. (Thermal oxidation rates for the 3C polytype may be found in Refs. 10 and 11). Plasma oxidation rates for both C- and Si-faced samples are larger for shorter times, while for longer exposure times, the thermal and plasma oxidation rates appear to converge. The cumulative exposure process used for the plasma oxidations may be a contributing factor, and for the Si-faced samples, the small layer thicknesses (measured using RBS techniques) also present some difficulty. The higher rates for plasma oxidation at 810°C compared to thermal oxidation at 1147°C may indicate a higher activation energy for the diffusion process which produces the oxide layer in the plasma process. However recent results for the atomic oxidation of thin copper films indicate that activation energies for the atomic and molecular oxidation processes are similar [12]. The increased growth rate for the atomic oxidation is attributed to the fact that molecular oxygen does not have to be dissociated at the metal surface prior to the start of the diffusion process for oxygen atoms.

Raman spectra were generated for C- and Si-faced 6H samples oxidized by both plasma and dry, thermal methods. In all cases (i.e. for both sample types using either oxidation method) the Raman spectra were similar. Argon and Krypton ion lasers were used to produce four excitation wavelengths - 4067Å (3.06 eV), 4131Å (3.01 eV), 4579Å (2.72 eV) and 4765Å (2.62 eV). These lines bracket the 6H-SiC bandgap which is approximately 3.0eV. Figures 3 and 4 show Raman spectra for

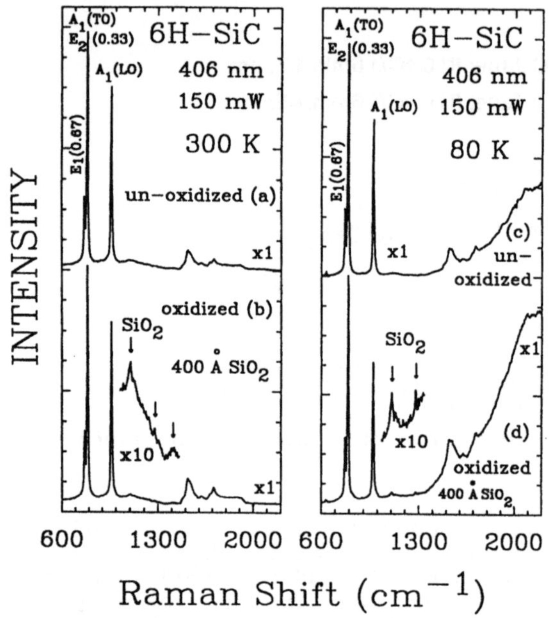

Figure 3. Raman spectra for the wave number range 600-2000 cm^{-1} for unoxidized and dry, thermally oxidized 6H-SiC.

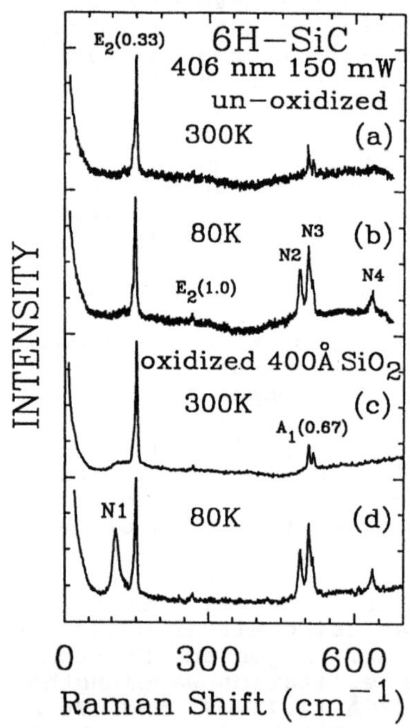

Figure 4. Raman spectra below 700 cm^{-1} for unoxidized and dry, thermally oxidized 6H-SiC.

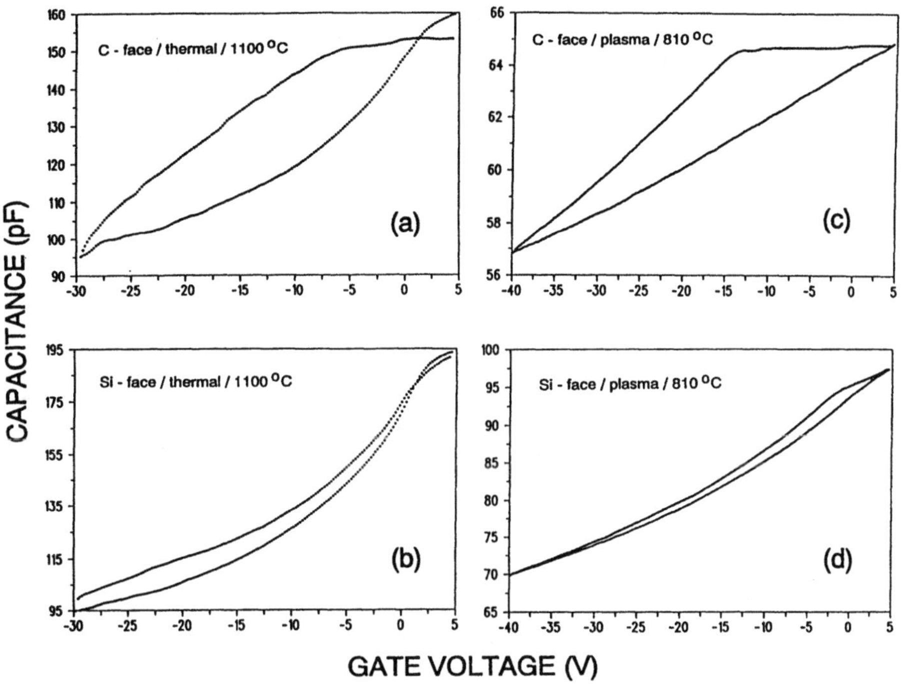

Figure 5. C-V measurements for C- and Si-faced 6H SiC samples oxidized thermally and in a microwave plasma system.

a thermally oxidized, C-faced sample. The strong lines labeled E_1, A_1/E_2 and A_1 represent well-known mode symmetries characteristic of 6H-SiC [13]. A reduced wave vector q/qmax is defined, and the points x=0, 0.33, 0.67 and 1.0 correspond to q=0 wave vectors in the Brillouin zone. LO and TO represent the longitudinal and transverse optical phonon modes at x=0. In Fig. 3(b), three weak bands at 1104, 1274 and 1410 cm^{-1} are superimposed on the SiC signal. These bands are characteristic of SiO_2. The low temperature (80°K) data in Figs. 3(c) and (d) show broad photoluminescence (PL) features above 1300 cm^{-1} which result either from recombination across the indirect bandgap or from defects. The SiO_2 modes at 1105 and 1277 cm^{-1} are sharpened and shifted slightly to higher wave numbers in comparison to the 300°K data. The SiO_2 feature near 1400 cm^{-1} is obscured by the PL signal. The PL band was not observed when the samples were excited using the Ar^+ laser with wavelengths 4579Å and 4765Å. In the low temperature spectra, the relative intensity of the PL band compared to the SiC Raman lines (E_1, A_1/E_2 and A_1) is larger for the oxidized sample. This indicates an increase in defect density due to oxidation. The types of defects introduced are uncertain.

In the wave number range 0-600 cm^{-1} (Fig. 4), oxidized and unoxidized spectra for same C-faced sample exhibit the axial acoustic phonon doublet $A_1(0.67)$ at 506 and 515 cm^{-1}, the planar acoustic phonons $E_2(1.0)$ at 266 cm^{-1} and the $E_2(0.33)$ doublet near wave number 150 [13]. Spectra 4(a) and (b) were taken on the unoxidized face of the oxidized sample, and the termination (either C or Si) of the unoxidized face was not specified. As the sample cooled, new Raman features N_2, N_3 and N_4 were observed at 488, 505 and 636 cm^{-1}, respectively. The N_3 line overlapped the $A_1(0.67)$ peak. Additionally, the feature N_1 at 108 cm^{-1} appeared for the cooled, oxidized sample [Fig. 4(d)]. The lines N1-N3 have been interpreted previously as $1s(A_1)$ to 1s(E) valley-orbit transitions at three inequivalent n-donor sites [14]. Present results indicate that the origin of the N1 line is affected by the oxidation process. Future studies will confirm whether this is actually the case.

The results of typical C-V measurements are shown in Fig. 5 for C-and Si-faced samples oxidized both thermally and in the plasma system. Regardless of the method of oxidation, the Si-faced samples appear to have better C-V characteristics - i.e., less hysteresis which is normally an indication of trapping centers which may possibly result from interface states in the present case. The hysteresis characteristics are no more pronounced for the plasma oxidations than for the thermal oxidations which indicates oxide layers of comparable quality.

4. Summary

RBS techniques have been used to determine oxide layer growth rates at 810°C for C- and Si-faced 6H-SiC samples oxidized using a microwave plasma system. Future work is planned at different temperatures in order to determine activation energies for the oxidation process. Raman spectroscopy and C-V measurements indicate comparable quality for oxide layers produced using plasma oxidation and those grown using standard thermal oxidation techniques. Oxide layers can be produced in less time at substantially lower temperature using a plasma process. This result has implications for several areas related to device fabrication. For example, n- and p-type dopants (N, P and Al, B, respectively) have been observed to undergo concentration redistribution during thermal oxidation [15]. Shorter oxidation times at lower temperatures offers the possibility of eliminating this problem.

This work was supported by NASA Grant NAGW-1192 through the Center for the Commercial Development of Space (CCDS) of the Auburn University Space Power Institute.

References

1. R.C.A. Harris and R.L. Call, **Silicon Carbide 1973**, eds. J.W. Faust and C.E. Ryan, University of South Carolina Press, Columbia, 1974.

2. J.A. Powell, J.B. Petit, J.H. Edgar, I.G. Jenkins, L.G Matus, W.J. Choyke, L. Clemen, M. Yoganathan, J.W. Yang and P. Pirouz, Appl. Phys. Lett. **59(2)**, 183 (1991).

3. B.J. Finlayson-Pitts, T.E. Kleindienst, M.J. Ezell and D.W. Toohey, J. Chem. Phys. **78**, 453 (1981).

4. J.H. Greiner, J. Appl. Phys. **42**, 5151 (1971); *ibid.*, **45**, 32 (1974).

5. P.K. Van Staagen, J.R. Williams and P.A. Barnes, Nucl. Instr. and Meth. in Phys. Res. **B56/57**, 785 (1991).

6. C.C. Tin, P.A. Barnes, J.R. Williams, C.S. Patuwathavithane and P.K. Van Staagen, J. Appl. Phys. **66(9)**, 4201 (1989).

7. J. L. Davidson, C. Ellis and R. Ramesham, J. Electron. Mater. **18**, 711 (1989).

8. L.R. Doolittle, Nucl. Instr. and Meth. in Phys. Res. **B9**, 344 (1985).

9. B.E. Deal and A.S. Grove, J. Appl. Phys. **36**, 3770 (1965).

10. J.B. Petit, J.A. Powell and L.G. Matus, Transactions of the First International High Temperature Electronics Conference, 198 (1991), eds. D.B. King and F.V. Thome.

11. J.W. Palmour, H.J. Kim and R.F. Davis, Mater. Res. Soc. Proc. **54**, 553 (1986).

12. B.C. Gibson, J.R. Williams, A.T. Fromhold, M.J. Bozack, W.C. Neely and A.F. Whitaker, "The Interaction of Atomic Oxygen with Thin Copper Films," to be published J. Chem. Phys. 1991-92.

13. D.W. Feldman, J.H. Parker, W.J. Choyke and L. Patrick, Phys. Rev. **170**, 698 (1970).

14. P.J. Colwell and M.V. Klein, Phys. Rev. **B6**, 498 (1972).

15. J.W. Palmour, R.F. Davis, H.S. Hong, S.F. Corcoran, and D.P. Griffis, J. Electrochem. Soc. **136(2)**, 502 (1989).

Electronic Band Structures of SiC Calculated from a Hybrid Pseudopotential and Tight-Binding Model

P. Srichaikul[1], *A.-B. Chen*[1], *and W.J. Choyke*[2]

[1]Department of Physics, Auburn University, Auburn, AL 36849, USA
[2]Department of Physics and Astronomy, University of Pittsburgh, Pittsburgh, PA 15260, USA

A hybrid pseudo-potential and tight-binding band model is used to extrapolate from β-SiC to the α-polytypes. Preliminary band structures, including band gaps, locations of the conduction band minimum E_c, and effective masses at E_c, have been calculated. Due to uncertainties in the extrapolation of pseudopotentials, these results have to be further analyzed. However, because of its computational efficiency and simplicity in interpretation, this hybrid method can effectively interact with experiments to provide a comprehensive study of band structures of SiC.

I. Introduction

SiC, because of its potential applications to electronics, has drawn considerable research interest recently. As growth capability [1,2] improves, a comprehensive experimental study of properties across different polytypes is now possible. Among the physical quantities that influence the device electrical characteristics, the band structure is perhaps the most fundamental. Yet for SiC, even the most important band states around the band gap are still uncertain. Here we briefly review the theoretical status of SiC band structures, and discuss our results based on a hybrid pseudo-potential and tight-binding model.

A number of ab-initio SiC band structures have been calculated [3-8]. Most of them [4-7] are only for the cubic structure (β-SiC). A majority [5-8] are obtained as byproducts of ground-state calculations based on the local density approximation [9] (LDA). LDA is known to be inaccurate to describe the conduction bands. An improved method [10] over LDA by incorporating the excitation self-energy [11] in the band calculation has yielded good band gaps. However, such a calculation requires extremely long computing time. Its ability to predict the effective masses and other detailed band features remains to be tested.

On the other hand, the empirical pseudo-potential method (EPM) [12] has been very successful for producing high-quality band structures of semiconductors. The success of EPM, however, depends on the availability of reliable band structure data. For SiC, good band structure data are still very limited. EPM was applied to SiC by Hemstreet and Fong (HF) [13] and by Junginger and van Haeringen (JV) [14] many years ago. They used pseudo-potential form factors extrapolated from previous Si and diamond (C) work. Although the band gaps for both β- and 2H-SiC from both calculations are in reasonable agreement with each other, the shapes of these two sets of band structures for β-SiC are very different. For 2H-SiC, the top of the valence band E_v is doubly degenerate in the bands by HF, but is a singlet in

Springer Proceedings in Physics, Vol. 71 Amorphous and Crystalline Silicon Carbide IV
Editors: C.Y. Yang · M.M. Rahman · G.L. Harris

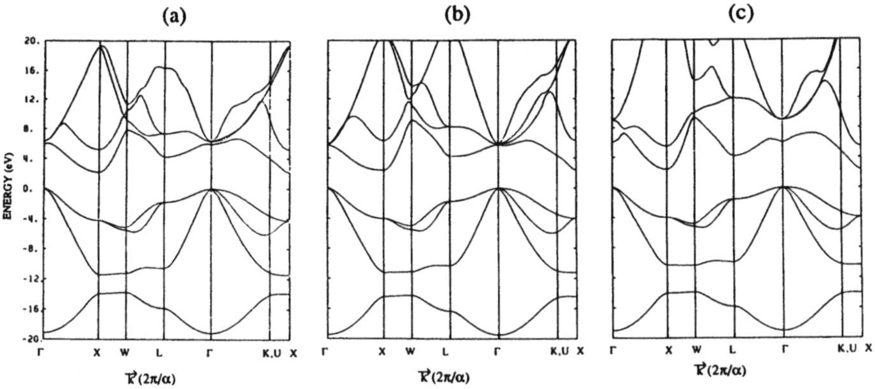

Figure 1. Band Structure of β-SiC. (a) Converging plane-wave calculation using the same local pseudo-potentials as in Ref. 13. (b) A direct tight-binding calculation using four Gaussian orbitals per atoms as described in the text. (c) Fine tuned band structure described in Sec. II and III.

JV's bands. These differences show the uncertainties in the extrapolation of pseudo-potentials. Moreover, the plane-wave calculation becomes very demanding as the size of the unit cell increases. For example, while a typical 500x500 matrix is needed to obtain a reasonable convergence in band energies for 6H-SiC, larger matrices are required to achieve convergence in effective masses.

Another simple approach, the empirical tight-binding method (ETB), has been widely used as a band structure interpolation scheme in semiconductors, including SiC [15,16]. Because long-range matrix elements are truncated, the ETB bands are often too flat to predict the correct effective masses. Furthermore, because of different layer arrangements in the cubic and hexagonal lattices, the extrapolation of TB matrix elements from β- to α-polytypes also runs into difficulty.

A hybrid scheme by Chen and Sher (CS) [17] combining the EPM and ETB has been applied with success to the transport properties of semiconductors and alloys [18]. This method is particularly appealing for the present study, because it not only keeps the small matrix size of ETB but also accurately retains the important EPM band structures (see Figure 1a and 1b). This is the method used in the present calculation.

II. Hybrid Pseudo-Potential Tight-Binding Method

In the EPM using local pseudo-potentials, the potential is written as

$$U(r) = \sum_{g} \exp(ig.r)\ V(g)\ ,$$

where V(g) is the Fourier transform of the crystal potential, and g are the reciprocal lattice vectors. V(g) can be further written in terms of atomic form factors $v_j(g)$ as

$$V(g) = \sum_{j} \exp\ (g.\tau_j)\ v_j(g),$$

where j runs over the atoms in a unit cell, and τ_j are the atomic positions with respect to the lattice vector **L**. If $v_j(g)$ is assumed to depend on the magnitude g and if $v_j(g)$ dies off fast as g increases, then only several values of these form factors from the first few g's are needed to perform the band structure calculation. In EPM, these form factors are determined empirically.

In the CS hybrid approach, a minimum set of four Gaussian (or Slater) orbitals (one s and three p) per atom are used as the basis functions. For a semiconductor containing n atoms per unit cell, 4n Bloch basis functions $|\mathbf{k}j\gamma\rangle$ can be formed from these orbitals for each **k** vector, where γ indicates the orbital types (s, p_x, p_y or p_z), and j, ranging from 1 to n, specifies the atoms in a given cell. It is then straightforward to calculate the 4nx4n overlap matrix S(**k**) and Hamiltonian matrix H(**k**) using empirical pseudo-potentials. Solutions to the secular equation $[H(\mathbf{k}) - E(\mathbf{k})S(\mathbf{k})]c = 0$ then give the band energies E(**k**). Figure 1a and 1b show that the first five bands for β-SiC produced by this hybrid model using an 8x8 matrix are essentially the same as those from converging plane-wave calculation using matrices larger than 100x100.

One important reason for using this approach is that it provides a way to fine tune the band structures. This can be achieved by adding a perturbation Hamiltonian h(**k**) to H(**k**), with h(**k**) taking the form of the first-neighbor TB Hamiltonian. There are, at a maximum 9 parameters in h(**k**), which can be adjusted to fit the desired band energies and effective masses.

The calculation of H(**k**) requires the pseudo-potential form factors. With limited experimental data for the α-polytypes, we can try to extrapolate these from the established values for silicon, diamond and β-SiC. In the work of JV, these form factors were multiplied by the unit-cell volumes Ω_c and are extrapolated as smooth functions of the square of the wave vectors q^2. In HF's work, smooth curves were drawn through the β-SiC form factors to obtain the needed values for the 2H-SiC. However, we found that smooth interpolation can be achieved for $\Omega_c v$ as a function of $\mathfrak{q}^2$, where $\mathfrak{q}$ is a dimensionless reduced wave vector defined by $q = \mathfrak{q}\,(2\pi/a)$, with the effective lattice constant a given by $\Omega_c = a^3/4$. Note that the equivalent cubic unit cell volume Ω_c is related to the actual unit-cell volume Ω in the α-polytypes by $\Omega_c = 2\Omega/n$, where n is the number of atoms per unit cell. These interpolated (extrapolated) form factors plus the TB parameters in h(**k**) for the β-SiC then allow us to calculate the band structures for the α-polytypes.

III. Results

β-SiC. Our calculated band structure for β-SiC is shown in Figure 1c. The experimental indirect band gap X_{1c}-Γ_{15v} of 2.417 eV [19], the 6.0 eV for the direct optical transition energy Γ_{1c}-Γ_{15v} [20], and the separation X_{3c}-X_{1c}=3.1 eV [21] are well accepted in the literature and are fitted precisely. The 4.2 eV tentatively assigned L_{1c}-Γ_{15v} [20] is also fitted. However, our bands give a value of 4.0 eV for X_{1c}-L_{3v}, as compared to the 3.55 eV experimental value [20]. The experimental longitudinal mass m_l=0.667 [22] was also fitted precisely, but the transverse mass m_t was not. Our calculated value for m_t=0.185 is smaller than the 0.247 experimental value [22]. Our masses give a combined transport mass m_c of 0.24, which is comparable to the 0.26 value in Si [23]. Our valence bands look very similar to the original HF bands. However, our higher conduction bands are much

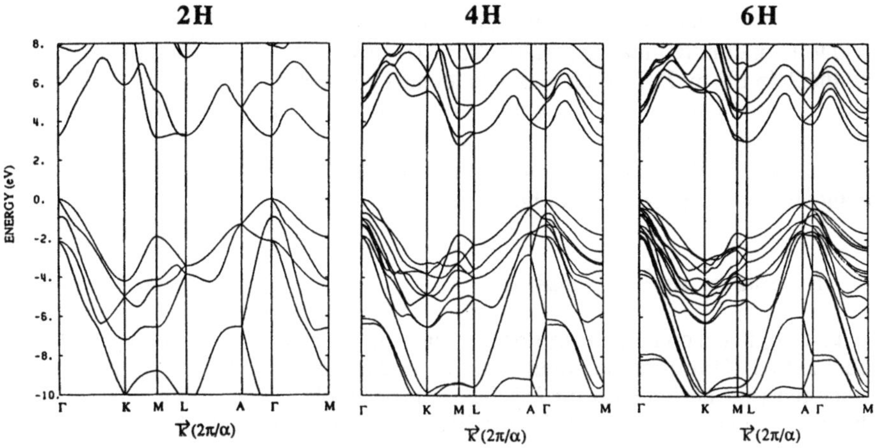

Figure 2. Band structures of three α-polytypes of SiC.

Table 1. Calculated band gaps E_g (in eV), the location of E_c, and effective masses for the three α-SiC polytypes studied. The masses in the three principal directions are m_1(Γ-M), m_2(M-L) and m_3(M-K).

	E_g	E_c location	m_1	m_2	m_3	m_c
2H	3.184	M	0.954	2.380	0.112	0.289
4H	2.800	M	1.039	0.293	0.136	0.256
6H	2.867	L-M	0.509	18.8	0.152	0.35

too high. The high conduction bands can be improved by increasing the number of basis functions. Since these states do not directly affect the transport properties, we have not tried to pin them down in the present study.

2H-SiC. The calculated bands are shown in Figure 2a. The band gap is 3.18 eV and E_c is located at M. Although this value is not very different from the well quoted 3.3 eV value from experiment [24], the location is different from the well accepted symmetry point K. The effective mass at E_c along the M-K direction is very small, but the other two masses are quite large. Their values are listed in Table 1. The combined transport mass is 0.29, which is larger than that in β-SiC. Because going from 3C to 2H-SiC represents the largest extrapolation, these results are probably the most uncertain among the different polytypes studied here.

4H-SiC. Our 4H-SiC bands are plotted in Figure 2b. Because of the doubling of the unit cell along the c-axis, band folding effects can be seen as compared to Figure 2a. The band gap is 2.800 eV with E_c remaining at M. This is the same as that by JV, but is lower than the experimental value [25] of 3.26 eV. The mass in the M-K direction remains small, while the other two masses are not as large as those in 2H. These masses combine to give a transport mass of 0.256, which is comparable to that in β-SiC.

6H-SiC. The bands are plotted in Figure 2c. It is surprising that the band gap is larger than that in 4H-SiC. Our E_c is 2.867 eV and is now located at a **k**

point close to L and about 7% between L and M. This gap is slightly smaller than the low- temperature measured value [25] 3.023 eV. Our value is certainly better than the 2.45 eV by JV, which was also located between M and L. Our transverse mass m_3 in the L-H direction remains small, but the other transverse mass m_2 (along L-M) becomes very large. The longitudinal mass m_1, however, is smaller than that in 4H-SiC. The combined transport mass is 0.35, which is the largest among all the four polytypes studied.

IV. Summary and Discussion

We have applied a combined empirical pseudo-potential and tight binding band model to SiC. However, due to uncertainty in the extrapolation, our results require further analysis, particularly for 2H-SiC. It is desirable to have more experimental data on the band to band transitions in 2H-SiC, so that our method can be used effectively to interpolate the band structures for all α-polytypes of SiC.

References

* Work supported by NASA Grant No. NAGW-1192 through the Center for the Commercial Development of Space (CCDS) of the Auburn University Space Power Institute.
1. "Amorphous and Crystalline Silicon Carbide", edt. G. L. Harris and C. Y. -W. Yang, (Springer-Verlag, Berlin, New York 1989).
2. J. A. Powell, L. G. Matus and M. A. Kuczmarski, J. Electro. Chem. Soc. 134, 1558 (1987).
3. F. Herman, J. P. Van Dyke and R. L. Kortum, Mat. Res. Bull. 4, 5176 (1969).
4. A. R. Lubinsky, D. E. Ellis and G. S. Painter, Phys. Rev. B11, 1537 (1975).
5. W. R. Lambrecht and B. Segall, Phys. Rev. B41, 2948 (1990).
6. N. Churcher, K. Kunc and V. Heine, J. Phys. C: Solid State Phys. 19, 4413 (1986).
7. K. J. Chang and M. L. Cohen, Phys. Rev. B35, 8196 (1987).
8. V. I. Gavrilenko, A. V. Postnikov, N. I. Klyui, and V. G. Litovchenko, Phys. Stat. Sol. (B) 162, 477 (1990).
9. P. Hohenberg and W. Kohn, Phys. Rev. 136, B864 (1964); W. Kohn and L. J. Sham, Phys. Rev. 140, A1133 (1965).
10. M. S. Hybertson and S. G. Louie, Phys. Rev. Lett. 55, 1418 (1985).
11. L. Hedin and S. Lundqvist, Solid State Phys. 23, 1 (1969).
12. M. L. Cohen and J. R. Chelikowsky, "Electronic Structure and Optical Properties of Semiconductors", 2nd. Ed. (Springer-Verlag, Berlin,1989)
13. L. A. Hemstreet and C. Y. Fong, Solid State Comm. 9, 643 (1971); Phys. Rev. B6, 1464 (1972); and in "Silicon Carbide 1973" (U. of South Carolina Press, 1974), P. 258.
14. H. G. Junginger and W. van Haeringen, Phys. Stat. Sol. 37, 709 (1970).
15. Y. Li and P. J. Lin-Chung, Phys. Rev. B36, 1130 (1987).
16. D. N. Talwar and Z. C. Feng, Phys. Rev. B44, 3191 (1991).
17. A.-B. Chen and A. Sher, Phys. Rev. B22, 3886 (1980).

18. S. Krishnamurthy, A. Sher and A.-B. Chen, Appl. Phys. Lett. 52, 468 (1988).
19. R. G. Humphreys, D. Bimberg and W. J. Choyke, Solid State Comm. 39, 163 (1981).
20. W. J. Choyke and L. Patrick, Phys. Rev. 187, 1041 (1969).
21. L. Patrick and W. J. Choyke, Phys. Rev. 186, 775 (1969).
22. R. Kaplan, R. J. Wagner, H. J. Kim and R. F. Davis, Solid State Comm. 55, 67 (1985).
23. S. M. Sze, "Physics of Semiconductor Devices" (John Wiley and Sons, New York, 1981).
24. L. Patrick, D. R. Hamilton and W. J Choyke, Phys. Rev. 143, 526 (1965).
25. W. J. Choyke, D. R. Hamilton and L. Patrick, Phys. Rev. A133, 1163 (1964).

Metallization Studies on Epitaxial 6H-SiC

J. Crofton[1], *J.M. Ferrero*[1], *P.A. Barnes*[1], *J.R. Williams*[1], *M.J. Bozack*[1], *C.C. Tin*[1], *C.D. Ellis*[2], *J.A. Spitznagel*[3], *and P.G. McMullin*[3]

[1]Department of Physics, Auburn University, Auburn, AL 36849, USA
[2]Department of Electrical Engineering, Auburn University, Auburn, AL 36849, USA
[3]Westinghouse Science and Technology Center, Pittsburg, PA 15235, USA

Abstract. A study of electrical and physical data for metals deposited on n and p—type epitaxial 6H—SiC is reported. The metals investigated were Ni, NiCr, W, TiW, and Al. Electrical characterization included IV and CV data for Schottky contacts and specific contact resistance for ohmic contacts. The surface preparation prior to metallization was found to be critical to the quality of the contacts formed. The near surface region of the metal—semiconductor interface was investigated using Rutherford Backscattering (RBS) and light ion channeling (LIC) to detect interdiffusion and lattice damage. The ion scattering experiments were complimented with Auger Electron Spectroscopy (AES) and X—ray Photoelectron Spectroscopy (XPS) measurements. Surface morphology was investigated using SEM analysis including Energy Dispersive X—ray (EDX) analysis.

1. Introduction

The hexagonal polytype of SiC (6H) is an indirect, wide bandgap semiconductor with enormous potential for device applications at both room and elevated temperatures. Devices fabricated with this new semiconductor will require ohmic contacts which have low specific contact resistances and are chemically stable at operating temperatures exceeding 300 ^{0}C.

This paper summarizes our recent successful attempts to develop ohmic contacts to 6H—SiC. The ohmic contacts were fabricated using a variety of metals and alloys to both n— and p—type epitaxial 6H—SiC purchased from Cree Research Inc. in Durham, North Carolina, USA. The epitaxial layers were grown on bulk 6H—SiC of the opposite conductivity type. The n—type layers were nitrogen doped while the p—type layers were aluminum doped. Capacitance—voltage (CV) measurements and specific contact resistance measurements using the circular transmission line method on contacts that were ohmic are reported. Auger and Rutherford Backscattering Spectrometry (RBS) data are also reported.

2. Contacts to n—type 6H—SiC

Titanium—Tungsten (TiW). Tungsten has been reported as a suitable ohmic contact metal to β—SiC[1]. However, our own work with W on 6H—SiC indicated adhesion problems between the W and SiC. We

Springer Proceedings in Physics, Vol. 71 **Amorphous and Crystalline Silicon Carbide IV**
Editors: C.Y. Yang · M.M. Rahman · G.L. Harris

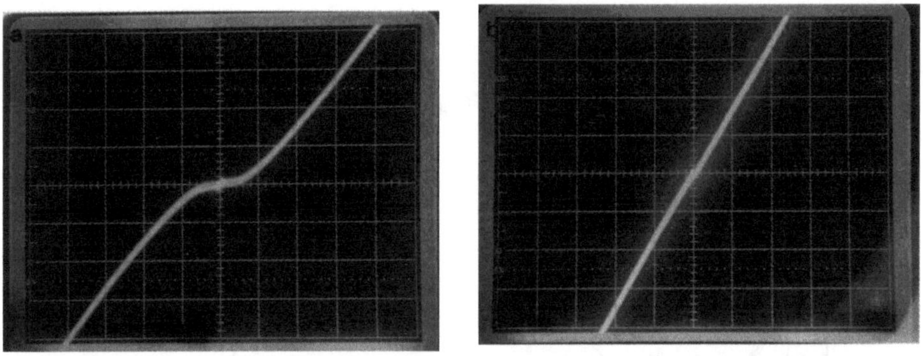

Figure 1. IV characteristic for TiW on n–type 6H–SiC. Scale: 2 V/cm horizontal, 2 mA/cm vertical. (a) as deposited, (b) after annealing at 600 ^{0}C.

therefore decided to pursue TiW (10% Ti–90% W by weight) as an alternative. The TiW–SiC contacts were prepared as follows: the SiC was degreased in TCA, acetone, and methanol followed by a rinse in DI water. After blowing dry with nitrogen, Shipley 1650 positive resist was spun on followed by a 90 ^{0}C bake. The resist was exposed and developed followed by a 10 second dip in buffered oxide etch (BOE). Finally, before sputter deposition of TiW, the patterned SiC was descumed in an oxygen plasma for 30 seconds. This descum step was found to be crucial to ohmic contact formation. After sputter deposition of approximately 0.83 μm of TiW, the unwanted metal was lifted off in acetone in an ultrasonic cleaner. The TiW films adhered very well, and after lifting off the undesired metal, many contact resistance patterns were available for probing.

As deposited TiW was a blocking Schottky contact yielding a non–linear IV characteristic as shown in Fig. 1a. The IV characteristic was measured between two 100 μm radius dots. (cf. Fig 4). After annealing at 600 ^{0}C for 5 minutes in 10% H_2–90% Ar environment, the contacts became ohmic (Fig. 1b). Contacts annealed at temperatures below 600 ^{0}C were non–ohmic. TiW contacts on SiC with n–type doping, $N_d = 4.7 \times 10^{18}$ cm^{-3} yielded ohmic contacts as did heavier doped samples whereas TiW contacts on SiC with n–type doping $N_d = 2 \times 10^{18}$ cm^{-3} were Schottky barriers. We therefore conclude that doping concentrations, N_d, greater than 5×10^{18} cm^{-3} are required for ohmic contact formation with TiW. However, if the descum step prior to TiW deposition was omitted, doping concentrations as high as $N_d = 7 \times 10^{18}$ cm^{-3} also annealed at 600 ^{0}C did not yield ohmic contacts to SiC. Surface preparation is therefore critical to TiW ohmic contact formation.

The contact resistance measured on the 4.7×10^{18} cm^{-3} sample was $r_c = 7.8 \times 10^{-4}$ Ω–cm^2. The circular transmission line method[2] used to measure the specific contact resistance also determines both the sheet resistance of the epitaxial layer under the metal contact, (R_{sk}) as well as the sheet

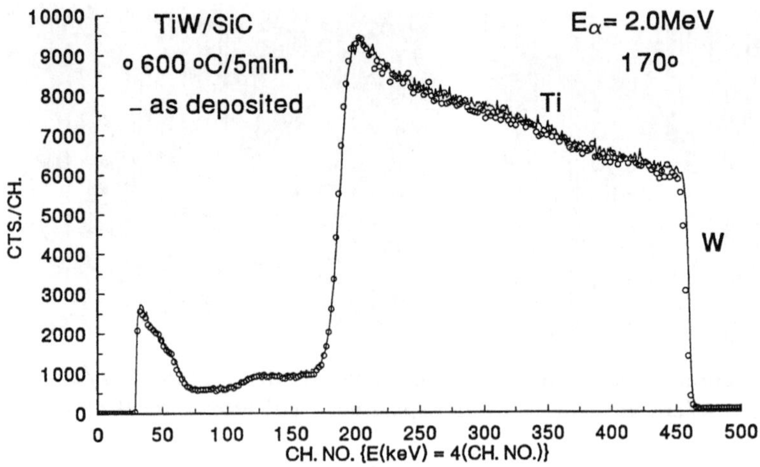

Figure 2. RBS spectra of TiW on 6H—SiC before and after annealing at 600 ^{0}C.

resistance of the epitaxial layer not under the metal (R_{sh}). Our measured results were $R_{sk} = 8.1 \times 10^2$ $\Omega/\square$ under the TiW and $R_{sh} = 9.5 \times 10^2$ $\Omega/\square$. If the annealing caused significant reaction between the metal and the underlying semiconductor, one would expect the sheet resistance under the contact to change, perhaps by as much as an order of magnitude[3]. The fact that R_{sk} and R_{sh} are so close in value indicates that reaction at the TiW—SiC interface is minimal and the contact is a non—alloyed contact. This is supported by RBS data of the TiW—SiC interface taken before and after annealing (Fig. 2). The RBS spectra shows no movement of the Ti, W, Si, or C to within a depth resolution of a few hundred Angstroms at the interface. We hypothesize that the annealing simply consumes any residual oxide present on the SiC prior to TiW deposition. This oxide consumption has been observed previously on Ti contacts to β—SiC[4].

A CV pattern was also fabricated on a second sample identical to the one described above. CV measurements were made to determine the carrier concentration and barrier height. The carrier concentration calculated from the slope of the $1/C^2$ verses V plot was $n = 1.2 \times 10^{19}$ cm^{-3} assuming a dielectric constant of 10.0. This is approximately twice the value of the doping concentration of 4.7×10^{18} cm^{-3} supplied by CREE with the material. The voltage axis intercept of the $1/C^2$ verses V plot gives a calculated barrier height of $\varphi_b = 1.45$ eV for TiW on 6H—SiC. This agrees well with the work of Wu and Campbell[5] who measured $\varphi_b = 1.40 \pm 0.05$ eV for Au Schottky barriers on chemically cleaned 6H—SiC. However, more recent data of Yasuda et al[6] indicates $\varphi_b = 1.7 \pm 0.1$ eV for Al Schottky barriers on the Si face of 6H—SiC. Accurate measurements of barrier heights on 6H—SiC require further careful work.

Nickel (Ni). Our attempts to deposit large area (1 cm^2) Ni films by electron beam evaporation on 6H—SiC failed due to the high stress present in the deposited films. Although e—beamed Ni did adhere nicely when deposited over small areas through a shadow mask, we could not pattern the Ni. Thus CV and contact resistance measurements using the circular transmission line method were not possible. One sample was made using a shadow mask containing a set of 200 μm radius dots. The samples were prepared as described above for TiW except that the descum step was not performed. The Ni was then e—beam deposited on SiC doped at $N_d = 4.7 \times 10^{18}$ cm^{-3} to a thickness of 0.3 μm. IV data taken between two of the Ni dots revealed that the as deposited Ni contacts were Schottky barriers; after annealing at 950^{0}C for 5 minutes in H_2—Ar environment, however, the contacts became ohmic. Using the simple method of Terry and Wilson[7], we estimate the contact resistance to be in the mid 10^{-2} Ω—cm^2 range.

Nichrome (NiCr). NiCr has already been patented as an ohmic contact to β—SiC[8]. NiCr is therefore a natural choice for an ohmic contact to 6H—SiC in light of the fact that Ni makes an ohmic contact with 6H—SiC. NiCr contacts were deposited on n—type epitaxial SiC having a donor concentration of $N_d = 4.7 \times 10^{18}$ cm^{-3} that was prepared using the method described above for TiW. The NiCr alloy was 60% Ni—40% Cr by weight and was sputter deposited to a thickness of approximately 0.425 μm. The NiCr circular transmission line patterns did not lift off as well as the TiW but nevertheless enough of the patterns were clearly delineated so that contact resistance measurements could be made. The as deposited IV measurements were made between two 100 μm radius dots. After annealing the sample at 950^{0}C for 5 minutes in H_2—Ar, the IV characteristic was linear indicating ohmic contact formation.

After ohmic contact formation, the circular transmission line patterns were probed to yield the contact and sheet resistances. The specific contact resistance was found to be $r_c = 1.8 \times 10^{-3}$ Ω—cm^2 while the sheet resistance

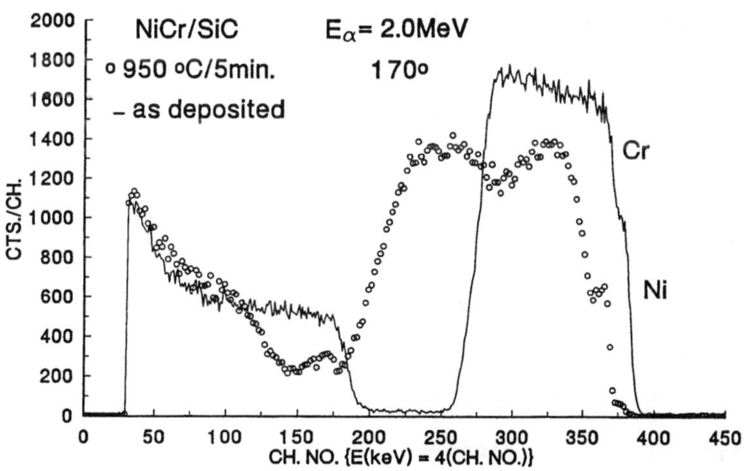

Figure 3. RBS spectra of NiCr on 6H—SiC before and after annealing at 950 ^{0}C.

under the contact was $R_{sk} = 1.0 \times 10^3\ \Omega/\square$ with a bulk sheet resistance of $R_{sh} = 6.7 \times 10^2\ \Omega/\square$. The change in sheet resistance (R_{sk}) under the contacting metal, indicates that a reaction has taken place between the metal and the underlying semiconductor that has consumed some of the epitaxial SiC. This makes the SiC layer thinner and causes the sheet resistance to increase.

RBS analysis of the NiCr–SiC system before and after annealing shows a pronounced segregation of the Ni and Cr during the annealing process (Fig. 3). The Ni segregates toward the metal–semiconductor interface. This is also supported by Auger analysis at the surface of the annealed NiCr film which shows the presence of Cr but not Ni.

3. Contacts to p–type 6H–SiC

Aluminum (Al). The fact that Al is used as a p–type dopant in SiC[9] makes it a natural choice for an ohmic contact metal to p–type SiC. Samples were cleaned using the method described earlier except that there was no descum prior to Al deposition. The Al was deposited by e–beam evaporation to a thickness of 0.3 μm. The as–deposited Al yielded a Schottky contact on SiC with doping concentration $N_a = 1.85 \times 10^{18}$ cm^{-3}. After annealing at 700^{0}C for 10 minutes in H_2–Ar, these contacts became ohmic. Considerable spiking or pitting was observed (Fig. 4) indicating that pure Al may have limited use as an ohmic contact for p–type SiC.

Contact resistance measurements on systems where pitting is present generally are inaccurate[10] due to the non uniformity of the contact. Nevertheless, the contact resistance measured can be considered to be an average resistance between areas of high and low conduction. Our measured contact resistance for Al on SiC was $r_c = 1.7 \times 10^{-3}\ \Omega$–cm^2 while the sheet resistance under the contact was $R_{sk} = 7.8 \times 10^2\ \Omega/\square$ and the bulk sheet resistance was $R_{sh} = 2.4 \times 10^3\ \Omega/\square$. That the low sheet

Figure 4. Optical micrograph showing pitting of Al on 6H–SiC after annealing at 700 ^{0}C. The innermost dot diameter is 100 μm.

resistance under the contact (R_{sk}), is so much lower than the bulk sheet resistance (R_{sh}), suggests that the Al has heavily doped the SiC under the Al contact. This heavy doping allows the carriers to tunnel through the barrier thereby yielding an ohmic contact.

The problem of Al spiking on Si has been solved[11] by depositing varying ratios of AlSi with 1 or 2 weight percent Si. We evaporated 200 Å of Si followed by 4000 Å of Al. The as deposited metals were Schottky contacts. Subsequent annealing at 850 ^{0}C for 5 minutes caused the contacts to become ohmic. Optical and SEM microscopy indicated significantly reduced pitting in comparison to the case of pure Al. Contact resistance measurements and Auger profile analysis are currently being performed.

4. Conclusions

We have shown that TiW, Ni, and NiCr form ohmic contacts on n—type 6H SiC and Al and AlSi form ohmic contacts on p—type 6H—SiC. It appears that in the case of TiW, the ohmic contact forms when the native oxide present between the SiC and TiW is consumed during the annealing process. If this is true, TiW deposited on SiC that is vacuum cleaved or perhaps ion cleaned in situ prior to TiW deposition should be ohmic as deposited and make TiW a truly non—alloyed contact. Ni and NiCr appear to form ohmic contacts due to a reaction between Ni and SiC at the metal—semiconductor interface which results in a reduced barrier height. We are continuing barrier height measurements on lightly doped SiC to verify this supposition.

The Al and AlSi contacts appear to heavily dope the p—type SiC under the contact resulting in a thinner barrier through which carriers can tunnel to form an ohmic contact. We are continuing our work with these systems in the hopes of finding a contact system that will not pit or spike while heavily doping the semiconductor immediately beneath the metal.

Results for both n— and p—type 6H SiC are summarized in Table 1. Ultimately, the success or failure of these contacting systems will be determined by their stability to high currents and temperatures. We are presently beginning our work in this area.

TABLE 1. OHMIC CONTACT FORMATION ON 6H—SiC

Metal	Doping (cm^{-3})	Anneal Cycle	r_c (Ω—cm^2)
TiW	N_d=4.7x10^{18}	5 min. @ 600^{0}C	7.8x10^{-4}
Ni	N_d=4.7x10^{18}	5 min. @ 950^{0}C	mid 10^{-2}
NiCr	N_d=4.7x10^{18}	5 min. @ 950^{0}C	1.8x10^{-3}
Al	N_a=1.85x10^{18}	10 min. @ 700^{0}C	1.7x10^{-3}

5. Acknowledgments

This work was supported by NASA Grant No. NAGW—1192 through the Center for the Commercial Development of Space (CCDS) of the Auburn University Space Power Institute and the Alabama Research Institute through Grant No. ARI—90—201.

6. References

1. K.M. Geib, J.E. Mahan, C.W. Wilmsen, in Amorphous and Crystalline SiC and Related Materials II, Edited by M.M. Rahman, C.Y.—W. Yang and G.L. Harris, Springer—Verlag, Berlin, 1989., pp.224—228.
2. G.K. Reeves, Solid State Electronics, **Vol. 23**, 487—490, 1980
3. W. Kellner, Siemens Forsch. u. Entickl. **Ber. Bd. 4**, Nr. 3, 1975
4. M.I. Chaudhry, W.B. Berry, M.V. Zeller, Proceedings of the Materials Research Society Symposium, **Vol. 162**, 1990
5. S.Y. Wu, R.B. Campbell, Solid State Electronics, **Vol 17**, pp. 683—687, 1974.
6. K. Yasuda, T. Hayakawa, M. Saji, IEEE Trans. Elec. Devices. **Vol ED—34**, pp. 2002—2008, 1987.
7. L.E. Terry, R.W. Wilson, Proceedings of the IEEE, **Vol. 5**, No.9, pp 1580—1586. 1969
8. A. Addamiano, U.S. Patent Office, No. 3,510,733. May 5, 1970
9. J.A. Edmond, D.G. Waltz, S. Brueckner, H.S, Kong, J.W. Palmour, C.H. Carter, Transactions First International High Temperature Electronics Conference. D.B. King, F.V. Thome, Editors Albuquerque, NM June, 1991
10. S.S. Cohen, G. Gildenblat, VLSI Electronics—Vol. 13. Academic Press 1986
11. S. Wolf, R.N. Tauber, Silicon Processing for the VLSI Era Vol. 1. Lattice Press, 1987

TEM Study of β-SiC Films Grown on (111) Silicon Substrates

K. Fekade[1], *M.G. Spencer*[1], *S. Nahm*[2], *and L.G. Salamanca-Riba*[2]

[1]MSRCE, School of Engineering, Howard University, Washington, DC 20059, USA

[2]Engineering Materials Department, University of Maryland, College Park, MD, USA

Abstract: A transmission electron microscopy study was carried out on β-SiC films grown on Silicon {111} by the CVD technique.[1] Both plan view and cross-sectional samples were observed. A high density of interfacial twins and stacking faults were observed in the epilayers. Cross-sectional high resolution micrographs showed the interfaces to be semi-coherent. In addition, it was found that only films grown at 1100°C with flow rates of 10/10, 20/20 and 30/30 were single crystalline.

Introduction

Electronic devices for high technological applications require the growth of perfect single crystalline films. For this reason, several investigators are engaged in growing the cubic polytype of silicon carbide (β-SiC) on Silicon (Si)[1-4]. The electrical and thermal properties of β-SiC make it attractive for high power, high temperature and/or high-frequency applications[5-6]. In recent years, there has been a considerable achievement made in the growth of large area single crystal films of β-SiC. Most studies were carried out, however, using CVD growth techniques on a{100} plane of Si and at temperatures in excess of 1300°C[1-4]. In this paper, we report preliminary TEM results of β-SiC films grown using the CVD technique, but on the (111) plane of Si and at temperatures under 1300°C.

Experimental

The β-SiC films studied were grown at the Ioffe Physical Technical Institute of the USSR. The epilayers were grown on Si substrates {111} 4° off the axis by CVD techniques. The growth of β-SiC was carried out at atmospheric pressures and temperatures ranging from 1000°C to 1250°C. Mono silane (SiH_4) and trichlorethylene (C_2HCl_3) were used as sources of silicon and carbon, respectively. Hydrogen was used as a carrier gas. The thickness of the epitaxial layers ranged from 0.5 to 10 mm.[7]

The thin foils for conventional TEM were prepared as follows. First the samples were mechanically thinned to about 150 mm and subsequently dimpled to about 50 mm. This procedure was followed by ion beam thinning using an ANA-TEC AIM 2.0 ion beam thinning instrument. A beam of argon ions with an energy of 3 KeV was used to thin the samples to electron transparency. A JEOL 2000 FX-II TEM operating at 200KV was used to examine the samples.

Result and Discussion

Figure 1 is a low magnification plan view image showing the defects present in the samples grown at 1100°C. The micrograph was typical for samples grown at 1100°C with flow rates of 10/10, 20/20 and 30/30. The structure shown in Fig. 1 represents the bulk of the material. This micrograph shows a high density of stacking faults (S.F.) lying on the four

Springer Proceedings in Physics, Vol. 71 Amorphous and Crystalline Silicon Carbide IV
Editors: C.Y. Yang · M.M. Rahman · G.L. Harris

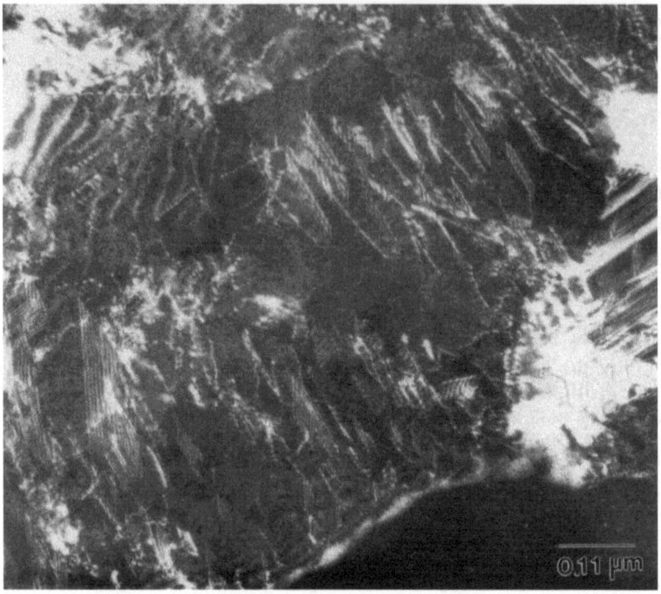

Fig. 1. Plan-view micrograph of a SiC film (typical for all samples grown at 1100°C with flow rates of 10/10, 20/20, 30/30) showing a high density of SF.

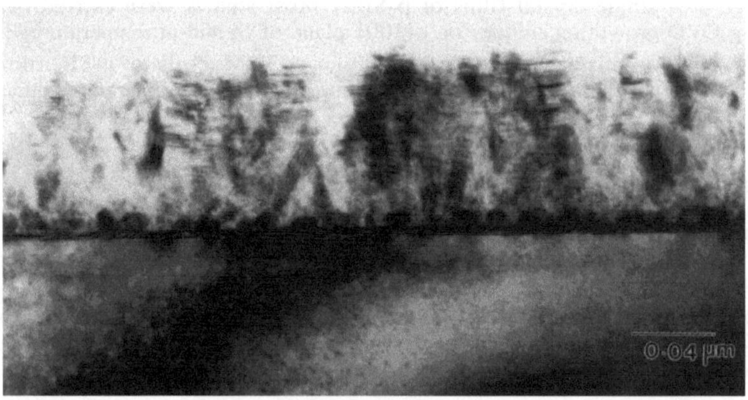

Fig. 2. Bright field image of Si/SiC interface showing a poly crystalline film of SiC.

{111} planes. Isolated perfect dislocations were not observed. The stacking faults were uniformly distributed throughout the foil. Our observation shows that the density of these planar defects is independent of flow rate. Also, the density of SF does not decrease as we move away from the interface.

A cross-sectional view of a Si/SiC interface is shown in Fig. 2. This figure shows that the SiC film is polycrystalline and that close to the interface there is island formation in the film. For most of the samples we studied the silicon substrate is defect free. However, the interfaces were not very sharp and in the case of the poly crystalline specimen, there seems

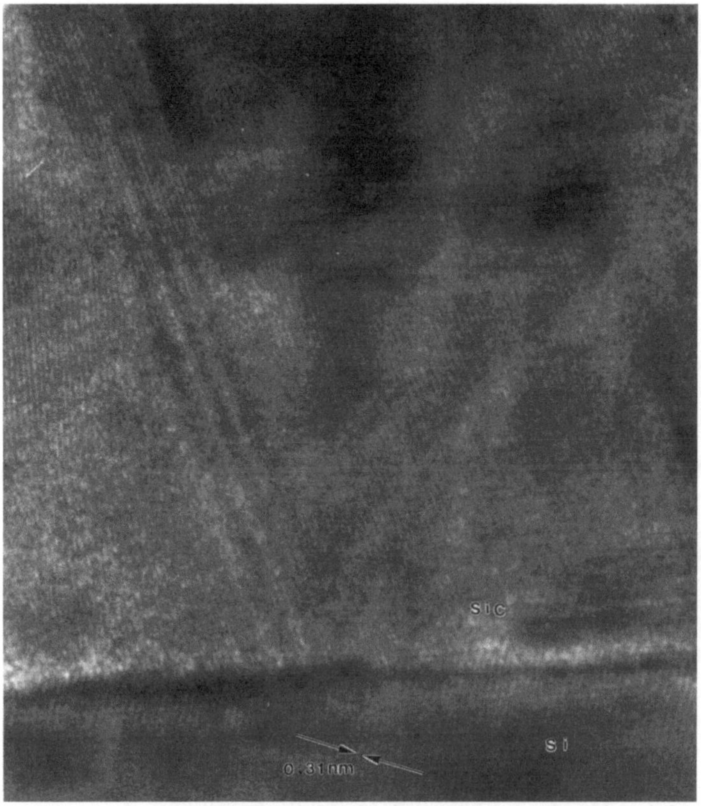

Fig. 3. HREM micrograph showing the Si/SiC interface of a sample grown at 1100°C and 10/10 flow rate.

to be island formation at the interface as shown in Fig. 2. We have observed straight dislocations on the Si substrate in one of the samples we examined (Sample #86-1-B).

Figure 3 shows a high resolution micrograph of a single crystalline specimen grown at 1100°C and with a 10/10 flow rate. This micrograph was typical for samples grown at 1100°C but with flow rates of 10/10, 20/20 and 30/30. The twining in the SiC is apparent in this figure. The twins nucleated on the Si surface as shown in Fig. 3.

A high resolution micrograph of the single crystalline β-SiC film is shown in Fig. 4. The inset to this figure shows the diffraction pattern from the area. This diffraction pattern contains streaks and satellite spots along the <111> direction due to the twins and stacking faults present.

Samples grown at temperatures either higher (1250°C) or lower (1000°C) than 1100°C were observed to produce poly crystalline SiC films. Also, samples grown at 1100°C but with flow rates of 40/40 and 60/60 produced poly crystalline films. Our results are summarized in Table 1.

Figure 5 shows a high magnification view of the interface of a specimen grown at 1000°C and 10/10 flow rate. The diffraction pattern from the interface region is shown as an inset to Fig. 5. The inner set of diffraction spots is due to Si and the outer set due to SiC. In

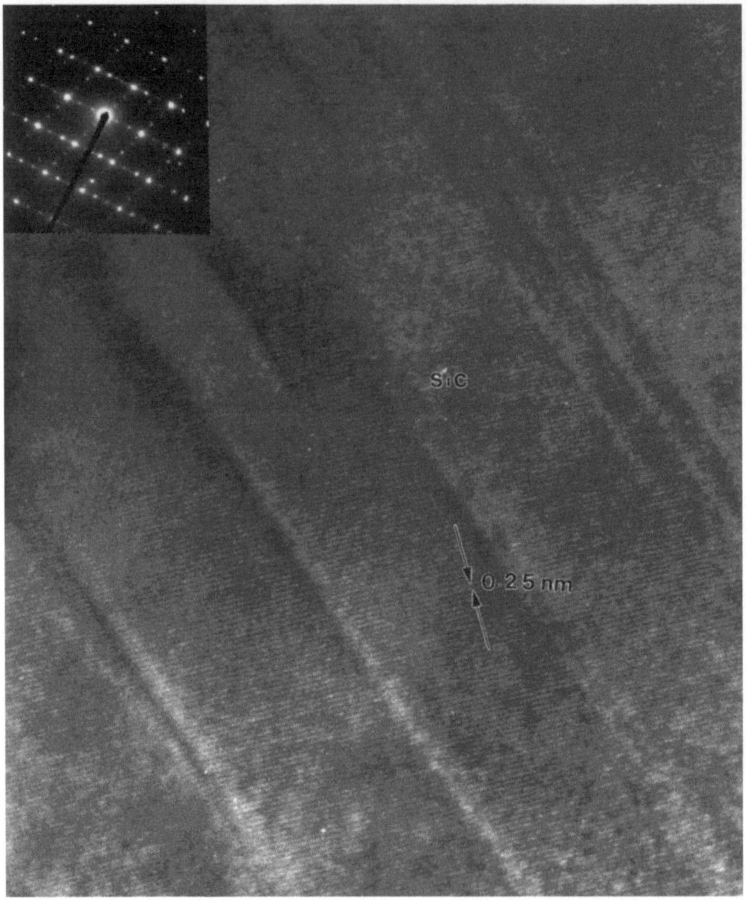

Fig. 4. (110) HREM micrograph of SiC film showing twins. The inset shows the twins and diffraction pattern of the area.

addition to showing poly crystallinity, the diffraction patterns also showed the epitaxial relation between Si and SiC. We can see the misorientation of the different grains which resulted in poly crystalline film. The arrows mark dislocations at the interface.

A high resolution micrograph of the specimen grown at 1100°C with 40/40 flow rate is shown in Figure 6. The diffraction pattern of the interface region is also shown. Again, the different orientation of the grains show that this film is poly crystalline.

It has been proposed by Pirouz, et al.[8], that if the orientation of the nuclei which are formed at the initial stages of growth, are parallel to the substrate the result will be a single crystalline film. However, in the case of the films grown at 1000°C and 10/10 flow rate (see Fig. 5), we believe that the main contributing factor for the films being poly crystalline is a low rate of surface diffusion rather than misorientation of nuclei. Because of the slow diffusion rate, the atoms will not be mobile enough to be able to rearrange themselves on the Si surface. Hence, instead of a single crystalline film a poly crystalline epilayer forms. However, in the case of the films grown at either high temperature, i.e., 1250°C or high

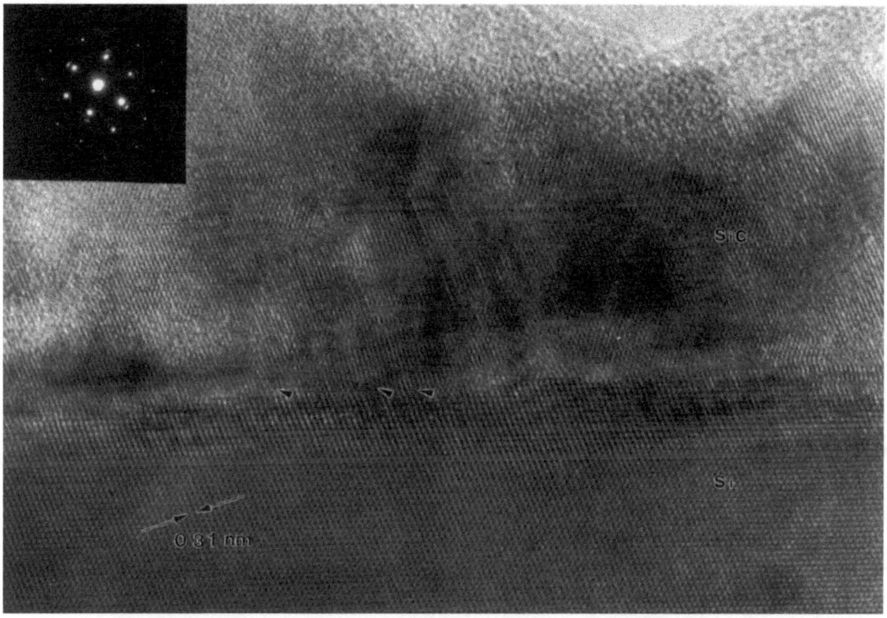

Fig. 5. (110) HREM micrograph of a poly crystalline SiC film grown at 1100°C and 40/40 flow rate showing dislocation and Si/SiC interface.

Table I.
Growth conditions and structure of the samples studied in this work

Sample #	Temperature	Flow Rate	Observation
85-1-B	1100°C	10/10	single crystal
86-1-B	1100°C	20/20	single crystal
83-5-B	1100°C	30/30	single crystal
134-B	1100°C	40/40	poly crystalline
136-B	1100°C	60/60	poly crystalline
88-4B	1000°C	10/10	poly crystalline
81-3-B	1250°C	30/30	poly crystalline

flow rate (40/40, 60/60), we believe that poly, crystalline films are due to the formation of a high density of nuclei which are misoriented with respect to the substrate (see Fig.6). Our observations are similar to those of other investigators, and we believe that 3-dimensional nucleation coupled with error in stacking produces a high density of SF and twins in the crystals we examined.

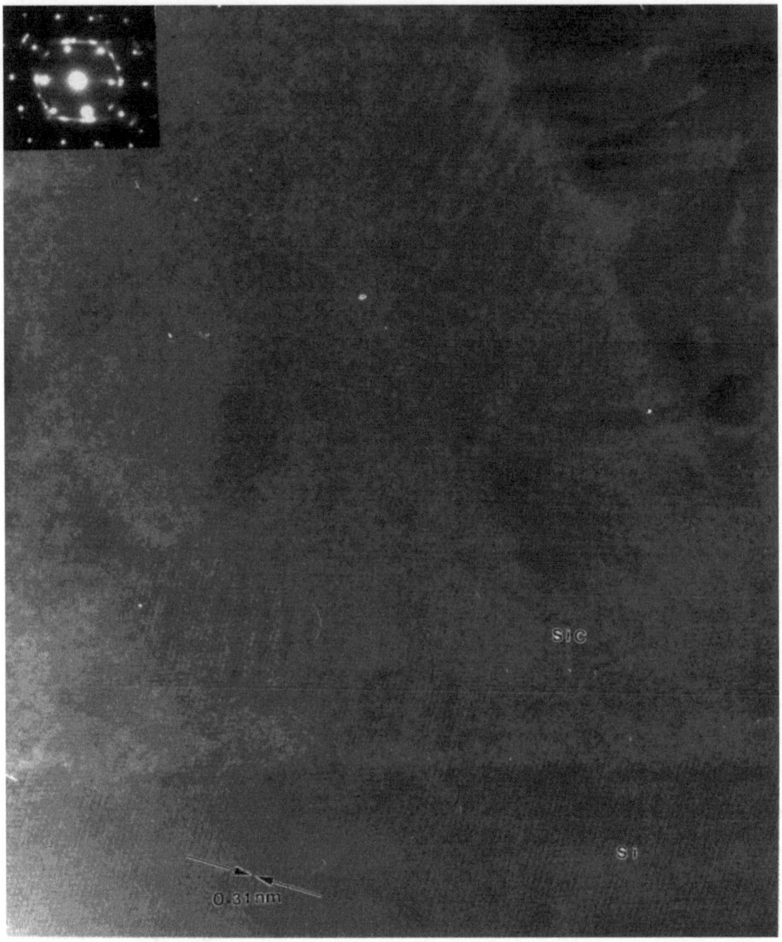

Fig. 6. (110) HREM micrograph of a poly crystalline SiC film grown at 1000°C and 10/10 flow rate. The insert shows the diffraction pattern from the area.

Conclusions

A comparison of the structure of films grown with different growth temperature and flow rate has been carried out. The lattice defects in these films have been observed to be mostly stacking faults and twins. Our results show that 1100°C and flow rates in the range 10/10 - 30/30 are necessary for obtaining single crystalline films. Either lower or higher growth temperatures as well as higher flow rates than these give rise to poly crystalline films. It has therefore been shown that CVD growth of β-SiC with a temperature under 1350°C, i.e. 1100°C could result in growth of single crystalline films. A more detailed study will be followed to determine the types of partial dislocations associated with the SF and the burgers vectors of the partial dislocations associated with the stacking faults.

References

1. K. Shibahra, T. Saito, S. Nishino, and H. Matsunami, Extended Abstracts of the 18th International Conference on Solid State Devices and materials, Tokyo, 1986, pp. 717-718.

2. J.A. Powell, L.G. Matus, M.A. Kuczmarski, C.M. Chorey, T.T. Cheng, and P. Pirouz, Appl. Phys. Lett. 51, 823 (1987)

3. L.G. Matus and J. A. Powell, Proceedings of the 1st International Conference on Amorphous and Crystalline Silicon Carbide and Related Materials, Dec. 10-11, 1987.

4. H.S. Kong, Y.C. Wang, J.T. Glass, and R.F. Davis, J. Mater. Res. 3, 521 (1988).

5. W.E. Nelson, F.A. Halden and A. Rosengreen, J. Appl. Phys. 37, 333 (1966).

6. H. Matsunami, S. Nishino and H. Ono, IEEE Trans. Electron. Devices ED-28 1235 (1981).

7. M.Baranov. V.A. Dmitriev, I.P. Nikitina, and T.S. Kondvateva. See this proceeding.

8. P. Pirouz, F. Ernst and T.T. Cheng. Mat. Res. Soc. Symp. Proc. Vol 116, (1988)

Thermal Oxidation of Single-Crystal Silicon Carbide: Kinetic, Electrical, and Chemical Studies

J.B. Petit[1], *P.G. Neudeck*[2], *L.G. Matus*[3], *and J.A. Powell*[3]

[1]Sverdrup Technology, Inc., MS 77-1, 2001 Aerospace Parkway, Brook Park, OH 44142, USA

[2]Ohio Aerospace Institute, MS 77-1, 2001 Aerospace Parkway, Brook Park, OH 44142, USA

[3]NASA-Lewis Research Center, MS 77-1, 21000 Brookpark Rd., Cleveland, OH 44135, USA

Abstract. This paper presents kinetic data from oxidation studies of the polar faces for 3C and 6H SiC in wet and dry oxidizing ambients. Values for the linear and parabolic rate constants were obtained, as well as preliminary results for the activation energies of the rate constants. Examples are presented describing how thermal oxidation can be used to map polytypes and characterize defects in epitaxial layers grown on low tilt angle 6H SiC substrates. Interface widths were measured using Auger electron spectroscopy (AES) with Ar ion beam depth profiling and variable angle spectroscopic ellipsometry (VASE) with effective medium approximation (EMA) models. Preliminary electrical measurements of MOS capacitors are also presented.

1. Introduction

Thermal oxidation is expected to play an enabling role in planar device processing technology for SiC, just as it has for silicon. Thermally grown oxides can be used as passivation layers on SiC device structures, as the insulator in MIS structures, and as a mask for selected area ion implantation. Numerous studies of thermal oxidation of SiC have been reported, but these have used many different types of material. Specimens have included Acheson furnace platelets [1,2], Lely platelets (grown by unseeded sublimation)[2-5], and (100) 3C-SiC epilayers grown on Si substrates [6]. No published oxidation studies have been conducted on 6H wafers grown by the seeded sublimation technique, nor is there any data on the polar faces of the 3C polytype. Previous work exhibits a large spread in the values reported for the rate constants, which may be due to variations in impurity content and defect density. Therefore, it would be useful to collect data specific to this semiconductor grade material, including the 3C films with low defect density grown on these substrates. In addition to kinetic studies, we have demonstrated that oxidation can be a very useful diagnostic tool for the structural characterization of SiC substrates and epitaxial films[7,8].

2. Experimental Procedure

The samples were prepared by dicing 6x5.5 mm^2 substrates from 25 mm diameter wafers which were cut from sublimation grown boules (produced by Cree research,Inc., Durham, NC). Epitaxial

Springer Proceedings in Physics, Vol. 71 **Amorphous and Crystalline Silicon Carbide IV**
Editors: C.Y. Yang · M.M. Rahman · G.L. Harris

layers were grown by chemical vapor deposition (CVD) on these substrates. The 6H-SiC substrates included (0001)Si and (0001)C samples of both low-tilt angle (0.1° to 0.6° tilt) and off-axis (3° to 4°) orientations. It was found that epitaxial layers grown upon low-tilt angle substrates could contain the 6H and/or the 3C polytypes, depending on tilt angle, surface quality, and the parameters of the pre-growth in-situ HCl etch (time, temperature, HCl concentration)[9]. The kinetic data for the 3C polytype was obtained from the films grown upon on-axis substrates, while the work on 6H included off-axis film samples and as-received substrates.

Thermal oxidations were conducted in a horizontal hot-wall quartz tube furnace. In all cases the oxygen flow rate was about 0.5 l/m. Wet oxidations were performed by bubbling oxygen through de-ionized water at 95°C. The cleaning consisted of organic solvents, HF etch, and DI water rinse prior to oxidation. Thickness measurements were obtained using a manual null ellipsometer (Gaertner) with a He-Ne laser (632.8 nm) at multiple angles of incidence. Variable angle spectroscopic ellipsometry (VASE) was performed with an instrument made by J.R. Woolam Inc.; wavelengths ranged from 270 to 800 nm and angles ranged from 65 to 75°. The data was fitted to models using the effective medium approximation (EMA) with software provided by the manufacturer. The Auger electron spectroscopy (AES) was performed with a Perkin-Elmer 590 system equipped with an Ar ion gun for depth profiling. To prevent reduction of the oxide by the electron beam the beam voltage was set at 3kV, the beam current kept under 200 nA, and the beam was defocused and rastered over the area of interest.

Metal-Oxide-Semiconductor (MOS) capacitors were fabricated on the silicon face of off-axis, as received 6H SiC substrates. A 75 nm SiO_2 insulator was formed by a 2-hour 1150°C wet thermal oxidation. Circular aluminum gates were sputter-deposited and patterned, and fabrication was concluded by the sputtering and definition of large-area backside AuTa "ohmic" contacts.

3. Results

A plot of oxide thickness vs. time for both polarities of 3C and 6H is given in Figures 1 and 2 for wet and dry oxidations, respectively. The difference in the oxidation rates of the

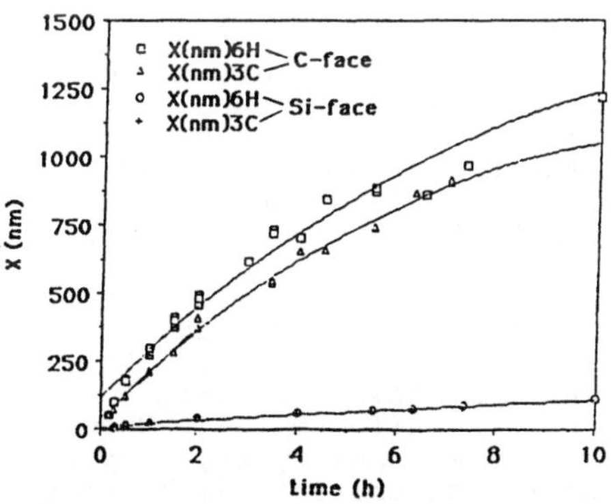

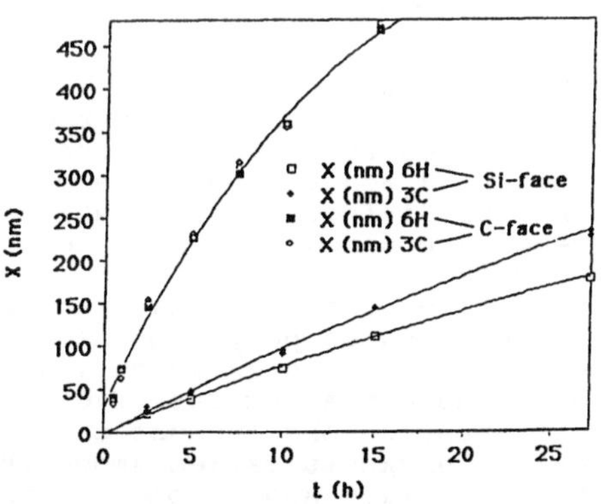

Figures 1 and 2: Thickness vs. time plots for wet (left) and dry (right) oxidations of polar faces of 3C and 6H SiC

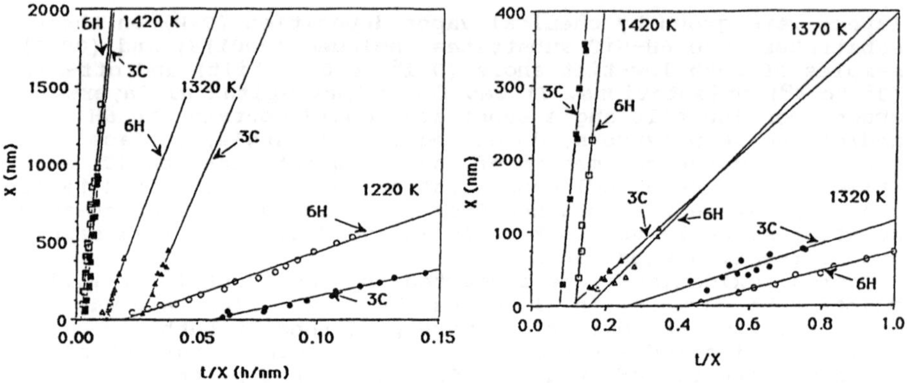

Figures 3 and 4: X vs. t/X for wet oxidation of C-face (left) and dry oxidation of Si-face (right)

different polytypes depends on the polarity and oxidizing ambient. For the Si-face, the difference is greater in dry oxidation, while for the C-face the difference is greatest in wet oxidation. There is very little or no difference in oxidation rates of the polytypes when the Si-face is oxidized in wet ambients or when the C-face is oxidized in dry ambients. Also note that 3C oxidizes faster in dry oxidation of the Si-face, while 6H is the "fast" polytype in the wet oxidation of the C-face. As will be discussed later, the data fits the general relationship for oxidation kinetics developed by Deal and Grove [10],

$$x^2+Ax=B(t+\tau_o),$$

which is derived from consideration of the diffusion and surface reaction processes. In this equation, x is the oxide thickness, t is time, A and B are temperature dependent rate constants, and τ_o is a constant that depends on the initial state of the surface (presence of native oxide, damage, etc.).

The oxidation rate equation may be written in the form,

$$x=B[(t+\tau_o)/x]-A,$$

which indicates that a plot of x vs. $(t+\tau_o)/x$ should exhibit a linear relationship where the constants A and B, were determined from the y-intercept and the slope, respectively. Figures 3 and 4 show these plots for the C-face (wet oxidation) and the Si-face (dry oxidation). Table 1 lists the values that were determined for the rate constants and their activation energies.

Figures 5 and 6 illustrate the use of thermal oxidation for the mapping of polytypes and delineation of defects in epitaxial layers grown on low-tilt angle 6H substrates. It is well known that oxides of different thicknesses exhibit different colors because of interference effects [11]. When the oxide thickness varies among the different polytypes, they may be identified and mapped according to the characteristic color of the native oxide. Figure 5 shows a Si-face sample after a 5 h dry oxidation. In this case the 3C region oxidized faster and

Temperature (K)	B/A (nm/h)		B (x 10^3 nm^2/h)	
	6H	3C	6H	3C
	wet oxidation, C-face			
1420	531 ± 74	296 ± 46	196 ± 11	198 ± 14
1320	90.5 ± 30	37.2 ± 6.6	43.9 ± 5.2	42.4 ± 3.1
1220	59.8 ± 12.4	16.3 ± 2.9	4.8 ± 0.2	3.9 ± 0.2
E_a kJ/mol	155 ± 63	207 ± 61	277 ± 20	
	dry oxidation, C-face			
1420	100 ± 13	94.8 ± 16.4	20.4 ± 0.7	21.9 ± 1.1
1370	71.2 ± 12.7	65.8 ± 7.1	14.1 ± 0.6	14.6 ± 0.5
1320	51.8 ± 11.3	39.6 ± 4.5	8.8 ± 0.5	10.8 ± 0.4
E_a kJ/mol	106 ± 2	137 ± 9	121 ± 2	
	wet oxidation, Si-face			
1420	33.8 ± 10.2	28.4 ± 18.8	1.76 ± 0.13	1.75 ± 0.21
1370	12.2 ± 8.5	12.0 ± 13.4	1.06 ± 0.15	0.88 ± 0.15
1320	19.6 ± 11.2	24.2 ± 6.8	0.12 ± 0.01	0.10 ± 0.01
E_a kJ/mol	---	---	443 ± 133	
	dry oxidation, Si-face			
1420	8.2 ± 1.9	12.6 ± 2.6	5.86 ± 0.54	6.4 ± 0.47
1370	6.3 ± 3.2	9.2 ± 3.9	0.51 ± 0.7	0.45 ± 0.05
1320	2.4 ± 1.2	3.8 ± 3.9	0.13 ± 0.01	0.15 ± 0.03
E_a kJ/mol	197 ± 59	187 ± 46	590 ± 131	

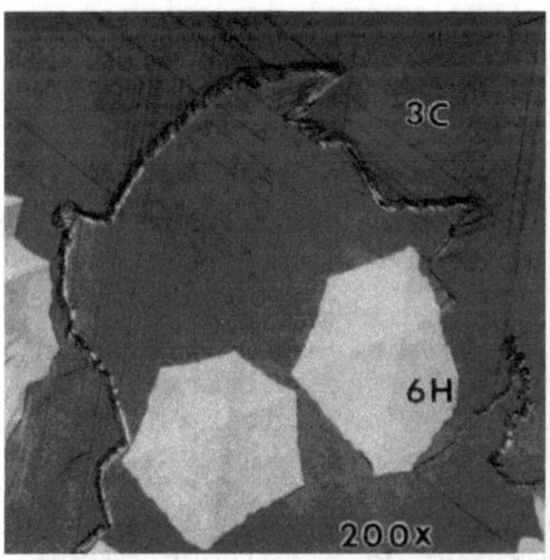

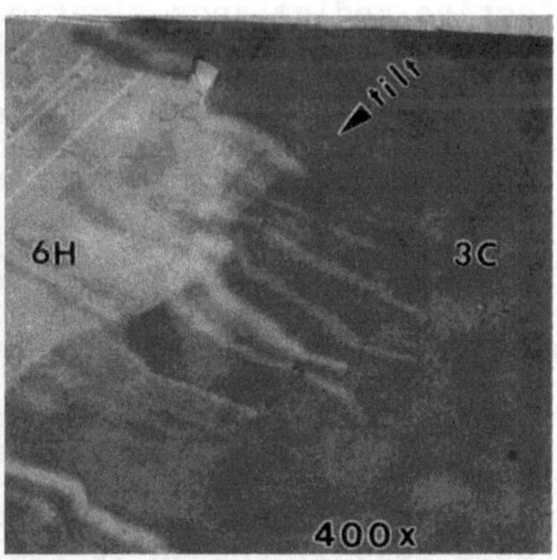

Figures 5 and 6: Si-face after 5h dry oxidation (left) and close-up of 3C/6H boundary (right)

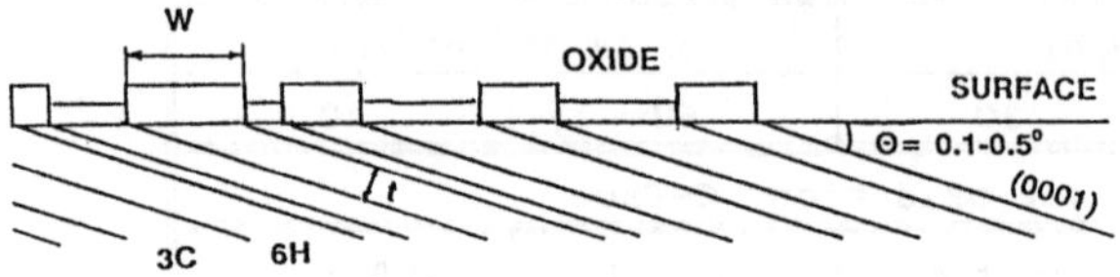

Figure 7: Schematic showing detection of heterostructures by oxide.

resulted in a brown colored oxide (70 nm), while the 6H yielded a light tan colored oxide (50 nm). The oxidation also enhanced the visibility of a double positioning boundary (DPB, jagged curve) and stacking faults (straight lines). Wet oxidation of the C-face could map polytypes in 30min; the oxide on the 6H was 180nm thick and appeared very light yellow-green, and the oxide on the 3C was 120nm thick and appeared royal blue. Defects on the C-face were less sensitive to delineation by oxidation and required the growth of thicker oxide layers (300-500 nm) to become visible. Figure 6 shows a close-up of a 3C/6H boundary where a variation in the abruptness of the boundary can be observed to be dependent on the direction of the boundary relative to the tilt direction [8]. Also visible are oxide thickness (color) fluctuations caused by oxidation of 3C/6H heterostructures (figure 7). The ratio of the width of these fluctuations to the thickness of the layers is given by,

$$W/t = 1/\sin\theta,$$

where θ is the tilt angle. For low tilt angle substrates this apparent lateral magnification factor varies from 115 to 575, as θ decreases from 0.5° to 0.1°. Thus, very fine (about 10 nm) heterostructures can be resolved by this technique.

The width of the SiO_2/SiC interface was measured by both Auger electron spectroscopy (AES) and variable angle spectroscopic ellipsometry (VASE). The measurement by AES was done by measuring the time required for the extrapolated linear change of the O_{KLL} peak at the interface and multiplying by a sputtering rate determined by the total time required to sputter through oxides of known thicknesses. Values obtained by this method ranged from 70 to 150 nm. The VASE data was fitted to an effective medium approximation (EMA) model in which a layer that consists of a SiO_2/SiC mixture is assumed to exist at the interface. Values for interface width obtained by VASE ranged from 80 to 120 nm. No correlations where found between interface width and polytype, polarity, or oxidation method. These values are in agreement with those reported in reference 4, but are less than those reported in reference 12; the values reported in ref. 12 used (100)SiC/Si samples, whose rougher surfaces may explain the broader interfaces.

The C-V characteristics of the MOS capacitors at 23°C and 100°C are shown in Figures 8 and 9, respectively. For gate voltages (Vg) greater than 5 V, the measured capacitance corresponds to the theoretical accumulation capacitance. Inversion is not observed in these curves due to the extremely small carrier generation rate of 6H-SiC. Minority carriers (holes in this case) necessary to form an inversion layer are not generated quickly enough to follow the gate voltage sweep rate of 0.1 V/second.

The C-V curves of Figures 8 and 9 clearly exhibit charge trapping phenomena indicative of a large density of defects at

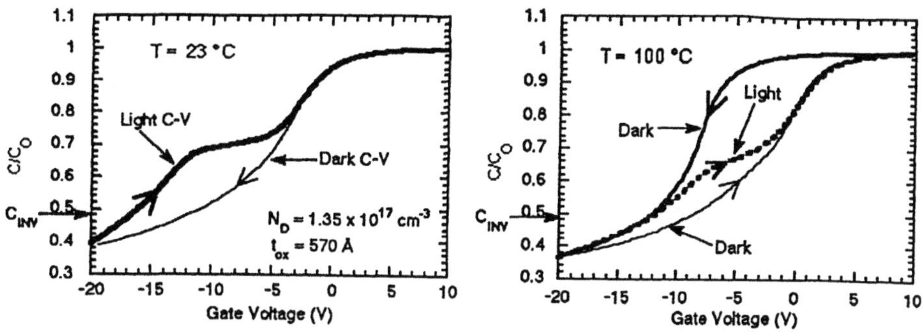

Figures 8 and 9: Capacitance-Voltage plots at 23°C (left) and 100°C (right).

the SiO_2/SiC interface (i.e., interface states). The trapping of excess minority carriers generated by the light is believed to be responsible for the ledges, and the hysteresis observed in dark C-V curves is typical of abundant interface state charge trapping in MOS structures [13,14]. From the amount of shifting observed, the interface state density is at least 3 x 10^{12} cm^{-3}. This number compares unfavorably with interface state densities in the SiO_2/Si system ($\sim10^{10}$ cm^{-2}). However, it is hoped that anneal treatments similar to those used to attain low interface state densities in the SiO_2/Si system will lead to a drastic reduction in SiO_2/SiC interface trap densities [13,14].

4. Conclusion

Kinetic data for the thermal oxidation of 3C and 6H SiC polytypes was obtained. Examples of polytype mapping and defect delineation were discussed. Measurements of interface width were made. Results of initial MOS capacitor studies were reported.

5. Acknowledgments

The work described in this paper was conducted at NASA-Lewis Research Center under internal funding support. The authors wish to thank Margaret Eastman for assistance with the ellipsometric measurements, Mary Zeller for help with the AES, and Carl Salupo for MOS fabrication.

6. References

1. Z. Zheng, R.E. Tressler, K.E. Spear, and J.A. Costello, J.Electrochem. Soc., **137**, 854 (1990)
2. J.A. Costello, Doctoral Thesis, Penn. State Univ. (1983)
3. W. von Münch, and I. Pfaffeneder,J. Electrocem. Soc.,**122**, 642 (1990)
4. A. Suzuki, H. Ashida, N. Furui, K. Momeno, and H. Matsunami,Jpn. J.Appl. Phys., **21**, 579 (1975)
5. Y. Laukhe, Y.M. Tairov, V.F. Tsvetkov, and F. Shchepanski, Izvest. Akad. Nauk. S.S.S.R., Neorgan. Mat., **17** (1983)
6. C.D. Fung, and J.J. Kopanski, Appl. Phys. Lett.,**45**, 757 (1984)

7. J.A. Powell, J.B. Petit, J.H. Edgar, I.G. Jenkins, L.G. Matus, W.J. Choyke, L. Clemen, M. Yoganthan, J.W. Yang, and P. Pirouz, Appl. Phys. Lett., **59**, 183 (1991)
8. J.B. Petit, J.A. Powell, and L.G. Matus, Proceedings High Temperature Electronics Conference, Albuquerque, NM, 16-20 June 1991
9. J.A. Powell, D.J. Larkin, J.B. Petit, and J.H. Edgar, this proceedings, and references therein
10. B.E. Deal and A.S. Grove, J. Appl. Phys., **36**, 3770 (1965)
11. W.A. Pliskin and E.E. Conrad, IBM J. Res. Dev., **8**, (1964)
12. R.W. Kee, K.M. Geib, C.W. Wilsen, and D.K. Ferry, J. Vac. Sci. Technol., **15**, 1520 (1978)
13. E.H. Nicollian, and J.R. Brews, MOS (Metal Oxide Semiconductor) Physics and Technology, Wiley, New York, (1982)
14. R.F. Pierret, Field Effect Devices, Modular Series on Solid State Devices Vol. 4, Addison-Wesley, Reading, MA, 1983

Determination of Stacking-Fault Abundances and Distributions in SiC Using XRPD and HRTEM

P.J. Schields

Department of Chemistry, Arizona State University, Tempe, AZ 85287-1604, USA

Abstract. Relative abundances of stacking faults in β-SiC were determined from high-resolution transmission electron microscopy (HRTEM) structure images. These stacking-fault abundances were correlated to the intensity of the {10} two-dimensional line in the x-ray powder diffraction (XRPD) pattern. The {10} line corresponds to rods of reciprocal-lattice intensity generated by the disorder of the faulted structure. Relative abundances of stacking faults are defined by the relative peak intensity of the {10} line. Systematic changes of XRPD relative line intensities for β-SiC are consistent with theoretical predictions based on the local hexagonal symmetry of the faulted crystal structure.

Distributions of interfault distances in heavily faulted regions of β-SiC were determined from HRTEM images. These distributions are consistent with a single, random, fault probability greater than 0.5 (relative to β-SiC). Conventional XRPD fault probability determinations can not be applied since line broadening, asymmetry and displacements are ambiguous to measure and interpret. The proposed XRPD stacking-fault analysis was used to interpret the transformation of β-SiC as a function of time and temperature in a carbon-rich atmosphere.

Structural Defects in Epitaxial Layers SiC-3C/Si Grown by CVD

R.N. Kyutt[1], *I.P. Nikitina*[1], *S.S. Ruvimov*[1], *and D. Baither*[2]

[1]A.F. Ioffe Physical-Technical Institute, Russian Academy of Sciences, Politechnicheskaya 26, 194021 St. Petersburg, Russia
[2]Max-Planck Institut für Mikrostrukturphysik, Weinberg 2, O-4010 Halle (Saale), Fed. Rep. of Germany

The present study reports research results of a dislocation structure of epitaxial layers SiC-3C/Si, grown by CVD without a buffer layer and by "buffer" technology. X-ray differential diffractometry and TEM were used to establish the structure of the grown SiC-3C layers. It was revealed that single crystalline epitaxial films can be produced on Si(111) substrate without buffer layer at 1100°C. These films have a mosaic structure with coherent scattering block size of 300-600A near the substrate. By increasing the layer thickness, the size of blocks becomes larger with transformation of defect structure into randomly distributed dislocation ensemble.

1. Introduction

The high value of gap energy (E_g= 2,2 eV) and the highest value of carrier mobility (μ = 1000 $cm^2/V \cdot sec$) give the opportunity to use the cubic polytype of silicon carbide in high-speed and high-power semiconductor devices and heterojunction structures with Si. But some parameters of such devices (breakdown voltage, leakage currents) are much worse than the expected [1-4]. The reason for such discrepancy is very likely to be the defects in the crystalline structure of epitaxial SiC-3C layers. The structural investigations of SiC-3C films carried out by X-ray diffraction and TEM showed the possibility of the single crystalline SiC-3C epitaxial layers on Si substrate by buffer technology [4-7]. A large density of the defects such as microtwins, stacking faults, antiphase boundaries and dislocations was observed [7,8]. These defects are caused by mismatch (≈20%) between the layer lattice and the silicon substrate lattice. Structural perfection of single crystalline epitaxial 3C-SiC layers grown on substrate Si(111)

Springer Proceedings in Physics, Vol. 71 **Amorphous and Crystalline Silicon Carbide IV**
Editors: C.Y. Yang · M.M. Rahman · G.L. Harris

and Si(100) by CVD without and with a buffer layer were investigated in the present work.

2. Experimentals

In order to characterize a dislocation ensemble in single crystalline epitaxial layers of a cubic polytype, we used the method of X-ray differential deffractometry and TEM (JEOL JEM-7A, 100 kV, JEM-1000, 1 MV). TEM specimens were prepeared by chemical etching of Si substrate followed by ion milling of SiC layers from one or both sides with 3-5 kV Ar ions. The differential difractometry gives information about scattering intensity distribution near a reciprocal lattice site along two mutually perpendicular directions: along vector **H** of a reciprocal lattice (radial disfribution) and along the direction perpendicular to **H** (azimuthal disfribution). Rocking curves, obtained at simultaneous scanning of a sample and a detector near Bragg angle (θ) with scanning speed ratio 1:2, demonstrate radial distribution of scattering intensity, (θ-2θ - mode) while rocking curves, obtained at scanning of a sample and fixed detector, which corresponds to an exact Bragg reflection, show azimuthal distribution (θ - mode). In line with the theoretical viewpoints developed by Krivoglaz [9], chaotically distributed dislocations, dislocation dipoles, dislocation loops of long radius and stacking faults, as well as the formation of dislocation boundaries dividing crystal into blocks - this causes diffraction effects which differ in great extent. Halfwidth ratio of intensity distribution in radial and azimuthal directions, radial distribution as a function of a Bragg angle and form of a rocking curve depend on crystal dislocation structure.

Rocking curves of the both types on reflections of first and second orders from planes {111} were received by double-crystal X-ray spectrometer using CuKα radiation. Monocrystal creaking cut of dislocation-free Si(111) and Si(100) were used as monochromators. In connection with the fact, that halfwidths of rocking curves were from some hundreds to some thousands seconds of arc, it appeared possible to use as an analyser a slot of 0.25 mm in front of the diffracted radiation detector at distance 18 cm from sample. Rocking curves were plotted point-by-point in intensity accumulation

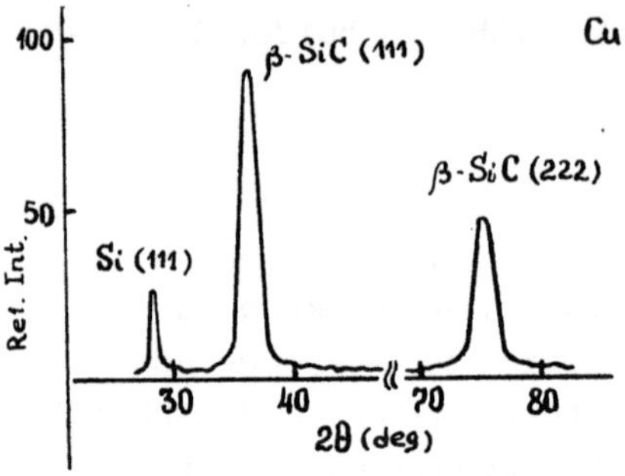

Fig.1. X-ray diffraction pattern for SiC-3C layer on Si(111) substrate (1100°C; 1 arb.unit).

mode over constant time. Accumulation time was equal to 100 seconds, step value of scanning angle varied from 0.025° to 0.1°.

Growth of SiC-3C was carried out at the atmospheric pressure and temperatures from 1000°C to 1250°C. Monosilane (SiH_4) and trichlorethylene (C_2HCl_3) were used as sources of silicon and carbon respectively. Hydrogen was a carrier gas. Variations of growth temperature and reagent gas flow rate allowed the determination of optimal conditions for producing single crystalline layers of silicon carbide cubic polytype on silicon substrates. Thickness of layers grown varied from 0.5 to 11 μm depending on the flow rate.

It was revealed, that crystalline epitaxial SiC-3C layers could be produced only at 1100°C on substrate Si(111) and different reagent gas flow rates without a buffer layer and at 1250°C on substrate Si(100) with a buffer layer [10].

Fig.1 shows an X-ray diffraction pattern for single crystalline SiC- 3C layer on a Si(111) substrate, obtained at 1100°C and reagent gas flow rate of 1 (in arbitrary units). On the diffraction pattern there are only peaks corresponding to the reflections of first and second order from planes of type {111}, besides the peak from substrate.

Fig.2 represents X-ray diffraction rocking curves measured in θ-2θ and θ-modes for the sample 1 (see Table 1). Similar curves are obtained for the samples with higher reagent gases flow rates. The θ-scanning curves give the peaks of the finite halfwidth which is independent of the reflection order. However, they are much broader than the θ-2θ peaks. The θ-2θ peak halfwidth increases with increasing of the Bragg angle. The obtained results of halfwidths of the rocking curves for

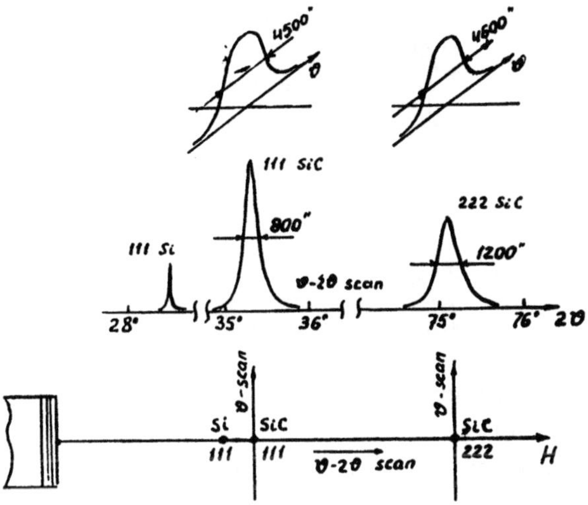

Fig.2. Rocking curves, measured in θ-2θ and θ-modes for SiC-3C layer (1100°C; 1 arb.unit).

Table I

Sample	Reagent gases flow rate (arb. units)	Thick-ness (μm)	Radial distribution (111) halfwidth (sec.of arc)	Azimuthal distribution halfwidth (sec.of arc)	Halfwidth ratio for radial distribution (222)/(111)
1	1	0.98	790	4270	1.48
2	1	1.02	800	4780	1.7
3	2	2.34	720	3750	2.2
4	3	5.6	700	3780	2.3
5	4	9.0	800	8470	2.4
6	6	11.2	900	9960	2.2

epitaxial layers, produced at 1100°C and different flow rates of reagent gases are given in Table 1.

$$\frac{\sec \theta_{222}}{\sec \theta_{111}} = 1.2 \qquad \frac{\mathrm{tg}\ \theta_{222}}{\mathrm{tg}\ \theta_{111}} = 2.4$$

Fig.3. represents a typical electron micrograph (a) and electron diffraction pattern (b) of SiC-3C layer near Si-substrate. One can see the sharp reflections of Si and enlarged SiC-reflections on the electron diffraction pattern. The elongations of latters are likely to be caused by mosaic

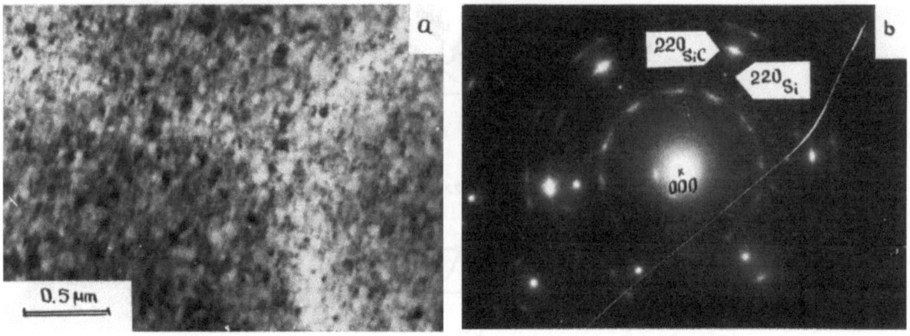

Fig.3. a) Typical electron micrograph of SiC-3C layer near Si-substrate (sample 1);
b) Electron difraction pattern, corresponding to a)

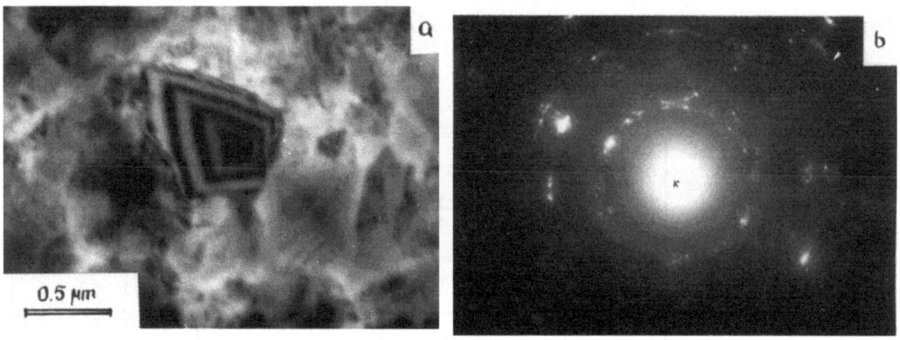

Fig.4. a - TEM micrograph of SiC-3C layer near the surface (sample 2);
b - electron diffraction of SiC-3C, correcponding to a)

SiC structure containing a high defect density. The size of coherent scattering SiC-areas is 300-600Å.

The mosaic structure transforms to another one on the surface of the layer with the increasing of block size and their misorientation (Fig.4). The block size on surface has been increased up to 0.5 µm and more. There are many stacking faults in each block as well as high density of dislocations on their boundaries. Similar features of the diffraction curves were obtained also for SiC-3C grown on Si(100) substrate with buffer layer. However the halfwidths of intensity distribution in both directions for (200) reflection are half of the value for the (111) reflection from (111) films.

3. Discussion

The study of the rocking curves obtained has shown that the halfwidth of rocking curves along the direction perpendicular to H for all samples examined was approximately 5-10 times as much as that of rocking curves along H. It is a feature of chaotically distributed dislocations of high density with domination of one-sign dislocations, i.e. at plastic bending. In this case halfwidth of radial distribution must have varied proportionally with tgθ. Nevertheless, an angular function of radial distribution varied from the law similar to sec θ to the tg θ law depending on reagent gases flow rate. As a rule, a dependence of Bragg angle occupied an intermediate position between sec θ law and tg θ law (see Table 1).

Abrupt asymmetry of intensity distribution in reciprocal space and intermediate dependence of Bragge angle of radial distribution for law flow rate of reagent gases is described well by a model which epitaxial layer consists of small blocks of coherent scattering, divided with dislocations boundaries. Dislocation density within boundaries is higher than that inside the blocks, while spaces (D_o) between dislocations in the boundaries are much less than the block sizes (D). The equality of the halfwidth in θ-modes shows that the misorientation is the main reason for the diffraction peaks broadening. The misorientation of the crystal sections with dislocation boundaries causes also the fact that the curve for radial distribution must be Lorentzian and for azimuthal distribution - close to Gaussian. Rocking curves form analysis along H and along the direction perpendicular to H showed a satisfactory coincidence of the forms of experimental rocking curves with Lorentian and Gaussian outlines, respectively, for epitaxial layers, produced at reagent gas flow rate of 1 arb.unit.

On the basis of radial distribution broadening on reflections (111) and (222) the average sizes of the blocks of coherent scattering and the values of relative deformation $\Delta a/a$ were calculated with the help of nomograms for separation of the broadening effects resulting from microstresses and dispercivity of the blocks [11]. The average size values of the block coincides with the TEM estimation of size near SiC-3C/Si interface. Using values of rocking curve halfwidths along the azimuthal direction, the angles of blocks disorientation from

Table II

Sample	Block size (Å)	Relative deformation $\Delta a/a \cdot 10^4$	Misorientation angle (sec.of arc)
1	470	5.0	380
2	580	9.3	430

each other were estimated. The calculation results are given in Table II.

When the reagent gas flow rate increases, the angular dependence of radial intensity distribution follows the tg θ law. The halfwidth of the θ-curves (azimutal distribution) increases. This can be indicative of the fact, that at fast flow rates of reagent gases the block structure "goes to pieces", dislocation ensemble being chaotically distributed dislocations of high density with the domination of one-sign dislocations. This causes a plastic bending of epitaxial layer. Really, in some cases when layers, grown at fast flow rates of reagent gases, are thicker than 10 μm, separation of epitaxial films from substrates is observed.

This diffraction pattern can be also explained by the presence of large high damaged misoriented blocks. Most probably this result is due to the fact that the layers grown at fast flow rates have larger thicknesses. These results are in accordance with the TEM observation which shows the increase of the block size on the film surface (Fig.4). The results obtained for SiC-3C films grown by "buffer" technology on the Si(100) substrate confirm the conclusion of the previous studies [4,6,8] about sufficiently good structural quality of these layers. The ratio of the θ-2θ curves halfwidth for (200) and (400) reflections corresponds to tg θ law, which can be explained by the presense of large density of randomly distributed dislocations.

4. Summary

In the present work the structural imperfections in epitaxial SiC-3C layers, grown by CVD on (111) and (100) Si-substrates without and with a buffer layer were investigated by X-ray diffraction and TEM. It was shown that the single crystalline epitaxial layers of sufficient quality can be produced on Si(111) substrate at T = 1100°C without buffer layer. It was found that epitaxial layers have a mosaic structure, coherent scattering block size of 300-600A near the substrate. When reagent gas flow rate increases the size of blocks becomes larger, the density of defects within blocks becomes the same order of magnitude as inside the dislocation boundaries, i.e. dislocation ensemble transforms into a "net" of chaotically distributed dislocations with surplus number of one-sign dislocations. These results can be used when analysing the discrepancy between real and design parameters of devices on the basis of heteroepitaxial layers of silicon carbide cubic polytype.

5. References

1. Y.Kondo, S.Misawa, H.Daimon, M.Yamanada and S.Yoshida, IEEE Electron Device Letters, EDL-7(1986) 404-406.
2. G.Kelner, S.Binari, K.Sleger, H.Kong, IEEE Electron Device Letters, EDL-8(1987) 428-430.
3. T.Sugii and T.Ito, Y.Furumura, D.Doki, F.Mieno and M.Maeda, J.of Electrochem.Soc., 134(1987) 2545-2549.
4. S.Nishino, J.A.Powell and H.A.Will, Appl.Phys.Lett., 42(1983) 460-462.
5. A.Addamiano, J.A.Sprague, Appl.Phys.Lett., 44(1984), 525-527.
6. S.Nishino, H.Suhara, H.Ono, H.Matsumuri, J.Appl.Phys., 61(1987), 4889-4893.
7. P.Pirouz, C.M.Chorey, T.T.Cheng and J.A.Powell., Microscopy of Semiconducting Materials, 1987 (IoP Conf.ser.87), 175-180.
8. S.R.Nutt, D.Smith, H.Kim, R.Davis, Appl.Phys.Lett., 50(1987) 203-205.
9. M.A.Krivoglaz, "X-ray and neutron diffraction in nonideale crystals", Kiev, "Naukova Dumka", p.407, 1983 [in Russian].
10. I.M.Baranov, N.A.Belov, V.A.Dmitriev, N.G.Ivanova, T.S.Kondratieva, I.P.Nikitina, V.E.Chelnokov, V.F.Shatalov, R.N.Erlih, Pisma w ZhTF, 15, N 12(1989) 50-52 [in Russian].
11. S.S..Gorelic, L.N.Rastorguev and U.A.Skakov, "X-ray diffraction and electron diffraction analysis", Moskow, "Metallurgia", p.367, 1970 [in Russian].

Observation of Linear Electro-Optic Effect in Cubic Silicon Carbide

X. Tang, K.G. Irvine, D. Zhang, and M.G. Spencer

Materials Science Research Center of Excellence, Department of Electrical Engineering, School of Engineering, Howard University, Washington, DC 20059, USA

Abstract. We report the first observation of linear electro-optic effect of cubic silicon carbide (β - SiC). Thin films of silicon carbide are grown on silicon (Si) substrates. A high electric field is produced in the material by a reverse biased Schottky barrier. The electro-optic coefficient is measured along the [100] crystal direction. At the wavelength of 633nm, the value of the electro-optic coefficient of β - SiC is determined to be (2.7 $\pm$ 0.5) x 10^{-12} m/V which is about 1.7 times larger than that in GaAs measured at 10.6μm. The half-wave voltage of 6.4 kV for β - SiC is obtained. Besides this favorable electro-optic property, we demonstrate the possibility of fabricating a waveguide as a light switch in epitaxial β - SiC. It is shown that β - SiC is a promising candidate in electro-optic applications for high optical intensity in the visible region.

1. Introduction

Silicon carbide has been considered a good material for high power and high temperature devices because of its attractive mechanical, thermal and electronic properties. Many of these electronic properties have been characterized and reviewed[1], but very little is known about the non-linear optical and electro-optical characteristics. Cubic silicon carbide (β - SiC) has a zinc blend crystal structure which belongs to $\bar{4}3m$ point group, and possesses a noncentral inversion symmetry, therefore, an electro-optic effect is anticipated. For this kind of crystal structure, when we choose coordinates x,y,z parallel to crystal axes [100], [010], [001], respectively, the electro-optic tensor components r_{41}, r_{52}, r_{63} have the same non-zero value, and the others are all zero. As a result of this electro-optic effect, it will be possible to fabricate discrete as well as integrated optical components. Optical devices fabricated from β - SiC can be operated at wavelengths as short as 560nm due to the large indirect band of 2.2eV. The high mechanical strength and thermal conductivity of the material will make it attractive for use at high optical power densities and elevated temperatures. An electro-optic coefficient is the most important parameter for the materials which are used to fabricate electro-optic devices. In this work, we report the first measurement of the electro-optic coefficient of β - SiC.

In our experiments, thin films of β - SiC are grown on silicon (Si) substrates using vertical Low Pressure Chemical Vapor Deposition (LPCVD)[2]. The carrier concentration of the epitaxial layer is approximately 10^{16} cm^{-3}, (n-type unintentionally doped). Free standing films of β - SiC are obtained by etching the silicon substrate away in a HF and HNO_3 solution. In order to produce a high electric field in the material without drawing significant current, two metal rectifying contacts were used. Semi-transparent gold Schottky contacts (of thickness of 100 angstroms) were evaporated on both sides of the film. The breakdown voltage of these contacts exceeds 20 volts with low leakage currents.

A waveguide geometry reported earlier[3] was employed as a light switch driven by an applied electrical field (Fig. 1), but it was difficult to analyze the date for obtaining

Springer Proceedings in Physics, Vol. 71 Amorphous and Crystalline Silicon Carbide IV
Editors: C.Y. Yang · M.M. Rahman · G.L. Harris

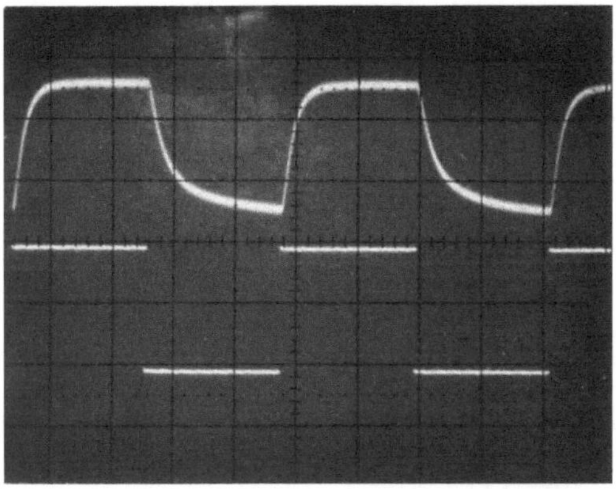

Fig. 1. A light signal (up) is switched by a pulsed voltage (down) x: 0.5ms/Div.

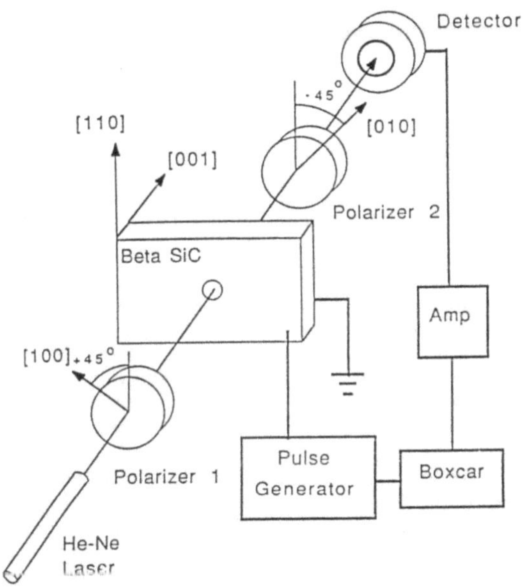

Fig. 2. Experimental set-up for measurement of electro-optic coefficient of beta silicon carbide.

precise value of the electro-optic coefficient. The measurement geometry used in this experiment is shown in Fig. 2. A He-Ne laser beam of wavelength λ of 633nm is incident onto the [001] surface of the crystal and is linearly polarized at +45 degreetoward the [100] crystal direction. The laser beam passes through the semi-transparent contacts on both sides of the sample while a train of electrical pulses of width 10ms is applied between the contacts. Thereafter, the light passes through polarizer 2 which is oriented -45 degree toward the [010] crystal direction. Finally, the light is detected and signal-averaged. In our experimental configuration, the total phase retardation is independent of the thickness of the sample. For small phase retardation, the electro-optic coefficient is related to the peak-to-peak amplitude of the detected light intensity I in the following way:

$$r41 = [(I/I_o)^{1/2}/V] \cdot [\lambda/\pi\, n_o{}^3] \quad (1)$$

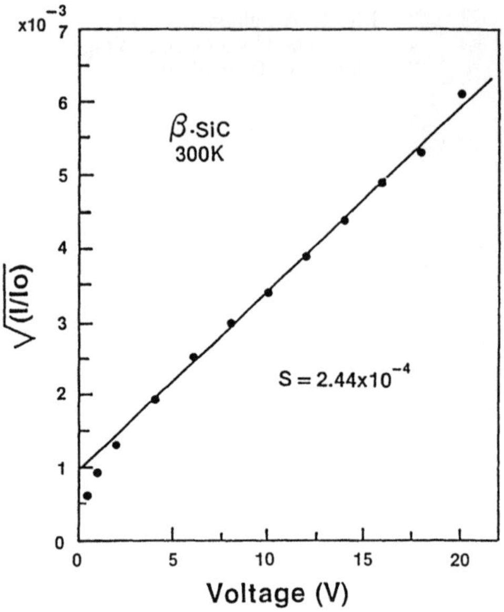

Fig. 3. Square root of the normalized electro-optic signal verses applied voltage for beta silicon carbide.

where I_o is the intensity measured without voltage applied and with polarizer 2 paralleled with polarizer 1, V is applied voltage, n_o is the zero field refractive index of β - SiC, respectively. The results of the measurements performed at room temperature are shown in Fig. 3. Throughout most of the region, the slope is constant, however, in the low voltage region the curve follows a different slope. It can be explained as follows: phase retardation due to the electro-optic effect is not observed at zero applied voltage although there is a large built-in field at both metal semiconductor interfaces. There is not a net phase retardation, because the opposite directions of the built-in fields cause the electro-optic effects produced at both junctions to cancel themselves out. As increased voltage is applied to the device, the two contacts no longer exactly cancel the elctro-optic effects and therefore, a net phase retardation is observed. After one of the junctions becomes forward biased the applied voltage drops entirely across the reversed biased junction (the value of applied voltage where the cross over occurs is about 2.2V or twice the barrier height of 1.1V)[4]. This analysis implies that the voltage where the experimental curve becomes linear is above 2.2V or approximately the value seen in Fig .2. Using the slope of the plotted data (for voltage greater than 2V) and the value $n_o = 2.639$ for the refractive index of β - SiC[5], the values of the electro-optic coefficient and the half wave voltage for β - SiC at a wavelength of 633nm are determined to be $(2.7 \pm 0.5) \times 10^{-12}$ m/V and 6.4 kV, respectively. The electro-optic coefficient of β - SiC is about 1.7 times larger than that of gallium arsenide measured at 10.6 microns. We also calculated the value of $n_o^3 r_{41}$ defined as a figure of merit to be 49×10^{-12} m/V, which is slightly smaller than the comparable value for gallium arsenide. Most of the experimental error is caused by the contact resistance and unsatisfactory uniformity of material conductivity. In Table 1, we present some of the non-linear coefficients and figures of merit of common electro-optical materials GaAs[6], KDP[7] and $LiNbO_3$[8] for comparison.

In summary, we report the first measurement of the electro-optic coefficient in cubic silicon carbide. The value of electro-optic coefficient compares favorably with GaAs. Although the electro-optic coefficients for $LiNbO_3$ and some poled organic materials[9] are very high, the lower optical damage threshold may limit their usefulness in case of high optical intensity. We will study the optical damage property of SiC and a high optical

Table I: Electro Optical Properties for Some Common Electro-Optic Materials

Material	Room Temperature Electro-Optic Coefficient (10^{-12}m/V)	Index of Refraction	Figure of Merit (10^{-12}m/V)
β – Sic at 633nm (this work)	$r_{41} = 2.7 \pm 0.5$	2.639	49
GaAs at 10.6μm	$r_{41} = 1.6$	3.34	59
KDP at 546nm	$r_{41} = 8.77$	1.51	29
	$r_{63} = 10.3$	1.47	34
LiNbO3 at 633nm	$r_{13} = 10$	2.1155	
	$r_{22} = 6.8$	2.1974	
	$r_{33} = 32.2$	2.1974	

damage threshold is expected due to its strong chemical bonding and high thermal conductivity. We believe that this material may be a promising candidate in electro-optic applications for high optical intensity in the visible region.

We gratefully acknowledge the support of the National Science Foundation under MSRCE Grant #NSF-RII8947 537 and NASA Grant #NAG-3887. One of us (M.G. Spencer) acknowledges support under Presidential Young Investigator Program.

REFERENCES

1. R. Freer (Ed), The physics and Chemistry of Carbides, Nitrides and Borides, Klumer Academic Publishers, Printed in the Netherlands, 1990, pp.563-587.
2. K. Irvine, B. Molnar, M.G. Spencer, Proceeding of Amorphous and Crystalline Silicon Carbide III, Springer-Verlag, to be published.
3. Xiao Tang, Kobchat Wongchotigul and Michael G. Spencer, Appl. Phys. Lett., **58**, 917 (1991).
4. Peizhen Zhou, Michael G. Spencer, Gary L. Harris and Konjit Fekade, Appl. Phys. Lett., **50**, 1384, (1987).
5. Peter T.B. Schaffer and Robert G. Naum, J. Opt. Soc. Am., **59**, 1498 (1969).
6. A. Yariv, C.A. Mead and J.V. Parker, IEEE J. Quantum Electronics, QE-2, 243 (1966).
7. J.H. Ott and T.R. Silker, J. Opt. Soc. Am., **54**, 1442 (1964).
8. J.D. Zook, D. Chen and G.N. Otto, Appl. Phys. Lett., 11, 159 (1967).
9. G.H. Cross, A. Donaldson, R.E. Gymer, S. Mann, and N.J. Parson, Proceedings of the SPIE Symposium on Integrated Optics and Optoelectronics, Boston, 1989 (Society Photo Optical Instrument Engineering, Washington, D.C. (1990), vol. 1177, pp. 79-91.

Optically Induced Near-IR Absorption Lines in 6H-SiC

Th. Stiasny[1], *R. Helbig*[1], *and R.A. Stein*[2]

[1]Institute für Angewandte Physik, Universität Erlangen-Nürnberg, Staudtstr. 7, W-8520 Erlangen, Fed. Rep. of Germany

[2]Siemens AG, Research Laboratories, Scharowskystr., W-8520 Erlangen, Fed. Rep. of Germany

By FTIR transmission spectroscopy of 6H–SiC samples sharp absorption lines in the spectral region of 1300 – 1400 nm have been reported. But in transmission measurements with monochromatic light only in 6H–SiC these lines could not be observed. The optical generation of the corresponding absorption centers can be done by additional monochomatic light focussed on the sample. The excitation spectra for different absorption lines were measured. The cut–off wavelength in the excitation spectra gives the ground state energy of 1.45 eV of the created center relative to the valence band. Comparision with EPR measurements shows that the corresponding absorption center is V^{4+}. Additionally the transient behaviour of the transmission was investigated for different excitation intensities in the temperature range from 15 – 80 K.

1 Introduction

The application of 6H–SiC as a semiconductor material in bipolar device structures depends strongly on the control of minority carrier lifetime which is determined by the concentration of deep impurities, acting as recombination centers (for a review see [1]). One of these deep centers is Vanadium in 6H–SiC which was investigated recently by EPR– and transmission– measurements [2][3]. The absorption, however, can be observed only when the sample is illuminated with white light additionally. This situation normally occurs in Fourier–Transform–Spectroscopy.

We have measured the excitation spectra of the induced absorption. Additionally the increase and the decay of the induced absorption due to these Vanadium related centers with time were investigated.

2 Experimental set up and samples

The experimental set up is shown in fig. 1. A continuous flow helium-cryostat was used for cooling the sample in the temperatur region between 300–20 K. The samples were excited by monochromatic light produced by a 0.27m–Jarrell–Ash Mark X monochromator illuminated by a Xenon XBO–1000W lamp in combination with two suitable colour filters.

To probe the induced transmission we used monochromatic light (spectral resolution 0.2 Å) from a Spex 1702 monochromator illuminated with a tungsten halogen lamp filtered with two RG1000 filters (Schott). The transmitted light was detected by a cooled germanium–detector [4]. To suppress atmospheric absorption the whole set up was flooded by dry nitrogen.

Springer Proceedings in Physics, Vol. 71 **Amorphous and Crystalline Silicon Carbide IV**
Editors: C.Y. Yang · M.M. Rahman · G.L. Harris

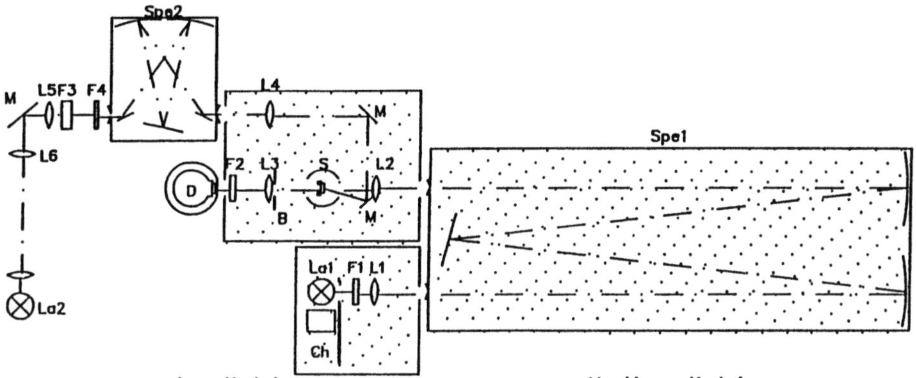

probe light		excitation light	
La1:	tungsten halogen lamp 100W	La2:	XBO 1000W lamp with condenser
L1, L2, L3:	lens f = 5cm Ø= 3cm	L4:	lens f = 8cm Ø= 7.3cm
F1, F2:	filter RG1000	L5:	lens f = 10cm Ø= 5.8cm
Ch:	chopper	L6:	lens f = 20cm Ø= 7.2cm
Spe1:	spectrometer Spex 1702	M:	mirror
S:	sample in cryostat	F3, F4:	filter
D:	detector in cryostat	Spe2:	spectrometer Jarrell–Ash Mark X
⁚⁚⁚ :	nitrogen flooded area		

fig. 1: Experimental set up.

The investigated samples were p–type 6H–SiC single crystals with an electrical active Al–concentration of $N_A = 7 \cdot 10^{17} cm^{-3}$ and compensated at a level of $5 \cdot 10^{16} cm^{-3}$. These concentrations were determined by Hall–effect–measurements. The concentration of V impurities was determined by an emission spectral analysis to approx. 40 ppm. The thickness of the samples was 700 μm and the polished surfaces were perpendicular to the c–axis. All spectra are therefore for light polarization perpendicular to the c–axis.

3 Experimental results

In fig. 2 the effect of an additional illumination on the transmission spectrum of the 6H–SiC sample is demonstrated. We have denoted the induced transmission peaks from 1 to 6 (in a recent publication of Schneider et al. [2] they are named in groups as α, β and γ). The same absorption peaks are shown in fig. 3 with greater spectral resolution. For the transmission lines 1 and 2 the normalized (to the incident excitation photons) transmission is shown for different excitation wavelengths in fig. 4. The same behaviour is observed for the other induced transmission peaks. The excitation spectra are summarized in fig. 5: all induced absorption peaks have the same cut–off wavelength of $\lambda \approx 855$ nm, corresponding to an energy of 1.45 eV. The time dependence of the transmission for peak 1 is shown in fig. 6 for different excitation intensities. To observe the time dependence for greater times and lower excitation intensitiy respectively, the illumination was done with white light (100 W tungsten halogen lamp).

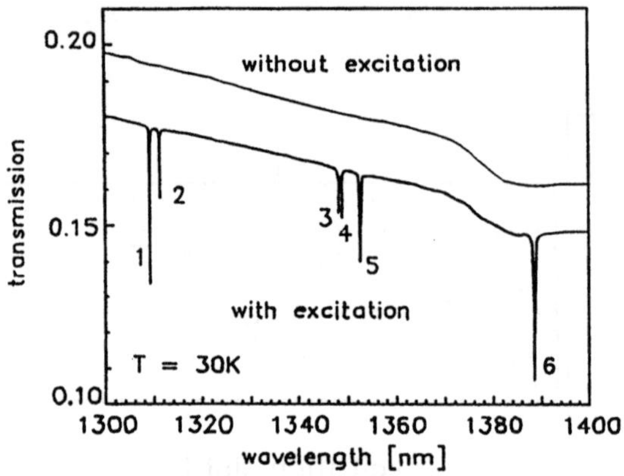

fig. 2: Absorption spectrum of a 6H-SiC crystal (p-type) with and without an additional illumination. The induced absorption could be bleached by heating above 210 K.

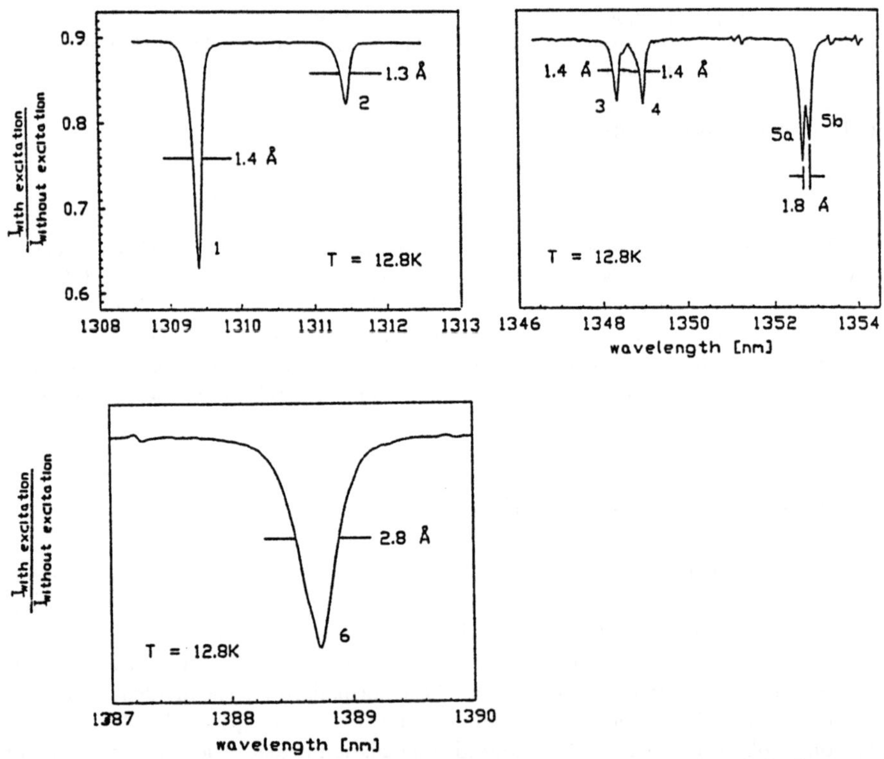

fig. 3: Transmission spectra of the induced absorption peaks 1-6 with higher spectral resolution (spectral resolution 0.2 Å). The observed absorption lines are due to electronic transition in the 3d-shell of the V^{4+}-ion on the different silicon sites. The energy levels in the 3d-shell are split by the crystal field and spin-orbit interaction, for detail see [2],[3].

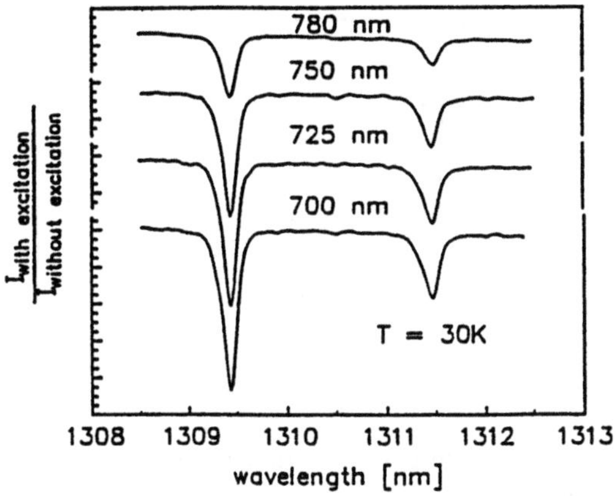

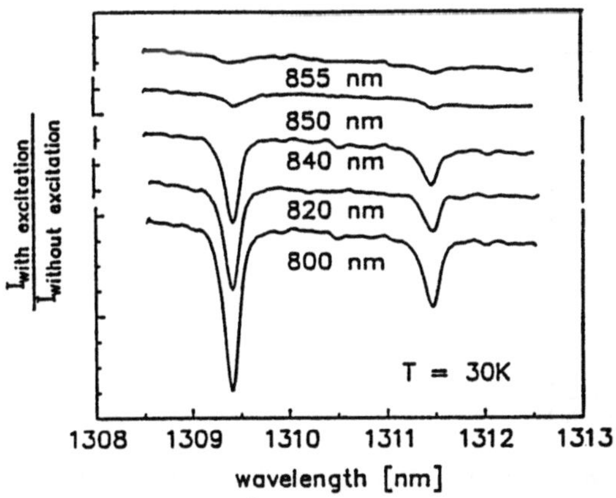

fig. 4: Induced transmission spectra of the absorption peaks 1 and 2 for different excitation wavelengths.

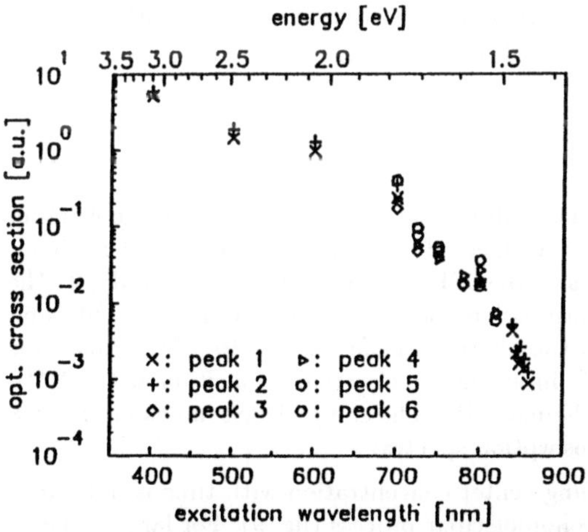

fig. 5: Excitation spectra for the induced absorption peaks 1–6. All absorption lines have the same cut-off wavelength at $\lambda \approx 855$ nm.

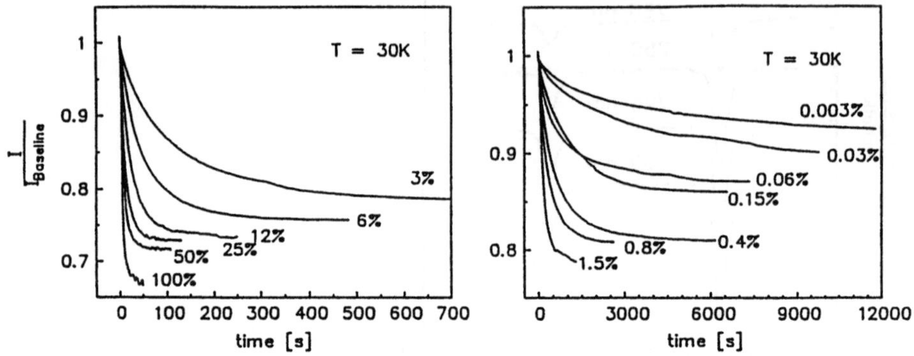

fig. 6: Time dependence of transmission for different excitation intensities of white light, shown is the transmission for peak 1.

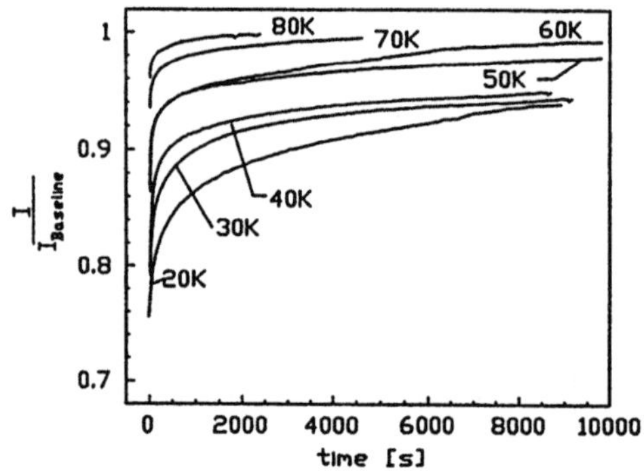

fig. 7: Increase of induced transmission with time after switch off the excitation light for different temperatures.

The increase of the induced transmission after switching off the excitation light was observed at different temperatures and is shown in fig. 7.

4 Discussion

From the excitation spectra for the induced absorption alone we cannot decide whether the corresponding electronic transition is connected with the conduction or the valence band. But if we compare our results with EPR–measurements of [2] and [3], the defect responsible for the induced transmission should be a Vanadium–impurity and the excitation process is correlated to the charge transfer of V^{5+} to V^{4+}. This leads us to the energy scheme given in fig. 8, in which the electronic transition at the V^{4+}–center (peak 1–6) are indicated very schematically. The energy levels indicated in fig. 8 explain the observed excitation and absorption spectra.

The observed decrease of absorbing center concentration with time is not exponentially as one should expect for a monomolecular process (fig. 9). For larger times a linear dependence is observed, indicating a bimolecular process. The shown dependence with time can only be explained if additional traps for the holes are involved.

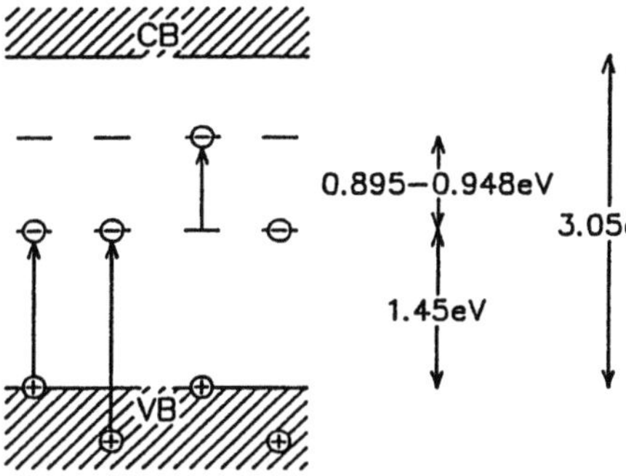

fig. 8: Proposed energetic schema of the different states of the defect responsible for the observed induced transmission.

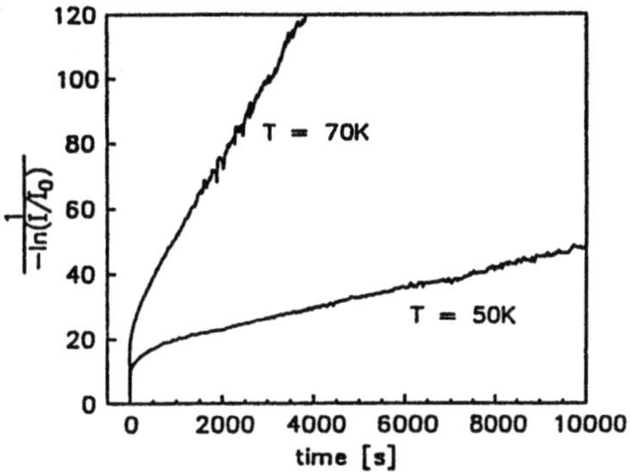

fig. 9: Increase of the induced transmission with time after switch off the excitation. Shown is the reciprocal logarithm of the ratio of the transmitted intensity I to the intensity I_0 in case of maximal absorption which exists for $t = 0$. This is proportional to the reciprocal of the concentration of the absorbing centers .

References

[1] G.Pensl and R.Helbig; Advances in Solid State Physics Vol 30, p.133 (1990).

[2] J.Schneider, H.D.Müller, K.Maier, W.Wilkening, F.Fuchs, A.Dörnen, S.Leibenzeder and R.A.Stein; Appl.Phys.Letters 56, 1184 (1990).

[3] K.Maier, H.D.Müller and J.Schneider; ICDS-16, Bethlehem 1991.

[4] R.Lindner, R.Helbig, R.F.Schulz, W.Rühle, P.Eckstein, P.A.Glasow and J.Walter; Proc of the 13th International Symposium of the Technical Committee on Photonic Measurements (IMEKO), Braunschweig 1987, p.187.

Interference Fringes in the Infrared Reflectance of 6H-SiC Films on 6H-SiC Substrates

M.F. MacMillan, W.J. Choyke, and R.P. Devaty
University of Pittsburg, Pittsburgh, PA 15260, USA

Abstract. We present the results of room temperature IR reflection measurements near the Restrahlen region for epitaxial 6H SiC single crystal films deposited on boule grown 6H SiC substrates. The data were collected at near normal incidence (5°) using a Nicolet 740 FTIR spectrometer. Fabry-Perot fringes presumably caused by the film-substrate interface are seen in many samples. The nature of this interference is difficult to understand since the index of refraction of the films and the substrate should be comparable. We have attempted to model this effect by inserting a thin layer of different materials between the film and the substrate. We conclude that a metallic-like layer of about 250Å between the film and the substrate will reproduce our data. However, at the present time we have no direct experimental evidence for the existence of such a layer.

1. Introduction

Optical reflectance is a simple, nondestructive technique that can yield valuable information about thin film samples. Fabry-Perot fringes caused by interference in a film layer are often seen for samples where the film material and substrate material are different[1][2][3]. These fringes can be used to determine the thickness of the film sample if the index of refraction of the film is known[2]. The contrast of the fringes is ideally determined by the index mismatch between the film and the substrate[4]. Therefore, a system in which the film and substrate are made of the same material ideally should not show any interference fringes.

In this paper we show the results of room temperature infrared (IR) reflectance measurements near the Restrahlen region for epitaxial 6H-SiC single crystal films deposited on boule grown 6H SiC substrates. Fabry-Perot fringes are seen in two samples and are presumably caused by the film substrate interface. This interference effect is difficult to understand since the index of refraction for the film and the substrate should be nearly identical.

Holm, et. al.[1] have reported enhanced contrast of fringes in SiC films deposited by CVD onto Si substrates. They observed fringes near the Restrahlen peak, and they observed that the fringe contrast was unusually large in the frequency region below the Restrahlen peak. They model this system by inserting a 500Å conducting layer of graphite between the film and the substrate. We have attempted to model our system in a similar way, but we have tried a wide variety of possibilities for the layer between the film and substrate.

2. Experiment

The reflectance data were collected on a Nicolet 740 Fourier Transform Infrared (FTIR) spectrometer using a KBr beamsplitter and a DTGS room temperature detector. The resolution of the spectra is 1 cm^{-1}. The data were collected at near normal incidence (5^{0})

Springer Proceedings in Physics, Vol. 71 Amorphous and Crystalline Silicon Carbide IV
Editors: C.Y. Yang · M.M. Rahman · G.L. Harris

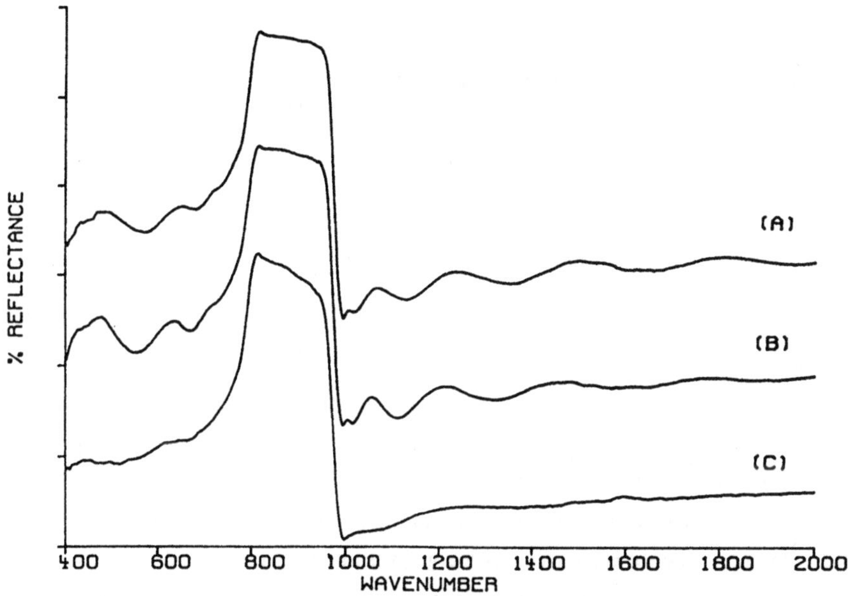

Fig 1: Measured room temperature reflectance spectra for several 6H SiC samples

A) Cree #1: (BO-270-8-NP 2020190-3): 3.0μm p type epi 6H film 1.7-4.0 x 10^{16} cm^{-3}
B) Cree#2 : (BO-270-8 2022490-2): 2.3μm undoped epi film
C) Cree#3 : (BO-270-8): 6H substrate for films A and B

using a Spectra Tech 500 reflectometer. A platinum film deposited on a sapphire substrate was used as the reflectance standard for the measurements[5].

The 6H SiC film samples used in this study were supplied by the Cree Corp.[6]. The first sample, denoted as Cree#1 (BO-270-8-NP 2020190-3), is a 3.0 μm thick, p-type film with doping of 1.7-4.0 10^{16} cm^{-3}. The second film, denoted as Cree#2 (BO-270-8 2022490-2), is 2.3 μm thick and undoped. Both film samples are deposited on the same 6H single crystal substrate, and the sample denoted as Cree#3 (BO-270-8) is a bare piece of that same substrate crystal. The thicknesses of the film samples were determined by comparing the deposition times for the samples to that of a standard film of known thickness. The measured reflectance of all three Cree samples is shown in Figure 1. The spectra have been shifted vertically, and Cree#1 and Cree#2 both show fringes in the reflectance due, presumably, to interference in their film layers. There are no fringes present in the bare substrate. The fringe contrast in the p-type Cree#1 sample is slightly larger than that of the undoped Cree#2 sample, and in both samples the contrast decreases rapidly with increasing frequency. This indicates that the film-substrate interface may be rough enough to cause scattering that would reduce the fringe contrast with increasing frequency[7]. Calculation of the film thicknesses from the fringes gives results that do not agree with the measured values stated above. Fringe thickness measurements yield a thickness of 5.65 $\pm$ 0.15 μm for Cree#1 and 5.83 $\pm$ 0.28 μm for Cree#2. We have used 5.4 μm as the thickness of Cree#1, and 5.6 μm as the thickness of Cree#2 in all calculated spectra because these thicknesses seem to best reproduce the data.

3. Models

We have carried out a series of calculations similar to that done by Holm et al., but we look at the effects of inserting a wide variety of possible substances between the film and substrate. The calculated spectra are generated by standard thin film techniques[8]. The complex, frequency dependent dielectric functions of the various materials are determined from the following oscillator model[9].

$$\varepsilon(\omega)=(n+ik)^2=\varepsilon_\infty\{1-\omega_p{}^2/\omega(\omega+i\gamma)+(\omega_L{}^2-\omega_T{}^2)/(\omega_L{}^2-\omega^2-i\Gamma\omega)\} \tag{1}$$

For 6H SiC, (ordinary ray), we use the following parameters in the oscillator model [10][11]: ω_L = 969.9 cm^{-1}, ω_T = 789.9 cm^{-1}, ε_∞ = 6.5, Γ = 4.76 cm^{-1}, m^* = 0.5m_0 , and γ = 53.1 cm^{-1} , where ω_L and ω_T are the longitudinal and transverse optic phonon frequencies, Γ is the phonon damping constant, m^* is the effective mass of the electron, ε_∞ is the high frequency dielectric constant, and γ is the free carrier damping constant. The parameter m^* together with the free carrier density, N, determine the plasma frequency as $\omega_p = 4\pi Ne^2/m^*\varepsilon_\infty$, where e is the electron charge in esu units. If one considers a conducting layer, then we use a Drude model with the following parameters [12]: m^* = 0.037m_0 , ε_∞ = 6.0, and γ = 18.0 cm^{-1} .

$$\varepsilon(\omega)=(n+ik)^2=\varepsilon_\infty\{1-\omega_p{}^2/\omega(\omega+i\gamma)\} \tag{2}$$

This layer with the parameters listed above is the same as the graphite layer that Holm uses in his calculations. We also consider Si or SiO_2 layers which behave like simple dielectrics in this region. They have indices of refraction of n=3.43 and n=1.43 respectively[13][14].

4. Discussion

We first attempted to model this interference effect by varying the doping levels of the substrate while considering the film undoped. The results of these calculations are compared to the reflectance data in figure 2. Small doping concentrations ($n<10^{17}$) do not have much of an effect on the spectra, and so are not shown here. The range of doping of the substrate, N = 2.0-4.0 10^{18} cm^{-1} ,was chosen because the samples have a green tint indicative of nitrogen impurities of this concentration. Reflectance spectra were also calculated for a doped film on an undoped substrate. We varied the thicknesses used in the calculated spectra, but this did not improve the fit to the data. Neither set of calculations reproduced the data well, and so it is clear that simple differences in the doping between the film and the substrate are not responsible for this interference effect.

We also tried inserting a dielectric layer between the film and substrate. We inserted a silicon layer (n=3.43) for one calculated spectrum and a silicon oxide layer (n=1.43) for the other spectrum. Both of these layers fail to reproduce the observed fringes. Varying the thickness of the dielectric layers does not improve the fit to the data. We conclude that a simple dielectric layer is not the cause of this effect.

The effects of inserting a highly doped SiC layer between the film and sample is shown in figure 3. The layer thickness is 250Å and N=1.0x10^{20}. This layer is essentially metallic.

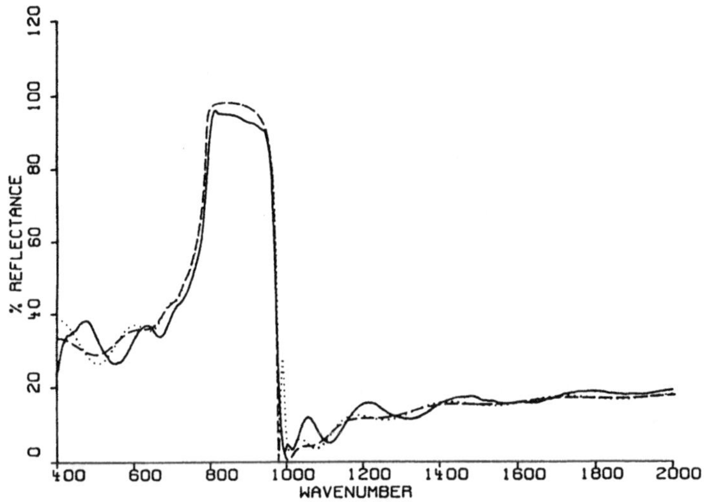

Fig 2: Reflectance of Cree#2, (solid line), compared to calculated spectra for an undoped SiC film on a doped SiC substrate. Dashed line: 5.6μm undoped SiC film on a SiC substrate with $N=2.0x10^{18}$ cm^{-3}. Dotted line: 5.6μm undoped SiC film on a SiC substrate with $N=4.0x10^{18}$ cm^{-3}.

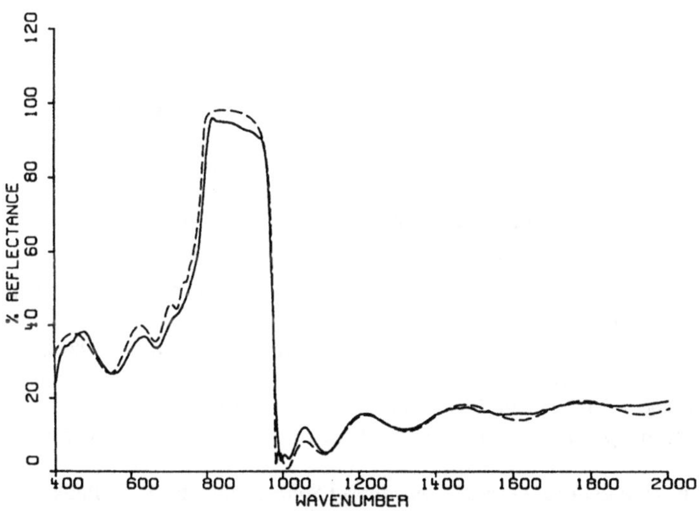

Fig 3: Reflectance of Cree#2, (solid line), compared to calculated spectrum, (Dashed line), for a 250Å SiC layer, ($N=1.0x10^{20}$ cm^{-3}), inserted between a 5.6 μm, undoped SiC film on a SiC substrate with $N=2.0x10^{18}$ cm^{-3}.

The oscillations above and below the Restrahlen region are reproduced by this model. We have done a similar comparison for a conducting layer of thickness 250Å and free carrier density of $N=1.0x10^{19}$ inserted between the film and substrate. This layer also reproduces the data in detail. A similar comparison done with the Cree #1 sample yields comparable fits for both the highly doped SiC layer and the conducting layer.

Although there is no direct experimental evidence for the existence of a metallic-like layer, the reflectance data seems to indicate that such a layer might exists and can have an important effect on the reflectance spectra we have measured. Electron microscopy should be used to determine if in fact an interfacial layer is the cause of the discontinuity in the optical properties.

5. Conclusions

We have measured the room temperature reflectance of single crystal 6H SiC films epitaxially deposited on boule grown 6H single crystal substrates. Fabry-Perot fringes are observed in two samples. Simple differences in the free carrier concentration between the film and substrate do not account for the interference effect. We have modeled our films by inserting layers of different materials between the film and the substrate. Both a metallic layer of graphite, and a highly doped, essentially metallic layer of 6H SiC will reproduce the fringes observed. There is no direct experimental evidence for this layer at the present time.

Acknowledgements

We thank ONR for partial support of this work under grant # N00014-89-J-1517.

References

1) R. T. Holm, P. H. Klein, and P. E. R. Nordquist Jr., J. Appl. Phys. 60 (4), 1479, (1986)
2) J. L. Bradshaw, R. P. Devaty, W. J. Choyke, and R. L. Messham, Appl. Opt. 29, 2367, (1990).
3) D. Huang, D. Mui, and H Morkoc, J. Appl. Phys. 66 (1), 358, (1989).
4) Fringe contrast is defined as $(R_{max}-R_{min})/(R_{max}+R_{min})$, where $R_{max}=\{(n_s-1)/(n_s+1)\}$ and $R_{min}=\{(n_s-n_f)/(n_s+n_f)\}$, and n_s and n_f are the indices of refraction for the substrate and film.
5) J. V. Mantese, W. A. Curtin, and W. W. Webb, Phys. Rev. B. 33, 9897, (1986).
6) We thank Dr. Calvin Carter Jr. for making these samples available for these studies.
7) H. E. Bennett and J. M. Bennett, Phys. Thin Films 4 1 (1967).
8) O. S. Heavens, in *Optical Properties of Thin Solid Films,* Butterworths, London, 1955.
9) S. S. Mitra, in *Handbook of Optical Constants of Solids,* edited by E. D. Palik Academic, New York, 1985, p. 213.
10) W. J. Choyke and Lyle Patrick, J. Opt. Soc. Am. 58 (3) 377, (1968).
11) W. G. Spitzer, D. Kleinman, and D. Walsh, Phys. Rev. 113 (1) 127, (1959).
12) J. W. McClure, Phys. Rev. 112, 715, (1958).
13) D. F. Edwards, in *Handbook of Optical Constants of Solids,* edited by E. D. Palik Academic, New York, 1985, p. 547.
14) H. R. Philipp, in *Handbook of Optical Constants of Solids,* edited by E. D. Palik Academic, New York, 1985, p. 749.

Low Temperature Photoluminescence of SiC: A Method for Material Characterization and the Influence of an Uniaxial Stress on the Spectra

Ch. Haberstroh[1], *R. Helbig*[1], *and S. Leibenzeder*[2]

[1]Institut für Angewandte Physik, Universität Erlangen-Nürnberg, Staudtstr. 7, W-8520 Erlangen, Fed. Rep. of Germany
[2]Siemens AG, Research Laboratories, Scharowskystr., W-8520 Erlangen, Fed. Rep. of Germany

Abstract. Photoluminescence spectra of SiC at 1.9K were investigated. Different polytypes and p- or n-type dopants in the crystals can be determined. Specific lines for N-, Ti- and Al-correlated defects can be observed. The line shape and the intensity ratio of special lines are correlated to the quality of the crystals. After ion implantation a series of characteristic lines for different polytypes are visible.
To obtain a better understanding of the recombination mechanism an external uniaxial stress (up to 3 GPa) was applied to the samples. The different recombination lines show a specific behavior: some of them shift to longer wavelengths, some other to shorter wavelengths and some of them split with an uniaxial stress in $[11\bar{2}0]$-direction.

1. Introduction

With the efforts of SiC boule growth and the fabrication of good quality epitaxial layers there is a need for a suitable characterizations method. Deep temperature photoluminescence (DTPL)-measurements of SiC-crystals are known as a powerful means for material characterization for a long time, hence it is highly sensitive, specific for different kind of impurities and it is a non-destructive technique. Due to the indirect band gap of all SiC polytype crystals most of the excited charge carriers become localized to a impurity or a crystalline defect for radiative recombination. Therefore an understanding of the DTPL-spectrum may reveal information about occurrence and concentration of these centers.

A lot of work was done in this field and today the origin most of the features of the spectra seems to be known [1]. In the present work we report measurements done both to characterize SiC-boule-material and epitaxial layers grown on SiC-substrates. Typical spectra for different polytypes, the influence of n- or p-dopants and the improvement of crystal quality of an epilayer in comparison to the substrate will be shown. Emission lines due to defects generated by quenching of 6H and 15R-SiC polytype samples are observed. In order to identify the different symmetry of different centers an uniaxial stress was applied to the samples under various crystallographic orientations.

2. Experimental procedure

To achieve an effective cooling even under high excitation intensity the samples were immersed in pumped down liquid helium (T=2K). For optical excitation the UV-lines of a Krypton-Ion-Laser were used ($h\nu$=3.51 and 3.54eV; $\approx$100W/cm^2) with a penetration depth of $\approx$18 μm for 6H-SiC e.g. The emission spectra were analyzed by a grating monochromator with a resolution of 0.5 Å. All spectra were recorded photoelectrically by a photomultiplier. Uniaxial stress up to 3 GPa could be pneumatically applied and altered by a piston, while the samples were immersed in superfluid helium.

Samples were grown in a sublimation furnace at Siemens AG, Erlangen [2]. Epitaxial layers were grown by chemical vapor deposition [3]. Crystallographic orientation of the crystals for uniaxial stress experiments was determined by natural sound faces of crystals grown by the Lely-technique [4].

3. Experimental results

Different polytypes of SiC show characteristic line spectra in low temperature photoluminescence. Typical spectra are shown for 3C, 4H, 6H and 15R polytypes (n-type material) in Fig. 1. The lines next to the excitonic band gap (labelled as O, S_o, R_o, P_o, and Q_o lines) are related to excitons bound to neutral nitrogen donors. The corresponding phonon satellites of these lines could

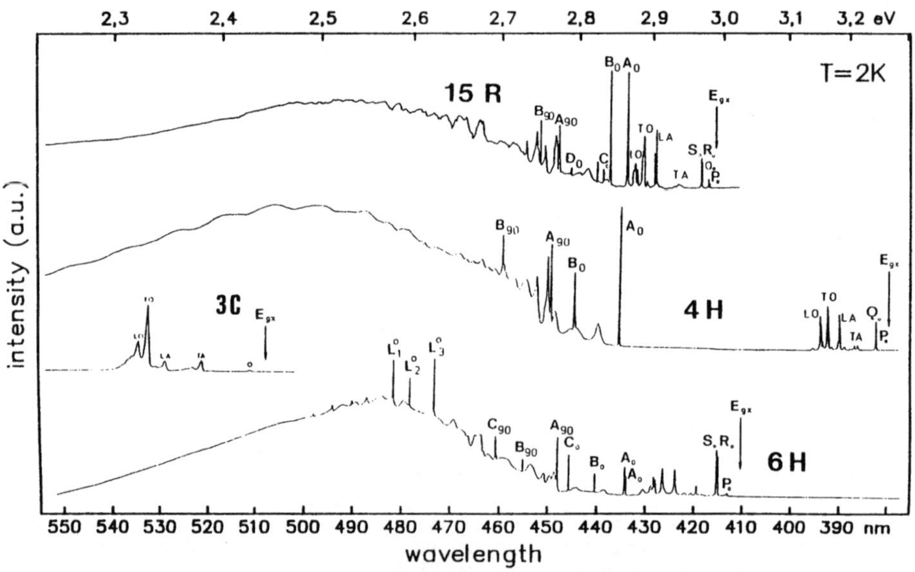

Fig. 1) Typical spectra of 3C, 4H, 6H and 15R-SiC polytypes (n-type material). The lines next to the excitonic band gap (labelled as O, S_o, R_o, P_o, and Q_o lines) are related to excitons bound to neutral nitrogen donors, followed by the corresponding phonon satellites. The lines A_o, B_o, C_o, D_o, A_{90}, B_{90}, and C_{90} indicate the presence of Titanium impurities. The lines $L_1{}^o$, $L_2{}^o$, and $L_3{}^o$ are observed e.g. after ion implantation and annealing (vacancy-complex).

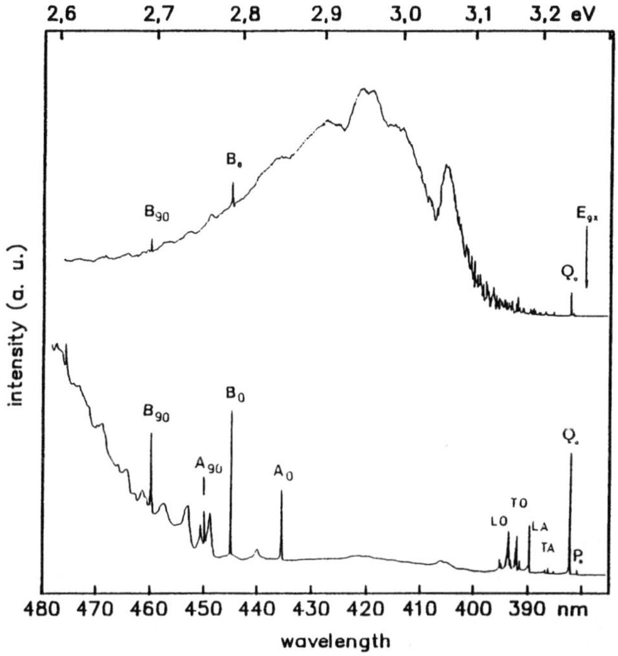

Fig. 2) Low Temperature photoluminescence spectra of 4H-SiC samples: due to different impurities and the concentrations different spectra could be observed.
upper curve: compensated p-type samples (Al-acceptor, N-donor) with pair spectrum;
lower curve: n-type sample with N-related bound exciton lines and Titanium lines.

be observed. The lines A_0, B_0, C_0, D_0, A_{90}, B_{90}, and C_{90} indicate the presence of Titanium impurities, an effective recombination center in SiC. The lines $L_1{}^0$, $L_2{}^0$, and $L_3{}^0$ are observed e.g. after ion implantation and annealing (vacancy-complex) [5].

To obtain p-type SiC usually Al is used as dopant. Together with the omnipresent nitrogen donor in p-type or highly compensated material donor-acceptor-pair recombination becomes the dominant radiative process. This is shown in Fig. 2 for 4H-SiC: in the p-type spectrum additional to the weaker nitrogen and titanium-related lines Q_0 and B_0, B_{90} an array of sharp lines arises at energies near the excitonic band gap E_{gx}, reflecting the discrete distances of donors and acceptors in the SiC-lattice. At higher energies it is merging into a broad continuum according to the low energy separation of large distance pairs. Because DAP recombination is comparatively effective, even small quantities of Al can be detected. This might be useful for the characterization of differently doped thin epitaxial layer were Hall-measurements are not possible, e.g. n^- layer on n^+-substrate (or p^- layer on p^+-substrate).

Intensity and line width of the zero phonon lines are sensitive to the crystal quality. In Fig. 3 near band gap emission-spectrum of both n-type 6H substrate and CVD-epitaxial layer are compared.

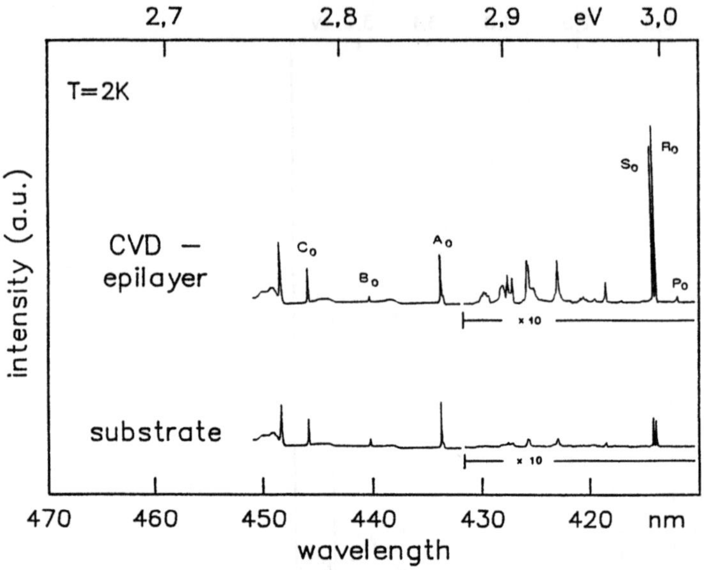

Fig. 3) Comparison of low temperature 6H-SiC photoluminescence of a Lely-grown-substrate (n-type) and a n-type epi-layer on that substrat: The improvement of crystal quality of the epi-layer is visible in higher intensity and line shapes of the S_0, R_0, P_0 N-bound exciton lines (epi-layer thickness $\approx 3\,\mu m$).

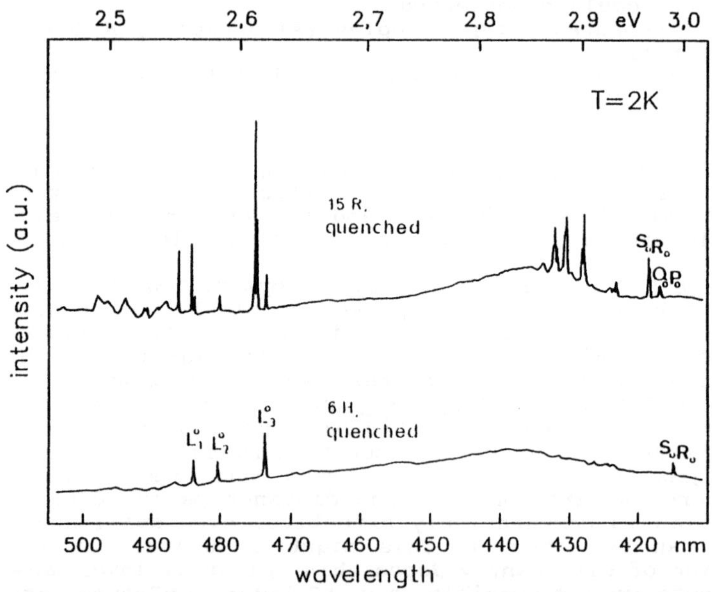

Fig. 4) Photoluminescence of two regions of the same Lely-grown SiC crystal. The crystal was heated to 2000°C and rapidly quenched to room temperature. In the 6H-SiC polytyp region the additional lines L_1^o, L_2^o and L_3^o appear. The spectrum for the 15 R-region shows a number of new lines at corresponding energy.

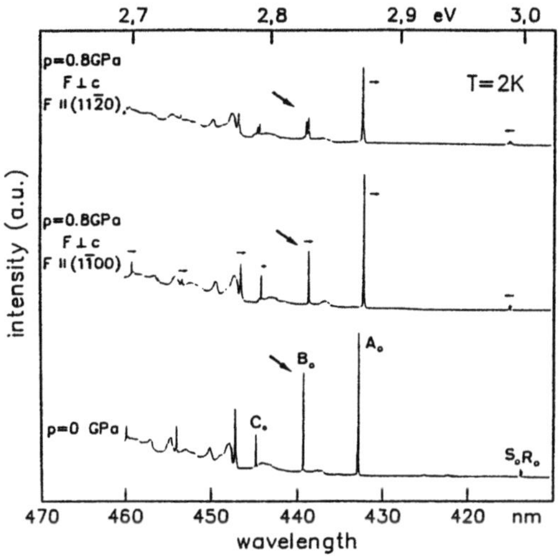

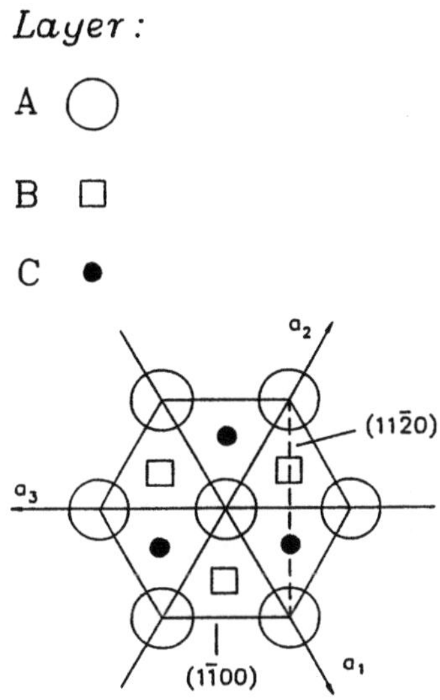

Fig. 5) Influence of an uniaxial stress on the photoluminescence spectra of a 6H-SiC sample:
for a stress ∥ c-axis no splitting could be observed, the emission lines shift only;
for a stress ⊥ c-axis but ∥ to the (1$\bar{1}$00) face only a line shift is oberserved too;
for a stress ⊥ c and ∥ to the (11$\bar{2}$0) face a splitting is observed especially for the B_o and C_o lines.

For the penetration depth of the exciting light is greater than the thickness of the epi-layer (d≈3μm) in the upper spectrum the substrate lines still are visible. Shape and intensity of the nitrogen related lines S_o,R_o and P_o reflect the improvement of crystal perfection of the epilayer.

In respect to SiC-application for semiconductor devices detection of crystalline defects will become a major issue. A defect visible in the DTPL-Spectrum is the complex termed D_1, in 6H-SiC at T=2K consisting mainly of three prominent zero-phonon-lines L_1^o, L_2^o and L_3^o. This complex, probably some form of divacancy, is observed e.g. after ion implantation (Z>1) and subsequent annealing. It is generated also by quenching: a 6H-SiC-crystal containing a 15R-polytype region was heated up to 2000°C and rapidly quenched to room temperature. In the spectra of both polytypes (Fig.4) strong defect luminescence is created the 15R-polytype defect complex according to the greater number of inequivalent lattice sites emits more different lines (the 6H-SiC L_i^o-lines are visible too, because the sample had only a thin layer of 15R-SiC).

In order to obtain a better understanding of the PL-spectrum the influence of an uniaxial stress to the spectrum was observed. Most interest was put on stress perpendicular to the c-axis, hence along the c-axis the crystal is distorted anyway by the crystal field for the different polytypes and no additional symmetry splitting should be expected.
Fig. 5 shows the influence of an uniaxial stress of 0.8 GPa perpendicular to the c-axis. Stress on the (11$\bar{2}$0)-face evokes

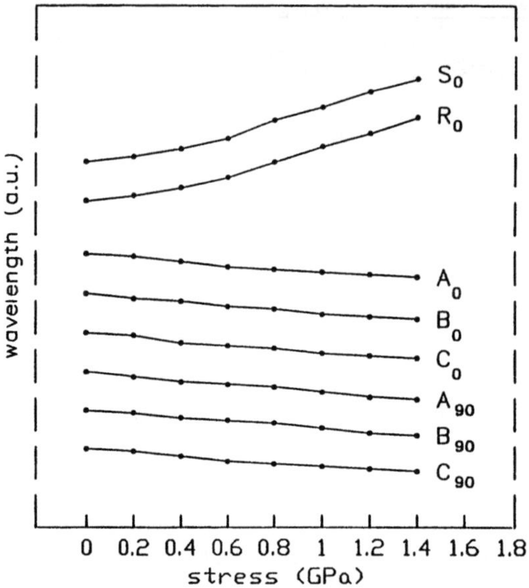

Fig. 6) Line shift versus uniaxial stress on the $(1\bar{1}00)$-face in linear orbitrary units. The Ti-related lines Ao, B_o, C_o, A_{90}, B_{90} and C_{90} shift towards higher energy in contrary the nitrogen-related lines S_oR_o, which show a great shift towards lower energy.

both a energy shift of the different lines and a splitting of the Ti-related lines B_o and C_o. Greater stress in this orientation effects an inversion of the S_oR_o-intensity ratio [6]. If stress is applied to the $(1\bar{1}00)$-face, only a energy shift is visible. All the Ti-related lines shift towards shorter wavelength. The shift of the nitrogen-related lines S_oR_o (line P_o not visible here) is about three times stronger towards greater wavelength (Fig.6). For the understanding of this behavior further work is necessary.

References

[1] W.J.Choyke, L.Patrick, Proc. of the Third Int.Conf. on Silicon Carbide, Miami Beach (1973) p. 261

[2] G.Ziegler et al., IEEE Trans. ED-30, 277 (1983)

[3] S.Karmann et al., Verhandl. DPG (VI) 26, 1156 (1991)

[4] W.v.Münch, J.Electrochem. Soc. 122, 642 (1975)

[5] P.J.Dean et al., Proc. Int. Conf. on Defects ... in Semicond., Nice 1978, publ. London 1979, p. 447

[6] C.Haberstroh, to be published

Part III

Growth and Characterization of Polycrystalline, Microcrystalline, and Amorphous Silicon Carbide

Novel Feedstocks for a-SiC:H Films

Y.-M. Li, B.F. Fieselmann, and A. Catalano

Solarex Corporation, Thin Film Division,
826 Newton-Yardley Road, Newtown, PA 18940, USA

Abstract. Several series of a-SiC:H alloys, with bandgaps of 1.8 - 2.0 eV, have been prepared from both methane and polysilylmethane feedstocks, with and without hydrogen dilution by the PECVD method. Improvements in structural, optical, and photo-transport properties have been observed in these a-SiC:H alloys compared to CH_4 based films without hydrogen dilution. Also observed are improvements in the initial performance of p-i-n single junction cells using a-SiC:H i-layer from the novel carbon feedstocks as well as in the stability of these cells under extended illumination.

1. Introduction

Wide bandgap, amorphous semiconductors such as hydrogenated amorphous silicon carbon alloys (a-SiC:H) are key to the development of high efficiency amorphous silicon based multi-bandgap, multijunction solar cells. Kuwano, et al. [1] have calculated that a conversion efficiency of 24% can be achieved in a triple junction structure using a 2.0 eV absorber in the top junction. Unfortunately, the transport properties of a-SiC:H films deteriorate rapidly as the carbon content (and bandgap) increases. Of even greater concern is the suggestion that the wide bandgap a-SiC:H films may degrade at a higher rate than a-Si:H [2,3].

We must regard the transport, optical and structural properties of a-SiC:H in the context of crystalline silicon carbide and the silicon-carbon phase diagram. Unlike the crystalline silicon-germanium alloys that form a continuous series of solid solutions over the entire range of composition, silicon-carbon alloys have only a single, stable compound SiC. An amorphous silicon carbon alloy must represent a substantial departure from the lowest energy state of the system which would be represented by their disproportionation into sp^3 bonded Si and sp^2 bonded carbon (i.e., elemental Si and C), except at the unique composition SiC. It should not be surprising to learn from Raman measurements, for example, that substantial distortions of bond angles occur in the amorphous alloy [4]. Such structural distortions, arising from the inherent structural instability, are likely the origin of the extensive band tail states that have been found to increase with carbon addition in the a-SiC:H alloys. These states certainly reduce minority carrier lifetime, carrier mobility, and limit further the ability to successfully dope the alloy. Improvements in the properties of a-SiC:H have been associated with the use of high hydrogen dilution of methane (CH_4), as reported for example in the work of Matsuda, et al. [5,6]. The use of hydrogen dilution has been found to improve the photoconductivity of the alloy. The PECVD or glow discharge decomposition of CH_4 leads to CH_n (n = 2,3) bonds in the film which are evident in the infrared absorption spectra. The difficulty in dissociating the CH_n bonds effectively prevents the formation of Si-C bonds and a random network of C-Si bonds. The polymerization of silane which occurs at lower power, relative to methane, also aids in the formation of an inhomogeneous alloy by promoting the formation of Si-Si bonds.

Alternative feedstocks have also been employed in order to improve structure and transport properties. For example, Matsuda, et al. [5,6] prepared films from silylmethane (CH_3-SiH_3) but the properties of these films, although better than CH_4 based films without H-dilution, are not as good as the diluted case (CH_4+H_2). W. Beyer, et al. [7] prepared photoconductive a-SiC:H with bandgap > 2 eV from disilylmethane (DSM, $[SiH_3]_2CH_2$). The concept of forming the C-Si bond in the feedstock is attractive, since it would bypass the unfavorable gas-phase plasma chemistry encountered with the more prosaic feedstocks [8]. Polysilylmethanes represent such a class of feedstocks, however the syntheses of the higher order compounds trisilylmethane ($(SiH_3)_3$ CH) and tetrasilylmethane ($(SiH_3)_4C$) are difficult [9-12].

In this paper, we review the results of a comparative study of a-SiC:H prepared using CH_4, disilylmethane (DSM), and trisilylmethane (TSM) as the source of carbon. In addition to insights into the structure as inferred from infrared measurements, we report on the transport properties such as ambipolar diffusion length as measured by the steady state photocarrier grating (SSPG) method [13], dark and photo-conductivity, and optical properties such as the Urbach Energy and mid-gap absorption determined by Photothermal Deflection Spectroscopy (PDS). The performance and stability of p-i-n devices is presented.

2. **Experimental**

Trisilylmethane (TSM) and disilylmethane (DSM) were synthesized as described in the literature (11-12). The gases were mixed with silane prior to use. The details of the dependence of film properties on deposition conditions will be described elsewhere. Control samples of a-SiC:H were prepared from methane-silane mixtures with and without hydrogen dilution. H-dilution, when employed, was near 25:1 (H_2: CH_4 + SiH_4). The substrate temperatures were 270°C (CH_4) and 300 - 310°C (CH_4 + H_2), respectively. The rf power density was near 120 mW/cm^2. A film thickness of ~ 1 μm was used to avoid surface effects. Single junction a-SiC:H p-i-n devices were fabricated on tin oxide, and coated with Al back contact. Only the i-layer (with bandgap near 1.90 eV) of the devices was prepared using the source gases under test; the diborane doped p-layers were all prepared from undiluted silane-methane mixtures. The optical absorption, measured by PDS, was used to obtain the defect absorption, the Urbach energy, and the optical bandgap which was empirically taken as the energy corresponding to an absorption coefficient of 2×10^3 cm^{-1}. Infrared spectra were taken for films on crystalline Si substrates. The mobility lifetime product, $\mu\tau$, of the majority carrier (electrons), was computed from photocurrent measured with a flux of 1-2 x 10^{17} $cm^{-2}S^{-1}$ of semi-penetrating photons at 25°C. Room temperature dark conductivity was either measured directly when possible, or extrapolated (for the very resistive samples) from high temperature data by an Arrhenius plot.

3. Results and Discussion

The infrared transmission spectra in Figure 1 show the substantial effect of the feedstock on the bonding structure of a-SiC:H with 1.9 eV bandgap. Although device quality a-Si:H films are characterized by a single stretch peak at 2000 cm^{-1}, the addition of CH_4 to the gas phase results in a film in which absorption at 2080 cm^{-1} dominates the spectra, indicating a large degree of dihydride bonding. The chemical shift due to adjacent carbon atom is believed to be small at the low C concentrations employed here as discussed in reference [14-15]. The 2000 cm^{-1} peak increases in intensity relative to the 2080 cm^{-1} peak in the order CH_4, DSM, CH_4+H_2, TSM, suggesting a trend of improving film quality. In a similar fashion the extent of CH_n (n = 2,3) bonding decreases as indicated by the absorption at ~ 2900 cm^{-1} and ~ 1400 cm^{-1}. This suggests that the paradigm of solid structural modification based on the bonding configuration of the feedstock is valid. This is an important first step in obtaining control of the structure and film properties.

The optical properties of the samples also show important improvements that are coincident with these suggested structural changes. Figure 2 plots the Urbach energy taken from the slope of the absorption edge measured by PDS. The bulk of the data for CH_4, CH_4+H_2, and DSM based films lie roughly along the line (a fit to the CH_4 data) shown in the figure (except for CH_4+H_2 films with gap near 1.91 eV), indicating an increase in the tail density of states with increasing bandgap, at least over the narrow range studied. The Urbach energy measured for films prepared from TSM is markedly lower indicating a narrower distribution of states; furthermore, only a slight increase in Urbach energy is evident as bandgap increases. The Urbach energy is often assumed to be largely a measure of state distribution associated with the valence band-tail in a-Si:H based alloy, so that a narrower distribution would be expected to lead to an improvement in the transport properties of holes. Similarly,

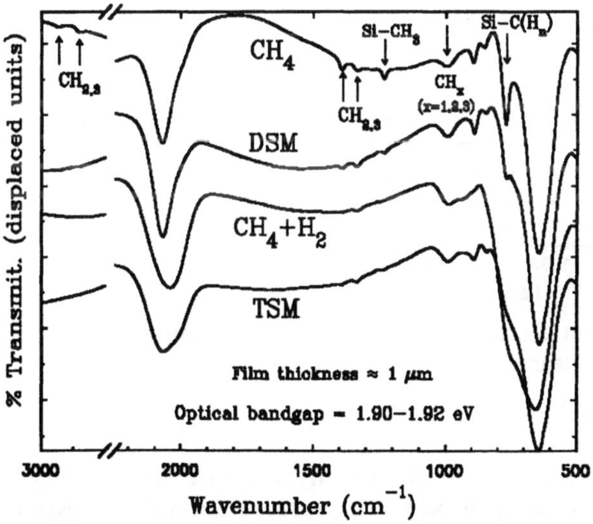

Figure 1. Infrared transmission spectra for a-SiC:H alloys with near 1.91 eV bandgap.

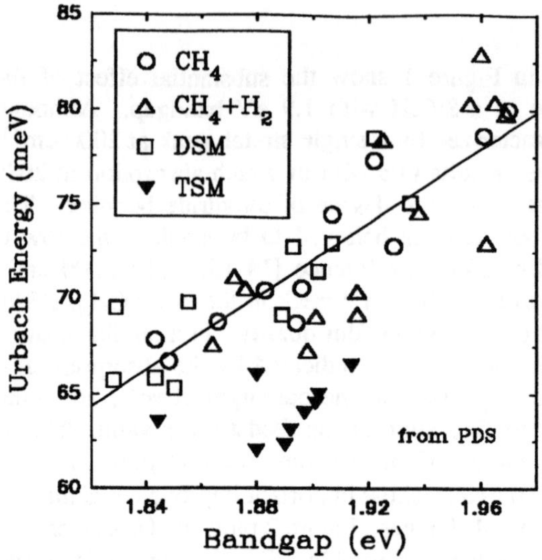

Figure 2. Urbach energy versus bandgap for a-SiC:H films based on different feedstocks.

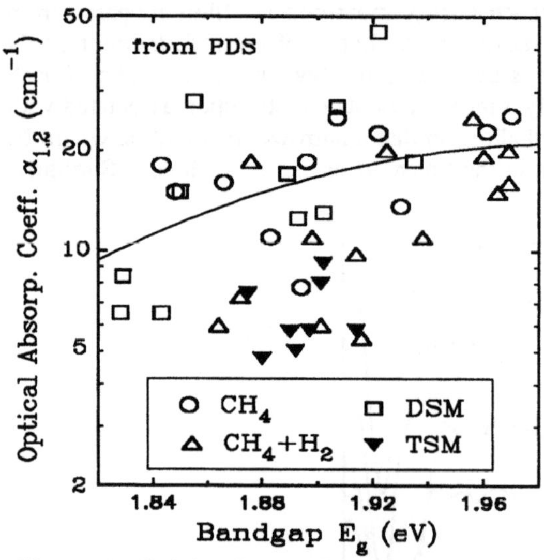

Figure 3. Sub-bandgap defect absorption at 1.2 eV versus bandgap for a-SiC:H prepared by different means.

the defect related absorption taken (somewhat arbitrarily) at 1.2 eV, shown in Figure 3, is lower for TSM than the other feedstocks. This reduced absorption is taken as a reduction in the level of mid-gap defects for the TSM. A reduction in $\alpha_{1.2}$ for CH_4+H_2 films (relative to CH_4 films) can also be seen in Figure 3.

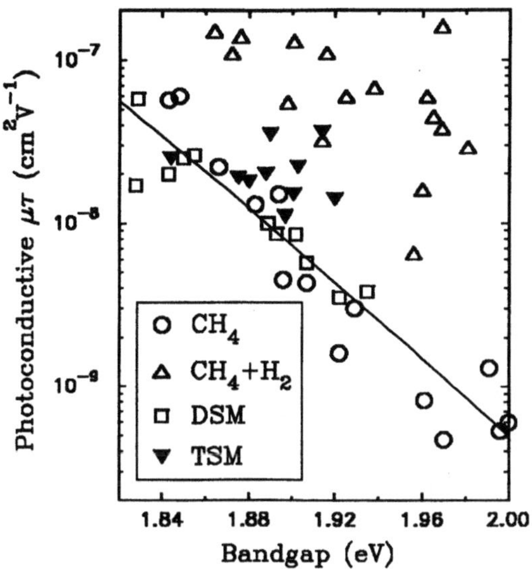

Figure 4. Electron mobility-lifetime product versus bandgap for several series of a-SiC:H alloys.

The transport properties confirm the improvement in the materials implied by the structural changes. Figure 4 plots the $\mu\tau$ product of the films as a function of optical gap for the films under study. Both DSM and CH_4 grown films show a linear drop off of log $\mu\tau$ with increasing bandgap. We have not been able to reproduce films with high photo-conductivity from DSM as reported in Ref. 7 for wider gap alloys. In contrast, CH_4+H_2 films are much better, achieving $\mu\tau \sim 10^{-7}$ (near the value for a-Si:H) even at a bandgap of 1.9 eV. TSM films show a lower $\mu\tau$ product than CH_4+H_2 films, but the bulk of the data are at a higher level than the corresponding CH_4 and DSM films. The 1.9 eV TSM films show a dark conductivity activation energy of 0.95 eV indicating the position of the Fermi level is close to mid-gap, compared to the CH_4+H_2 film where the activation energy is found to be 0.1 eV or so lower. This suggests the Fermi level is 0.1 eV closer to the conduction band than in the TSM (and DSM) films. In fact, the CH_4+H_2 films are the least resistive among the four groups. Figure 5 shows the variation with bandgap of photosensitivity, signified by the ratio $\mu\tau/\sigma_d$ (dark conductivity), for these alloys. Because of their much lower σ_d, the TSM and DSM based films have higher photosensitivity in spite of their lower $\mu\tau$ compared to the CH_4+H_2 films.

The ambipolar diffusion length determined by SSPG is shown in Figure 6. A comparison of the values, which represent the transport properties of primarily holes, reveals a far higher diffusion length for TSM films than any of the other films, although the comparison is most favorable to TSM below 1.9 eV. Below 1.88 eV, the diffusion length, in the neighborhood of ~ 1150Å, is comparable to that of unalloyed a-Si:H.

Although the CH_4+H_2 based alloys show very high electron $\mu\tau$ in Figure 4, their hole diffusion length has not improved as remarkably by comparison. In fact, some

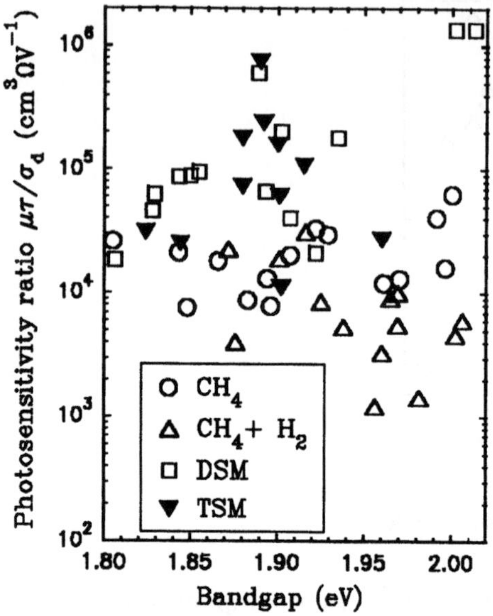

Figure 5. The variation with bandgap of the ratio of $\mu\tau$ to dark conductivity for a-SiC:H films from various recipes.

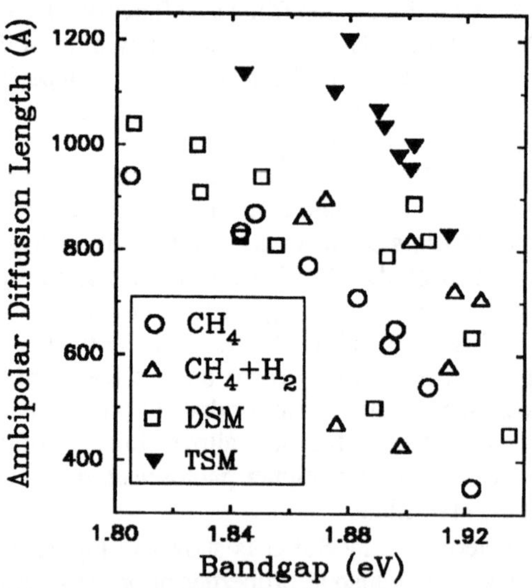

Figure 6. Hole diffusion length versus bandgap for a-SiC:H alloys from different feedstocks.

films with high $\mu\tau$ exhibit rather poor hole diffusion length. This, combined with their relatively low ratio of $\mu\tau/\sigma_d$ indicated in Figure 5, suggests that their improved $\mu\tau$ may be partly due to Fermi level or some kind of sensitization effect rather than entirely due to gap state reduction. Overall, the DSM films have longer ambipolar diffusion length than the CH_4 films, an observation relevant to our discussion on device performance below.

Representative device results obtained using simple p-i-n devices with a-SiC:H i-layer (bandgap near 1.9 eV) from different recipes are summarized in Table I. An unalloyed a-Si:H cell is included for reference. The i-layer thickness for the cells studied was varied from 1000 to over 2000 Å. It should be noted that no attempt has been made to optimize the a-SiC:H device structures through interface grading, etc., nonetheless some interesting observations can be made. DSM yields devices with the best initial performance, particularly fill factor, followed by TSM. It is difficult to understand why the DSM based cells (initially) perform so well given the rather poor properties of the DSM based films (with the exception of mild improvement in hole diffusion length). Perhaps a better p/i interface is formed due to the lower rf power used or the absence of CH_3 radicals found in CH_4 plasma. Thin cells using TSM show good parameters when made with a CH_4 or DSM buffer (without H_2 dilution), at the p/i interface. More work on TSM based cells will be needed to see if the good transport properties can translate into superior device performance. The CH_4+H_2 based devices are particularly poor (low V_{oc} and disappointing FF) despite the otherwise excellent transport properties. As expected, CH_4 based i-layer yields

Table I Representative a-SiC:H Single Junction Cells

Cell I.D.	Carbon Source	I-Layer Thick-ness (Å)	Band-gap (eV)	V_{oc} (V)	J_{sc} (mA/cm^2)	FF	η (%)
10515ZD[(a)]	None	2,200	1.75	0.83	11.0	0.72	6.6
S10516ID	CH_4	1,300	1.90	0.90	5.4	0.53	2.6
S10725JD[(a)]	CH_4	2,000	~1.91	0.91	5.9	0.56	3.0[(b)]
S10515BD[(a)]	CH_4+H_2	1,800	1.90	0.77	7.4	0.56	3.2
S10724HD[(c)]	CH_4+H_2	1,300	1.90	0.86	7.6	0.57	3.6
S10515CD[(a)]	TSM	2,000	1.90	0.86	7.4	0.51	3.2
S10516GD[(c)]	TSM	1,100	1.89	0.90	6.0	0.70	3.8
S10523YD	DSM	2,300	~1.93	0.92	7.8	0.67	4.8
S10710UD[(a)]	DSM	2,300	~1.89	0.89	9.6	0.59	5.0[(b)]

(a) Cells used in stability comparison in Figure 7.
(b) ITO/Ag rear contact (higher current) and thinner p-layer.
(c) CH_4 buffer (without H) at p/i interface ~ 100Å.

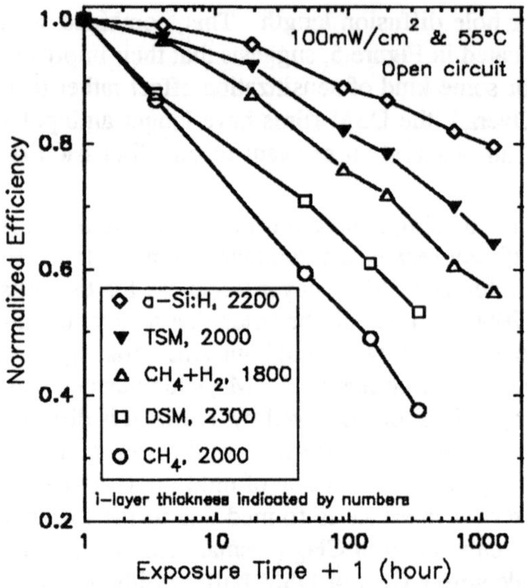

Figure 7. Efficiency dependence on illumination time for p-i-n single junction cells using a-SiC:H i-layer with 1.9 eV bandgap prepared by different means (see Table 1 for initial cell parameters).

the worst devices among the four groups. We believe the poor initial efficiency of the $CH_4 + H_2$ based cells (and the TSM cell S10515CD in Table I which had hydrogen dilution at the p/i region) may be due to hydrogen bombardment at the p/i interface. The resulting damage or etching, possibly facilitated by the high substrate temperature, may lead to a high effective recombination rate. Damage to the p-layer may also reduce majority carrier mobility leading to "blocking" behavior. There also exists the possibility that the $CH_4 + H_2$ based i-layer studied here is inhomogeneous and has a low-bandgap component (or phase) which, among other things, would limit the V_{oc} below the value expected from the nominal bandgap of the i-layer.

The light induced changes in devices of comparable thickness (near 2000 Å) from different recipes are summarized by a plot of the normalized conversion efficiency vs. exposure time as shown in Figure 7. The initial parameters of the devices are listed in Table I. Only degradation data from cells without any p/i buffer are shown since the purpose here is to compare the relative stability of the i-layers rather than the performance of the cells. The relative degradation rate of thinner cells (1100-1300 Å) from various feedstocks are qualitatively similar to that shown in Figure 7. Not surprisingly, the a-Si:H cell exhibits the least amount of degradation despite the higher photogeneration rate due to its narrower bandgap. What is notable is the reduction in instability of the TSM and $CH_4 + H_2$ devices, which are nearly within the experimental error of the measurement, compared to the far poorer stability of the CH_4 based cell. The DSM based cells appear to show only slight improvement in stability over CH_4 based cells as judged by the data from several cells. The better stability of the TSM and $CH_4 + H_2$ devices appears to be related to the low levels of SiH_2 bonding (and reduced CH_n density) evident in the infrared spectra when compared to the CH_4 and DSM based materials which exhibited rather strong SiH_2 absorption at 2080 cm^{-1}. The broad Urbach edge in the CH_4 and DSM based a-SiC:H alloys may be indicative of higher density of potential sites for photo-induced metastable defects. It appears hopeful that thin a-SiC:H (about 1000 Å) with 1.9 eV bandgap can be made reasonably stable to be used in the top junction of a-Si:H based triple junction solar cells.

4. Summary

We have studied the optical and transport properties of a-SiC:H alloys prepared from novel carbon sources TSM and DSM, and from CH_4 with high H_2 dilution, in comparison with those obtained from the reference films derived from conventional CH_4 + SiH_4 mixture without H_2 dilution. The TSM based alloys show significantly improved bonding structure, lower sub-bandgap optical absorption, and better electron and hole phototransport. The CH_4 + H_2 based films also show the above improvements to various degrees. The enhancement in $\mu\tau$ reported earlier (5,6) is the most notable. However, the data (e.g., Urbach energy and hole diffusion length) show considerable scattering, and the photosensitivity $\mu\tau/\sigma_d$ is relatively low, for the CH_4 + H_2 specimens. The only improvements observed for DSM based films are somewhat greater ambipolar diffusion length and higher $\mu\tau/\sigma_d$. Preliminary results from simple p-i-n cells using a-SiC:H i-layer show that DSM and TSM lead to good initial cell parameters, while the CH_4 + H_2 recipe has not yet worked well presumably due to hydrogen damage at or near the p/i interface. The stability of the cells improves remarkably when using TSM and CH_4 + H_2, and slightly when using DSM. Further work is needed to optimize the device structure and the i-layer deposition conditions for better performance of a-SiC:H based cells.

Acknowledgements

This work was partially supported by NREL under Subcontract #ZM-0-19033-1.

References

[1] Y. Kuwano, M. Ohnishi, H. Nishiwaki, S. Tsuda, T. Fukatsu, K. Enomoto, Y. Nakashima, and H. Tarui, Proc. of the 16th IEEE PVSC, San Diego, CA, 1982, p. 1338.

[2] M. Ohsawa, A. Asano, T. Yoshida, T. Hama, M. Nagano, Y. Ichikawa, and H. Sakai, Tech. Dig. of PVSEC-5, Kyoto, Japan, 1990, p. 989.

[3] M.S. Bennett, S. Wiedeman, and K. Rajan, Mat. Res. Soc. Symp. Proc., Vol. 149 (MRS, Pittsburgh, 1989), pp. 577-582.

[4] E. Bustarret, J. Non-Cryst. Sol. 114 (1989), p. 13.

[5] A. Matsuda, T. Yamaoka, S. Wolff, M. Koyama, H. Imanishi, H. Kataoka, H. Matsuura, and K. Tanaka, J. Appl. Phys. **60** (11), 1986, p. 4025.

[6] A. Matsuda and K. Tanaka, J. Non-Cryst. Sol. 97 & 98 (1989), p. 1367.

[7] W. Beyer, R. Hager, H. Schmidbaur, and G. Winterling, Appl. Phys. Lett. **54** (1989), p. 1666.

[8] U.S. Patent # 4,690,830 by C.R. Dickson and D.E. Carlson (1987).

[9] R. Hager, O. Steigelmann, G. Muller, and H. Schmidbaur, Chem. Ber. 122 (1989), p. 2115.

[10] R. Hager, O. Steigelmann, G. Muller, H. Schmidbaur, H.E. Robertson, and David, W.H. Rankin, Angoe. Chem., Int. Ed. Engl. 29 (1990), p. 201.

[11] R. Muller and G. Seitz, Chem. Ber. 91 (1958), p. 22.

[12] E. Amberger and H.D. Boeters, Chem. Ber. 97 (1964), p. 1999.

[13] D. Ritter, K. Weiser, and E. Zeldov, J. Appl. Phys. **62** (1987), p. 4563.

[14] See, e.g., R.R. Koropecki, F. Alvarez, and R. Arce, J. Appl. Phys. 69 (1991), p. 7805.

[15] A.H. Mahan, R. Raboisson, D.L. Williamson, and R. Tsu, Solar Cells 21 (1987), p. 117.

Studies of 1,3-Disilacyclobutanes as Single-Source CVD Precursors to Silicon Carbide

D.J. Larkin[1] *and L.V. Interrante*[2]

[1]NASA Lewis Research Center, Cleveland, OH 44126, USA
[2]Rensselaer Polytechnic Institute, Troy, NY 12180, USA

Abstract. Substituted 1,3-disilacyclobutanes were studied as potential low temperature single-source CVD precursors to silicon carbide. These studies were carried out in a cold-wall LPCVD system maintained at a total reactor pressure of 1.0 Torr using a mass flow controlled argon carrier gas. The precursor was decomposed on resistively heated Si(100) substrates at temperatures ranging from 700°C - 1100°C. Gas chromatography-Fourier transform infrared spectroscopy and quadrupole mass spectrometry were used to identify the volatile pyrolysis byproducts from each precursor. The decomposition onset temperature was separately determined for each precursor by monitoring the major pyrolysis byproducts, by using quadrupole mass spectrometry, while slowly increasing the substrate temperature. Scanning and transmission electron microscopy were used to determine the cross-sectional and surface morphology and also the deposition rate, which ranged from ca.0.035 μm/h to 12.0 μm/h.

1.0 Introduction

Research on the development of silicon carbide as a high temperature and high power semiconductor is continuing to gain momentum internationally.[1,2] With the onset of commercially available 6H-SiC 1" wafers and the demonstration of blue LED operation at 350°C,[3] it is evident that this technology will continue to develop as a viable source of semiconductor material for electronic device manufacture. A critical aspect of successful device manufacture is the ability to consistently produce high quality epitaxial layers. Stoichiometry control is typically accomplished with the use of separate mass flow controlled source gases, one for Si and another for C. Although there have been some reports[4] of single-source precursors to SiC, a flow of hydrogen is usually required for efficient removal of excess carbon or chlorine, in the case of chlorocarbosilanes, and substrate temperatures typically exceed 1100°C.

Alternatively, we have undertaken a systematic study of various derivatives of 1,3-disilacyclobutane as a single-source CVD precursor to silicon carbide. These compounds have been shown to produce stoichiometric SiC in an inert atmosphere of argon at a relatively low deposition temperature (as low as 850°C).[5] The correct SiC stoichiometry is "designed" into the four-membered $-(SiC)_2-$ ring structure, which contains a considerable amount of ring-strain and aids in obtaining a relatively low decomposition temperature. Upon decomposition, this ring opens with the possible formation of reactive species, where both the silicon and carbon can serve as reactive atomic centers. This assists in the incorporation of both the silicon and carbon into the growing film, thereby utilizing more effectively both the Si and C of the disilacyclobutane ring. We report here on selected aspects of this investigation of disilacyclobutane, and derivatives thereof, which includes a description of the molecular structures for each derivative, pyrolysis byproducts detected by GCFTIR and QMS, and film analysis by SEM, TEM, and FTIR.

Editors: C.Y. Yang · M.M. Rahman · G.L. Harris

2.0 Experimental

The disilacyclobutane derivatives I, II, and III were chemically synthesized either by directly using known literature preparations[6-9] or modifications thereof. Each compound was purified by fractional distillation at atmospheric or reduced pressure, in a nitrogen environment. The purified compounds were characterized by using ^{1}H, ^{29}Si, and ^{13}C Fourier transform NMR (FTNMR); FT infrared spectroscopy (FTIR); and quadrupole mass spectrometry (QMS). All three compounds are colorless liquids which are non-pyrophoric, but slowly oxidize when exposed to oxygen.

A schematic of the all stainless, turbo-pumped, cold-wall, hot-stage, low pressure CVD (LPCVD) apparatus that was used in the pyrolysis studies of these compounds is shown in Figure 1. Attached to the LPCVD apparatus is a QMS system which aided in the determination of the initial decomposition (onset) temperature for each compound, as described previously.[10] For this determination, the mass peaks (m/e) which are characteristic of initial precursor decomposition (methane m/e = 15, 16; hydrogen m/e = 2, as determined in preliminary experiments) were monitored as the silicon substrates were slowly heated from room temperature.

The substrates (silicon wafer sections) were resistively heated via a DC current. The "as received" Si(100) substrates (Wacker Siltronic Co.; ρ=1.1 Ω cm; p-type; 525 μm thick) were scored on the unpolished face and then cleaved into 1.4 cm X 6.5 cm rectangular shaped pieces. Typically, two to three pieces were then connected in a series circuit by using tungsten clips and tantalum leads and subsequently loaded into the stainless steel reactor. A current-limited DC power supply was used as the power source to resistively heat the silicon and maintain the surface at temperatures ranging between 650°C-1150°C. The substrate surface temperature was determined by using the combination of an optical pyrometer and a previously calibrated thermocouple fastened to the silicon substrate.

Prior to each experiment, the entire LPCVD system was evacuated and backfilled using purified argon. A base pressure of ca. 5 X 10^{-8} Torr was reproducibly achieved with the apparatus walls maintained at 80°C and the silicon substrates at > 800°C. The isothermal CVD experiments were typically carried out with the silicon substrates at the desired deposition temperature prior to the introduction of precursor into the reactor. The growth parameters include; a 100 sccm flow of argon, a 1 Torr of total

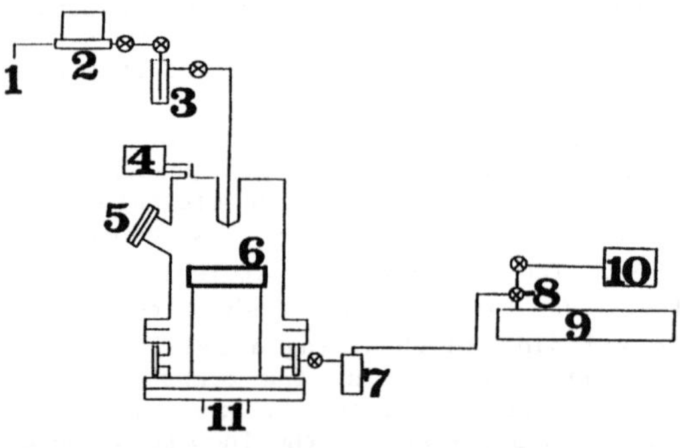

Figure 1: A schematic of the LPCVD system with an attached QMS; 1.) Carrier gas inlet, 2.) Mass flow controller, 3.) Stainless bubbler containing precursor, 4.) Capacitance manometer, 5.) Sapphire view port, 6.) Silicon substrates, 7.) Series of 3 cold traps, 8.) Sampling leak valve, 9.) Quadrupole mass spectrometer, 10.) Mechanical pump used to maintain the gas sampling flow, 11.) to the main L-N_2 cooled trap and pumping system.

reactor pressure, and a bubbler temperature corresponding to a 2 Torr precursor vapor pressure (21°C for I; -13°C for II; -42°C for III).

The volatile byproducts were condensed in a stainless steel, liquid-N_2 cooled trap, located downstream of the heated substrates, for later analysis by using Gas Chromatography-FTIR (Perkin-Elmer 1800 FTIR with an MCT detector; Varian 1500 GC using a VZ-10 column).

The resulting silicon carbide coated Si(100) pieces were analyzed for residual Si-H and C-H absorptions, indicative of partial decomposition of the precursor, by using FTIR in both the specular reflectance and transmission mode. An uncoated, unheated Si(100) piece was used for background subtraction.

Scanning electron microscopy was routinely used to examine the surface morphology of the deposited SiC films, whereas cleaved SiC/Si(100) samples were used to determine the film thicknesses and cross sectional morphologies. TEM was also employed to examine the SiC-film/Si(100) interface of a thin film (ca. 80 nm) which resulted from the decomposition of precursor (I) at the previously reported[10] deposition onset temperature (755°C).

3.0 Results And Discussion

Chemical Structure of Precursors:

The molecular structure for 1,3-disilacyclobutane (III) and for the derivatives (I) and (II) are shown in Figure 2. In particular, note the four-membered $-(SiC)_2-$ "core-structure" common to all three of the precursors. This four-membered ring is of particular interest because of 1). the 1:1::Si:C stoichiometry "designed" into the ring structure 2). the significant amount of ring strain energy contained in the ring 3). the availability of both Si and C as very reactive centers, as a result of ring opening upon precursor decomposition.

The correct Si/C ratio for SiC is designed into the four-membered ring, which potentially eliminates the requirement for stoichiometry control through the use of mass flow controllers. By choosing both the proper substituents contained on the four-membered ring (i.e. methyl, hydrogen, chloro, etc.), which can be eliminated under the appropriate deposition conditions (e.g., methyl eliminated as methane), it is possible to produce stoichiometric SiC films.[5,10] The relatively lower decomposition

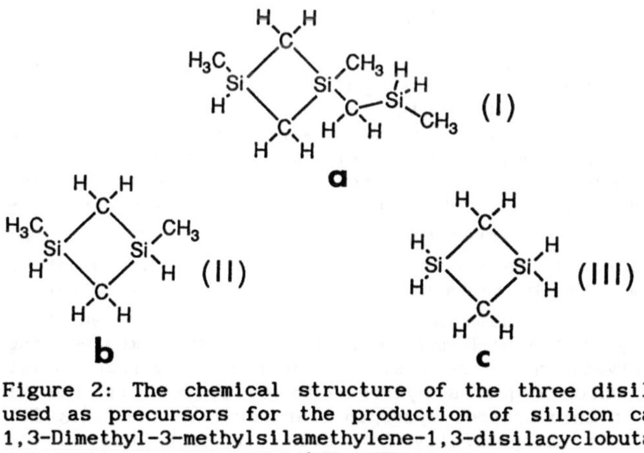

Figure 2: The chemical structure of the three disilacyclobutane compounds used as precursors for the production of silicon carbide thin films: a.) 1,3-Dimethyl-3-methylsilamethylene-1,3-disilacyclobutane; b.) 1,3-Dimethyl-1,3-disilacyclobutane; c.) 1,3-disilacyclobutane.

temperatures associated with the disilacyclobutanes is, in part, related to the strain contained in the four-membered ring. This ring strain arises because the four atoms contained in the ring are forced to adopt bond angles smaller, and therefore less stable, than the tetrahedral (109.5°) bond angle. For example, the ring strain energy has been calculated for the tetra-methyl substituted disilacyclobutane derivative as 24.6 ± 2.6 Kcal/mol.[11]

The remaining important feature of this class of compounds is the reactivity of both the Si and C atoms contained in the ring. Results of work by Auner, et al.[12] have indicated that the disilacyclobutane ring can thermally decompose to yield reactive intermediate fragments such as the π-bonded silene ($R_2Si{=}CH_2$)[12,13] and the diradical species, silylene ($R_2\dot{S}i{-}\dot{C}H_2$). In both cases the Si and C atomic centers are reactive, which aids in the incorporation of both the Si and C into the growing film. This helps to preserve the stoichiometry designed into the $-(SiCH_2)_2-$ four-membered ring structure.

Pyrolysis Byproducts:

Each compound was separately used as a LPCVD precursor in a set of similar experiments. After each set of experiments, involving one of the three precursors and carried out at different deposition temperatures, the byproducts that had condensed in the L-N_2 cooled trap (downstream of the reactor) were analyzed by GCFTIR. The non-condensable byproducts were analyzed during the course of the LPCVD experiment by using the QMS. The major pyrolysis byproducts from compounds (I) and (II) were methane and hydrogen, compared to compound (III) which produced only hydrogen as the major pyrolysis byproduct. The minor byproducts from compounds (I), (II), and (III) consisted of a mixture of various methylsilanes ($H_{4-x}SiMe_x$; x = 1-4). The amount of methyl substitution on the methylsilanes, for each precursor, was found to be directly realted to, and limited by, the number of Si-C bonds contained in the molecular structure of each precursor. For example, compound (III) produced only methylsilanes of the form ($H_{2+x}SiMe_{2-x}$; x = 0 - 1), whereas compound (I) yielded (H_xSiMe_{4-x}; x = 0 - 3). This suggests that the pyrolysis byproducts are a result of discrete mechanistic decomposition pathways and are not produced from random recombination of radical species. Furthermore, methane is believed to be produced primarily from the methyl substituents and not from the $-(CH_2)-$ contained in the ring, which is supported by the lack of methane detected from the pyrolysis of compound (III).

The identities of the major pyrolysis byproducts were used to determine the decomposition onset temperature for each precursor. This was accomplished by monitoring the LPCVD reactor contents, as a function of an increasing substrate temperature, by using the attached QMS. The resulting plots of m/e peak intensities indicative of methane (for precursors I and II) and hydrogen (for precursor III) versus the Si(100) surface temperature were used to determine each precursor decomposition onset temperature. The series of decomposition onset temperatures (I{760°C}; II{740°C}; and III{680°C}) were determined to be inversely proportional to the amount of hydrogen substitution on the Si contained in the ring. This trend is consistent with a decomposition scheme described by Auner, et al.[12] which is based on the availability of the Si-H moiety within the four membered ring of disilacyclobutane derivatives. Auner, et al. describe a hydrogen shift decomposition mechanism which requires hydrogen to be present on the Si contained in the ring. In this scheme, hydrogen transfer from the "ring-Si" to an adjacent "ring-C" is energetically favorable, in part, due to the release of the ring strain energy.

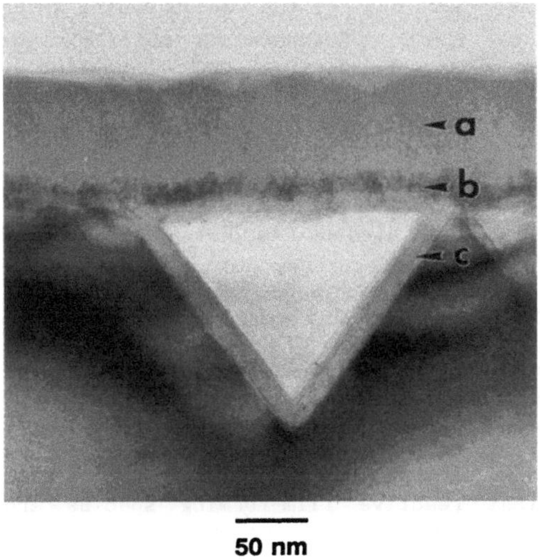

Figure 3: Cross-sectional TEM of the film produced from the decomposition of compound (I) at the decomposition onset temperature of 760°C. This micrograph reveals the a.) amorphous region, b.) polycrystalline SiC layer at the SiC/Si(100) interface, and c.) the continuous amorphous layer lining the pit that formed in the silicon substrate.

Film Analysis:

The films produced from each of the three precursors were routinely analyzed by using SEM, FTIR, and powder XRD and selected films were examined at the Si/SiC interface by using TEM. SEM was used to examine the surface and cross-sectional morphology of the films deposited at different substrate surface temperatures, which ranged from ca. 700°C to 1150°C. The surface and cross-sectional morphology appeared smooth and featureless at the lower deposition temperatures and became columnar in appearance at higher temperatures. The onset of the columnar growth occurred at a different temperature for each precursor, which in each case corresponded to the maximum deposition rate of ca. 12.0 µm/h. These observations are consistent with a transition from a kinetically controlled to a gas phase diffusion limited deposition.[14]

Fourier transform IR (FTIR) was used in both the specular reflectance and transmission mode to analyze for IR absorptions in the 1900 cm^{-1} - 2000 cm^{-1} and in the 2900 cm^{-1} - 3200 cm^{-1} regions. These spectroscopic regions correspond to $Si-H_x$ and $C-H_x$, respectively, and absorptions in these regions would indicate incomplete decomposition of the precursor. Since no absorptions were detected in either of these regions then the concentration of $A-H_x$ (A = Si or C) is less than the detection limit for FTIR or there is complete decomposition of the precursor on the substrate.

The film deposited from compound (I), at the decomposition onset temperature (760°C), was examined in cross-section at the Si/SiC interface by using high resolution TEM. The total thickness of the film is ca. 90 nm, consisting of a top amorphous layer (≅ 75.0 nm) and a polycrystalline β-SiC layer (determined by SAD) located at the Si/SiC interface. As shown in Figure 3, the silicon substrate contains a series of pits which are defined by the Si(111) planes. These pits are believed to have formed prior to the start of CVD, during an initial "bake-out" of the vacuum system which includes heating the substrate to ca. 800°C. Evidence for pit formation prior to the start of CVD is given in Figure 3(c.), which shows a continuous coating lining the inner surface of the pit. A TEM lattice image of the pit area revealed that this portion of the film was completely amorphous.

4.0 Conclusions

Derivatives of 1,3-disilacyclobutane were studied as potential single-source, low temperature (<1000°C) LPCVD precursors to polycrystalline silicon carbide films on Si(100) substrates. These compounds are non-pyrophoric and, under appropriate CVD conditions, pyrolyze to stoichiometric silicon carbide films in an argon environment. These compounds do not require the use of a co-reactant, such as hydrogen, as most other sources of CVD SiC commonly employ for complete removal of excess carbon, chlorine, or similar byproducts. Each precursor had a maximum deposition rate of 12 μm/h, which was consistent with a mass transport limitation. The correlation of the decomposition onset temperature for each compound with initial film formation suggests that reactive film-forming species are produced at these decomposition onset temperatures. This class of compounds show good potential as CVD precursors and will be studied further as potential CVD sources of SiC single crystal epitaxy at low (< 1000°C) temperatures.

Acknowledgments

The authors would like to thank J. Yang and Prof. P. Pirouz for their TEM expertise. This work was supported at RPI by the Chemistry Division of the Office of Naval Research.

References

[1] Proceedings: <u>First International High Temperature Electronics Conference</u>, ed. by D. B. King, F. V. Thome (Albuquerque, N. Mex., June 1991).
[2] <u>Amorphous and Crystalline Silicon Carbide II</u>, ed. by M. M. Rahman, C. Y.-W. Yang, Springer Proceedings in Physics, Vol. 43 (Springer-Verlag Berlin, Heidelberg, 1989).
[3] Cree Research Inc., Durham, NC 27713.
[4] S. P. Tay, J. P. Ellul, S. B. Hewitt, N. G. Tarr, A. R. Boothroyd, <u>Wide Band-Gap Semiconductors</u>, Vol. 242 of the Materials Research Soc. Symp. Proc. Series (MRS Boston Meeting, Dec. 1991), in press.
[5] D. J. Larkin, L. V. Interrante, Chem. Mater., in press.
[6] W. A. Kriner, J. Org. Chem., **29**, 1601(1961).
[7] R. M. Irwin, J. M. Cooke, J. Laane, J. Amer. Chem. Soc., **99**(10), 3273(1977).
[8] N. Auner, J. Grobe, J. Organomet. Chem., **222**, 33(1981).
[9] N. S. Nametkin, V. M. Vdovin, V. I. Zav'yalov, P. L. Grinberg, Izv. Akad. Nauk. SSSR, Ser. Khim., **5**, 929(1965).
[10] D. J. Larkin, L. V. Interrante, J. B. Hudson, B. Han, Proc. M.R.S. Meeting; <u>Chem. Perspec. Microelectr. Mater. II</u>, ed. by L. H. Dubois, L. V. Interrante, M. E. Gross, K. F. Jensen (Mater. Res. Soc. Symp. Proc., *204*, Pitts. PA 1991) p. 141.
[11] H. E. O'Neal, M. A. Ring, J. Organomet. Chem., **213**, 419(1981).
[12] N. Auner, I. M. T. Davidson, S. Ijadi-Maghsoodi, F. T. Lawrence, Organomet., **5**, 431(1986).
[13] D. M. Brown, J. D. Parsons, US Patent No. 4 923 716 (8 May 1990).
[14] J. M. Blocher Jr., in <u>Vapor Deposition</u>, ed. by C. F. Powell, J. H. Oxley, J. M. Blocher Jr. (John Wiley & Sons, Inc., New York 1966) p.11.

Frequency Dependence of Conductivity of Hydrogenated Amorphous SiC Films Prepared by PCVD

A. Tabata, K. Tomiita, Y. Suzuoki, and T. Mizutani

Department of Electrical Engineering, Nagoya University, Furo, Chikusa, Nagoya 464-01, Japan

Ac conductivity of hydrogenated amorphous silicon carbide (a-Si_xC_{1-x}:H) films prepared by plasma CVD method from silane and methane gases was measured. Ac conductions in the films with silicon content $0.7<x<0.9$ were explained by the classical hopping of electrons among the defect states, the densities of hopping sites were estimated to be 10^{17}-$10^{18}cm^{-3}$. Ac conductions in the films with $x>0.9$ above 300K were discussed in terms of the trap-release model. The Fermi level was in the center of the bandgap and the densities of states at the Fermi level were estimated to be of the order of $10^{16}cm^{-3}eV^{-1}$.

1. Introduction

Hydrogenated amorphous silicon carbide (a-Si_xC_{1-x}:H) has been expected to be an optoelectronics material, because its optical bandgap can be potentially controlled from 1.8eV to 3eV as the silicon-carbon content changes. However, its properties have not been completely explained yet. It is important to understand the localized gap states for preparation of high quality film and for device application.

The measurement of frequency dependence of conductivity is one of the simple and useful techniques for investigating the localized states within the band gap, because carriers hop among the localized states under the applied ac field. The ac conductivities of chalcogenide glasses [1][2], SiO_2 glass [3][4][5], a-Ge [6] and a-Si(:H) [7][8] were measured and it was reported that $\sigma_{ac}=A\times\omega^s$, where A and s(<1) were temperature dependent parameter [9]. This was explained in terms of correlated barrier hopping (CBH) model and the densities and energy levels of the localized states were estimated [1]-[8].

In this study, the frequency dependence of conductivity of a-Si_xC_{1-x}:H was measured and the transport process and the localized states were discussed.

2. Experimental

The a-Si_xC_{1-x}:H films were prepared by RF(13.56MHz) plasma CVD method (ANELVA PED-301). The reactive gases were silane and methane gases, gas flow rate of which were 10SCCM. The dilution gas was hydrogen gas, gas flow rate of which was 90SCCM. The total gas pressure was 800mTorr, RF power was 90W ($290mW/cm^{-2}$), and the substrate temperature was 573K. The film was deposited on a glass substrate (Corning 7056) with an evaporated magnesium electrode and after deposition a magnesium electrode was evaporated to form a sandwich structure. The film thickness was about 1μ m. The silicon content x was determined by X-ray photoelectron spectroscopy (XPS).

The capacitances and dielectric loss tangents were measured by an LCR meter (YHP 4247A) at frequencies between 10^2 and 10^5Hz, and the capacitances and conductances were measured by a capacitance bridge (General Radio 1616) at frequencies between 10 and 10^4Hz. For measurement of dc conductivity, a vibrating reed electrometer (VRE) (Takeda-Riken TR-84M) was used. All measurements were carried out under vacuum in the dark at temperatures between 80 and 370K.

3. Results and Discussions

3.1 Hopping between defect states

Figure 1 shows ac conductivity of a-Si_xC_{1-x}:H (x=0.49). Similarly to those in many amorphous materials [1]-[9], it obeys the power law $\sigma_{ac}=A\omega^s$, where A and s(<1) were temperature dependent parameters. The correlated barrier hopping (CBH) model was proposed to explain these results [1].

There are several assumptions to derive the ac conductivity. First assumption is the pair approximation in which carriers hop only between two sites and multiple hopping is excluded. Secondly, the hopping between the two sites is regarded as rotation of a dipole moment, whose response is assumed to be Debye-type. The ac conductivity then can be regarded as a sum of such individual losses. In the CBH model, besides the abovementioned assumptions, it is assumed that carriers hop not by tunneling but by thermal activation over the potential barrier W lowered by Coulombic interaction between the two sites. W is related to the distance R between the two sites,

$$W=W_M-ne^2/\pi\varepsilon_0\varepsilon_r R \quad (1)$$

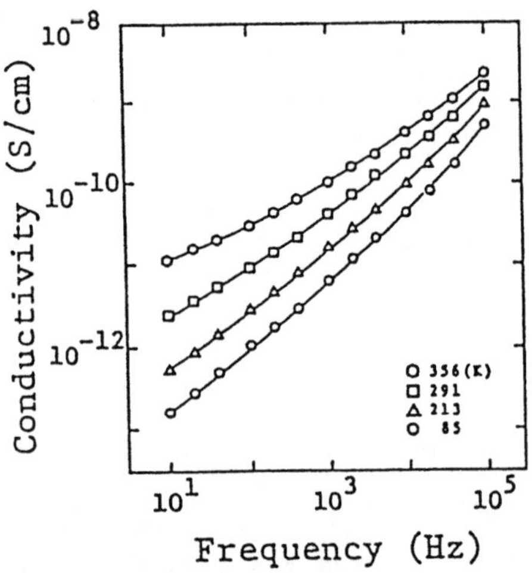

Figure 1
Frequency dependence of the conductivity of a-Si_xC_{1-x}:H. (x=0.49)

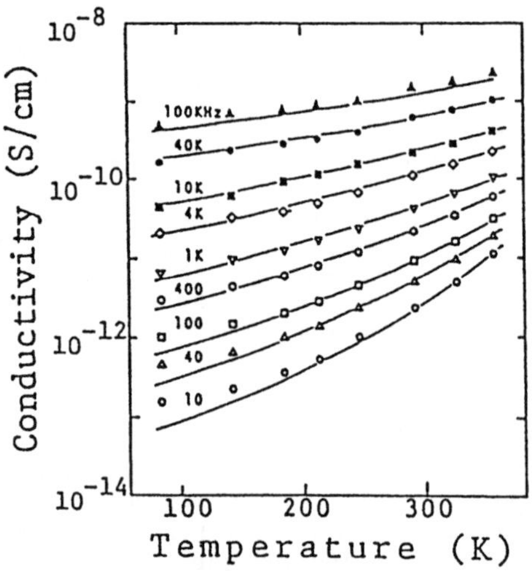

Figure 2
Temperature dependence of the conductivity of a-Si_xC_{1-x}:H. (x=0.49) Solid lines are calculated from CBH model.

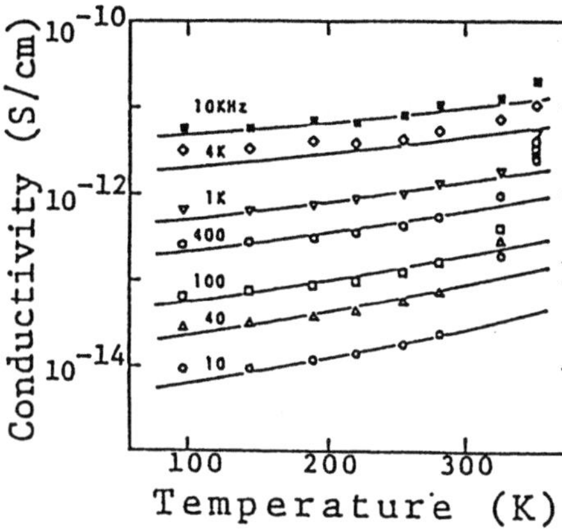

Figure 3
Temperature dependence of the conductivity of a-Si_xC_{1-x}:H. (x=0.72) Solid lines are calculated from CBH model.

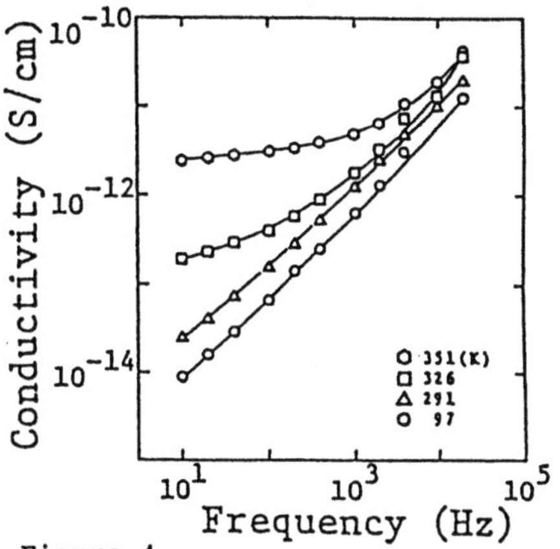

Figure 4
Frequency dependence of the conductivity of a-Si_xC_{1-x}:H. (x=0.72)

where W_M is the maximum barrier height, ε_r is the relative dielectric constant and n is the number of carriers that hop (n=1 for single-polaron, n=2 for bipolaron). The relaxation time is $\tau = \tau_0 \exp(W/kT)$. Ac conductivity in the CBH model is expressed by

$$\sigma(\omega)= \frac{n N^2 \pi^3 \varepsilon_0 \varepsilon_r}{12} \omega \left(\frac{n e^2}{\pi \varepsilon_0 \varepsilon_r}\right)^6 [W_M - kT \ln(1/\omega\tau_0)]^{-6} \quad (2)$$

where N is the density of localized states. If $W_M >> kT$, the ac conductivity is approximated by

$$\sigma(\omega)= \frac{n N^2 \pi^3 \varepsilon_0 \varepsilon_r}{12} \left(\frac{n e^2}{\pi \varepsilon_0 \varepsilon_r W_M}\right)^6 \tau_0{}^{\beta} \omega^{1-\beta}, \quad \beta = \frac{6kT}{W_M} \quad (3)$$

and this obeys the power law.

Figure 2 shows the temperature dependence of the ac conductivity replotted from figure 1. The solid lines in figure 2 were calculated from the CBH model with parameters $N=8.4\times 10^{17}cm^{-3}$, $W_M=0.77eV$, $\tau_0=2.0\times 10^{-9}sec$, n=2 and $\varepsilon_r=6.1$. ε_r is calculated from the capacitance measured simultaneously with measurement of the ac conductivity. The experimental data agree with the theoretical curves in the whole temperature range of the measurement and can be explained by the CBH model.

Figure 3 shows the temperature dependence of the ac conductivity of a sample with higher silicon content, i.e. a-Si_xC_{1-x}:H (x=0.72), together with the theoretical curves ($N=2.4\times 10^{18}cm^{-3}$, $W_M=1.33eV$, $\tau_0=3.9\times 10^{-9}sec$ and $\varepsilon_r=7.3$). Below 300K the experimental data agree with the theoretical curves. Above 300K, however, the experimental data deviate from the theoretical curves. To make this point clear, the ac conductivity in figure 3 was replotted against frequency in figure 4. It is clear in figure 4 that above 300K the conductivity does not obey the power law and that the frequency dependence becomes smaller as the frequency decreases. This behavior results from

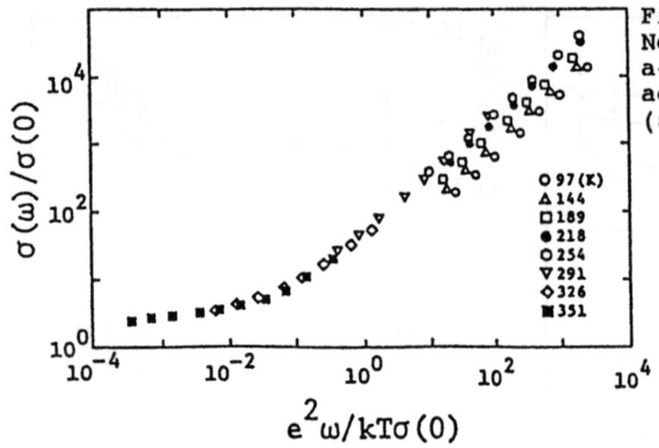

Figure 5
Normalized conductivity for $a\text{-}Si_xC_{1-x}:H$ ($x=0.72$) plotted against reduced frequency. (scaling law)

breakdown of the pair approximation and can be explained by the extended-pair-approximation (EPA) model [10]-[12]. The EPA model is based on the Miller-Abrahams equivalent circuit [13] derived from the rate equation. In the EPA model two arbitrary sites are considered and the effect of the external network on this pair is described by two admittances taken into account in an averaged way. $\sigma(\omega)$ approaches dc conductivity when $\omega \rightarrow 0$, and within the limits of large ω the EPA model is consistent with the pair approximation. The EPA model results in scaling law [14][15]

$$\sigma(\omega)/\sigma(0)=f(e^2\omega/\sigma(0)kT),$$

$$f(x)=1+x^n, \qquad (4)$$

where $\sigma(0)$ is the dc conductivity. Figure 5 shows scaling law and the normalized conductivity $\sigma(\omega)/\sigma(0)$ of $a\text{-}Si_xC_{1-x}:H$ ($x=0.72$) are well represented by a universal curve. It is suggested that the transport mechanism in the whole temperature range is the electron hopping among the defect states. The CBH model based on the pair approximation is valid in the lower temperature region and the multiple hopping process becomes predominant at the low frequencies in the high temperature region.

The parameters calculated from the CBH model for $a\text{-}Si_xC_{1-x}:H$ with three different silicon contents are shown in Table 1 with the optical band gap $E_{g.opt.}$ estimated from Tauc plot. The densities are reasonable because the spin densities N_s in $a\text{-}Si_xC_{1-x}:H$ with silicon content $0.5<x<0.8$ were reported to be 10^{17}-$10^{18}cm^{-3}$ [16].

Table 1 Parameter calculated from CBH model.

x	$N(cm^{-3})$	$W_M(eV)$	$\tau_0(sec)$	ε_x	$E_{g.opt.}(eV)$
0.72	2.4×10^{18}	1.33	3.9×10^{-9}	7.3	2.09
0.63	9.4×10^{17}	0.93	1.4×10^{-9}	6.3	2.22
0.49	8.4×10^{17}	0.77	2.0×10^{-9}	6.1	2.44

3.2 Interfacial relaxation

Figure 6 shows the ac conductivity of a-Si:H films. Above 290K the dispersion peak are observed. Figure 7 shows the

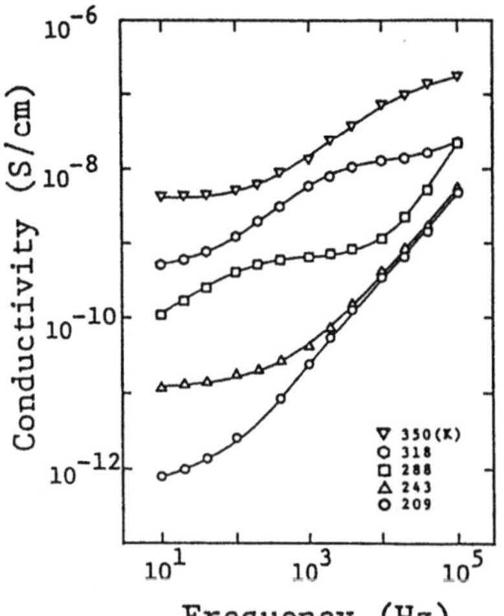

Figure 6
Frequency dependence of the conductivity of a-Si:H.

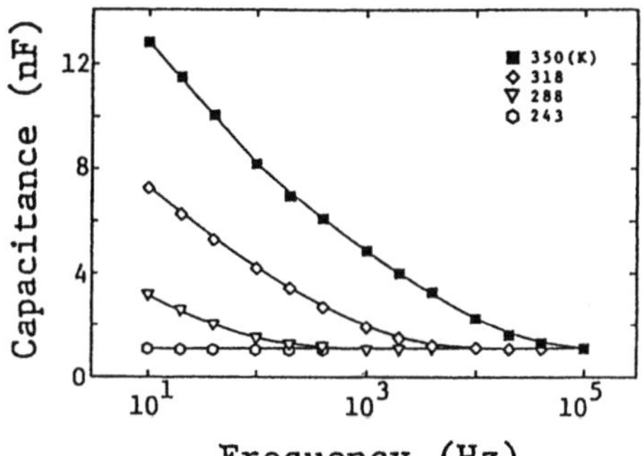

Figure 7
Frequency dependence of the capacitance of a-Si:H.

frequency dependence of capacitance measured simultaneously with measurement of the conductance. The capacitance is independent of frequency below 250K. On the other hand, above 280K the capacitance increases as the frequency decreases. This behavior can not be explained by the effect of the bulk but can be explained by the trap-release model based on Schottky barrier at the interface between magnesium electrode and a-Si_xC_{1-x}:H. The ac response can be calculated from electron transfer process between the deep level within the depletion layer and conduction band. The relaxation time is considered to be distributed around a certain mean value τ . The conductance $G(\omega)$ is given by

$$\frac{G(\omega)}{\omega} = \frac{e}{2} \, (\varepsilon_0 \varepsilon_r N_f)^{1/2} S \, \frac{\sin(\beta \pi/2)}{\cos(\beta \pi/2)+\cosh(\beta \ln \omega \tau)} \qquad (5)$$

where N_f is the density of states at the Fermi level E_F, β is the Cole-Cole parameter ($0<\beta<1$) and S is the area of electrodes. The mean relaxation time τ can be given by $\tau=\tau_p \exp(E_r/kT)$, where $E_r=E_C-E_F$.

Figure 8 shows $G/\omega - \omega$ plot for a-Si:H. The solid lines were calculated from the theoretical equation (5) with the parameter $N_f=4.2\times 10^{16}cm^{-3}$, Er=0.82eV, $\varepsilon_r=7.3$ and $\beta=0.74$. β was estimated from the Cole-Cole plot and ε_r was from the capacitance at high frequencies. It is apparent in figure 8 that above 280K the experimental data in the dispersion region agree with the theoretical curves and can be explained by the trap-release model.

Similar dispersion was observed in a-Si_xC_{1-x}:H with silicon content x>0.8 above 300K (figure 9). In this temperature region the theoretical curves from the trap-release model agree with the data. The parameters used for the curve fitting are shown in Table 2. Fermi level is about the middle of the band gap. The estimated values of N_f are reasonable, as the density of

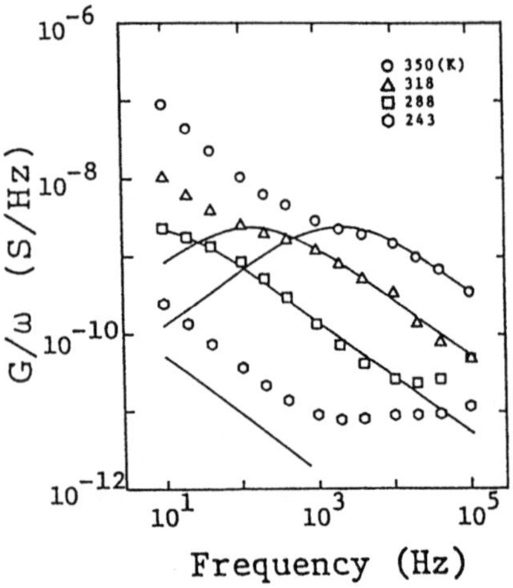

Figure 8
Frequency dependence of G/ω for a-Si:H. Solid lines are calculated from trap-release model.

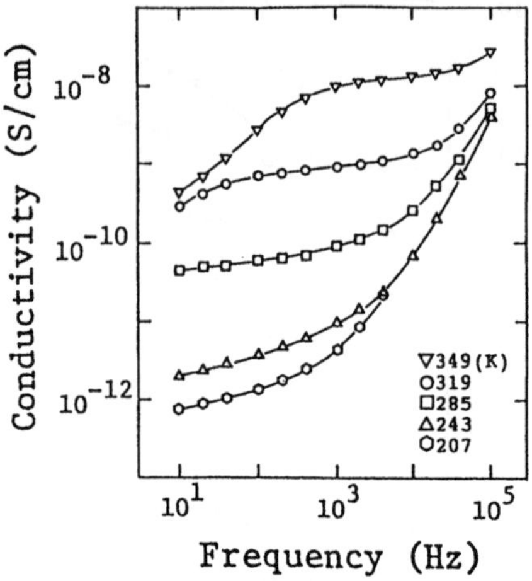

Figure 9
Frequency dependence of the conductivity of a-Si_xC_{1-x}:H. (x=0.93)

Table 2 Parameter calculated from trap-release model.

x	N_f ($cm^{-3}eV^{-1}$)	E_r (eV)	τ_p (sec)	ε_r	β	$E_{g.opt.}$ (eV)
1.00	4.2×10^{16}	0.82	1.2×10^{-16}	9.2	0.74	1.73
0.93	5.0×10^{16}	0.88	2.4×10^{-16}	8.1	0.88	1.83
0.82	5.9×10^{16}	0.89	3.4×10^{-16}	8.9	0.86	1.88

the states at Fermi level was reported to be of the order of $10^{16}cm^{-3}eV^{-1}$ [18].

In the low temperature region, however, the values predicted by the trap-release model were very small as compared with the experimental data, so that the obtained ac conduction in a-Si_xC_{1-x}:H with x>0.9 in the low temperature region could not be explained by the trap-release model, suggesting it results not from the interfacial relaxation but from the bulk. Although the CBH is a bulk process, the CBH model can not explain the ac conduction in the low temperature region, because the theoretical curves from the CBH model could not be fitted to the experimental data.

4. Conclusion

Frequency dependence of conductivities of a-Si_xC_{1-x}:H films prepared by plasma CVD method from silane and methane gases were measured. Ac conductivities of a-Si_xC_{1-x}:H with silicon content 0.5<x<0.8 showed the power-law frequency dependence and can be explained by the classical hopping of electrons among the defect states. The densities of defect states were estimated to be of the order of 10^{17}-$10^{18}cm^{-3}$.

Ac conduction in a-Si_xC_{1-x}:H with silicon content x>0.8 in the high temperature region,showed dispersion peak and can be explained in terms of the trap-release model based on Schottky barrier. The densities of the states at Fermi level were of the order of $10^{16} cm^{-3} eV^{-1}$ and Fermi level was located in the center of band gap.

References

[1] S.R.Elliott: Phil.Mag. 36 (1977) 1291; Phil.Mag.B 37 (1978) 153; ibid. 37 (1978) 553; ibid. 40 (1979) 507

[2] K.Shimakawa: Phil.Mag.B 46 (1982) 123

[3] R.Meaudre and M.Meaudre: Phil.Mag.B 40 (1979) 401; ibid. 46 (1882) 647

[4] K.Shimakawa and A.Kondo: Phys.Rev.B 27 (1983) 1136; ibid. 29 (1984) 7020

[5] A.Kondo, K.Shimakawa and Y.Inagaki: J.Phys.C Solid State Phys. 16 (1983) 5222

[6] A.R.Long and N.Balkan: Phil.Mag.B 41 (1980) 287

[7] K.Shimakawa, K.Watanabe and K.Hattori: Phil.Mag.B 54 (1986) 391

[8] S.R.Elliott: Phil.Mag.B 38 (1978) 315

[9] N.F.Mott and E.A.Davis; "Electronic Processes in Non-Crystalline Materials (2nd ed.)" (1979) Clarendon Press: Oxford

[10] P.N.Butcher: Phil.Mag.B 42 (1980) 799-802

[11] P.N.Butcher and S.Summerfield: J.Phys.C Solid State Phys. 14 (1981) L1099-L1102

[12] S.Summerfield and P.N.Butcher: J.Phys.C Solid State Phys. 15 (1982) 7003-7016; ibid. 16 (1983) 295-307; Phil.Mag.B 49 (1984) L36-L65

[13] A.Miller and E.Abrahams: Phys.Rev. 120 (1960) 734-755

[14] S.Summerfield: Phil.Mag.B 52 (1985) 9-22

[15] A.R.Long, J.McMillan, N.Balkan and S.Summerfield: Phil.Mag.B 58 (1988) 153-169

[16] A.Morimoto, T.Miura, M.Kumeda and T.Shimizu: J.Appl.Phys. 53 (1982) 7299

[17] I.G.Gibb and A.R.Long: Phil.Mag.B 49 (1984) 565

[18] J.Robertson: Phil.Mag.B 51 (1985) 183

Formation of SiC for Microelectronic Applications by C Implantation into Doped a-Si

B.M. Manning[1], *S.B. Hewitt*[1], *N.G. Tarr*[1], *and T.W. MacElwee*[2]

[1]Department of Electronics, Carleton University, Ottawa, Canada K1S 5B6

[2]Northern Telecom Electronics Ltd., Ottawa, Canada K1Y 4H2

Abstract. Heterojunction diodes have been formed by high-dose C implantation into thin n^+ amorphous Si layers deposited on lightly-doped p-type Si substrates. FTIR and ellipsometry both indicate the conversion of the implanted layer to SiC following rapid thermal annealing at or above 1050°C. Resistivities as low as 300 mΩcm were obtained after RTA for SiC films over oxide. Diode J-V characteristics suggest the presence of a large depletion region recombination current component.

1. Introduction

The use of silicon carbide as a wide-bandgap emitter for heterojunction bipolar transistors (HBT's) compatible with silicon integrated circuit technology has recently received considerable attention. Most work to date on SiC emitter HBT's has relied on the use of CVD or PECVD techniques to form the emitter. Formation of crystalline β-SiC requires deposition temperatures of 1000°C or greater, which may pose problems of thermal budget incompatibility with conventional VLSI processing. Amorphous or microcrystalline SiC films can be deposited at lower temperatures, but may have lower emitter efficiency and suffer long-term stability problems.

SiC synthesis by carbon implantation into silicon [1] may provide an alternative means of forming an SiC emitter, but has so far received relatively little consideration. Lezheiko *et al* fabricated heterojunction diodes by C implantation into p-type silicon substrates, followed by doping of the surface layer with a nitrogen implant [2]. Unfortunately, the resulting junctions exhibited very poor ideality factors in the forward direction, and extremely high reverse leakage. Kimura *et al* investigated the recrystallization of carbon-implanted layers and the redistribution of carbon during post-implant annealing [3]. More recently Reeson *et al* showed that buried SiC layers could be grown epitaxially on silicon substrates by post implant annealing at temperatures very close to the melting point of silicon [4].

This paper examines the formation of n^+p heterojunction diodes by high-dose carbon implantation into a deposited amorphous silicon (a-Si) layer. Doping of the implanted layer is accomplished by *in-situ* doping of the a-Si film rather than ion implantation. Rapid thermal annealing is used to attempt to remove implant damage while keeping the thermal budget at a level compatible with conventional silicon processing.

Springer Proceedings in Physics, Vol. 71 **Amorphous and Crystalline Silicon Carbide IV**
Editors: C.Y. Yang · M.M. Rahman · G.L. Harris

2. Experiment

The starting substrates for this experiment were lightly-doped ($\approx 5x10^{15}$ cm^{-3}) p-type silicon wafers of (100) orientation. Following thermal growth of a 100 nm passivating oxide, windows were opened and an amorphous silicon film heavily *in-situ* doped with phosphorus deposited by LPCVD at 560°C. Dektak measurements on control wafers showed a film thickness of 60 nm. Native oxide was removed from the windows by etching in 1% hydrofluoric acid immediately prior to deposition. SIMS indicated a phosphorus concentration of approximately 10^{21} cm^{-3} in the film. Wafers were next implanted with ^{12}C to a dose of $2.5x10^{17}$ cm^{-2} at an energy of 10 keV. The implant profile predicted by LSS theory is shown in Fig. 1; the implant should give a broad peak of essentially stoichiometric SiC. Due to the high dose required, implants were restricted to a region approximately 1 cm on a side at the center of each sample. This technique also allowed for side-by-side comparison of the properties of implanted and unimplanted material.

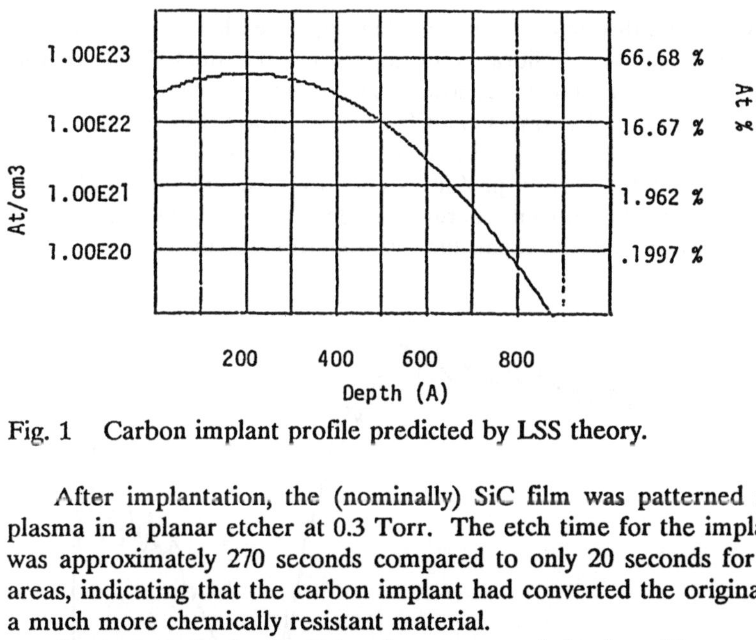

Fig. 1 Carbon implant profile predicted by LSS theory.

After implantation, the (nominally) SiC film was patterned in a CF_4/O_2 plasma in a planar etcher at 0.3 Torr. The etch time for the implanted regions was approximately 270 seconds compared to only 20 seconds for unimplanted areas, indicating that the carbon implant had converted the original a-Si film to a much more chemically resistant material.

Contacts to the implanted regions were made by depositing a second layer of heavily *in-situ* doped a-Si. Wafers were then furnace annealed at 650°C for 60 minutes to recrystallize the a-Si, and subsequently rapid thermal annealed for times ranging from 5 to 30 seconds at temperatures from 850 to 1250°C. Final contact was made using a sputtered $Ti_{0.1}W_{0.9}$ 0.2 μm barrier layer overlaid with 0.5 μm of aluminum; this metal layer was patterned by wet chemistry. A 0.5 μm Al back contact was then applied, and the samples sintered at 400°C for 20 minutes in H_2.

3. Results

3.1 Material Analysis

FTIR. Fourier Transform Infrared Spectroscopy (FTIR) revealed an absorption peak at 780 cm^{-1} in areas which had been implanted with carbon, indicating the presence of silicon-carbon bonds [1,3]. A peak was seen at the same wavelength in reference SiC material produced by CVD. The peak was absent in unimplanted regions. The peak sharpened and grew in height by a factor of approximately four after annealing at 1100°C (Fig. 2).

Ellipsometry. Ellipsometry was performed at a wavelength of 632.8 nm on implanted layers over the silicon substrate. Analysis was performed assuming an abrupt transition between the surface layer and the bulk, although in fact the transition must be gradual. Despite this problem, a refractive index of 2.7 to 2.8 was found after annealing at 1100°C, close to the value expected for bulk β-SiC. The refractive index of unannealed implanted regions was 3.2.

SIMS. A SIMS profile through an unannealed implanted layer is shown in Fig. 3. The depth axis was calibrated by measuring the final depth of the sputtered crater, while the atomic concentrations were determined by comparison with control silicon wafers of known C and P content. The SIMS profile shows a Gaussian distribution of carbon with peak concentration close to that of stoichiometric SiC, but the carbon profile extends to a far greater depth than expected from the known implant energy and dose. This result cannot be explained on the basis of material expansion resulting from the high implant dose,

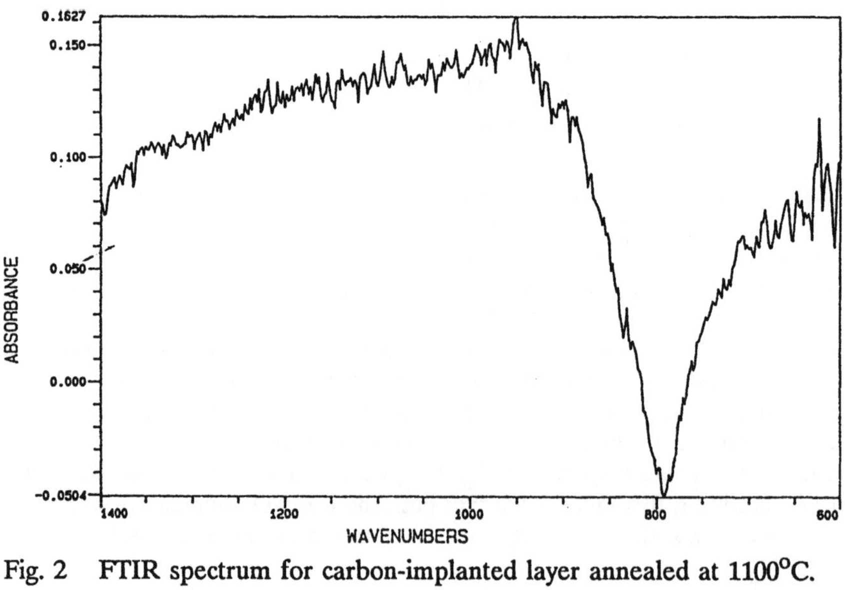

Fig. 2 FTIR spectrum for carbon-implanted layer annealed at 1100°C.

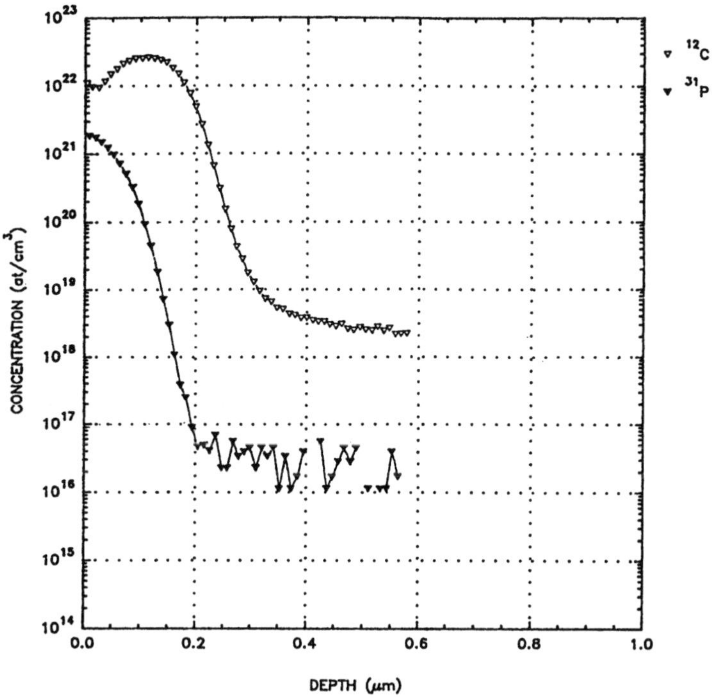

Fig. 3 SIMS profile through unannealed carbon-implanted layer.

since the fact that the lattice constant of β-SiC is 20% smaller than that of Si almost exactly cancels the expansion that might be expected from the added carbon atoms. It may be that the sputtering rate for SiC was lower than that for Si, leading to inaccurate calibration of the depth axis.

3.2 Electrical Characterization

Resistivity. The resistivity of the implanted layers was measured using a van der Pauw cross over oxide. The resistivity was found to drop from 7-10 Ωcm in as-implanted material to 300-500 mΩcm after annealing at 1050°C, with relatively little further improvement for higher RTA temperatures. Data are summarized in Fig. 4. Although this resistivity value is roughly three orders of magnitude higher than that expected for heavily-doped polysilicon, it is comparable to the resistivity of CVD emitter material in recently-reported high-gain HBT's [5]. The resistivity of unimplanted a-Si films reached a minimum of 0.8-1.0 mΩcm after annealing at 1150°C, but rose at higher RTA temperatures, presumably as the result of dopant evaporation.

Diodes. The current-voltage characteristics of 300 μm square diodes with implanted emitters annealed at 1100°C are shown in Fig. 5. Very similar results were obtained for devices annealed at 1050°C. For comparison, the

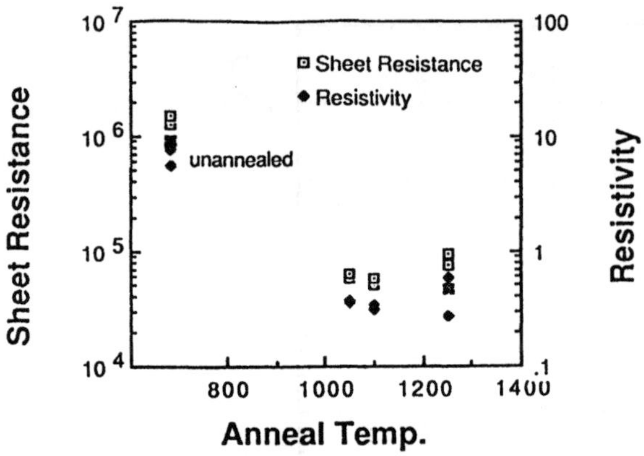

Fig. 4 Resistivity (Ωcm) and sheet resistance (Ω/□) of carbon-implanted films as a function of RTA temperature (°C).

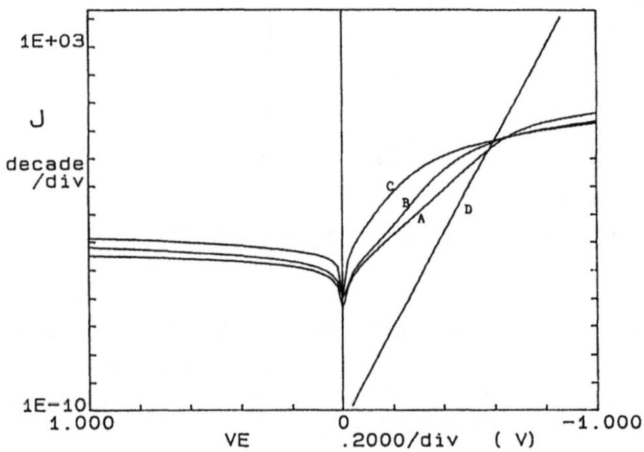

Fig. 5 Diode current-voltage characteristics (current density J in A/cm^2).
(a) homojunction control
(b) periphery of carbon implant
(c) center of implanted region
(d) ideal injection-diffusion current into substrate

characteristics of a control device lacking the carbon implant are also shown. All diodes exhibited a sharp breakdown at a reverse bias slightly greater than 50 V. Reverse leakage is high in all devices, including the homojunction control. Comparison of diodes with different perimeter-to-area ratios suggested that the reverse leakage had a large edge component, especially on small devices. The source of this large leakage current is uncertain at present; since we have found recrystallized amorphous silicon emitters can provide reverse leakage of approximately 1 nA/cm^2 in other experiments, the leakage current may result from damage to the passivating oxide associated with charge build-up during the high-dose carbon implant.

Near the periphery of the implanted regions, where the carbon dose is low, the forward characteristics show two regions of exponential dependence on bias with different ideality factors. Writing the current-voltage law as $J \propto e^{qV/nkT}$, at biases below approximately 0.3 V the ideality factor n is close to 2. Above 0.3 V the ideality factor becomes as small as 1.5. More importantly, in this bias range the characteristic comes close to coinciding with that of the homojunction control. At current densities greater than approximately 100 $mAcm^{-2}$ the curve deviates from an exponential dependence on voltage, presumably as the result of emitter resistance. For these devices the current is apparently dominated by recombination in the emitter-base depletion region at low forward bias, with electron injection-diffusion into the substrate becoming significant at high forward bias. The large depletion region recombination/generation component likely results from unannealed implant damage in the substrate.

For devices near the center of the implanted region, the forward current shows a single regime of exponential dependence on bias with $n \approx 1.5$. At a given bias in this regime the current is roughly two orders of magnitude greater than in the homojunction control. The relatively small n value for these devices suggests that recombination may be occurring close to the emitter edge of the depletion region, perhaps within the SiC layer itself.

One might expect that annealing at temperatures above 1100°C would be more effective in removing implant damage. Unfortunately, junctions annealed at 1250°C were barely rectifying. The reason for this loss of diode action after high temperature annealing is not known.

Lateral Transistors. The mask set used to fabricate the diodes included a lateral BJT with a 20 μm wide emitter stripe surrounded on both sides by collector stripes spaced 20 μm away. Measurements of the gain of these transistors supported the conclusions drawn above regarding the source of the diode current components. For transistors near the periphery of the carbon implant, where "two-slope" forward J-V characteristics were found for diodes, common emitter current gains of ≈ 0.1 were obtained for the lateral transistors. Near the center of the implant the gain dropped to ≈ 0.001. (It should be noted that with such coarse dimensions this lateral transistor structure normally provides gains less than 10 even for good-quality silicon homojunctions).

4. Conclusions

High-dose carbon implantation into heavily *in-situ* doped a-Si has been shown capable of forming heterojunction diodes with reasonably good rectifying characteristics. FTIR, ellipsometry and etch rate data all support the formation of an SiC layer (presumably non-stoichiometric) in the emitter. SiC resistivities as low as 300 mΩcm have been obtained after rapid thermal annealing at or above 1050°C. However, the best SiC emitter diodes still had far higher leakage current at small reverse bias than homojunction controls, and higher forward current until emitter resistance effects set in. Any reduction in hole back injection resulting from the conversion of the emitter to SiC appears to be more than compensated by an increase in depletion region recombination current.

Acknowledgement

The authors wish to thank L.P. Berndt and C.A. Adams for their assistance in device processing, and Dr. M. Simard-Normandin for help with material analysis.

References

[1] J.A. Borders, S.T. Picraux and W. Beezhold, Appl. Phys. Lett. **18**, 509 (1971).

[2] L.V. Lezheiko and E.V. Lyubopytova, Sov. Phys. Semicond. **10**, 1039 (1976).

[3] T. Kimura, S. Kagiyama and S. Yugo, Thin Solid Films **81**, 319 (1981).

[4] K.J. Reeson, P.L.F. Hemment, J. Stoemenos, J. Davis and G.E. Celler, Appl. Phys. Lett. **51**, 2242 (1987).

[5] T. Sugii, T. Yamazaki and T. Ito, IEEE Trans. Electron Dev. **ED-37**, 2331 (1990)

Polycrystalline SiC Films Prepared by a Plasma Assisted Method at Temperatures Lower than 1000°C

K. Kamimura[1], *K. Koike*[1], *H. Ono*[1], *T. Homma*[1], *Y. Onuma*[1], *and S. Yonekubo*[2]

[1]Department of Electrical and Electronic Engineering, Shinshu University, 500 Wakasato, Nagano 380, Japan
[2]Nagano Precision Industry Research Institute, 9959 Okaya, Nagano 394, Japan

Polycrystalline SiC films have been prepared by plasma assisted decomposition of tetramethylsilane (TMS) at temperatures lower than 1100 ℃.
The films were characterized by X-ray diffraction. The films prepared at higher than 840 ℃ consisted of polycrystalline silicon carbide. To obtain a large grain size, the substrate temperature must be higher than 1000 ℃. The poly-SiC/Si heterojunction diode showed the diode factor of 1.8.

1. Introduction

Polycrystalline SiC films have been prepared by plasma assisted chemical vapor deposition. Polycrystalline SiC films are useful for emitter of heterojunction bipolar transistors and electronic devices such as piezoresistive elements or thin film transistors used at high temperature [1,2,3].

By conventional methods, the substrate temperature must be higher than 1200 ℃ to prepare a single-crystalline SiC film [4]. The high substrate temperature is one of the limiting factors for the application of crystalline SiC films to electronic devices. Recently, some authors have reported the crystal growth of SiC at lower temperature [5]. The low growth temperature resulted in the improvement in device characteristics such as HBT [6,7].

In this report, polycrystalline SiC films were prepared by plasma assisted decomposition of tetramethylsilane. This method is rather safety as compared to the conventional method in which silane or disilane is used as the source gas. The source gas used in this work is only tetramethylsilane, which is stable at room temperature and is safe as compared to silane or disilane.

Films were characterized by X-ray diffraction, and ESCA. Hall effect was measured to characterize the electrical properties of the films.

2. Preparation

The SiC films were prepared by plasma assisted decomposition of tetramethylsilane (TMS). Figure 1 shows the schematic representation of the deposition apparatus. The reaction chamber was made from quartz tube with inside diameter of 37mm ϕ.

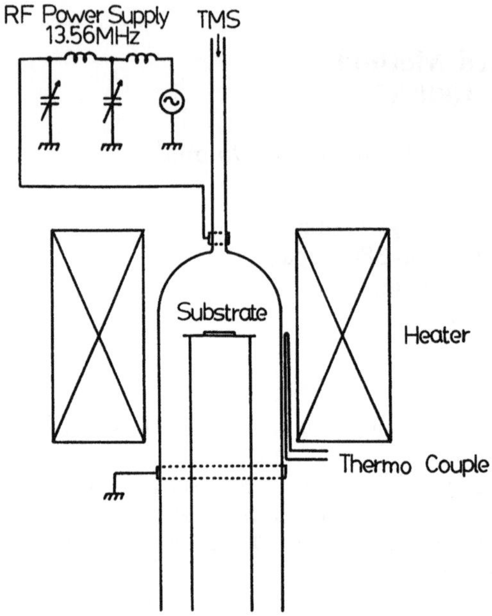

Fig.1 Schematic representation of experimental apparatus CVD system. The reaction chamber is made from quartz tube with inside diameter of 37mm ø.

The vapour pressure of TMS was controlled by controlling the temperature of TMS. The flow rate was measured by float meter and controlled by needle valve. Pure TMS (99.9999%) and pure H_2 (99.99999%) were introduced to the reaction chamber. The dilution ratio of TMS by H_2 was determined from the flow rate of each gas measured by flow meters. The TMS concentration in the source gas ($TMS/(TMS+H_2)$) was from 20 % to 100 %. Diluting the source gas with H_2 is expected to be effective to control the carbon concentration in the film, because the hydrogen radical should easily react with the carbon atoms in TMS and take out the excess carbon atoms.

The substrate was heated by a resistance furnace. The substrate temperature was deduced from the temperature of the furnace. The relation between the substrate temperature and the temperature of furnace had been measured prior to the deposition of the films.

The glow discharge was excited by RF power of 13.56 MHz to enhance the decomposition of TMS and the deposition of films. RF power was applied to the reaction chamber with a capacitively coupled ring electrode.

The substrates were single crystal Si and thermally oxidized SiO_2. The color of the films was black or dark gray to the naked eye.

3. Characterization

3.1 X-ray diffraction

Crystal structure of the films was characterized by X-ray diffraction. Figure 2 shows the X-ray diffraction patterns of the samples prepared at substrate temperatures from 740 ℃ to 1040 ℃. All the samples in Fig.2 were prepared from TMS without dilution with H_2. The films were deposited on (111) surface of single crystal silicon.

Diffraction peaks were observed for the films prepared at the substrate temperature of higher than 840 ℃. When the sample was prepared at temperatures from 840 ℃ to 940 ℃, the film consisted of polycrystalline SiC with small grain size. The grain size increased rapidly at the substrate temperature above 1000 ℃.

Figure 3 shows the X-ray diffraction patterns for the samples prepared from the source gases with different concentration. The diffraction peaks became higher and sharper with decreasing source gas concentration. The samples were prepared at 1040 ℃ with the RF power of 50 W under a pressure of 0.5 Torr. The concentration of TMS ($TMS/(TMS+H_2)$) was from 100 % to 20 %. The thickness of the films was controlled to be about 1 - 2 μm. Diluting TMS with H_2 was effective in increasing the grain size of the polycrystalline SiC films. The hydrogen radicals seemed to be useful for the formation of SiC films by cleaning the surface, assisting the decomposition of TMS or reducing the excess carbon atoms.

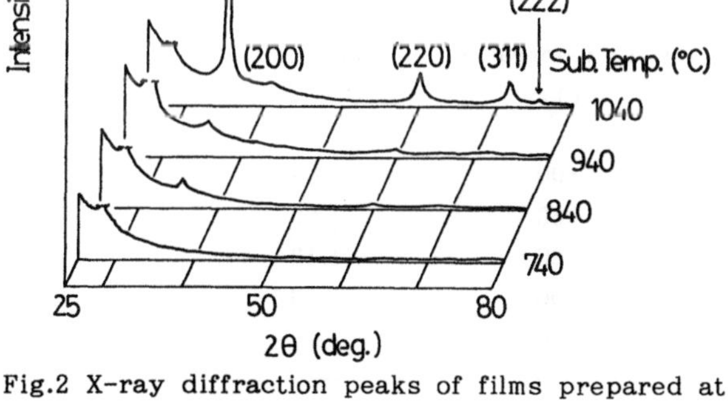

Fig.2 X-ray diffraction peaks of films prepared at the substrate temperature from 740 to 1040℃. The films were deposited on (111) surface of single crystal silicon.

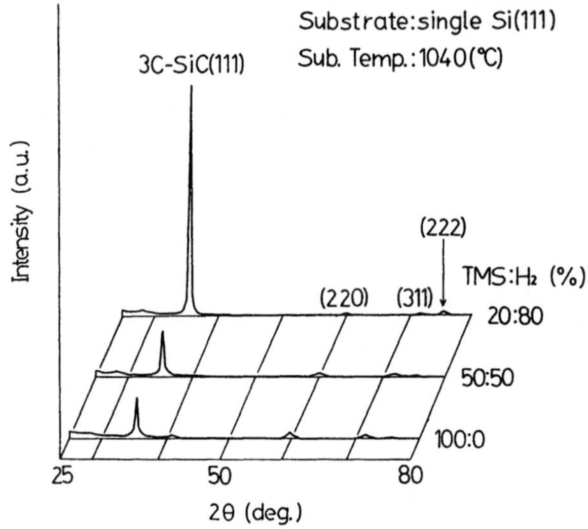

Fig.3 X-ray diffraction peaks of films. Patterns were depend on the flow rate of ($TMS:H_2$=100:0～20:80). The samples were prepared at 1040℃ with the RF power of 50 W.

3.2 Stoichiometry

The atomic ratio of C/Si in the films was estimated by ESCA measurement. Table 1 shows the result for some samples prepared at substrate temperatures from 840 to 1040 ℃, from the source gas with the TMS concentration from 20 to 100 %. The data in Table 1 were calibrated to the standard sample of bulk SiC.

Diluting the source with hydrogen had no obvious effect on decrease in the carbon concentration in the films. The substrate temperature was a more

Table 1. Atomic Ratio in the Films Estimated from EPMA Measurement.

Substrate Temperature (℃)	TMS Concentration (%)	Atomic Ratio (C/Si)
1040	100	——
1040	50	1.00
1040	20	1.12
940	100	2.30
940	50	——
940	20	——
840	100	2.10
840	50	1.80
840	20	2.02
740	100	2.40

important parameter to achieve the stoichiometry of the films. The excess carbon was observed in the films prepared at 840 ℃ and 940 ℃, and the films became almost stoichiometric when the substrate temperature was 1040 ℃.

3.3 Electrical properties

The electrical properties of the film were estimated by Hall measurement. The mobility, carrier concentration and resistivity of the films are listed in Table 2. Films were deposited on thermally grown SiO_2.

The carrier concentration was from the order of 10^{16} to 10^{19} cm^{-3}. No impurity was doped intentionally, and the origin of donor or acceptor have not identified yet.

The mobility was the order of 10^0 to 10^2 cm^2/Vs. The low mobility was observed in the sample with high carrier concentration. The donors or acceptors seemed to form scattering centers.

Table 2. Electrical Properties of Polycrystalline SiC Films.

Substrate Temp. (℃)	TMS Concentration (%)	Conduction Type	Carrier Concentration (cm^{-3})	Hall Mobility ($cm^2/V \cdot s$)	Resistivity ($\Omega \cdot cm$)
1040	100	p	2.64×10^{16}	349.0	7.03×10^{-1}
1040	50	p	3.90×10^{19}	0.6	2.68×10^{-1}
1040	20	n	1.18×10^{19}	6.8	1.02×10^{-1}
940	100	p	3.51×10^{18}	61.4	2.76×10^{-2}
940	50	p	5.85×10^{17}	147.9	7.22×10^{-2}
940	20	n	2.50×10^{19}	3.3	9.02×10^{-2}
840	100	p	8.89×10^{17}	94.3	7.62×10^{-2}
840	50	p	1.35×10^{19}	4.5	1.07×10^{-1}
840	20	n	1.41×10^{17}	338.8	1.34×10^{-1}
740	100	n	6.31×10^{18}	115.0	8.95×10^{-3}

4. Application

A heterojunction bipolar transistor is one of the promising applications of polycrystalline SiC film [1,5]. A heterojunction diode was fabricated by depositing the n-type polycrystalline SiC film onto (111) surface of a p-type single crystal Si wafer. The vacuum evaporated Au film was used as the ohmic

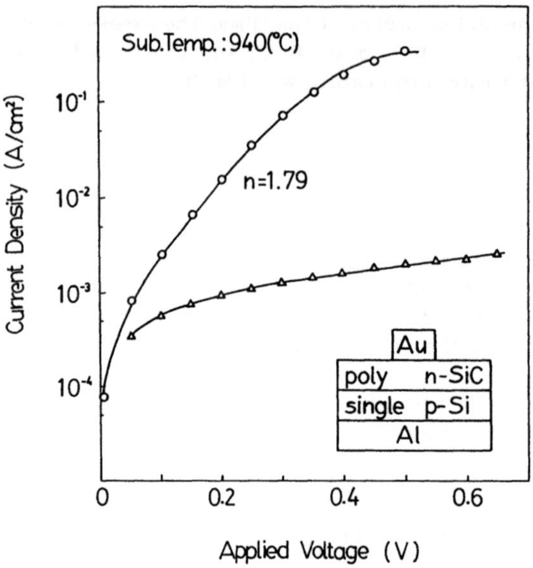

Fig.4 Current-Voltage characteristic of polycrystalline SiC / single crystal Si Heterojunction diode. The polycrystalline SiC was deposited at 940℃.

contact to polycrystalline SiC. The Au/polycrystalline SiC contact showed good ohmic characteristics when the carrier concentration in SiC was high enough.

Figure 4 shows the typical current-voltage characteristics of a polycrystalline SiC / single crystal Si heterojunction diode. Polycrystalline SiC was deposited at 940 ℃. The diode factor was estimated to be 1.8 from the linear portion in Fig.4. This value is almost the same as the samples prepared with the polycrystalline SiC film deposited by electron beam evaporation [1].

5. Conclusion

Polycrystalline SiC films were prepared by plasma assisted decomposition of tetramethylsilane at the substrate temperature of 840 to 1040 ℃. Although this work had been intended to get high quality films at lower than 1000 ℃, the grain size was fairly small when the films were deposited at lower than 1000 ℃. At this point, the substrate temperature was most important parameter to get high quality films, under the conditions in this work.

Diluting the TMS with hydrogen was effective to increase the grain size of films. The dilution also seemed to have some effects to reduce the excess carbon in the films, but it was not so obvious.

As compared to the conventional method using silane and hydrocarbon, the method in this work is rather safe because the source gas is tetramethylsilane which is very stable at room temperature.

References

1. K. Kamimura, Y. Nishibe and Y. Onuma, "SiC/Si HBT Using Polycrystalline SiC Layer Prepared by Electron Beam Evaporation", Springer Proceedings in Physics, Vol.43, (Springer-Verlag Berlin, Heidelberg, 1989) pp.207-211.
2. Y. Onuma, S. Miyashita, Y. Nishibe, K. Kamimura and K. Tezuka, "Thin Film transistors Using Polycrystalline SiC", Springer Proceedings in Physics, Vol.43, (Springer-Verlag Berlin, Heidelberg, 1989) pp.212-216.
3. Y. Onuma, K. Kamimura, Y. Nagura, Cai Hao Yi and M. Kikuchi, "Polycrystalline Silicon Carbide Films for Piezoresistive Elements", Sensors and Materials, Vol.2, No.4 (1991) pp.207-216.
4. S. Nishino, H. Suhara, H. Ono and H. Matunami, "Epitaxial Growth and Electrical Characteristics of Cubic SiC on Silicon", J. Appl. Phys. Vol.61, No.10 (1987) pp.4889-4893.
5. Y. Furumura, M. Doki, F. Mieno, T. Suzuki and M. Maeda, "Heteroepitaxial β-SiC on Si", J. Electrochem. Soc., Vol.135, No.5 (1988) pp.1255-1260.
6. T. Sugii, T. Ito Y. Furumura, M. Mieno and M. Maeda, "Heterojunction Bipolar transistors with Single-Crystalline β-SiC Emitters", J. Electrochem. Soc., Vol.134, No.10 (1987) pp.2545-2549.
7. T. Sugii, T. Aoyama and T. Ito, "Polycrystalline SiC for a Wide-Bandgap Emitter of Si-HBTs", J. Electrochem. Soc., Vol.136, No.10 (1989) pp.3111-3115.

Si_xC_{1-x} Alloys Deposited on Silicon Using a Low-Cost, Hot-Wall, LPCVD Reactor

F. Hébert

Advanced Bipolar Products, Avantek, Inc.,
39201 Cherry St., Newark, CA 94560, USA

Abstract. The deposition of Si_xC_{1-x} alloys on silicon using a low-cost, hot-wall, horizontal low pressure chemical vapor deposition (LPCVD) batch reactor, is investigated. Silane is used as the source for silicon atoms while two non-toxic and non-flammable carbon sources are studied for the carbon source, namely carbon tetrafluoride and 1,1,1-trichloroethane. Films with less than 2% carbon are obtained when carbon tetrafluoride is used at temperatures of 900°C or below. When 1,1,1-trichloroethane is used, SiC alloys with controlled carbon content greater than 15% are successfully deposited even at temperatures as low as 650°C.

1.0 Introduction

During the past decade or so, numerous techniques for obtaining device quality amorphous, micro-crystalline, poly-crystalline and epitaxial SiC films and Si_xC_{1-x} alloys have been reported in the literature. Chemical vapor deposition (CVD) techniques are typically used: for example, cold-wall CVD (e.g. [1-3]) and plasma enhanced CVD (e.g. [4,5]). Such equipment is rather specialized and not commonly available in many silicon semiconductor device fabrication facilities.

For the gas system, silane (SiH_4), disilane (Si_2H_6) and trichlorosilane ($SiHCl_3$) are used for the silicon source while hydrocarbon gases such as methane (CH_4), acetylene (C_2H_2) and propane (C_3H_8) are typically used for the carbon source. While the silane based gases are generally available in semiconductor device fabrication facilities, e.g. for polysilicon and silicon nitride deposition, the above mentioned carbon sources are not. Since the installation of flammable gas mixtures is costly and lengthy, due to special safety requirements such as the use of gas cabinets, sensors, alarms, etc., less hazardous carbon sources are desirable.

In this paper, deposition of SiC alloys on Si using source gases and equipment readily available in most silicon device manufacturing facilities is described. Depositions are carried out in an inexpensive reactor, generally used for polysilicon or silicon nitride depositions: a resistively heated, hot-wall, low-pressure chemical vapor deposition (LPCVD) batch reactor. Two relatively "safe" carbon containing gases are investigated as carbon sources for the deposition of SiC alloys. These are carbon tetrafluoride, or CF_4, typically used for plasma etching (e.g. [6]) and 1,1,1-trichloroethane, or TCA (CH_3CCl_3), generally used as an additive during the oxidation of silicon (e.g. [7]). For the Si source, silane is used as it is the most commonly used source gas for the deposition of polysilicon films.

2.0 Experimental Set-up

A schematic diagram of the reactor, ordinarily used for *in-situ* doped and undoped polysilicon thin-film depositions, is shown in Figure 1. The reaction chamber consists of a 115 mm internal diameter quartz tube inside a three-zone furnace. All gas flows were controlled using mass flow controllers (MFC). An automatic throttle valve controlled the pressure during deposition.

Springer Proceedings in Physics, Vol. 71 **Amorphous and Crystalline Silicon Carbide IV**
Editors: C.Y. Yang · M.M. Rahman · G.L. Harris

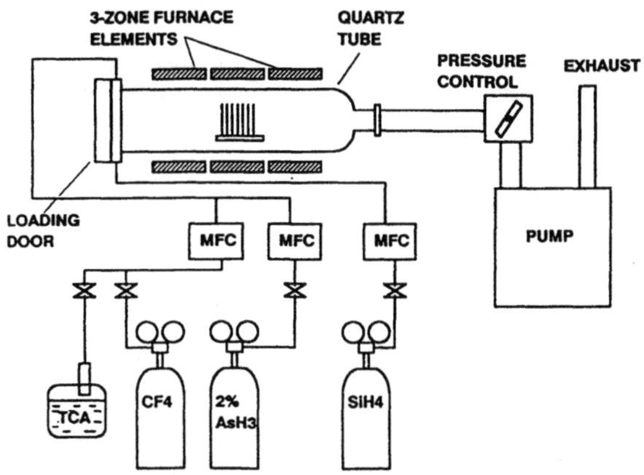

Figure 1 - Schematic of the LPCVD reactor

One-hundred percent silane was used as the Si source. One-hundred percent CF_4 gas, or pure TCA liquid contained in a temperature controlled quartz container [8], were separately used as sources for carbon. No carrier gas was used during the experiments: in the case of TCA, this means that the temperature of the quartz container limited the maximum TCA vapor draw. Two-percent arsine (AsH_3) in hydrogen (H_2) mixture was used for *in-situ* doping. Argon was used to purge the MFCs prior and after any depositions. Deposition temperatures in the range of 650°C to 900°C were investigated. The deposition pressure was maintained at 0.25 Torr.

3.0 Experimental Procedure

Films were deposited on thermally oxidized (100nm) and bare silicon 3-inch dia. wafers, with <100> and <111> orientation. The deposited film thickness was evaluated using an optical film thickness measurement system (NanoSpec) and after-etch step-height measurements using a surface profilometer (Dektak). Sheet resistance measurements were carried out using a six-point probe (Four Dimensions). Electron Spectroscopy for Chemical Analysis (ESCA) [9] was used to measure the chemical content.

4.0 Depositions Using Carbon Tetrafluoride

The effect of adding CF_4 to the silane gas flow was investigated at temperatures up to 900°C. Irrespective of the CF_4 to SiH_4 gas ratio, less than 2% carbon was found in the deposited polycrystalline films using ESCA. ESCA measurements revealed no contaminants other than some surface carbon contaminants and surface oxides, both of which were quickly removed using *in-situ* argon sputtering. The only observed effects of adding CF_4 to the silane gas flow were a reduction in deposition rate and an increase in the film grain size. For example, when the CF_4/SiH_4 ratio is increased from zero to 1.5, the grain size more than doubled while the deposition rate reduced by more than 2.5 times. Both of these effects are attributed to a dilution [10] caused by the CF_4 flow.

The low measured carbon content in the deposited films indicates that temperatures greater than 900°C, or the use of plasma enhanced equipment [11,12] are necessary for the dissociation the CF_4 molecule. It is speculated that this is due to the relative thermodynamic stability of CF_4 and the strong nature of the C-F bonds, when compared to hydrocarbon gases. Compared to methane, CF_4 is rather thermodynamically stable: Gibbs energy of CF_4 = -210 kcal/mole compared -12 kcal/mole for CH_4 [13]. Also, the C-F bond strength (128 kcal/mole) is greater than the C-H bond strength (80.6 kcal/mole) [13].

Certain hydrogen containing fluoro and chlorocarbon compounds, also non-flammable and relatively non-toxic, may be better candidates for use as carbon sources. Examples of such gases are CHF_3, $CHClF_2$ and CH_2F_2. These compounds are more reactive than CF_4 since they have less negative Gibbs energy and they contain C-H and/or C-Cl bonds which are weaker than C-F bonds [13]. These gases have not yet been investigated.

5.0 Depositions Using 1,1,1-Trichloroethane

When 1,1,1-trichloroethane was used in conjunction with silane, different results were obtained. The effect of TCA flow rate and temperature on the deposition rate is shown in Figure 2. The reduction in deposition rate which results from increased TCA flow, is not attributed to a dilution effect but rather, to a limited transport of TCA or its decomposed byproducts within the chamber.

The film structure is also observed to change with TCA flow rate. With no TCA flow, the films are grainy and hazy as discussed above. As the TCA flow is increased, deposited films on oxide or bare Si become specular with no visible structure. This is very much unlike the grain growth observed when CF_4 is added to the gas flow. The addition of TCA appears to result in films with amorphous and/or micro-crystalline structure.

ESCA measurements confirmed the successful realization of Si_xC_{1-x} alloys. For temperatures as low as 650°C, Si_xC_{1-x} films with greater than 15% carbon were obtained for TCA/SiH_4 flow ratios as low as 0.14. Up to 33% carbon content was measured in films deposited at 800°C with a TCA/SiH_4 flow ratio of 0.14. Optical bandgap measurements, described in [15], reveal that the SiC alloys are indeed wide-gap semiconductors.

Controlled grading of the carbon content has been found to be possible. ESCA measurements of such an example is shown in Figure 3. The deposition temperature and silane flow were maintained at 750°C and 36 sccm respectively while the TCA flow rate varied from 5.2 sccm to zero, in increments of approximately 1.3 sccm. The argon sputter etch rate of the film is approximately 18 to 20 A/min.

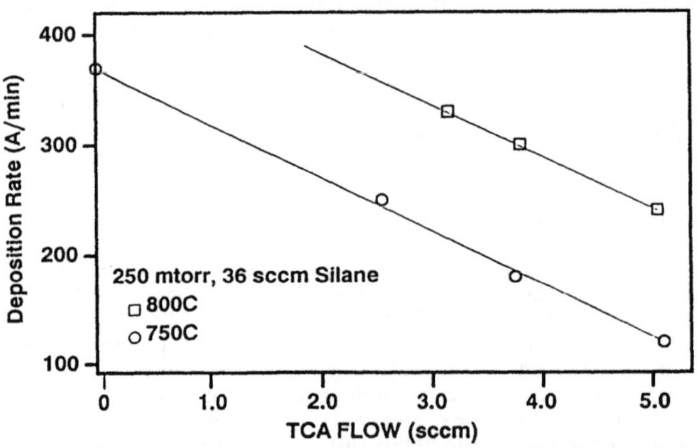

Figure 2 - Deposition rate versus TCA flow for a fixed SiH_4 flow of 36 sccm, 250 mtorr, at 750°C and 800°C.

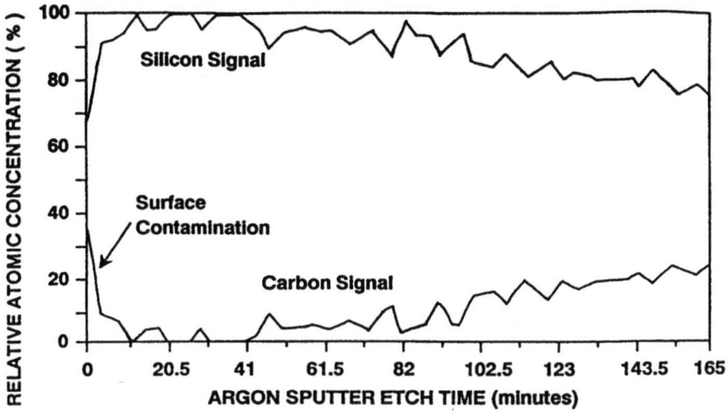

Figure 3 - Relative film composition versus depth for a graded SiC alloy deposited at 750°C.

6.0 Depositions Using 1,1,1-Trichloroethane and Arsine

Since the realization of emitters for silicon NPN heterojunction transistors is of interest, N-ype in-situ doping of the films using arsine (AsH_3) was investigated. When AsH_3 is added to the gas stream, N-type doping is realized. The deposition rate reduces to approximately 40 A/min at 750°C and 60 A/min at 800°C. This is attributed to a reduction in the density of available surface absorption sites [14] as well as a dilution effect caused by the H_2 portion of the dopant gas.

The measured as-deposited resistivity of films on bare <111> P-type Si wafers, as a function of AsH_3 flow, is shown in Fig. 4. The increase in resistivity with increasing TCA flow is attributed to an increase in the number of grain boundaries [16] within the film.

The resistivity is observed to be 1.5 to 3 times higher when the films are deposited on <100> P-type wafers instead of <111> wafers. Such a phenomena, observed in the case of thermally annealed ion implanted SiC films [17], may indicate that better quality films are obtained when <111> oriented Si is used.

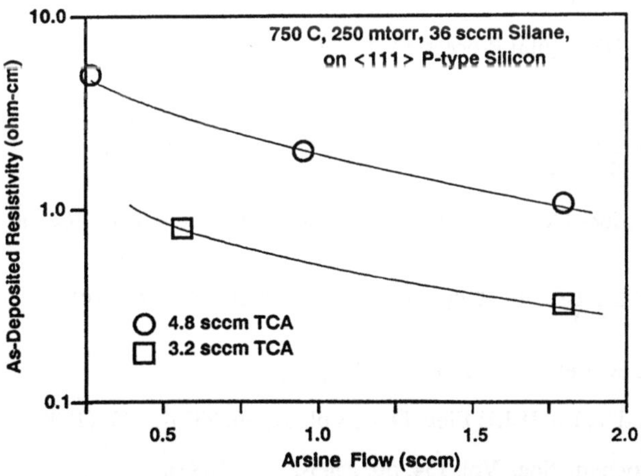

Figure 4 - Measured resistivity of the as-deposited N-type films on P-type <111> Si, as a function of arsine flow.

7.0 Conclusions

The deposition of SiC alloys with controlled carbon content, has been demonstrated using a hot-wall LPCVD reactor. Silane and 1,1,1-trichloroethane were used as the silicon and carbon sources respectively. Greater than 15% carbon content has been achieved at temperatures as low as 650°C. In-situ doping using arsine was also successfully achieved. Lower film resistivities are obtained when <111> rather than <100> oriented Si is used. Carbon tetrafluoride was also investigated as a carbon source. CF_4 was found to be less reactive since less than 2% carbon was measured in deposited films.

Acknowledgments. The author would like to thank Dr. Craig Snapp and Marc King for supporting this effort.

References

[1] T. Sugii, et.al., J. Electrochem. Soc., Vol. 136, pp. 3111-3115, (1989).

[2] M.I. Chaudry and R.L. Wright, J. Mat. Research, Vol. 5, pp. 1595-1598, (1990).

[3] Y. Ohshita, Appl. Phys. Lett., Vol. 57, pp. 605-607, (1990).

[4] K. Sasaki, M.M. Rahman and S. Furukawa, IEEE Elec. Device Lett., Vol. EDL-6, pp. 311-312, (1985).

[5] M. Kuwagaki, K. Imai and Y. Amemiya, Japan. J. Appl. Phys., Vol. 28, pp. L754-L756, (1989).

[6] S.M. Irving, et.al., US Patent No. 3,615,956, filed (1969).

[7] E. Janssens and G.J. Declerck, J. Electrochem. Soc., Vol. 125, pp. 1696-1703, (1978).

[8] EXTREMA-grade TCA, a product of J.C. Schumacher Co. (Oceanside, CA, USA).

[9] Charles Evans & Associates, Redwood City, CA, USA.

[10] T. Kamins, Polycrystalline Silicon for Integrated Circuit Applications, Kluwer Academic Publishers, Chapter 2, (1988).

[11] J.L. Vossen and W.Kern, "Thin Film Processes", Academic Press, part IV-1, p. 354, (1978).

[12] G. Ganguly, S.C. De, S. Ray and A.K. Barua, "Polycrystalline silicon carbide films deposited by low-power RF plasma decomposition of SiF_4-CF_4-H_2 gas mixtures", J. Applied Physics, Vol. 69, pp. 3915-3923, (1991).

[13] R.C. Weast (editor), Handbook of Chemistry and Physics, CRC Press, Boca Raton, FL, USA, (1982).

[14] T. Kamins, Polycrystalline Silicon for Integrated Circuit Applications, Kluwer Academic Publishers, p. 32, (1988).

[15] F. Hébert, IEEE Elec. Dev. Let., Vol. 12, pp. 477-479, (1991).

[16] T. Sugii, T. Yamazaki and T. Ito, IEEE Elec. Dev., Vol. 37, pp. 2331-2335, (1990).

[17] T. Sugii, et. al., J. Electrochem. Soc., Vol. 134, pp. 2545-2549, (1987).

Low Temperature PECVD Growth and Characterization of a-SiC:H Films Deposited from Silacyclobutane and Silane/Methane Precursor Gases

M.J. Loboda

Dow Corning Corp., Midland, MI 48686, USA

Research at Dow Corning includes the development of a-SiC:H films as protective coatings for electronics applications where the deposition temperature cannot exceed 250 °C. Results from a detailed study comparing a-SiC:H films deposited by electron cyclotron resonance (ECR) PECVD at 175 °C substrate temperature are presented. Two source gases were used: silacyclobutane, an organosilicon material produced in our laboratories, and a silane/methane mixture. Deposition properties, film structure and atomic composition have been characterized by ellipsometry, profilometry, internal reflection infrared spectroscopy, micro-Raman spectroscopy, X-ray photoelectron spectroscopy (XPS), Auger electron spectroscopy (AES) and Rutherford Backscattering/Hydrogen Forward Scattering spectrometry (RBS/HFS). The low temperature growth process on silicon substrates results in a-SiC:H films that are under compressive stress. All samples had a typical hydrogen content of 50 atomic percent and the average carbon/silicon ratio varied from 1.0-1.9. The silacyclobutane precursor gave higher film growth rates and higher carbon concentrations, while more oxygen is observed in the films deposited from the silane/methane mixture. XPS and Raman data show silicon carbide-like bonding and amorphous carbon/DLC bonding, respectively. The results indicate that the precursor molecular structure can be correlated with different film growth properties and composition, and also that the process conditions can influence film properties to a similar degree.

1.0 Introduction

Research on thin films of amorphous hydrogenated silicon carbide (a-SiC:H) continues to expand, from its beginnings as an extension of studies on amorphous hydrogenated silicon (a-Si:H) [1-3], into more recent applications such as semiconductor devices [4,5] and passivating/wear resistant coatings [6,7]. Deposition of a-SiC:H films is most commonly performed using a plasma enhanced chemical vapor deposition process (PECVD) or by reactive ion beam sputtering [1-7]. Typical of standard PECVD processes are precursors that are mixtures of silane and an alkane or alkene such as methane or ethylene, although the use of single precursors containing one or more silicon-carbon bonds has been investigated. Reports of the use of these organosilicon materials include the molecules containing four Si-C bonds investigated by Gerault, et. al., [8] to deposit a-SiC films by hot-wall CVD, and films grown by PECVD from methylsilane [9], and tetramethylsilane [10]. Among the advantages of organosilicon materials are their inherent safety relative to pure silane, the simplicity of using a single precursor, and the potential to tailor and create new radicals in the plasma, including deposition species with an intact Si-C bond.

Recently, Chandra reported on a new precursor for use in the low temperature (T<250 °C) PECVD growth of a-SiC:H films. This precursor, silacyclobutane (SiC_3H_8), was used to deposit a-SiC:H films using a parallel plate PECVD apparatus. The resulting films have been successfully used to enhance the reliability of integrated circuits [11]. Reported herein are results of continued research on PECVD of silacyclobutane (SCB). Microwave electron cyclotron resonance (ECR) PECVD was used to deposit films in the 175 °C temperature range. This technique was chosen on the basis of its capability to deposit high quality insulating films with the use of mild plasma conditions [12]. To help shed light on the a-SiC:H film growth process, a direct comparison was made against films deposited from a silane-methane mixture with silicon to carbon ratio equal to that in SCB. The purpose of this study is two-fold; first, to attempt to define correlations between the CVD precursor, the plasma decomposition, and the deposited film; second, to further demonstrate the efficacy of SCB as an alternative precursor for the deposition of a-SiC:H films.

2.0 Experiments and Results

2.1 a-SiC:H Film Deposition

Amorphous hydrogenated silicon carbide films were deposited at the applications laboratory of Astex, Inc., Woburn, MA. Figure 1 shows the microwave ECR deposition system used. The mass flow coefficient (relative to nitrogen) for SCB was determined from leak rate experiments to be approximately 0.2, while the value for argon (1.44) was used for the 10% silane/argon mixture. The deposition conditions are listed in Table 1.

Table 1 - a-SiC:H Deposition Conditions

i) Precursor Mixes and Flow:			
Mixture A:	Silacyclobutane (SCB)=3.3 sccm,		Ar=95 sccm
Mixture B:	SiH_4 (10% in Ar)=12 sccm,	CH_4 =3.6 sccm,	Ar=84 sccm
Mixture C:	SiH_4 (10% in Ar)=50 sccm,	CH_4=15 sccm,	Ar=35 sccm
ii) Chamber Pressure:	Base Pressure	3-5 μTorr	
	Deposition Pressure	3-4 mTorr	
iii) Microwave Power:	300-500 Watts		
iv) Wafer Temperature:	175-190 °C		
v) Substrate:	100mm Silicon wafers, <100> and <111> orientations		

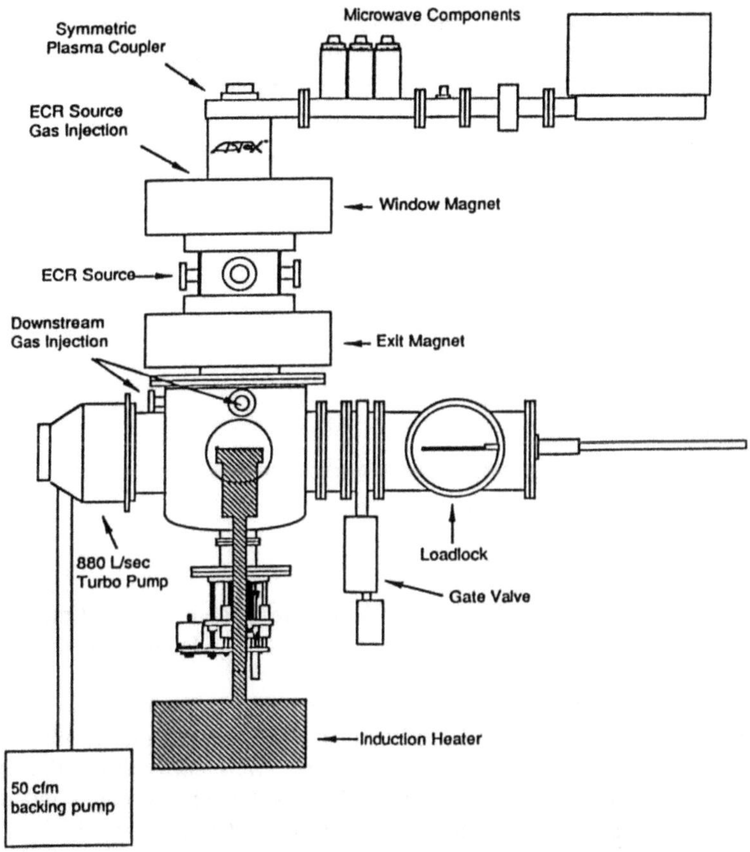

Figure 1. The ECR based PECVD system used to deposit the a–SiC:H films.

As shown in Figure 1, silicon wafers were placed onto a graphite susceptor whose temperature was controlled by an induction heating system. In order to be confident of the wafer temperature during processing, a thermocouple instrumented wafer was placed into the chamber. Using a gas flow of 100 sccm of argon and a pressure of 3 mTorr, the wafer was exposed to argon plasmas generated over a microwave power range of 300-500 W, with no power from the induction heater. The plasma was turned off after 10 minutes, the wafer temperature was measured, and then the power increased. Figure 2 plots the measured wafer temperature and the susceptor temperature, as a function of plasma power. It was found that for a fixed power input, the wafer temperature reached equilibrium in about 15-20 minutes. The steady-state temperatures were nominally 20 °C higher than those shown in Figure 2. The data show that for the conditions used, the wafer is thermally decoupled from the susceptor. For the low temperature applications desired, the presence of the plasma makes significant contribution to the wafer heating and cannot be neglected if wafer temperatures in the range of 175 °C are to be maintained. Based on these results, 500 W was determined to be the maximum microwave power allowable.

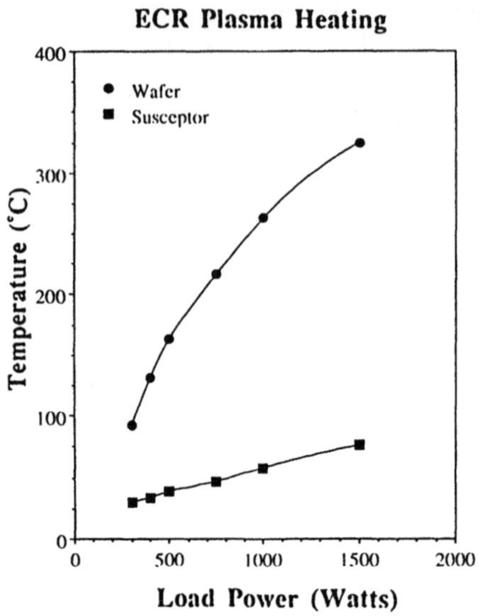

Figure 2. ECR plasma heating of the silicon subtrate and the graphite susceptor due to 10 minute exposure times.

Further measurements at power levels of 300-400 W were made in conjunction with induction heating of the susceptor to determine the conditions leading to ~175 °C wafer temperature.

After deposition, film thickness was measured using a Rudolph Auto-EL ellipsometer (632 nm) and a Tencor P-1 profilometer. The ellipsometer results, an average of 5 points spread over a 75mm diameter circle, were correlated with step height measurements made on a masked section of the wafer. The average standard deviation in film thickness over the wafer surface was 4.33±2.11%, and thickness uniformity showed no correlation to gas mixture used. In some cases film thickness was also measured using RBS [13]. Typical agreement with the profilometry/ellipsometry data was within 2-3%, with the worst disagreement being 6%. The average deposition rates obtained for the three gas mixtures used are shown in Table 2.

Table 2 - a-SiC:H Deposition Rates

Precursor Gas Mix	Power (W)	Deposition Rate (angstroms/min)
A	300	197.9
B	300	55.3
C	300	219.8
A	400	220.7
C	400	147.3
A	500	234.7
C	500	212.8

2.2 Film Stress

Film stress was measured using the long scan capability of the Tencor P-1 profilometer and the Tencor film stress analysis package. Each wafer was scanned over the center 90% of its diameter. Many samples were scanned twice, in mutually perpendicular directions. In all cases the wafer bow was uniform, and was indicative of compressive film stress. Table 3 compares the raw stress data and the data re-normalized to a standard 0.5 mm film thickness.

Table 3 - a-SiC:H Film Stress - All Values are Compressive Film Stress

	Measured Stress (10^9 dyne / cm^2)		*0.5 μm Normalized Stress (10^9 dyne / cm^2)*	
Power (W)	Gas "A"	Gas "C"	Gas "A"	Gas "C"
300	3.464	3.691	1.895	1.875
400	3.355	4.164	4.357	3.681
500	3.701	4.714	5.199	5.778

(Gas "A"=SCB Gas "C"=3:1 CH_4/SiH_4)

2.3 Molecular Bonding

Molecular bonding in the a-SiC:H films was characterized using infrared and Raman spectroscopy, and XPS.

Infrared spectra were measured using the internal reflection technique [14]. A Harrick Scientific (Ossining, NY) "Seagull" reflectance accessory mounted into a Nicolet 5SXB FT-IR spectrometer was used. Infrared light polarized perpendicular to the plane of incidence was coupled into the film through a germanium hemisphere at an incident angle of 80°. Figure 3 illustrates the measurement configuration and

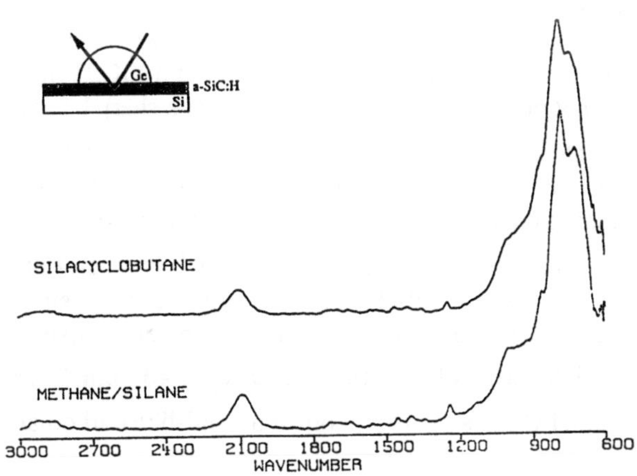

Figure 3.
Infrared spectra of the a-SiC:H films deposited from Mixtures A and C at 300 W and 175 °C. A schematic of the internal reflection technique used is shown in the upper left corner.

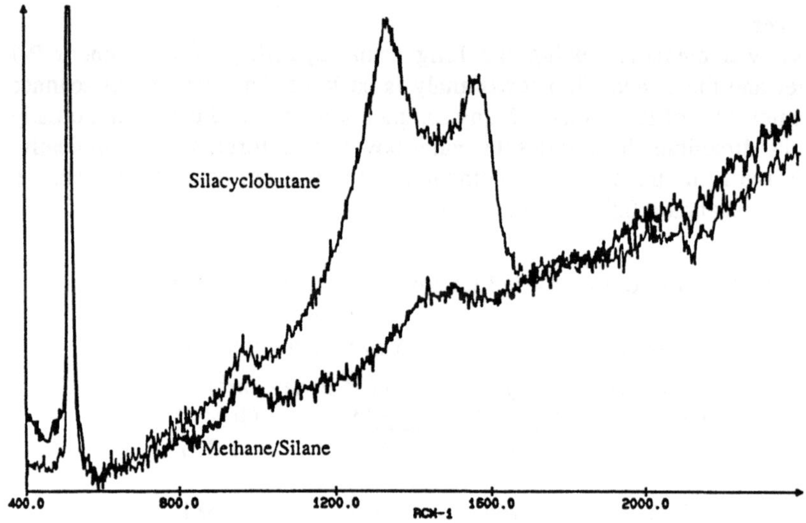

Figure 4. Micro-Raman spectra of the a–SiC:H films deposited from Mixtures A and C at 500 W and 175 °C.

shows typical spectra obtained on the a-SiC:H films deposited at 300W using the different precursors (mixtures A&C). The spectra measured with this technique are free of the interference phenomena which often distort transmission IR spectra of a–SiC:H films. Clearly observable in the spectra are the C-H stretching region (~2900 cm^{-1}), the Si-H stretching region (~2090 cm^{-1}), the symmetric bond bending of CH_3 groups attached to Si (~1250 cm^{-1}), and the prominent absorption centered near 780 cm^{-1}. This peak has been assigned to a number of overlapping bands, including the rocking of CH_n (n=2,3) groups attached to Si, Si-C stretching vibrations from continuous Si-C networks, and at the low frequency edge, the wagging of Si-H bonds [see 1,3 and references therein].

Raman spectra were measured using an Instruments S.A./Jobin-Yvon U-1000 MOLE Raman microscope. A 30 mW, 514.5 nm Ar ion laser line was used for sample excitation. Figure 4 shows the Raman spectra obtained on the a-SiC:H films deposited at 500W using the different precursors (mixtures A&C). The sharp band at ~520 cm^{-1} is due to the Si substrate. The remaining part of the spectra, consisting of very weak bands is typical of an amorphous network. Hicks, et. al., [15], has reported assignments of the Raman bands measured on a-SiC:H films deposited by glow discharge of silane/methane mixtures. The band near 960 cm^{-1} is mostly due to the two-photon absorption of the Si substrate, but may contain small contributions from Si-C networks in the film. Since these higher power deposition films were particularly carbon rich (see below), the bands at 1340 cm^{-1} and 1560 cm^{-1} associated with the D and G lines in amorphous/diamond-like carbon are also observed [16].

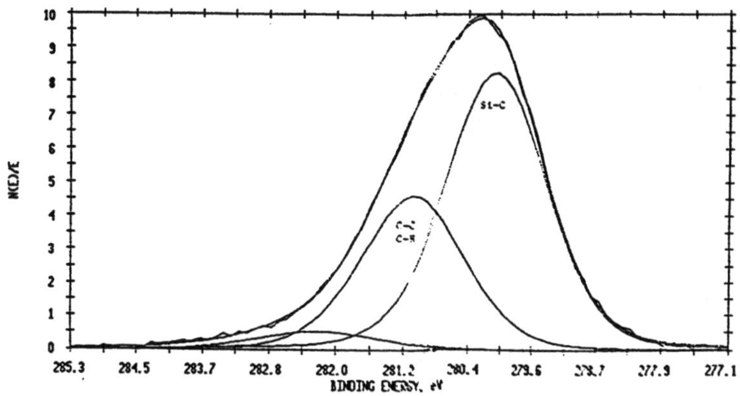

Figure 5. Typical carbon 1s photoelectron peak measured using XPS. This sample a–SiC:H film was deposited using SCB at 500 W and 175 °C.

Molecular bonding information obtained from XPS measurements on a-Si:C:H films is often reported as part of a-SiC:H film characterization [15,17,18]. XPS measurements were performed using a Perkin-Elmer 5400 ESCA system, the details of these measurements are reported elsewhere [13]. Deconvolution of the carbon 1s photopeak is often used to delineate carbide bonded carbon from homonuclear and hydrogen bonded carbon. Figure 5 shows a typical carbon 1s photoelectron peak measured on a sputter cleaned a-SiC:H film deposited at 300W from SCB, with a deconvolution algorithm applied to the data to separate the peaks [18]. Similar data measured on other samples deposited from both precursors showed only small variations from Figure 5. The XPS data obtained on the films described herein is of the same variety as those reported elsewhere [15,17,18].

2.4 Elemental Composition

Electron spectroscopy (XPS/AES) and high energy ion beam spectrometry (RBS/HFS) were used to analyze the elemental composition of the deposited a–SiC:H films [13]. The AES measurements were done using a Perkin-Elmer 600 SAM with 10kV accelerating voltage. Quantitation was performed using a hydrogen free, high temperature silicon carbide sample with known composition from RBS measurements. XPS measurements are described above, and elemental sensitivities were used for composition calculations based on integrated peak values. In both the XPS and AES measurements, the samples were sputter cleaned with an argon ion gun until the oxygen signal reached a consistent minimum value. RBS/HFS was performed using 2.3 MeV helium ions, and backscattering angles of 160° and 105° for RBS, and a scattering angle of 30° for HFS. Table 4 shows the C:Si ratio calculated by averaging the results from the XPS, AES and RBS data, the hydrogen concentration from HFS, and the film density as calculated from the RBS data.

Oxygen incorporated into the a-SiC:H films during and after growth is of concern in many applications [5,7] as it will have a direct impact on the passivating and

Table 4 - a-SiC:H Elemental Composition, Density and Refractive Index

Power (W)	Precursor Mix	Average C:Si Ratio	Hydrogen Content (at.%)	Density (g/cc)	Refractive Index (632nm)
300	A	1.52±.06	53±5	1.91	2.202
300	C	1.07±.06	52±5	1.93	2.508
500	A	1.86±.12	59±6	1.68	2.199
500	C	1.34±.10	53±5	1.79	2.204

Table 5 - a-SiC:H Film Oxygen Content by XPS

		(atom %)	
Power (W)	Precursor Mix	Surface Oxygen	Bulk Oxygen
300	A	23.6	4.5
300	C	23.2	6.7
500	A	20.7	3.0
500	C	21.2	3.8

corrosion resistance properties of the coatings. Table 5 shows the atomic concentrations of oxygen at the film surface and in the bulk as measured by XPS.

3.0 Discussion

As previously stated, one of the primary goals was to search for correlations between the CVD precursor, the plasma decomposition, and the deposited film. The choice of precursors and concentrations was made with this in mind. Silacyclobutane (SCB) contains two Si-C bonds, and is a strained ring molecule. It should be easily decomposed, and plasma species containing Si-C bonds may be produced. The two methane/silane mixtures were chosen such that mix B would treat the methane/silane combination as a single Si-C source, diluting it to 5% of the total flow, while mix C provided a flow of silicon species into the plasma comparable to the SCB. In both B and C, the Si/C ratio is the same as SCB, but of course there are no initial Si-C bonds.

The deposition rate analysis shows that the growth of the a-SiC:H film is mostly controlled by the silicon flow concentration, since films deposit at a comparable rate for mixtures A and C. Changing the power delivered to the plasma appears to have a much stronger influence on the deposition behavior of the methane/silane mixture. Review of the composition data shows that the plasma also has a profound effect on the carbon to silicon ratio in the film. It is interesting to note that for both precursors, an increase in load power caused an increase in C/Si ratio, 22% for SCB and 25% for methane/silane. Since the deposition rate remained relatively constant with power for the SCB, and in both cases the majority plasma species was argon, the increased carbon concentration in the 500 W depositions is probably due to sputter etching of weak Si-Si bonds by energetic argon ions and not increased production of depositing carbon species in the plasma. Such an explanation is consistent with the observed decrease in film density with increased power. Examination of the film stress data suggests the sputtering effects which lead to increased carbon concentration influence stress more than the precursor material.

The SCB consistently produced more carbon rich films. Although the hydrogen flow concentration is lower in the SCB depositions, we believe this quality is due to the decomposition of the SCB molecule itself, as Hicks, et. al., [15] reported a weak influence of hydrogen dilution on the stoichiometry of films deposited from silane and methane at high power levels. For the deposition conditions investigated the relative flow of hydrogen into the plasma, or the nature of its bonding in the precursor, appear to have little or no influence on hydrogen concentration in the film.

For both precursors, the molecular bonding is observed in the a-SiC:H films is similar to that previously reported [1,3,15,17,18]. Comparison of the films deposited from the different precursors shows that subtle differences are visible in the IR spectra of Figure 3, almost entirely in the 1100-600 cm^{-1} region, and the Raman spectra of Figure 4. These slight differences are probably related to the higher carbon concentrations found in the SCB/a-SiC:H films. The information obtained from XPS, shows both carbide and hydrocarbon bonding, but at the same time does not provide information on the polymeric bonding network in the film. Some insight into precursor dependent film properties may come from the slightly higher oxygen content observed in the films deposited from methane/silane. Since the deposition system achieved a 1000 times lower base pressure prior to deposition, and used a load lock, most of the oxygen was probably incorporated after deposition, thus implying a higher oxygen diffusion rate and reactivity may exist in the methane/silane films. Such an observation is consistent with the recent oxidation studies performed in our laboratory, where the methane/silane films show a larger decrease in the hydrogen related IR absorptions upon annealing in air at 450 °C. These results may suggest that differences in the a-SiC:H film structure that can traced back to the precursor are more likely to be found in microstructure of the film.

4.0 Conclusion

Amorphous hydrogenated silicon carbide films have been deposited from silacyclobutane, an organo-silicon ring molecule. A comparison of this process was made against methane/silane mixtures deposited under identical conditions. Correlations between the plasma power, the carbon concentration, film density and the film stress were identified. The experimental results indicate that the differences in the precursor bonding can be correlated with different growth properties, elemental composition, and oxygen diffusion rates. Relative to precursor influence on film properties, it is also observed that that plasma process (power, carrier gas, etc.) can have comparable influence on the film properties. Studies in our laboratory continue including comparisons of mechanical properties and oxidative stability.

5.0 Acknowledgements

The author acknowledges the valuable assistance of Dr. L. Borget, Astex, Inc.; Mr. J.A. Seifferly, Dow Corning; Dr. M. Edgell and Mr. S. Baumann, of Charles Evans and Associates; Dr. K. Stolt, Analytical Answers, Inc.; and Dr. D.S. Knight, MRL-Pennsylvania State University.

6.0 References

[1] Y. Catherine, G. Turban, Thin Solid Films, **60** (1979), p. 193-200

[2] Y. Katayama, T. Shimada, K. Usami, S. Ishioka, J. Journal of Appl. Phys., **19** (1980), Suppliment 19-2, p. 115-118

[3] H. Wieder, M. Cardona, C. Guarnieri, Phys. Stat. Sol. (B), **92** (1979), p.99-112

[4] M. Rahman, C. Yang, D. Sugiarto, A.S. Byrne, M. Ju, K. Tran, K. Lui, T. Asano, W. Stickle, J. Appl. Phys., **67** (1990), p. 7065-7070

[5] W. J. Lu, A.J. Steckl, T.P. Chow, W. Katz, J. Electrochem. Soc., **131** (1984), p.1907-1914

[6] M.A. Bayne, Z. Kurokawa, N.U. Okorie, B.D. Poe, L. Johnson, R.W. Moss, Thin Solid Films, **107** (1983), p.201-206

[7] J. Eldridge, J. Moore, G. Olive, V. Dunton, J. Electrochem. Soc., **137** (1990), p.2266-2271

[8] J. Gerault, R. Morancho, G. Constant, P. Mazerolles, J.J. Ehrhardt, M. Alnot, Thin Solid Films, **101** (1983), p.83-96

[9] M. Delplancke, J. Powers, G. Vandentop, M. Salmeron, G. Somorjai, J. Vac. Sci. Tech. A, 9 (1991), p.450-455

[10] S. Rynders, A. Scheeline, P.Bohn, J. Appl. Phys., **69** (1991), p.2951-2960

[11] G. Chandra , Mat. Res. Soc. Symp. Proc. Vol. 205 - Electronic Packaging Materials Science V, 1991, p. 97-108

[12] Y. Sugiyama, S. Fujimori, H. Yamazaki, I. Hatakeyama, Appl. Phys. Lett., **56** (1990), p.1403-1405

[13] M. Loboda, S. Baumann, M. Edgell, K. Stolt, to be published.

[14] N.J. Harrick, Internal Reflection Spectroscopy, 1979, Harrick Scientific Corporation

[15] S.E. Hicks, A.G. Fitzgerald, S.H. Baker, T.J. Hines, Phil. Mag. B, **62** (1990), p.193-212

[16] D. Knight and W. White, J. Mater. Res. , **4** (1989), p. 385-393

[17] W-Y. Lee, J. Appl. Phys. , **51** (1980), p.3365-3372

[18] A.S. Byrne, M. Ju, T. Asano, M.M. Rahman, C.Y. Yang, Springer Proc. in Physics, **Vol. 43**, Amorphous and Crystalline Silicon Carbide and Related Materials II, eds. M.M. Rahman, C.Y. Yang, and G.L. Harris. 1989, p.81-85

Study of Optimum Condition for Microcrystalline SiC Film Formation by ECR Plasma CVD Method

T. Miyajima, K. Sasaki, and S. Furukawa*

Department of Applied Electronics, Graduate Schoof of Science and Engineering, Tokyo Institute of Technology, 4259 Nagatsuta, Midori-ku, Yokohama 227, Japan

*On leave of absence from Nippondenso Co., Ltd.

Abstract. Microcrystalline SiC (μc-SiC) thin films have been obtained by using the electron cyclotron resonance (ECR) plasma CVD method. The result from X-ray diffraction measurement shows those films contain crystalline phase SiC without crystalline phase Si. And, the use of $Si(CH_3)_2H_2$ as a source gas has been found effective for μc-SiC formation by ECR plasma CVD method.

1. Introduction

It is needed to develop low temperature formation of highly conductive wide bandgap emitter materials, i.e., a-SiC [1], μc-SiC [2], μc-Si [3], in order to realize high-speed Si based heterojunction bipolar transistors (Si-HBTs). But, there are some problems as follows : By the conventional R.F. plasma CVD method, 1) only relatively high resistance SiC films were usually obtained, so that electronic property in HBT degrade due to high emitter resistance, on the other hand, 2) crystalline Si tend to be grown on the substrate when low resistive μc-Si is attempted grown instead of SiC [2], so that heterojunction property degrades. Therefore, if we could grow low resistive SiC films at low temperature, such films would be useful for application to Si-HBTs.

For such purpose μc-SiC films consisting of only crystalline phase SiC is necessary. The C-H bonds of source gas for carbon species are difficult to break by conventional R.F. plasma, therefore high excitation plasma is desired. Plasma generated by high power R.F. may give a damage to the emitter-base interface. ECR plasma CVD method which is capable to produce a high excitation plasma and low damage to the interface is successfully employed in this study, in order to prepare genuine μc-SiC films in such meaning. In this paper the effects of various deposition parameters, e.g. hydrogen dilution ratio, choice of source gases, and microwave power to produce μc-SiC by ECR plasma CVD will be discussed.

2. Experimental

Phosphorus-doped μc-SiC films were prepared by ECR plasma CVD method by using SiH_4-CH_4-H_2 or $Si(CH_3)_2H_2$-CH_4-H_2 gas mixture. PH_3 gas was 1% doped to the source gases. Some of the films were prepared by adding Ar gas to them. The gas pressure was kept around 1 mTorr and microwave powers were 300 ~ 900 W in depositions. Figure 1 shows a schematic diagram of the ECR plasma CVD apparatus. H_2 and CH_4 gases were introduced into the excitation chamber and the other gases were introduced into the deposition chamber. The 2.45 GHz microwave was

Editors: C.Y. Yang · M.M. Rahman · G.L. Harris

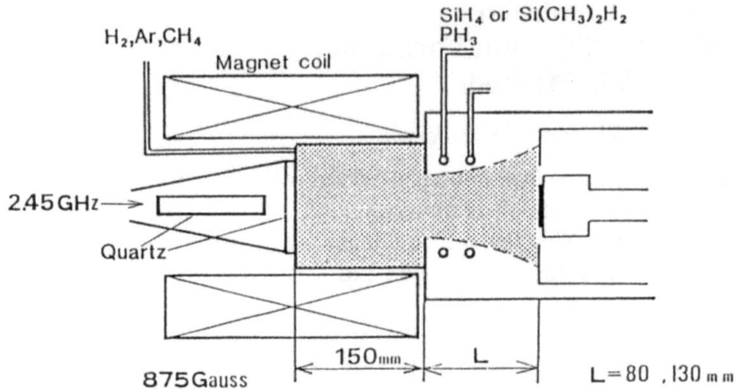

Figure 1 Schematic diagram of ECR plasma CVD apparatus.

Table 1 Deposition conditions of films by ECR plasma CVD.

Sample	Power (w)	Temp. (°C)	Source Gases	C/Si	H/(Si+C)	L (mm)
A	300	350	SiH_4+CH_4	1	13.0	130
B	300	600	SiH_4+CH_4	1	120.0	130
C	500	600	SiH_4+CH_4	2.25	105.6	130
D	500	600	SiH_4+CH_4	2.25	105.6	80
E	500	600	$Si(CH_3)_2H_2+CH_4$	3	69.5	80
F	800	600	$Si(CH_3)_2H_2+CH_4$	3	69.5	80
G	800	600	SiH_4+CH_4	3	70.8	80

introduced through the quartz window into the excitation chamber in which magnetic flux density of 875 Gauss was applied. The excitation chamber was 150 mm long and the distance from a substrate to the aperture of the excitation chamber was 80 or 130 mm. The μc-SiC films were deposited on a quartz and silicon substrate. Some of the deposition conditions are summarized in Table 1. The films were characterized by X-ray diffraction, Fourier Transform Infrared (FTIR), Electron Probe Micro Analysis (EPMA) and electrical conductivity.

3. Results and Discussion

Hydrogen Dilution Effect

First, we shall consider the case of using SiH_4 and CH_4 as source gases. Figure 2 shows the dependence of conductivity of the deposited films on H_2 dilution ratio. When carbon fraction of source gases, e.g. CH_4/SiH_4 was 1, low conductive a-SiC films (sample(A)) were obtained in

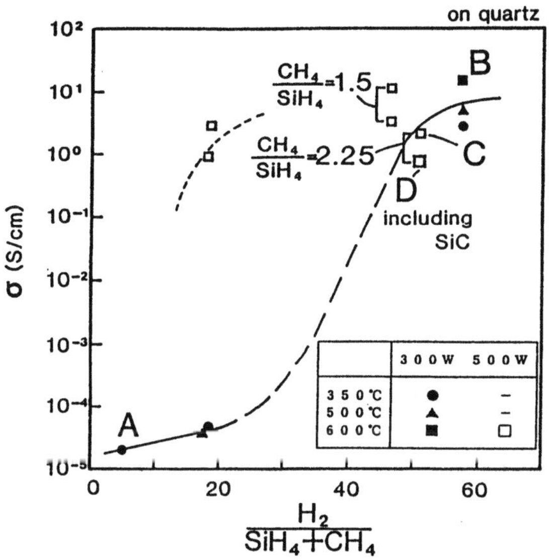

Figure 2 Relation between conductivity on quartz substrate and hydrogen dilution ratio. The data without stating CH_4/SiH_4 ratio are obtained under $CH_4/SiH_4=1$.

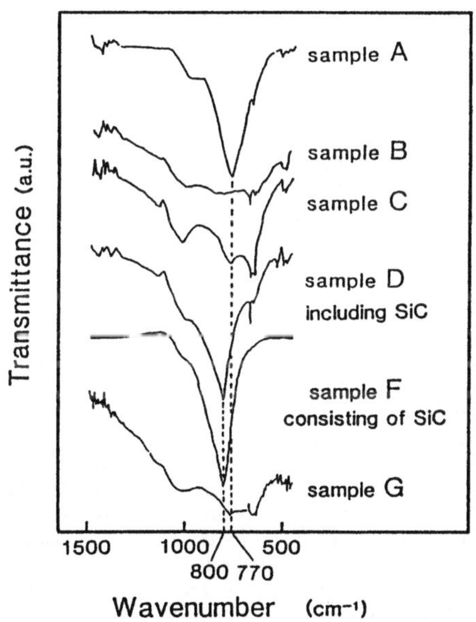

Figure 3 FTIR spectra of SiC films.

lower hydrogen dilution condition, while μc-Si films were obtained if not adding CH_4. As increasing the hydrogen dilution ratio, conductivity of films increased up to around 10 S/cm (sample(B)) but carbon atoms were not included in films. As shown in Figure 3, IR spectrum peak at around 770 cm^{-1}, which assigns $Si\text{-}CH_3$ bonds, was observed in sample(A) and disappeared in sample(B). From the X-ray

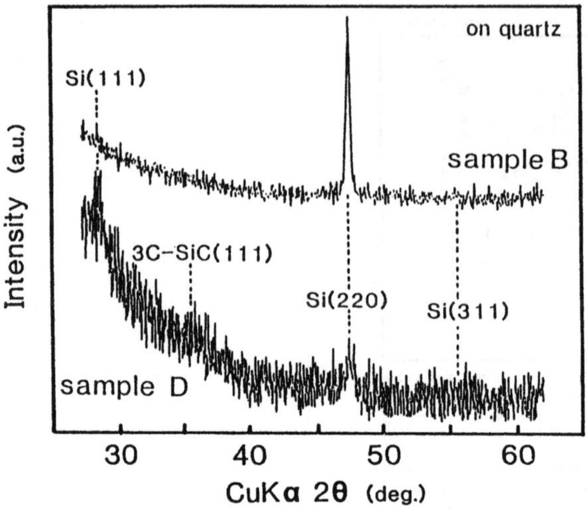

Figure 4 X-ray diffraction patterns of films deposited on quartz substrates.

diffraction measurement as shown in Figure 4, Si has been found grown. Only around 1 % carbon atoms content was measured by EPMA measurement in this sample. Besides, poly-Si can be formed even at low substrate temperature of 350 ℃. Similar phenomena were observed when the microwave power was increased up to 500 W, even though the hydrogen dilution ratio was decreased. These results are interpreted as follows : In the high hydrogen dilution condition, a large amount of hydrogen radicals produced eject carbon atoms from the film surface. In addition CH_3 species from the plasma combine with hydrogen atoms covering the film surface and they desorb from the surface as CH_4 molecules. That is, the hydrogen radicals promote crystallization of silicon, when the source gases are diluted by hydrogen gas.

Source Gas Ratio Effect

Next, carbon fraction of source gases was increased to introduce higher content of carbon atoms in films. Highly conductive films were obtained at carbon fraction of source gases 1.5. But they were not μc-SiC but poly-Si. It was found that highly conductive films (sample(C)) could be obtained even by further increasing carbon fraction of source gases up to 2.25 at substrate temperature of 600 ℃. In this case, carbon atoms were introduced in the film as shown in Figure 3. But only weak X-ray diffraction peaks for crystalline Si were observed, suggesting μc-Si containing a-SiC. It is noted that not only the presence of Si-C bonds of around 800 cm^{-1} has been observed by IR spectra but also the presence of 3C-SiC (111) has been confirmed by X-ray diffraction measurement as shown in Figure 3 and 4 for sample(D) when the substrate was close to the excitation chamber. This phenomenon is interpreted by change of the condition of radicals and substrate potential. These results suggest that substrate location, which has been discussed in few reports, is one of the important parameters to prepare genuine μc-SiC films by ECR plasma CVD. The X-ray diffraction peaks for Si were present and the peak for 3C-SiC was weak as shown in Figure 4. In order to increase the crystalline phase SiC, carbon fraction of source gases was increased. As the result, it is found that more carbon atoms could be introduced in the films, but that only a-SiC films could be obtained and they were low conductive films.

Source Gas Effect

We have not obtained highly conductive and wide bandgap μc-SiC under SiH_4 and CH_4 gas combination even if we had changed various parameter. We examined $Si(CH_3)_2H_2$ gas instead of SiH_4 as the source gases under the hydrogen dilution condition (H/(Si+C)) of 69.5. At 500 W microwave power, the weak X-ray diffraction peak for 3C-SiC was observed (sample(E)). As increasing microwave power up to 900 W at substrate temperature of 600 ℃, the large X-ray diffraction peak for 3C-SiC was clearly observed. Moreover interestingly, the peaks for Si were completely disappeared (sample(F)) as shown in Figure 5. The crystallite size estimated from FWHM of 3C-SiC (111) peak in X-ray diffraction pattern was 3 nm. The film contained 50 % carbon atoms. Though crystalline SiC phase was obtained, those films showed low conductivity (around 1×10^{-3} S/cm). This result suggested phosphorus atoms were not activated in the film and/or electrons are trapped by a large number of localized states. In order to make clear the effectiveness of $Si(CH_3)_2H_2$ gas for producing crystalline phase SiC, films were deposited by using SiH_4 and CH_4 gases under the same condition as the case of $Si(CH_3)_2H_2$ as shown in Table 1. As the result as shown in Figure 5 for sample(G), only poly-Si was obtained. This film contained only 1 % carbon atoms. The reason why the SiC crystalline phase is tend to be produced by using $Si(CH_3)_2H_2$ gas is considered as follows: $Si(CH_3)_2H_2$ gas has Si-C bond initially. These bonds may exist regularly in the film if these bonds are preserved even in the films. As we have mentioned, the ECR plasma CVD method as a deposition technique is useful for producing crystalline phase SiC at low temperature. It can be said that the use of gas sources having Si-C bonds such as $Si(CH_3)_2H_2$ gas is a key technique for producing μc-SiC. Though we have not obtained a sufficiently high conductivity for emitter material, we think its conductivity will be improved if we chose appropriate dopant for example nitrogen.

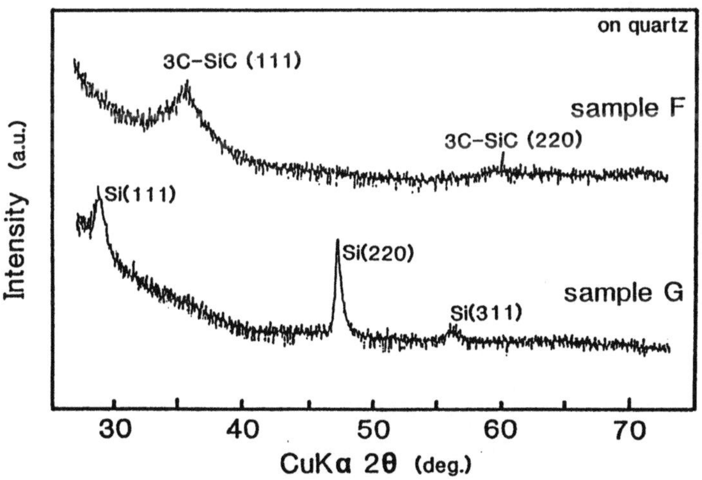

Figure 5 X-ray diffraction patterns of films deposited on quartz substrates.

4. Conclusion

μc-SiC films which contains crystalline SiC phase without crystalline phase Si have been obtained by using the ECR plasma CVD method.

Key techniques for producing μc-SiC by ECR plasma CVD method as follows : 1) Under heavy hydrogen dilution condition, which promotes crystallization, carbon is not introduced into the deposited films. 2) Location of substrate is one of the important parameters to prepare crystalline SiC phase. 3) The use of $Si(CH_3)_2H_2$ as source gas is effective for μc-SiC formation.

We consider that the ECR plasma CVD method is useful to obtain μc-SiC.

References

[1] K.Sasaki, M.M.Rahman and S.Furukawa : IEEE ED Lett. **EDL-6**, pp.311-312, 1985.

[2] T.Fukazawa, K.Sasaki and S.Furukawa : Springer Proc. in Phys. **43**, pp.49-53, 1989.

[3] K.Sasaki, T.Fukazawa and S.Furukawa : 1987 Intl. Electron Devices Meeting, Tech. Dig. pp. 186-189.

Part IV

Applications

Applications for 6H-Silicon Carbide Devices

J.W. Palmour, J.A. Edmond, H.-S. Kong, and C.H. Carter, Jr.

Cree Research, Inc., 2810 Meridian Parkway,
Durham, NC 27713, USA

Abstract. A variety of devices with promising characteristics have recently been demonstrated that confirm the potential for 6H-SiC. There are four primary application areas for 6H-SiC devices: 1) optoelectronics, 2) high temperature electronics, 3) radiation hard electronics, and 4) high power / high frequency devices. Optoelectronic devices include blue light emitting diodes and UV photodiodes. High temperature devices cover almost any device in SiC, from pn junctions to FET-based integrated circuitry. Radiation hard and EMP hard 6H-SiC devices can greatly improve the survivability of military systems and enhance nuclear reactor electronics. High power and high frequency devices in 6H-SiC can increase the power density of solid state devices by at least a factor of four over what is currently available and can also increase the operating temperature of these devices.

1. Introduction

Silicon carbide has long been known to have excellent physical and electronic properties which should allow production of electronic devices that can operate at significantly higher temperatures and power levels than devices produced from Si or GaAs. The most important of these properties are its wide bandgap (3.0 eV for 6H-SiC at room temperature), high electric breakdown field of $4x10^6$ V/cm [1], and high thermal conductivity of 4.9 W/cm-K [2]. These properties along with the high threshold displacement energy (45-90 eV) [3] of SiC should make these devices much more resistant to electromagnetic pulse (EMP) and radiation damage [4]. The high saturated electron drift velocity of $2.0x10^7$ cm/sec [5] also indicates that SiC is a very promising material for high power, high frequency operation.

Additionally, the optoelectronic properties of 6H-SiC have been known for years. Photodiodes in 6H-SiC were first demonstrated by Glasow, et al. [6] to have a peak response in the ultraviolet (UV) range due to the 2.9 eV bandgap. Despite the indirect bandgap of SiC, blue light emitting diodes (LEDs) have been reported since 1969 [7]. These devices emit in the 2.6 eV energy range because they utilize a donor to acceptor transition within the bandgap.

Although these properties have been known for many years, commercial development of these devices has been virtually non-existent. The primary reason is that a reliable production technique for growing large area 6H-SiC crystals for use as wafer substrates did not exist. Likewise, because high quality substrates were not available, suitable production techniques for controllable growth and doping of epitaxial layers were also very slow to develop.

However, significant advances have been made in the last few years in both crystal growth and epitaxial growth of 6H-SiC. This has resulted in the first commercial release [8] of 6H-SiC wafers and epitaxy in 1991. Along with these advances, device processing techniques such as ion implantation, oxidation, ohmic and Schottky contact formation, and reactive ion etching have also been developed.

As such, many promising device results have been produced in all of the device application areas mentioned previously. These advancements have lead to the development and commercial production of 6H-SiC blue LEDs, which are being used in a wide variety of applications. Some of the important device results for each area (optoelectronic devices, high temperature devices, radiation hard devices and high power / high frequency devices) will be discussed, along with some specific applications.

2. Optoelectronic Devices

The two primary optoelectronic devices in 6H-SiC are the blue LEDs and the UV photodiodes. The blue LEDs are by far the most established market for 6H-SiC devices to date.

Most of the major applications for blue LEDs involve combining them with commonly available GaP-based red and green LEDs to produce full color displays. By mixing red, green and blue light, solid state white light can be achieved. By varying the power to each discrete color LED, any color in the visible spectrum may be obtained. This is of particular interest to the manufacturers of LED-based displays that are used for both commercial signs and military full color LED monitors. Other applications are for discrete blue LED indicators for instrumentation and consumer electronics where an additional color provides enhanced differentiation or aesthetic appeal. Additionally, a very stable wavelength blue light source is required for some special blood and tissue analysis and for standards for colorimeters and spectrophotometers.

Blue LEDs fabricated in 6H-SiC utilize a donor-acceptor pair recombination in an epitaxial layer containing both Al and N dopants. The transition causes photon emission with a peak wavelength of about 470 nm, which is the deep blue spectrum. The typical spectral halfwidth is in the range of 70 nm.

The blue LEDs manufactured and sold by Cree Research have been shown to have radiant flux values as high as 17 μW at 20 mA with a forward bias of 3.0 V. This corresponds to an external quantum efficiency of 0.028%. By operating at the maximum rated current of 50 mA, a radiant flux of 30-35 μW can be achieved.

Photodetectors fabricated in 6H-SiC show peak quantum efficiencies in the range from 250-280 nm. Quantum efficiencies of nearly 100% have been measured. In addition to the very high efficiency, the extremely low leakage currents measured on 6H-SiC pn junctions (10^4-10^5 times less leakage than Si junctions) allow a sensitivity more than 10,000 times higher than common Si-based UV photodiodes.

The primary applications for SiC UV photodiodes are for UV detection, dosimetry, and analysis. Specifically, the detection of UV from flames can be used for applications ranging from solid state detection for combustion control to detection of missile plumes. Dosimetry applications can range from solid state total dose UV monitoring for industrial processes (UV curing, etc.) to a personal UV detector for monitoring sun exposure. Analysis applications include air quality monitoring equipment and UV spectroscopy.

For a more in-depth discussion of SiC LEDs and photodiodes, see Edmond, et al.[9].

3. High Temperature Devices

The wide bandgap of 6H-SiC gives the added advantage of high temperature operation for almost any device fabricated in this material. Most of the high temperature applications require a "family" of 6H-SiC devices for digital, analog, and power functions. In order to improve efficiency, new jet engine designs require higher operating temperatures. Since further fuel cooling for engine electronics is not desired, high temperature electronics will be required. These electronics include devices for the ignition systems as well as sense and control electronics.

Since radiation is the only method of cooling electronics in space and since the energy radiated is a T^4 function, the ability to operate electronics at higher temperatures greatly reduces the size and weight of the radiator panels on spacecraft. Thus 6H-SiC devices could greatly reduce the launch weight of space-based systems. Other applications areas are for automotive electronics, where the trend is to operate at higher temperatures, and for transmitters for deep well drilling.

A number of 6H-SiC devices have been demonstrated at high temperature since the mid 1960's. Recently, packaged high temperature 6H-SiC pn junction rectifiers have been reported that have very good characteristics up to 623K [9]. These devices have been tested under bias for more than 1000 hrs at 623K and showed no significant change in electrical characteristics. Unpackaged rectifiers have shown excellent properties at temperatures up to 873K.

A variety of FETs have also been measured at high temperature. Both MESFETs [10] and buried-gate JFETs [10,11] have been demonstrated to have very good characteristics at elevated temperatures. Both of these structures show similar behavior with temperature. The on-current and transconductance stays fairly stable up to about 323K, at which point they begin to decrease with temperature due to decreasing electron mobility. At 623K these parameters are roughly half of their room temperature values. The pinch-off voltage of these devices stays very stable throughout this temperature range. These devices were measured to have good characteristics to 773K.

The on-current and transconductance also decrease as temperature is decreased below room temperature. This decrease is due to decreasing carrier concentration with temperature (the N dopant is not fully ionized at room temperature). At 218K, the I-V characteristics are very much like those seen at 623K, with the on-current and transconductance being roughly half of the room temperature values.

Recent 6H-SiC MOSFETs have also shown promising results. The I-V curves shown in Fig. 1 are for an n-channel enhancement-mode MOSFET with a 7 μm gate length and 1 mm gate width. The room temperature drain current at $V_G = +16$ V is 18.6 mA and the transconductance at that gate voltage is 2.8 mS/mm, as is shown in Fig. 1(a). The threshold voltage is about 1.9 V. When the temperature is raised to 423K, as shown in Fig. 1(b), both the current and transconductance at $V_G = +16$ V increase to 29 mA and 3.6 mS/mm, and the threshold voltage decreases to about -0.2 V. This trend continues up to 573K, as shown in Fig. 1(c). At this temperature the current and transconductance at $V_G = +16$ V were 35 mA and 3.9 mS/mm, respectively. Although the threshold voltage of the device fell to $V_G = -0.8$ V, the drain current at $V_G = 0$ V was only 47 μA.

Some MOSFET devices measured had very good characteristics to temperatures as high as 923K, as shown by the I-V curves of a different 6H-SiC MOSFET in Fig. 2. This device had a threshold voltage of 0.0 V at 923 K. Good current saturation was obtained to drain voltages of 25 V with a drain current of 4.75 mA at $V_G = +15$ V. The g_{max} was 0.43 mS/mm at $V_G = +6$ V. The subthreshold leakage current was 22 μA at $V_D = 25$ V and $V_G = 0$ V.

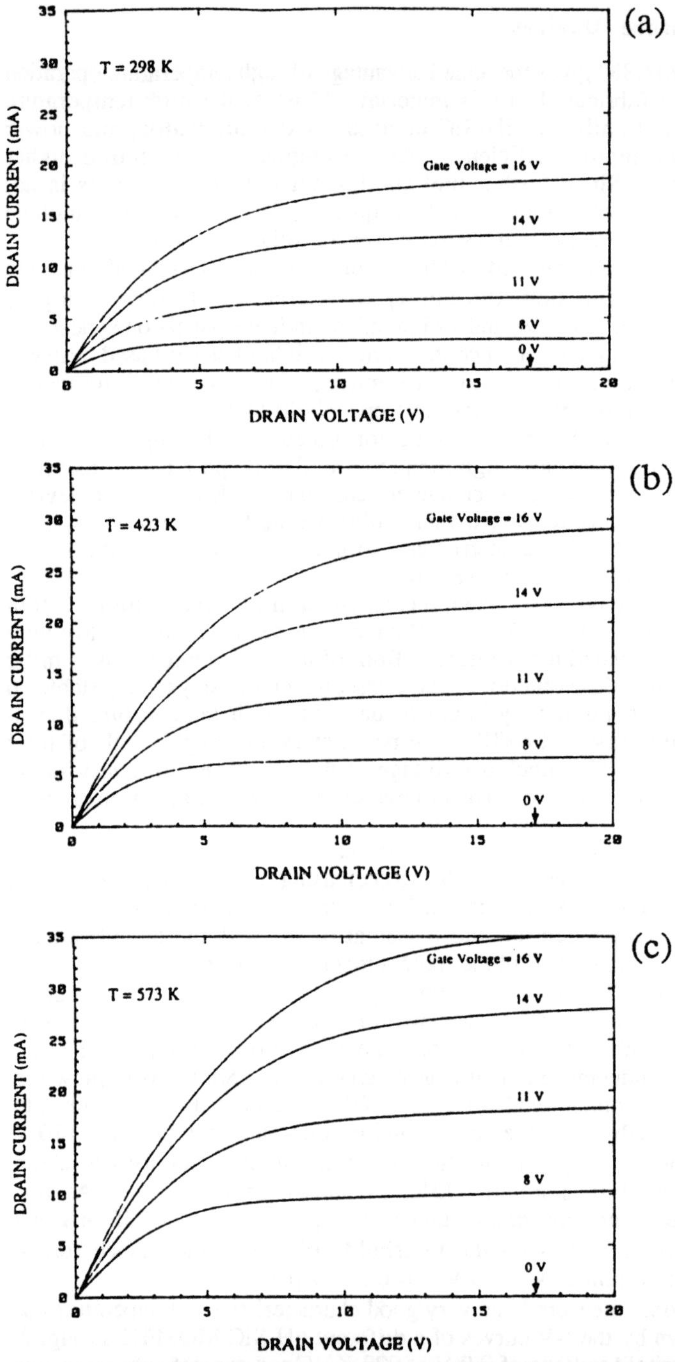

Figure 1. Drain current-voltage characteristics of an n-channel 6H-SiC enhancement-mode MOSFET at (a) 298K, (b) 423K, and (c) 573K. The gate length and width were 7 μm and 1000 μm, respectively.

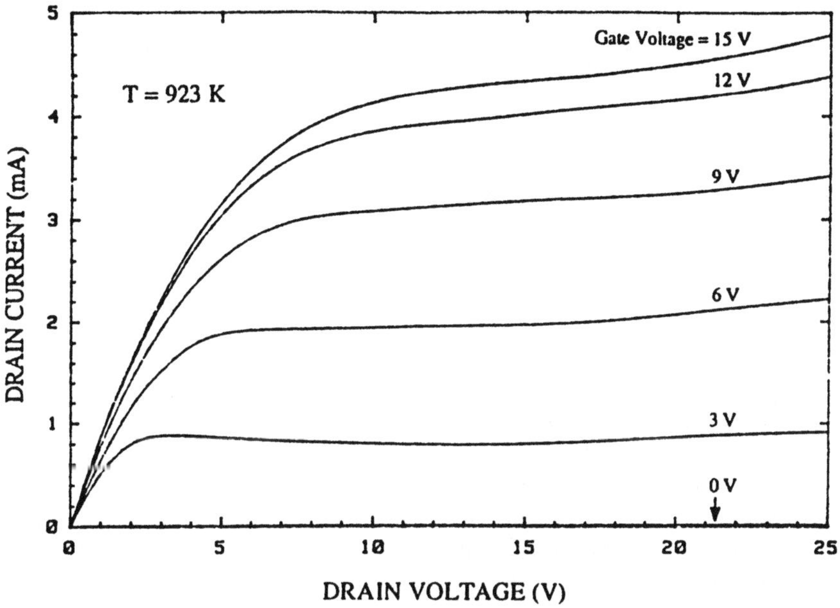

Figure 2. Drain current-voltage characteristics of an n-channel 6H-SiC enhancement-mode MOSFET at 923K. The gate length and width were 7 µm and 1000 µm, respectively.

4. EMP and Radiation Hard Devices

These applications are primarily in the military, the nuclear power industry and the space power industry. The primary military interest in EMP hard devices is for survivability of electronics during a nuclear weapons blast. The very high electromagnetic pulse experienced during this blast can cause solid state device failures hundreds of miles away. The pulse causes a large generation of electron-hole pairs in the semiconductor which in turn causes very high junction temperatures due to the sudden power pulse that can permanently damage or destroy the device. The major application for EMP-hard 6H-SiC devices would be for terminal protection devices such as surge suppressors. The amount of power required to cause device failure depends on the area of the junction and the pulse duration. Since 6H-SiC has a much higher thermal conductivity, heat capacity and melting temperature than Si, SiC parts should be inherently harder to EMP effects than equivalent Si devices.

Testing of EMP effects on 6H-SiC pn junction rectifiers has proven this to be true. The rectifiers were subjected to reverse bias avalanche power pulses for a variety of pulse times ranging from 1 µs to 100 µs. The power density was increased until the device showed a significant degradation in electrical characteristics. The experimental data is shown in Fig. 3 along with a calculated theoretical curve for 6H-SiC. While it is clear that these results are not as good as the theoretical values, the experimental data for 6H-SiC diodes shows at least twice the power pulse capability that is commonly achieved for Si rectifiers (also shown in Fig. 3).

Junction-based devices made in 6H-SiC have also been shown to be very resistant to damage from gamma radiation. Buried-gate JFETs were exposed to

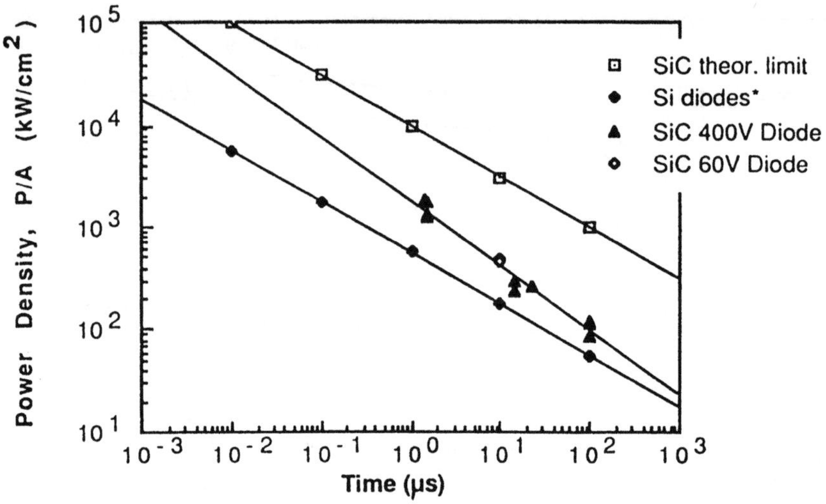

Figure 3. Average power per unit area as a function of pulse time required to cause junction failure in 6H-SiC and Si diodes. An SiC theoretical limit is also shown.

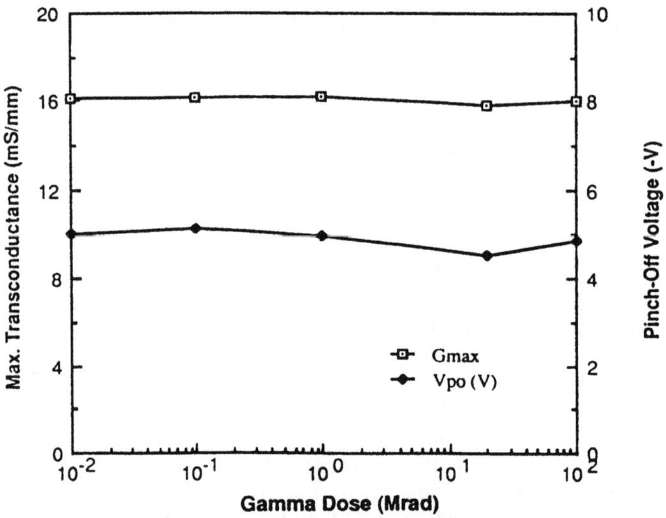

Figure 4. Average transconductance (at $V_G = 0$ V) and pinch-off voltage for 6H-SiC JFETs as a function of gamma radiation total dose.

increasing total doses of gamma radiation up to 100 Mrad, as shown in Fig. 4. It can be seen that there was no large change in the pinch-off voltage or the transconductance at $V_G = 0$ V up to 100 Mrad total dose [12].

Neutron irradiation has also been performed on 6H-SiC devices. Batches of both 125 V and 410 V 6H-SiC pn junction rectifiers were exposed to neutron fluences ranging from 10^{13} to 10^{15} n/cm^2. The forward voltage drop for 400 mA current and the avalanche voltage were measured before and after each irradiation. The 125 V rectifiers, which had higher doping, showed virtually no change in 400 mA forward voltage drop (3.0 V) after a 10^{15} n/cm^2 fluence, and the avalanche voltage increased by only 0.8%. The lower doped 410 V rectifiers had an 8.6% increase in 400 mA forward voltage drop and a 4% increase in avalanche voltage.

Radiation hardness is required for both space-based and ground-based nuclear reactor electronics. Power conditioning devices are of the most interest for space-based applications, while smart electrically controlled valves are of interest for ground-based reactors. Other applications are for electronics in crucial military systems that must survive radiation effects from a nuclear weapon.

5. High Power and High Frequency Devices

As was stated earlier, one of the main applications for SiC is in high power RF and microwave devices. Trew and co-workers [13] have performed physical modeling of SiC MESFETs and shown that RF output powers as high as 65 W could be achieved at 10 GHz (4 W/mm). Applications for these devices would be in Electronic Warfare and Electronic Countermeasure Systems such as jamming and threat warning systems. High frequency SiC MESFETs could have a major impact on the development of solid-state phased array radar. Very high power SiC MESFET devices would offer much higher reliability than current traveling wave tube (TWT) technology. Communication systems and high frequency power supplies will also benefit from the enhanced operating range of SiC.

As such, research on 6H-SiC MESFETs for high power / high frequency operation has been performed. These devices utilize a 0.6 µm gate length and 1 mm gate width. The n-type channel layer was epitaxially grown a low doped p$^-$ epilayer; the current was confined by etching a mesa around the device. A typical AC (120 Hz) current-voltage plot of one of these 6H-SiC MESFETs is shown in Fig. 5. This device showed good I-V characteristics to $V_D = 40$ V. The transconductance at $V_G = 0$ V and $V_D = 40$ V was about 30 mS/mm. The gate leakage current at $V_G = -3$ V and $V_D = 35$ V was 258 µA. The on-current for this device was 310 mA/mm and the pinch-off voltage was $V_G = -17$ V.

The device shown in Fig. 5 was measured at high frequency using an HP 8510 automatic network analyzer with a Cascade Microprober for standard S-parameter measurements. The plot in Fig. 6 shows that this device has a threshold frequency (f_t) of about 4.5 GHz, where the H_{21} parameter crosses 0 db gain. The device has an f_{max} of 3.0 GHz, where the power gain (G_{max}) crosses 0 db gain. At 1.0 GHz, this device had a power gain and a current gain of 13.5 db and 12.5 db, respectively. While no RF power measurements have yet been performed on these devices, it is anticipated that significant RF power output will be obtained for these devices.

In addition to the microwave applications, lower frequency 6H-SiC power devices will outperform currently available Si devices in terms of recovery speed and power density. Because of its short minority carrier lifetime, SiC pn junctions have inherently fast reverse recovery times (t_{rr}). Rectifiers with 1000 V breakdown have been measured to have t_{rr} of 17 ns at room temperature and 25 ns at 350°C. Primarily due to its high electric breakdown field, SiC power MOSFETs and Schottky diodes have been modeled to have much lower specific on-resistances than equivalent Si devices (see Baliga [14]).

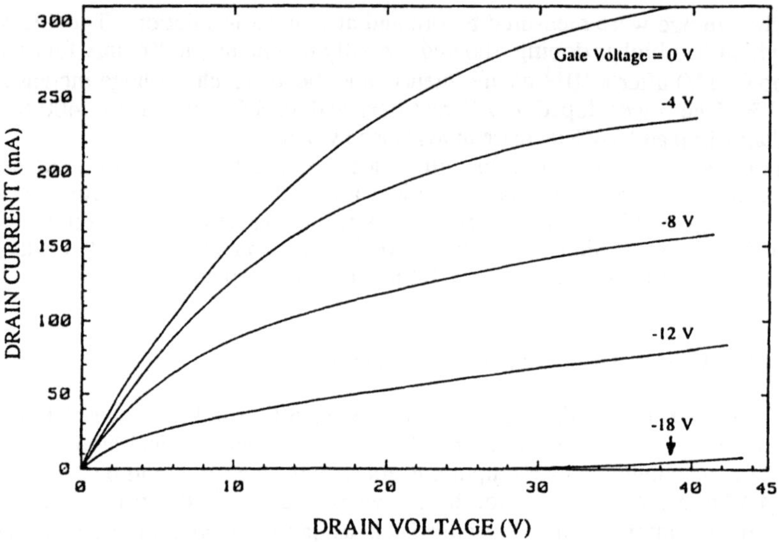

Figure 5. Drain I-V characteristics of a 6H-SiC MESFET at 120 Hz. The gate length and width were 0.6 μm and 1 mm, respectively. The source-drain length was 2.9 μm.

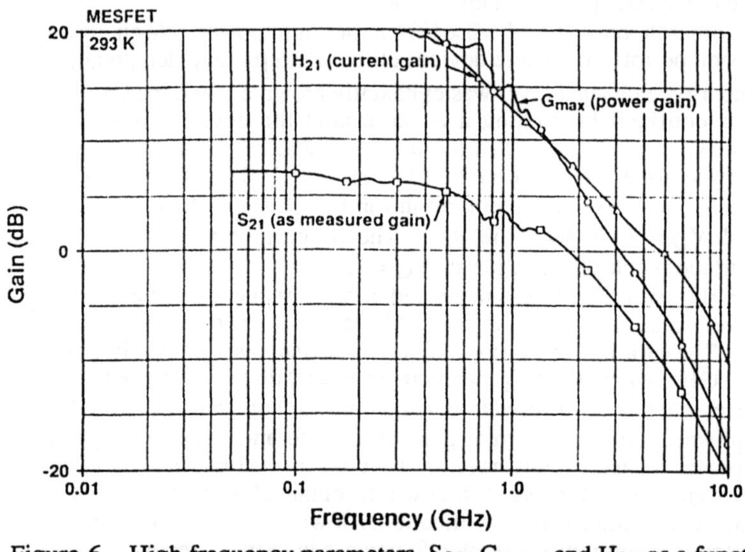

Figure 6. High frequency parameters, S_{21}, G_{max}, and H_{21} as a function of frequency for the MESFET shown in Fig. 5. Measurement conditions were $V_D = 35$ V, $I_D = 260$ mA, $V_G = -3.0$ V, and $I_G = 258$ μA.

6. Conclusions

Blue LEDs and UV photodetectors in 6H-SiC have been demonstrated, made commercially available, and are being designed into many optoelectronic applications. High temperature operation of pn junctions, n-channel 6H-SiC MESFETs, JFETs, and enhancement-mode MOSFETs has been demonstrated. Good I-V characteristics were observed for 6H-SiC MOSFETs at temperatures up to 923 K. Thus, the potential to develop a "family" of high temperature devices required for jet engine, spacecraft, and automobile applications has been proven. Both pn junctions and JFETs have been shown to have hardness to EMP and radiation effects superior to that of Si. Submicron gate length MESFETs have shown current gain up to 4.5 GHz, giving promise to 6H-SiC for high power / high frequency devices.

Acknowledgments. This research was performed at Cree Research, Inc. under both internal research funds and Small Business Innovation Research programs funded by the Department of the Air Force, the Strategic Defense Initiative Office (administered through the Office of Naval Research), and the National Aeronautics and Space Administration. The high frequency MESFET research was funded by the Office of Naval Research and additional funding for MOSFET research was obtained from the U.S. Army Research Office. The authors express their thanks to Dr. J. McGarrity of Harry Diamond Laboratories for the gamma radiation. The authors also wish to thank J. Spangler, M. Nieri, and M. Parrish for their technical assistance, and to Dr. R.J. Trew and A. Morris of N.C. State University for their valuable discussions and measurements on the high frequency MESFETs.

References

1. C. van Opdorp and J. Vrakking, J. Appl. Phys., **40**, 2320 (1969).
2. G.A. Slack,J. Phys. Chem. Solids, **34**, 321 (1973).
3. F.W. Clinard, Jr., and L.W. Hobbs, *Radiation Effects in Non-metals*, R.A. Johnson and A.N. Orlov, eds., (North Holland Press, 1986).
4. P.C. Canepa, P. Malinaric, R.B. Campbell, and J. Ostroski, IEEE Transactions on Nuclear Science, **NS-11**, 262 (1964).
5. W. von Muench and E. Pettenpaul, J. Appl. Phys., **48**, 4823 (1977).
6. P. Glasow, G. Ziegler, W. Suttrop, G. Pensl and R. Helbeg, SPIE, **868**, 40 (1987).
7. R.W. Brander and R.P. Sutton, Brit. J. Appl. Phys., 2, 309 (1969).
8. Cree Research, Inc., Durham, NC (1991).
9. J.A. Edmond, H.S. Kong, and C.H. Carter, Jr., this volume.
10. J.W. Palmour, H.S. Kong, D.G. Waltz, J.A. Edmond, and C.H. Carter, Jr., *Proc. First Intl. High Temp. Electronics Conf.*, D.B. King and F.V. Thome, eds., (U.S. Government Printing Office, Albuquerque, NM, 1991), p. 511.
11. G. Kelner, S. Binari, M. Shur and J. Palmour, Electron. Lett., **27**, 1038 (1990).
12. J. McGarrity, Harry Diamond Laboratories, Adelphi, MD.
13. R.J. Trew, J.B. Yan, and P.M. Mock, Proceedings of the IEEE, **79**, 598 (1991).
14. B.J. Baliga, this volume.

HBTs Using a-SiC and μc-Si

S. Furukawa and K. Sasaki

Graduate School of Science and Engineering, Tokyo Institute of Technology, 4259 Nagatsuda, Midori-ku, Yokohama 227, Japan

Heterojunction structures between non-crystalline materials including amorphous/microcrystalline phase and crystalline Si substrates are intensively investigated. The correlation of their electrical energy band structure and crystallographical structure is discussed. When μc-Si is deposited directly onto the crystalline Si substrate, crystallization of μc-Si takes place in the vicinity of the interface, so that a Si homojunction is formed. By inserting an a-SiC buffer layer as thin as a few nm, the crystallization or the formation of homojunction is successfully suppressed. As the result, it has been found that an abrupt and a uniform amorphous/crystalline Si heterointerface is obtained. A current gain as high as 523 is achieved by using this multilayered emitter structure.

1. Introduction

Si based heterojunction bipolar transistors (Si-HBTs) are being developed toward next generation high speed electron devices. We have offered that the Si based amorphous materials such as amorphous (a-)SiC and microcrystalline (μc-)Si were promising as heteroemitter materials for Si-HBTs[1),2)]. The key issues of the amorphous heteroemitter are reviewed as follows:

1. Heteroemitter materials need not always be a single crystal.
2. Si based amorphous materials have sufficiently wider bandgap than crystalline (c-)Si.
3. Interface states are terminated by hydrogens in depositing hydrogenated amorphous materials.
4. Preparation is carried out at low temperature process.
5. Si based amorphous materials do not bring contamination into the Si process.
6. Heteroemitter configuration has complete process compatibility with present bipolar self-align technologies.
7. Heteroemitter materials need very low resistivity.

Based on these motivations, amorphous heteroemitter research have been promoted. Recently, it has been demonstrated that a-SiC_x emitter Si-HBT fabricated by super self-align technology achieved cut-off frequency of 24GHz[3)]. This value has been a little lower than a poly-Si emitter homojunction transistor be cause of parasitic emitter resistance R_E, which decreases cut-off frequency, f_T by the delay time of R_EC_{BC} from intrinsic f_{Tin}, i.e., $1/2\pi f_T = 1/2\pi f_{Tin} + R_EC_{BC}$, where C_{BC} is the base-collector capasitance. Nevertheless, we can say that this result shows the feasibility of amorphous heteroemitter in early stage.

Springer Proceedings in Physics, Vol. 71 Amorphous and Crystalline Silicon Carbide IV
Editors: C.Y. Yang · M.M. Rahman · G.L. Harris

However, strict investigation of correlation between the amorphous/crystalline heterointerface structure[4)-7)] and its band profile has not been done. For further progress of the device performance, it is necessary to understand the real structure of the amorphous/crystalline heterojunction. In this paper, the electrical and crystallographical structures on the amorphous/crystalline heterojunction and their correlation will be discussed in the following.

2. Film Preparation

a-SiC and μc-Si films were prepared by L-coupled RF plasma CVD system using SiH_4-CH_4-PH_3/He/Ar gas mixture at pedestal temperature of 350°C and RF power of 60W. For a-SiC films, their optical bandgap linearly changed from 1.7eV to 2.1eV by changing CH_4 fraction, i.e., $CH_4/(SiH_4+CH_4)$, from 0 to 0.4. Electrical conductivity of the films changed from order of 10^{-5}S/cm to 10^{-8}S/cm within the CH_4 fraction range. These energy bandgaps were sufficiently wide, but conductivities were too small for a practical heteroemitter. In order to prepare a highly conductive SiC film, other new film preparation techniques such as ECR plasma CVD method will be necessary[8)].

In order to solve the high resistance problem, we have attempted to use μc-Si films instead of a-SiC films. μc-Si films can be prepared by only adding Ar to source gases in L-coupled plasma CVD method mentiond above. The μc-Si film resistivities are plotted for Ar flow ratio in Fig.1. For no Ar addition, the film deposited on SiO_2 substrate showed high resistivity. This film was found to be a conventional a-Si film. By adding Ar of around 20%, film resistivity drastically decreased as much as 5 decades and the value of around 0.1Ωcm was achieved with good reproducibility. These films showed the optical bandgap of 1.75eV, which value is sufficiently wide for use of a heteroemitter.

The reason why the μc-Si film has been prepared by Ar addition is speculated as follows; Ar atoms are easily ionized in the glow discharge and densify the plasma density, so that decomposition of SiH_4 gas is enhanced. Furthermore, Ar ion bombardment on the growing surface induces nucleation of Si crystallites and enhances the migration of SiH_n species. As a result, microcrystallization is occurred.

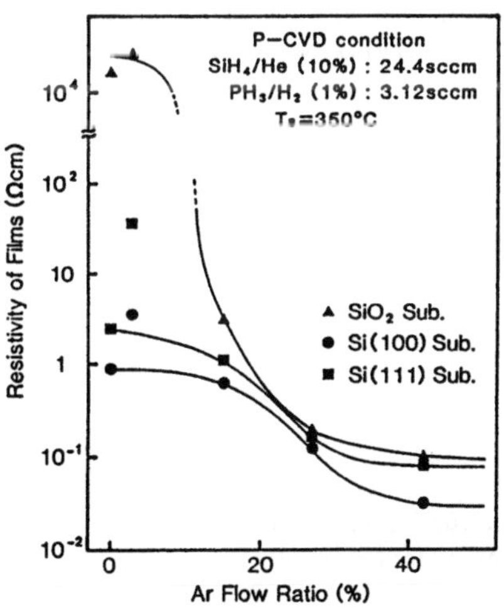

Fig. 1. Dependence of Ar flow ratio on resistivity of μc-Si films deposited on c-Si and SiO_2 substrates.

On the other hand, resistivity of the films on c-Si substrates showed low values in spite of no addition of Ar. This is because crystallization of μc -Si was induced by using c-Si substrates. This phenomenon, which is considered to significantly affect the HBT device performance, will be intensively analyzed and discussed below.

3. Analysis of Junction Properties

Junction properties including electrical band structure of the amorphous/c-Si heterostructures was investigated by employing the inner photoemission effect. The experimental principle is shown schematically in Fig. 2. Au/μc-Si structure Schottky diodes were fabricated on c-Si substrate. The diode photocurrents were measured with changing the incident photon energy. If there are no bandgap difference between the deposited film and c-Si substrate, photocurrent starts flowing at the same threshold energy as determined by c-Si. On the other hand, if the film has wider energy bandgap than the c-Si substrate, the threshold energy is expected to shift to higher energy side since current flow is blocked by the valence band discontinuity.

Fig. 3 shows the photocurrent spectra measured for the μc-Si/c-Si structure with changing the μc-Si film thickness. Dashed line in the figure shows reference spectra measured on n/n^+ c-Si and its threshold energy is observed around 1.2eV. This value is in fairly good agreement with the bandgap of c-Si. Samples with μc-Si films thinner than 250nm show the same threshold energy as c-Si. These results suggest that there are some portions having the same energy bandgap as c-Si in the deposited films. For the thicker film cases, the main photoresponse is observed to shift in higher energy side. However, there is a tail region reaching to the same threshold energy of c-Si.

This behavior can be explained by the structural model as shown in Fig. 4. When the film thickness is thin, the photocurrent can flow through crystallized portions, resulting in showing the same threshold energy as c-Si. As the film thickness increases, the crystallized portions reaching the Schottky contact decrease. Therefore the photocurrent decreases and the photoresponse has the tail region. Again, the photocurrent increases at the photon energy where the photoexcited holes obtain enough energy to overcome the barrier in the valence band. The fact that the extrapolated threshold energy agrees with the optical bandgap of μc-Si suggests that the band discontinuity occurred mainly in the valence band.

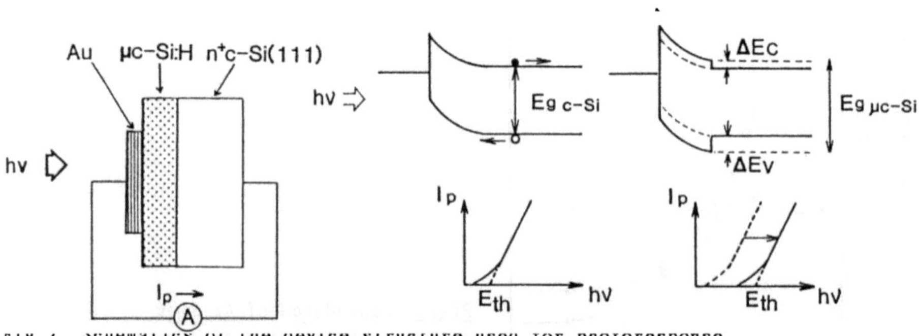

Fig. 2. Schematics of the device structure used for photoresponse experiments and the analysis principle of heterojunction conditions by the inner photoemission method.

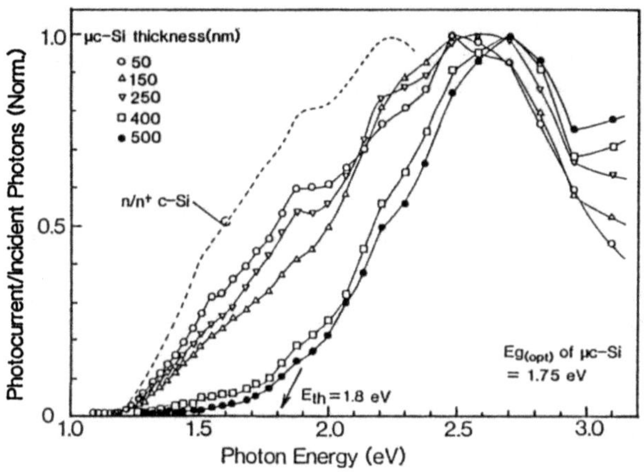

Fig. 3. Photocurrent spectra from Au/μc-Si/c-Si structure Schottky diodes fabricated with changing the μc-Si film thickness.

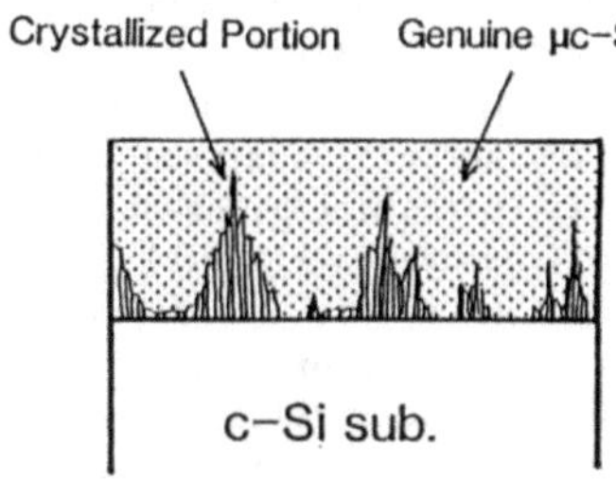

Fig. 4. Structural model for the μc-Si/c-Si structure.

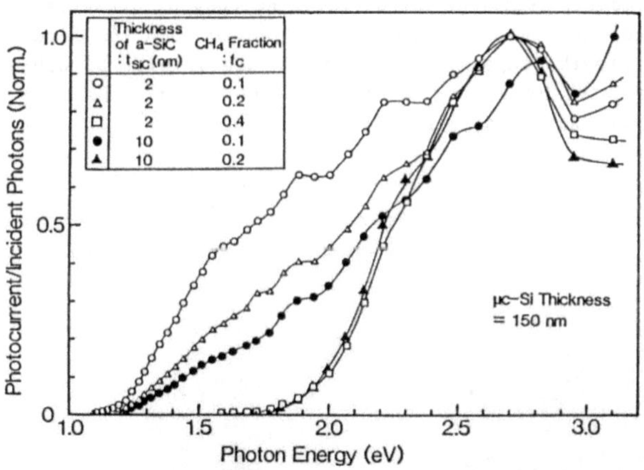

Fig. 5. Photocurrent spectra for samples with various a-SiC buffer layers.

In order to suppress such crystallization of μc-Si and obtain an abrupt and uniform heterojunction, we have attempted to introduce thin a-SiC films as a buffer layer. The photocurrent spectra for the samples with a-SiC buffer layers are shown in Fig. 5. The deposition conditions of a-SiC layer including the thickness t_{SiC} and CH_4 fraction f_C, are varied as shown in the table inserted in the figure. Here, the thickness of the upper μc-Si film is kept constant at 150nm. As can be seen in this figure, the spectra respond sensitively to the a-

SiC layer forming conditions. For the cases of $t_{SiC}(nm)/f_C$ = 2/0.1, 2/0.2 and 10/0.1, the threshold energies were same as c-Si, which means that the crystallization can not be completely suppressed under such a-SiC layer conditions. It can be seen that the slope of the spectra become gradual with increasing t_{SiC} or f_C. This change can be interpreted to show the change of extent of partial crystalline growth.

On the other hand, for the cases of $t_{SiC}(nm)/f_C$ = 2/0.4 and 10/0.2, the photocurrent spectra obviously changed and the threshold energy is observed at around 2.0eV. Furthermore, little tail region is observed. From these results, it can be considered that the crystallization of μc-Si near the interface has been completely suppressed, so that the ideal microcrystalline/c-Si heterojunction has been formed.

4. Microstructure of the Junction

In order to verify the above experiments and interpretations, the crystallographical structure of junctions were investigated by TEM observation. The cross-sectional TEM images of the μc-Si film deposited on c-Si(100) substrate are shown in Fig.6. Complete view of the around 300nm thick film is shown in Fig.6(a). Crystallized portions which are seen dark are markedly observed near the interface. Near the surface region, amorphous or microcrystalline phase appears. Further magnified TEM image near the interface is shown in Fig.6(b). The Si lattice image continuing from substrate or epitaxial growth is clearly observed in the corn shape. From these images, the crystalline growth of μc-Si seems to be induced by c-Si substrate. This may be because the crystallinity of underlying c-Si restricts the site of SiH_n species forming the film. Furthermore, this result shows that the initial substrate surface has not been oxidized. According to the recent research of Si surface, HF-treated Si surface is terminated by hydrogen, and the hydrogen remain stably up to 500℃[9]. Since in this experiment, the film growth temperature was 350℃, the initial Si substrate surface have been terminated by hydrogen, so that the surface is considered not to have been oxidized as shown in Fig.6(b).

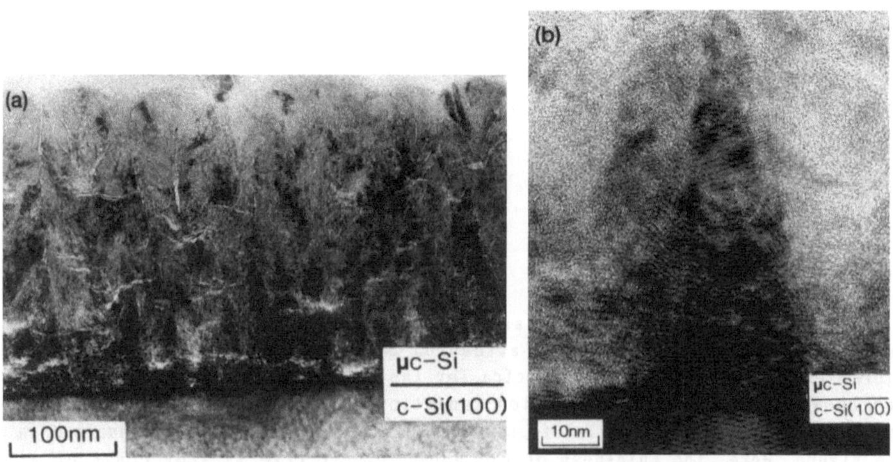

Fig.6. Cross-sectional TEM images of μc-Si/c-Si(100) structure.

Fig. 7. Cross-sectional TEM images of μc-Si/a-SiC buffer layer(f_C=0.5)/ c-Si(111) structure.

The TEM images of a sample with a-SiC buffer layer are shown in Fig. 7. Comparing with the Fig. 6, structural difference near the interface is clear. Seeing Fig. 7(a), amorphous phase is observed near the interface, while in the upper region, microcrystalline phase appears. In Fig. 7(b), around 3nm thick continuous amorphous layer is observed as a bright band. Amorphous-crystalline phase transition in the interface is quite sharp. Analyzing this region by EPMA and SIMS technique, carbon atoms were confirmed to be localized sharply at the interface. Also, no diffusion of the phosphorus atoms into the substrate was observed. From these analysis, an abrupt and a uniform heterostructure was found to be realized by only inserting a thin a-SiC buffer layer. Furthermore, it can be said that these results are consistent with the results of the junction analysis by the photocurrent response.

This a-SiC layer is desired to be thin on the viewpoint of decreasing parasitic resistance. In principle, this layer could be replaced by 1 monolayer of carbon. This kind of technology, that is, controlling surface in atomic scale will be important even in practical Si process technology in the near future.

5. HBT Characteristics

I_C-V_{CE} characteristics of the devices with and without the a-SiC buffer layer are shown in Fig. 8 as an example. It is clearly seen that current gain is drastically improved by using a-SiC buffer layers. High current gain exceeding 500

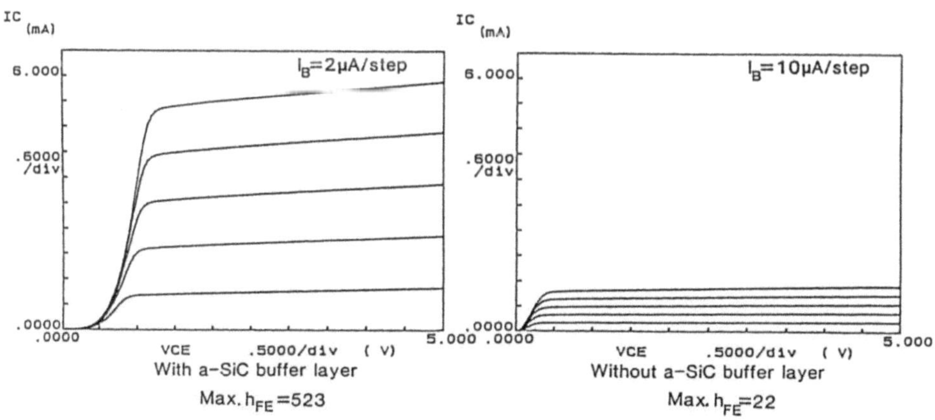

Fig. 8. I_C-V_{CE} characteristics of the HBTs with a-SiC buffer layer (f_C=0.5, t_{SiC}=2nm) and without buffer layer.

is achieved, while for the device without the a-SiC layer, the current gain is only 22. These results show that the heterojunction is formed by inserting the a-SiC buffer layer.

6. Conclusion

Physics of the μc-Si(a-SiC)/c-Si heterojunctions were intensively investigated. Their electrical band structure and crystallographical structure have been discussed. In this investigation, it has been shown that a-SiC is very useful to form an abrupt and uniform heterojunction and obtain a high current gain in ideal non-crystalline heteroemitter HBT. Though this buffer layer is very thin, it is clear that a-SiC has played a very important role.

References

1)S.Furukawa: Proc. of 1st ICACSC, Springer Proc. in Phys. **34**, p.58(1989).
2)K.Sasaki, T.Fukazawa and S.Furukawa: Tech. Dig. of IEDM87, p.186(1987).
3)M.Ugajin M.Kuwagaki and S.Konaka: Ext. Abs. 23rd Int. Conf. SSDM, p.159(1991).
4)M.Kondo, T.Saitoh, M.Tamura, S.Matsubara and M.Miyao: Jpn. J. Appl. Phys., **28**[9], p.1531(1989).
5)J.Symons, J.Nijs, and R.Mertens: IEEE Trans. on Elec. Dev., **ED-36**[12], p.2889(1989).
6)H.Fujioka, S.Ri, K.Takasaki, K.Fujino and Y.Ban: Tech. Dig. of IEDM87, p.190(1987).
7)T.Asano, D.Sugiarto, K.Tran, M.M.Rahman, C.Y.Yang and F.A.Ponce: Proc. of 2nd ICACSC, Springer Proc. in Phys. **43** p.153(1989)
8)T.Miyajima K.Sasaki and S. Furukawa: Proc. of 4th ICACSC, (this issue).
9)N.Yabumoto, K.saito, M.Morita and T.Ohmi: Ext. Abs. 22nd Int. Conf. SSDM, p.1067(1990).

Impact of SiC on Power Devices

B.J. Baliga

Power Semiconductor Research Center, Electrical and Computer Engineering Department, North Carolina State University, Raleigh, NC 27695-7294, USA

Abstract. This paper provides a review of the performance advantages of Silicon Carbide based power devices. It is shown here that devices made from either 3C-SiC or 6H-SiC will be significantly superior to silicon based devices. The high breakdown electric field strength in SiC is the major factor that results in a very low specific resistance for the drift region in high voltage devices. This reduces the specific on-resistance for power MOSFETs by a factor of more than 100. It also reduces the forward voltage drop of Schottky rectifiers to less than 2 volts even when designed to support 3000 volts. The development of these SiC devices can, therefore, be expected to result in high switching speed components with low on-state losses for power electronic applications.

I. Introduction

Over the last four decades, silicon technology has been extensively used to create a variety of power semiconductor devices for power electronic applications. In the 1950's and 1960's, bipolar power devices, such as the power bipolar transistor and the thyristor, were developed. An impressive growth in the ratings of these devices has steadily occurred over the years. At present, power bipolar transistors are available with blocking voltages of up to 1000 volts and current ratings of 300 amperes; while power thyristors are available with blocking voltages of 6500 volts and current ratings of 2000 amperes. In the latter device, it is important to note that a single device is fabricated from an entire wafer of silicon with a diameter of 100 mm.

In the 1970's, MOS technology was first introduced into the power semiconductor industry. The power MOSFET, whose structure is illustrated in Figure 1, became popular with developers of power electronic systems because of its high input impedance which greatly simplified the device control circuit. In addition, the power MOSFET had much faster switching speed with turn-off times in the range of nanoseconds, as compared with the bipolar power transistor with turn-off times in the range of microseconds [1].

During the last 15 years, extensive theoretical analysis of the power MOSFET structure has been done to optimize its blocking voltage capability and its on-resistance [1]. The specific on-resistance of the device is an important electrical characteristic of the device because it determines its current handling

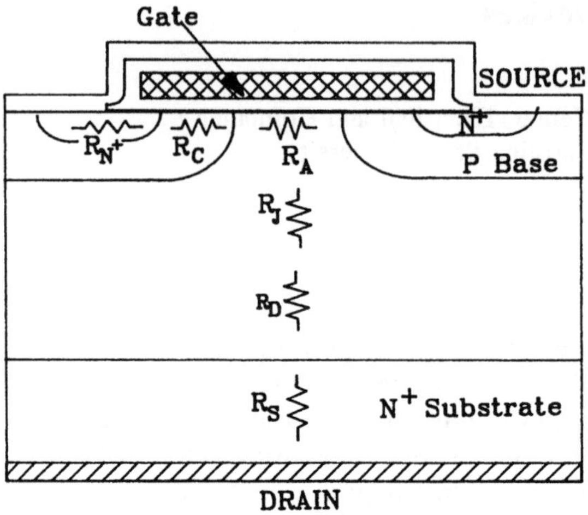

Figure 1 Cross-section of power MOSFET (DMOS structure) with the internal resistances indicated. Here R_{N+} is the resistance of the source diffusion, R_C is the channel resistance, R_A is the accumulation layer resistance, R_J is the JFET region resistance, R_D is the drift region resistance, and R_S is the substrate resistance.

capability. Using improved device design and process technology, it has been possible to continually reduce the specific on-resistance of the silicon devices. An example of this progress is given in Figure 2, where the specific on-resistance is plotted as a function of time for the case of devices with blocking voltage of 50 volts [2]. The components of the on-resistance of the power MOSFET structure were shown in Figure 1. Among these components, the channel, JFET region, and drift region resistances are the major components. Since the drift region is essential for supporting the voltage during the blocking state, its resistance is considered to be a fundamental value that represents the minimum achievable value for the power MOSFET. The other components are created by the need to control the current flow within the device. It can be seen from Figure 2 that improvements in silicon device technology have resulted in the specific resistances approaching the theoretical limit for silicon devices, namely, the resistance of the drift region. From this, it can be concluded that further advancement in the intrinsic performance of the power MOSFET cell can only be achieved by using other semiconductor materials.

Another important factor for the application of power MOSFETs has been a very rapid increase in the specific on-resistance with increase in the breakdown voltage [1]. As illustrated in Figure 3, the on-resistance increases by a factor of 300 times when the blocking voltage is increased by a factor of 10 times. As a consequence, the specific on-resistance of the power MOSFET becomes too

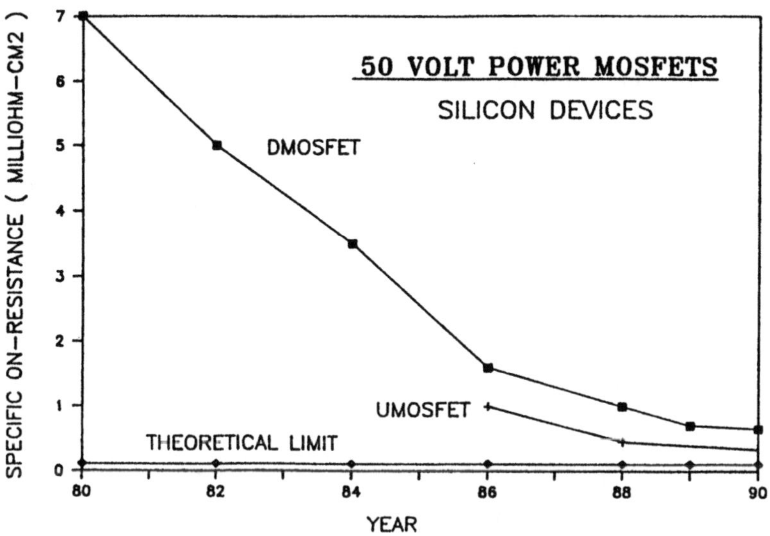

Figure 2 Progress made in reducing the specific on-resistance in silicon power MOSFETs by design and process technology advances.

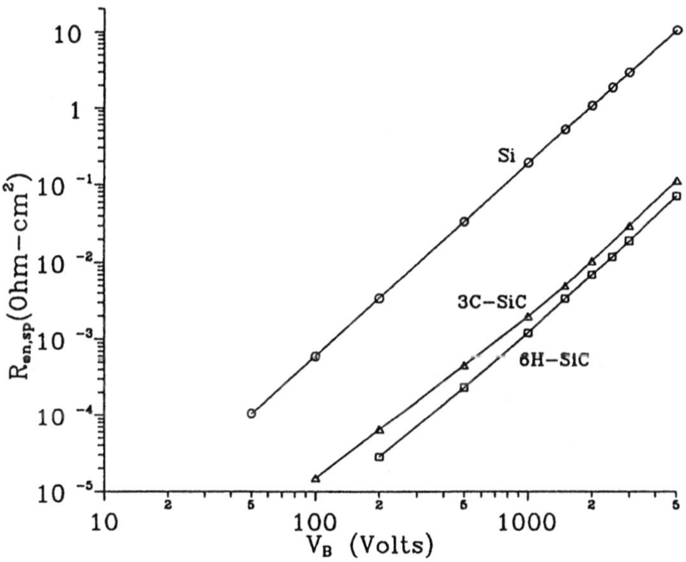

Figure 3 Comparison of the specific on-resistances of the drift region in power MOSFETs for 6H-SiC and 3C-SiC with those for silicon devices.

large for high voltage devices to enable operation at high current levels. One solution to this problem was the creation of the Insulated Gate Bipolar Transistor (IGBT) in the 1980's [1]. With this MOS-gated bipolar transistor structure, it has been possible to manufacture devices with current ratings of 300 amperes and 1500 volt blocking capability. These devices have rapidly displaced the power bipolar transistor in power electronic applications such as motor control. In spite of its great commercial success, the IGBT is not an ideal solution because its switching speed and safe-operating-area are limited by bipolar current flow. It would be advantageous to create power MOSFET structures from other semiconductor materials with low specific on-resistances that would be suitable for high voltage applications.

Power electronic applications require not only three terminal switching devices but also two terminal power rectifiers. In the case of silicon technology, Schottky barrier diodes are available with blocking voltages below 100 volts with excellent switching characteristics for high frequency applications. Devices with higher blocking voltages are not possible due to the large series resistance of the drift region which makes the on-state voltage drop very large and due to the soft breakdown characteristics observed for silicon devices. Consequently, for high voltage applications, the silicon P-i-N rectifier is used. Although this bipolar device has a low on-state voltage drop due to conductivity modulation of the drift region, it has poor switching characteristics as a result of the large amount of stored charge in the drift region [1]. Consequently, it would be advantageous to create improved high voltage Schottky barrier diodes using alternate materials with smaller drift region resistances.

2. Analysis of Drift Region Resistance

In the case of an abrupt one-dimensional junction fabricated in a uniformly doped semiconductor layer, the voltage is supported in a depletion layer with a linearly decreasing electric field. This field has its maximum value at the junction, and is equal to the critical electric field E_C at breakdown. From this field distribution, the doping concentration and the depletion layer thickness required to support the voltage can be derived [1]:

$$N_B = \epsilon \cdot (E_C)^2 \; / \; (2 \cdot q \cdot V_B) \qquad (1)$$

$$W = 2 \cdot V_B \; / \; E_C \qquad (2)$$

where V_B is the breakdown voltage. The specific on-resistance of this layer of material required to support the voltage is then given by:

$$R_{on,sp} = 4 \cdot (V_B)^2 \; / \; \left[\epsilon \cdot \mu \cdot (E_C)^3\right] \qquad (3)$$

where ϵ is the dielectric constant for the material and μ is the free carrier mobility.

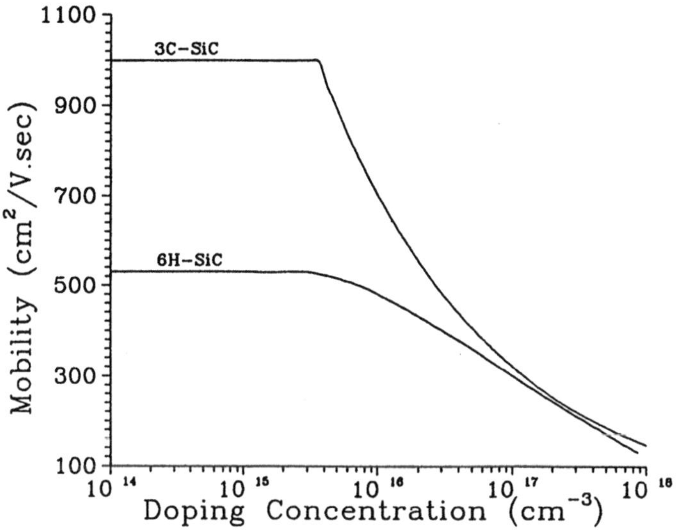

Figure 4 **Experimental data reported for the mobility of electrons in 6H-SiC and 3C-SiC.**

In general, μ and E_C are both functions of the doping level. In the case of silicon power devices, the doping level in the drift region is below 1E16 per cm3, and the mobility is independent of the doping level [1]. However, the critical electric field for breakdown is dependent upon the doping level [1] and is given by:

$$E_C = 4010 \cdot (N_B)^{1/8} \quad (4)$$

Based upon this relationship, the specific on-resistance for silicon drift regions is given by [1]:

$$R_{on,sp} = 5.93 \cdot 10^{-9} \cdot (V_B)^{2.5} \quad (5)$$

This variation of the specific on-resistance with breakdown voltage is plotted in Figure 3 for breakdown voltages ranging from 50 to 5000 volts. It can be seen that the specific on-resistance for silicon devices becomes too large at breakdown voltages above 1000 volts to make devices practical in this material.

In the case of silicon carbide, experimental data has been published on the variation of the mobility with doping concentration for both 6H-SiC and 3C-SiC [3,4]. This data is reproduced in Figure 4. It is worth pointing out that the mobility values for 6h-SiC are smaller than for 3C-SiC. In addition, the mobility becomes a function of the doping level for the SiC devices for the breakdown voltage range of 50 to 5000 volts because much higher doping levels can be used for the SiC drift region.

With regard to the breakdown field strength for SiC, measurements have only been reported for 6H-SiC [5,6]. There is a significant disparity between

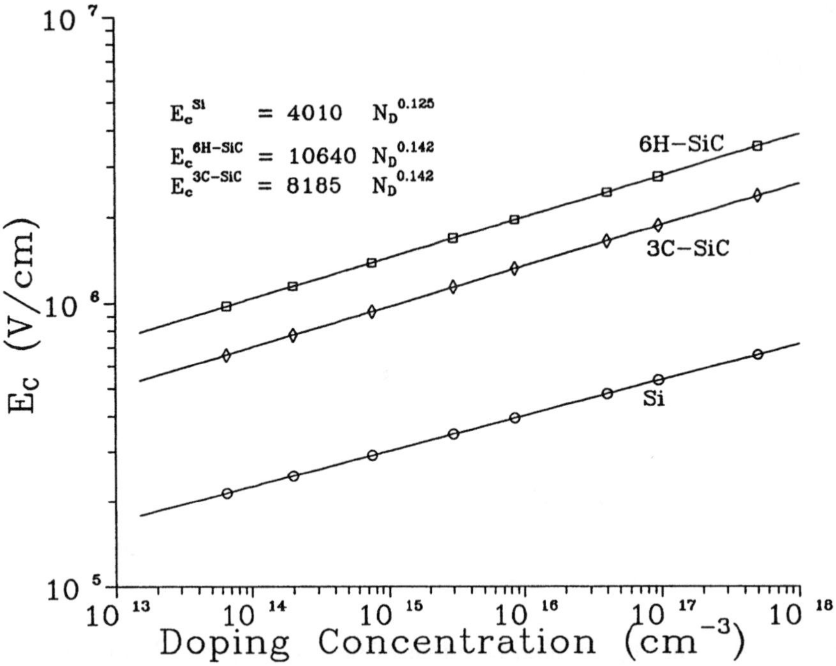

Figure 5 Critical electric field for breakdown for 6H-SiC, 3C-SiC, and silicon used for analysis in this paper. The values for silicon and 6H-SiC are based on experimental data published in the literature. The values for 3C-SiC are scaled from 6H-SiC using the ratio of the energy band gaps.

the values measured in the two studies. For the analysis of the specific on-resistance in this paper the higher value for the breakdown electric field strength has been used. This is based upon the assumption that the higher values represent material with superior quality. In addition, since no breakdown electric field strength values were found in the literature for 3C-SiC, these values were obtained by scaling them in proportion to the energy band gaps for these materials. Based upon this assumption:

$$E_C\,(3C\text{-}SiC) = 0.77\ .\ E_C\,(6H\text{-}SiC) \qquad (6)$$

using an energy band gap of 2.2 eV for 3C-SiC and 2.86 for 6H-SiC. The critical electric field for breakdown in 6H-SiC and 3C-SiC are compared with those for silicon in Figure 5.

3. Silicon Carbide Power MOSFETs

In the case of silicon carbide, lateral enhancement-mode n-channel MOSFETs have been experimentally demonstrated [7]. These devices were fabricated by the formation of the gate dielectric using thermally grown oxide layers. The same technology should be applicable to power MOSFETs. As indicated in Figure 1, the internal resistance in the power MOSFET structure contains many components. Since insufficient information exists to calculate the values of all these components, it will be assumed that technological advances will allow their reduction to values that are much smaller than for the drift region. This is justified by the trend shown in Figure 2 for silicon devices. The specific on-resistance of the drift region in silicon carbide power MOSFETs using 6H-SiC and 3C-SiC are compared with those for the silicon devices in Figure 3. It can be seen that the specific on-resistance for the silicon carbide devices is much lower than that for the silicon devices. It is worth pointing out that under the assumptions made for these calculations, 6H-SiC devices are expected to be superior to 3C-SiC devices in spite of the lower mobility for electrons in 6H-SiC.

4. Silicon Carbide Rectifiers

As mentioned earlier, power electronic systems require not only high performance transistors but also rectifiers with low on-state voltage drop and fast switching speed. Schottky rectifiers with low on-state voltage drops can be obtained using silicon carbide as shown in Figure 6, where the calculated

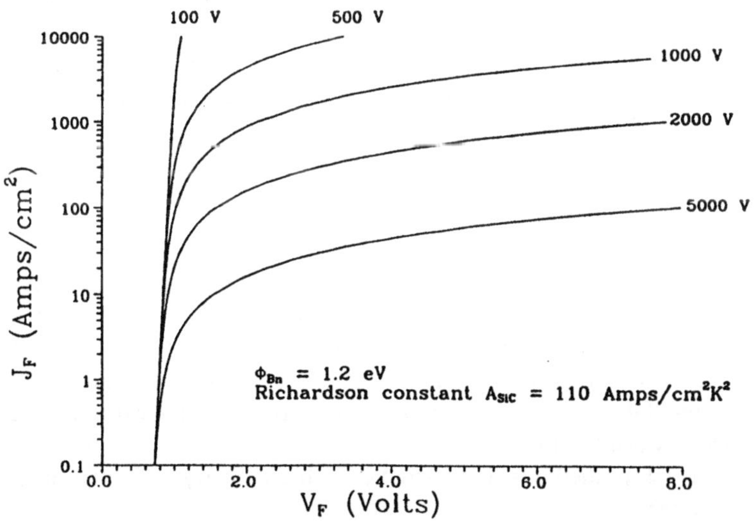

Figure 6. Calculated forward conduction characteristics of Schottky barrier diodes for 6H-SiC.

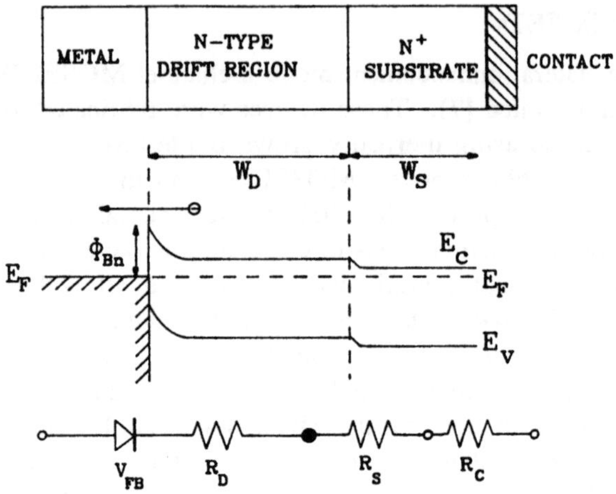

Figure 7 Cross-section of a high voltage Schottky barrier diode with series resistances indicated for the drift region, the substrate, and the ohmic contact.

on-state characteristics have been plotted for devices with breakdown voltages up to 5000 volts. In case of high voltage Schottky barrier diodes, the forward drop is determined not only by the Schottky barrier height but also by the voltage drop across the drift region as illustrated in Figure 7. The series resistance of the drift region increases when the breakdown voltage is increased as already discussed for power MOSFETs. As a consequence, the forward voltage drop of the diodes increases at higher breakdown voltages as shown in Figure 6. Since the forward voltage drop of the silicon carbide Schottky barrier diodes is calculated to be comparable to that for a silicon P-i-N diode even for the highest blocking voltage, the fast switching speed of the SiC Schottky barrier diode will favor its application in power systems.

5. Conclusions

The analysis performed in this paper indicates that silicon carbide power devices have the potential to displace all silicon power transistors and rectifiers up to a breakdown voltage of 5000 volts. The development of such devices is expected to take many years of intensive research to address basic technological requirements for the fabrication of these structures from silicon carbide. In particular, research is needed to:

(a) develop methods for growth of high purity epitaxial drift layers,

(b) develop techniques to produce p-n junctions and Schottky barriers with high breakdown voltages,

(c) develop processes to obtain very low resistance ohmic contacts, and

(d) develop methods for controlling inversion layer properties in MOS structures.

With these technologies in hand, it will be possible to demonstrate the performance advantages of silicon carbide for power electronic applications.

6. References

1. B.J. Baliga, "Modern Power Devices," John Wiley & Sons (1987).

2. B.J. Baliga, "Impact of VLSI Technology on Power Devices," International Conference on Solid State Devices and Materials, Abstract a-1-2, pp. 5-9 (1990).

3. H. Matsunami, "Semiconductor SiC - Expectations for Power Devices," Proceedings 2nd International Symposium on Power Semiconductor Devices and ICs, pp. 13-18 (1990).

4. R.B. Hillborn and H. Kang, "Charge Carrier Concentration and Mobility in n-type 6H Polytypes of SiC," in Silicon Carbide 1973, edited by R.C. Marshall, J.W. Faust, and C.E. Ryan, pp. 337-339.

5. C. van Opdrop and J. Vrakking, "Avalanche Breakdown in Epitaxial SiC p-n Junctions," J. Applied Physics, Vol. 40, pp. 2320-2322 (1969).

6. W. von Muench and I. Pfaffender, "Breakdown Field in Vapor Grown Silicon Carbide p-n Junctions," J. Applied Physics, Vol. 48, pp. 4831-4833 (1977).

7. K. Shibahara, T. Saito, S. Nishino, and H. Matsunami, "Fabrication of Inversion type n-Channel MOSFETs using Cubic SiC on Si," IEEE Electron Device Letters, Vol. EDL-7, pp. 692-693 (1986).

SiC_x:F Hetero-Emitter and Epitaxial-Base Bipolar Transistors

T. Sugii[1], *T. Yamazaki*[1], *Y. Arimoto*[1], *T. Ito*[1], *I. Namura*[2], *H. Goto*[2], *and A. Tahara*[2]

[1]Fujitsu Laboratories Ltd., 10-1 Morinosato-Wakamiya, Atsugi 243-01, Japan
[2]Fujitsu Limited, 1015 Kamiodanaka, Nakahara-ku, Kawasaki 211, Japan

This paper discusses the technology issues on an emitter and base for sub 0.5 μm bipolar transistors. The new opportunities offered by heterojunction structures and a low-temperature-grown epitaxial base increase the freedom of device design. The SiC_x wide-gap emitter with thin, highly doped epitaxial base shows superior characteristics for LSI application owing to the freedom. Low base sheet resistance was achieved using 1×10^{19} /cm^3 concentration base, while retaining an acceptable current gain and suppressing forward-bias tunneling current. We obtained a 30 GHz cutoff frequency for a 35-nm base width.

1. Introduction

The key devices in a fast CPU are high-speed gate arrays and SRAM, which are best implemented using bipolar technology. The bipolar transistor's exponential I-V characteristic gives it the highest transconductance of semiconductor devices and results in the bipolar transistor's excellent wiring capacitance drive capability and threshold voltage controllability.

The progress of high-speed bipolar transistors is clearly illustrated using the switching delay of emitter-coupled-logic (ECL) ring oscillators. Figure 1 plots gate delay data collected from several industrial laboratories and shows that gate delay has reached the sub-30-ps range. These short gate delays are due to advanced techniques.

A typical sub-30 ps t_{pd} bipolar transistor (Figure 2) features a poly-Si emitter, self-aligned base contact, low-energy boron ion-implantation, a pedestal collector,

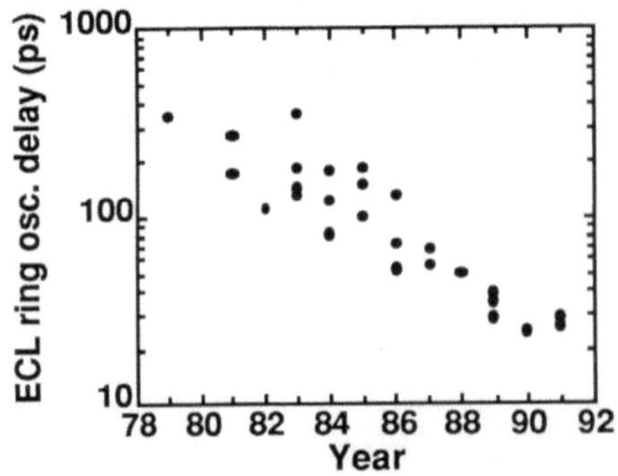

Figure 1 Gate delays in ECL ring oscillators

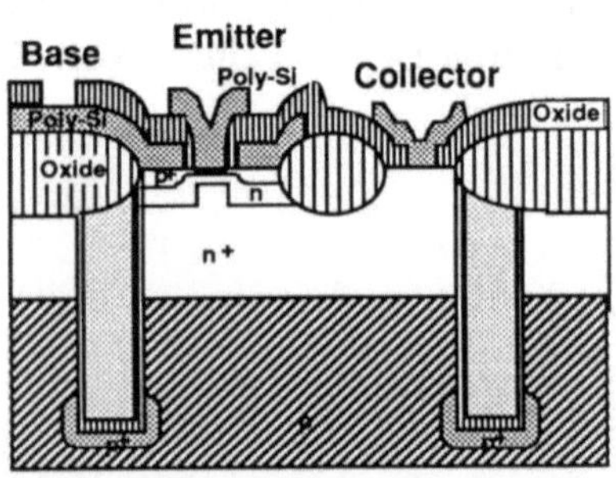

Figure 2 Cross-section of a typical high-speed bipolar transistor

Springer Proceedings in Physics, Vol. 71 Amorphous and Crystalline Silicon Carbide IV
Editors: C.Y. Yang · M.M. Rahman · G.L. Harris

and deep trench isolation. Fabrication behind this advanced structure is fully mature at R&D centers. A 0.5-µm- bipolar transistor is expected to be put to use in high-speed systems by the mid-1990s. The potential in the field is fast being outstripped by the need for new, more sophisticated techniques, however.

This paper describes an SiC_x:F wide-gap emitter and a low-temperature-grown epitaxial base expected to be a pioneer in next-generation devices[1,2].

2. Tradeoff between high f_T and low $r_{bb'}$

One target in device development is the reduction of gate delay. A self-aligned base contact virtually eliminates extrinsic base resistance. Current interests lie in an intrinsic base. A shallow intrinsic base produces a high cutoff frequency, but increases the intrinsic base resistance, resulting in a larger $r_{bb'}$. Ring oscillators using transistors with higher cutoff frequencies sometimes show lower t_{pd}[3,4], in what is basically a tradeoff problem solved by combining a high base concentration with a narrow base width. High base concentration degrades current gain and induces a large forward bias tunneling current, however. A heterojunction such as a wide-gap emitter and a narrow-gap base must be used at an emitter-base junction to suppress the degradation of current gain and forward bias tunneling current.

3. Single-crystal SiC HBT limitations

We started on Si-based HBTs using single-crystal SiC (3C-SiC)[5,6]. Thermal stability and large saturation drift velocity made it promising for use in high-speed electronic devices.

Figure 3 is the double-crystal X-ray diffraction spectrum for 400-nm-thick SiC film grown at 1000°C using $SiHCl_3$ and C_3H_8[7]. Using this film, we fabricated the SiC HBT shown in Figure 4. We obtained a maximum current gain of about 800 with an emitter carrier concentration of only 1×10^{19} /cm^3. The emitter carrier concentration is over one order smaller than that for conventional poly-Si emitters.

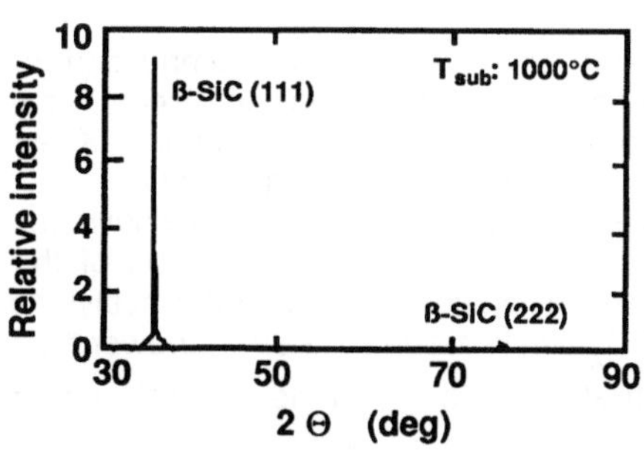

Figure 3 X-ray diffraction spectrum of the ß-SiC film

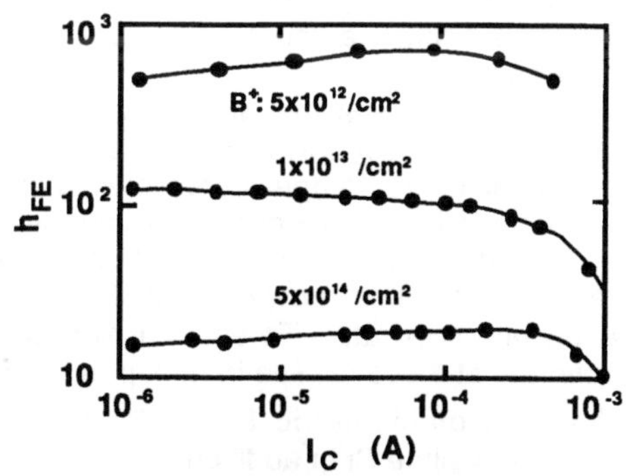

Figure 4 Current gain of the ß-SiC emitter HBT as a function of the collector current

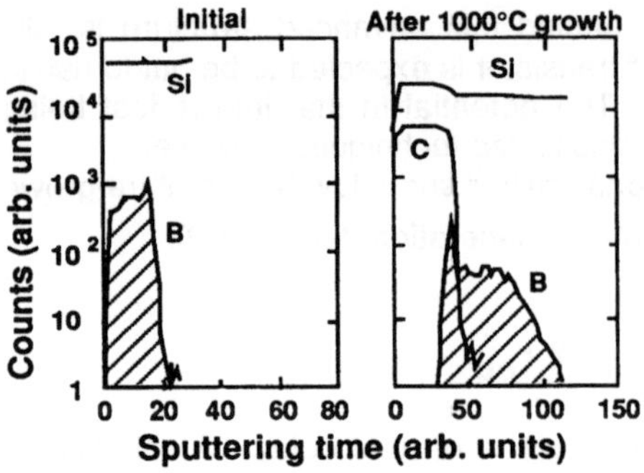

Figure 5 Impurity depth profiles before and after SiC growth at 1000°C for 10 minutes

The 3C-SiC can be grown by various methods[8-10], but commercially available devices using 3C-SiC have not yet been fabricated. One problem has been the difficulty in growing high-quality single-crystal SiC reproducibly over a large area. Another has been the high growth temperature, above 1000°C. For high-speed bipolar device applications, a lower growth temperature is needed to reduce dopant redistribution. Although the single-crystal SiC is effective as a wide-gap emitter, the SiC growth temperature of 1000°C is too high to fabricate the thin base, which must be less than 50 nm thick and have a base carrier concentration above 1×10^{19} /cm^3. Figure 5 shows the problems of high-temperature growth. SiC epitaxy at 1000°C induces dopant redistribution, which results in a thick base width.

4. Fluorine-doped SiC_x emitter

The SiC emitter fabrication temperature can be lowered by growing non-single-crystal SiC_x film. The emitter material need not be single-crystal, but the interface must be good enough to suppress the recombination current.

We tried terminating the dangling bonds in SiC_x by adding a terminator[11]. Fluorine appears the most suitable because the bonding energy of Si-F bonds is large and fluorine does not introduce new gap states. A fluorinated heat-resistant amorphous Si whose dangling bonds were terminated by fluorine has been used for solar cells[12]. We used difluorosilane as the fluorine doping gas. To introduce a large amount of fluorine into the SiC_x film, we lowered the growth temperature.

Figure 6 shows the depth profile of films grown at different temperatures, obtained by secondary ion mass spectroscopy. The film was 40 nm thick. Films grown at lower temperatures clearly incorporate more fluorine atoms. X-ray photoelectron spectroscopy revealed that fluorine atoms bond with Si or C.

The acetylene gas flow rate strongly effects film resistivity. Figure 7 shows the resistivity of films grown at 700°C on sapphire substrates as a function of the acetylene flow rate. To avoid a high emitter resistance, we chose the acetylene flow rate at 1 sccm. The bandgap of the SiC_x film is around 1.8 eV, narrower than that of stoichiometric SiC. We think that the film contains amorphous or microcrystalline Si in addition to SiC.

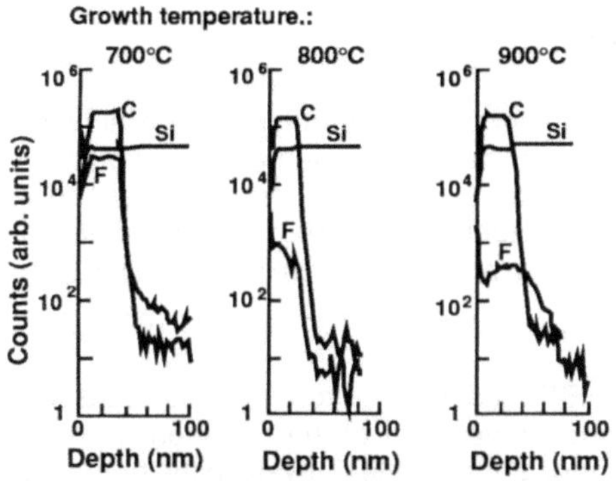

Figure 6 Impurity depth profiles of fluorine-doped SiC_x films grown at different temperatures

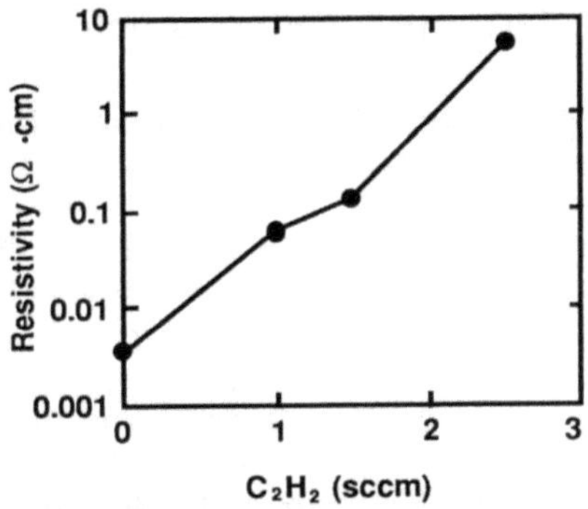

Figure 7 Relationship between the SiC_x film resistivity and C_2H_2 flow rate

5. Low-temperature-grown epitaxial base

The desirable HBT base is very thin and highly doped. Conventional boron ion implantation cannot be used because of boron ion channeling. We made the base using low-temperature Si photoepitaxy[13] based on CVD. Ultraviolet (UV) light from high-pressure Hg lamps vertically irradiated the wafer surface during epitaxial growth[14]. Disilane (Si_2H_6) was used as the source gas, and diborane (B_2H_6) as the dopant gas. Because UV irradiation enhances the dissociation of disilane, photoepitaxy enables sufficient growth rate at 600°C. Due to this low temperature, the base doping concentration and the base width can be optimized separately. The box profile of the base dopant is desirable because it minimizes the base width and maintains the required base Gummel number to achieve an acceptable punchthrough voltage.

Figure 8 shows the boron profile of a photoepitaxial layer 35 nm thick and having a 1×10^{19} /cm^3 concentration. Note the profile for ion implantation with 10 keV acceleration energy. Ion implantation could not form a shallow profile due to the channeling effect.

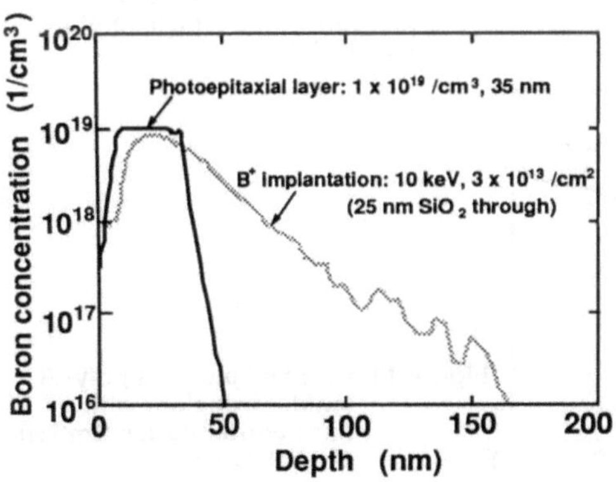

Figure 8 Boron profiles of a photoepitaxial layer 35 nm thick and 1E19 /cm³ concentration and a layer boron implanted with 3E13 /cm² at 10 keV through 25-nm-thick SiO_2

6. SiC_x-HBT (DC characteristics)

Figure 9 shows the HBT structure. The Si substrate was an n-type epitaxial wafer with (100) orientation. After growing a thick oxide, we opened a base region for epitaxial growth. A boron-doped single-crystal base and a boron-doped poly-crystal base electrode were grown by Si photoepitaxy at 600°C. The epitaxial base was 50 nm thick and had a carrier concentration of $1x10^{19}$ /cm^3. The SiC_x emitter was grown at 700°C. The carrier concentration was $1x10^{19}$ /cm^3, determined by Hall measurement.

Figure 10 shows Gummel plots of the HBT measured at room temperature and at 190K. The current gain was 80 , about 6 times that of the nonfluorine-doped SiC_x HBT[15]. The room temperature plot shows little forward-bias tunneling current. Because the tunneling current is less dependent on temperature than diffusion or recombination currents, it should become more prominent at lower temperatures, but, as shown in the figure, there is not significant tunneling current even at 190K.

Figure 11 shows a Gummel plot of the conventional poly-Si emitter transistor with the same base configuration. The base current characteristics apparently differ from those of the HBT. Due to the high carrier concentration in the poly-Si emitter (10^{20} to 10^{21} /cm^3) and the epitaxial base (10^{19} /cm^3), the forward-bias tunneling current is very large.

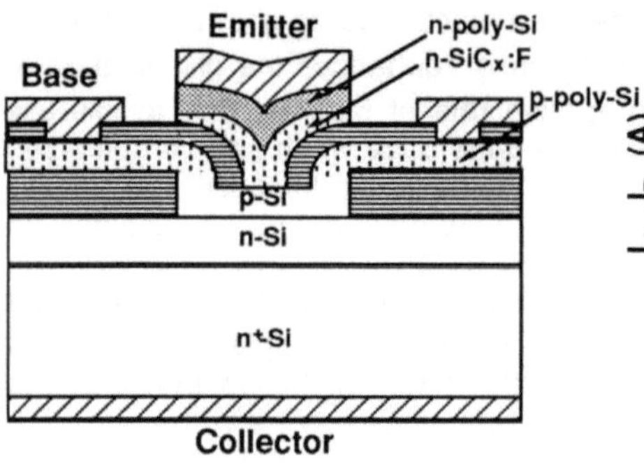

Figure 9 Cross-section of the SiC_x HBT for DC measurement

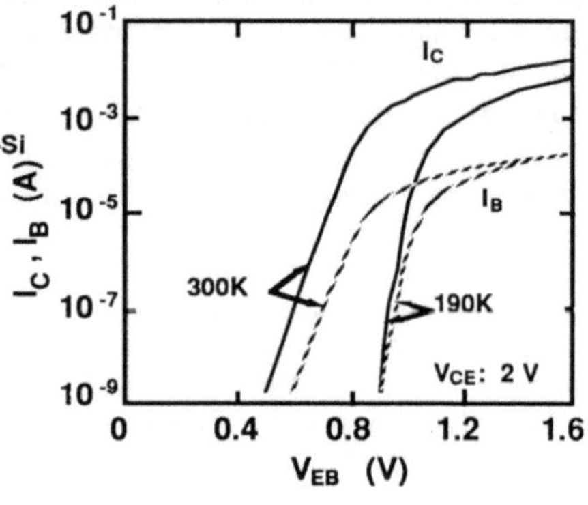

Figure 10 Gummel plots of the SiC_xHBT measured at room temperature and at 190 K

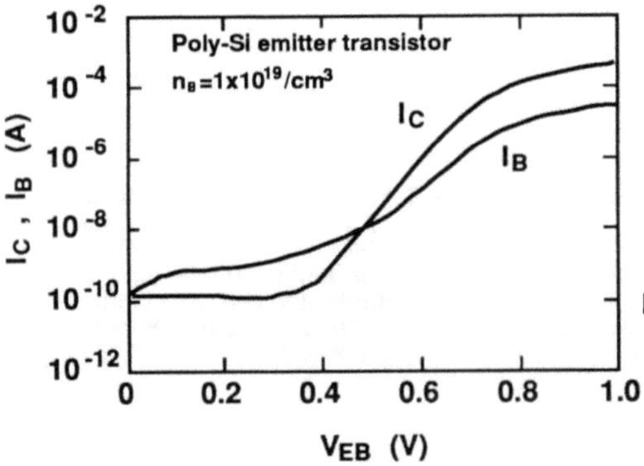

Figure 11 Gummel plot of a poly-Si emitter transisor with a base carrier concentration of 1E19 /cm^3

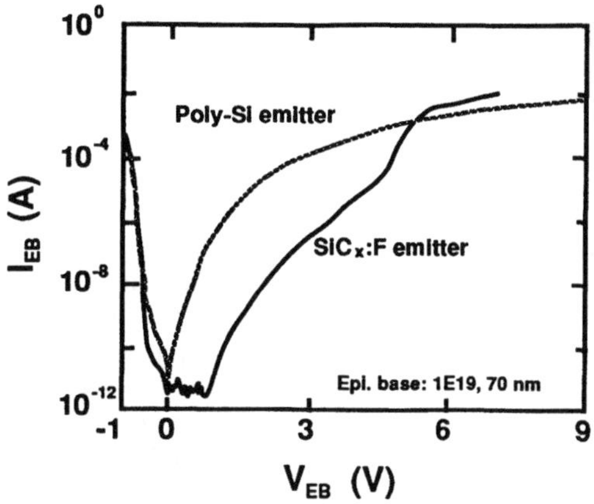

Figure 12 Emitter-base junction characteristics of SiC_x emitter and poly-Si emitter transistors

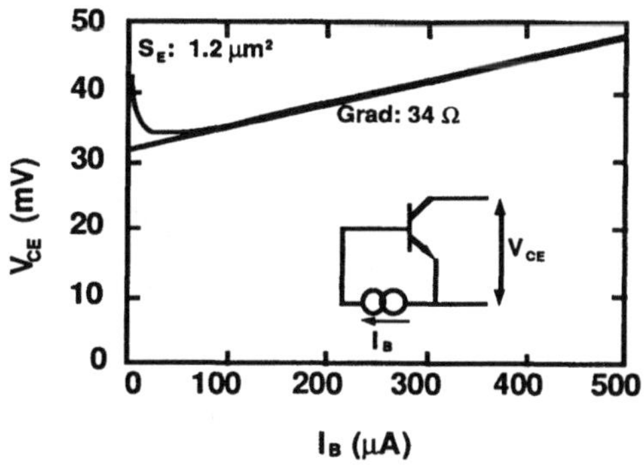

Figure 13 Typical result of emitter resistance measurement by the open collector method

The Early voltage of the HBT exceeded 100 V due to the high carrier concentration in the base. The measured pinch resistance was extremely low, 1.8 kΩ/square. A current gain of 80 and a pinch resistance of 1.8 kΩ/square is attainable only by using a heterojunction at the emitter-base junction.

Figure 12 shows the emitter-base junction characteristics. The lower carrier concentration in the SiCx emitter produces a larger reverse breakdown voltage, over three times larger at 10 μA.

Emitter resistance coupled with a base-collector junction capacitance seriously limits cutoff frequency and gate delay. A large emitter resistance also affects the DC transfer characteristics of the ECL gate. Emitter resistance does not scale with device geometry. To a first approximation, as device dimensions decrease, emitter resistance increases with the reciprocal of the emitter size. Scaling may be meaningless without emitter resistance reduction.

As far as current gain is concerned, the effect and existence of the interfacial oxide between emitter and base must be taken into account, because such interfacial oxide enhances current gain but seriously increases emitter resistance. It is difficult to distinguish the effect of the wide-gap emitter from that of the interfacial oxide in an evaluation using only current gain. Both emitter resistance and current gain must be measured with an identical transistor.

Figure 13[16] shows the typical result of the emitter resistance measurement by open collector method[17]. From the gradient of the straight line, the emitter resistance was 34 Ω, which corresponds to a contact resistance of 41 $\Omega \cdot \mu m^2$. This value is comparable to those for high-speed poly-Si emitter transistors.

7. SiCx-HBT (AC characteristics)

To measure AC characteristics, we fabricated SiC_x-HBTs using trench isolation, a double poly-Si process, and 0.8-μm lithography[18]. The collector had a 1×10^{17} /cm^3 concentration. We grew the heavily boron doped epitaxial layer on the field oxide, and covered the active base region with an oxide/nitride/oxide pad. A 0.5-μm-wide emitter was formed using side walls without self-alignment.

Figure 14 shows the dependence of the cutoff frequency on the collector current for a SiC_x HBT with a 35-nm-thick base. A maximum cutoff frequency of 30 GHz is the highest reported for wide-gap emitter Si HBTs. This decreased from 30 GHz to 25 GHz as the base concentration increased from 6×10^{18} /cm^3 to 1×10^{19} /cm^3. This degradation was mainly due to the decreased electron mobility in the base. Intrinsic delay in the SiC_x HBT (Figure 15) drastically decreased from 7.8 to 4.7 ps with base width and concentration due to the reduced base transit time. However, even a base transit time of 4.7 ps cannot be neglected. A thinner base improves the cutoff frequency.

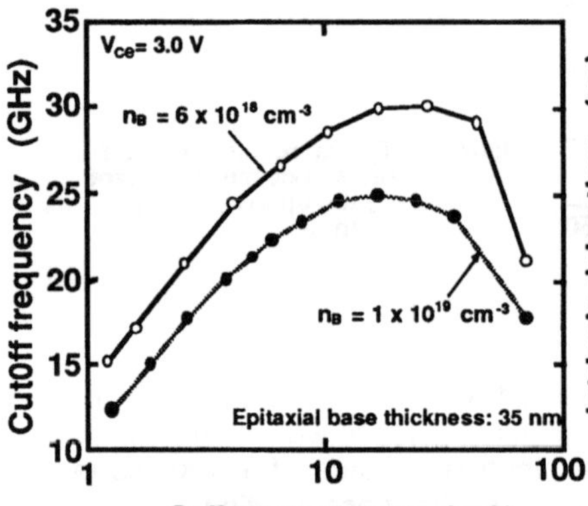

Figure 14 Dependence of cutoff frequency on collector current for the SiC_x HBT with a 35-nm-thick base

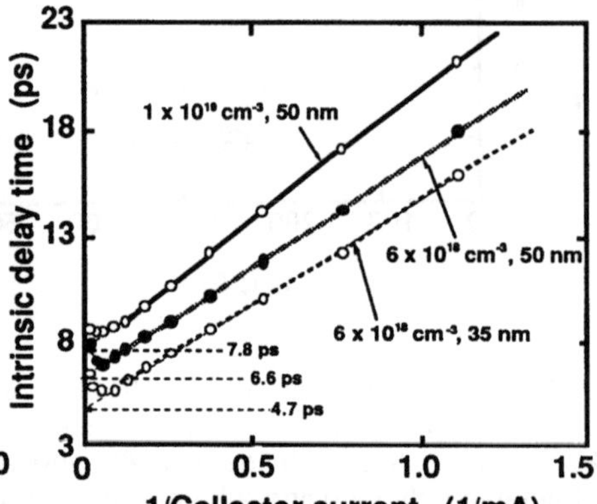

Figure 15 Relation between intrinsic delay and inverse collector current

8. Conclusion

We fabricated Si HBTs using an SiC_x wide-gap emitter and a low-temperature-grown epitaxial base. The SiC_x HBT had DC characteristics suitable for LSI application. The low base sheet resistance was achieved with a doping of 1×10^{19} /cm^3, while retaining an acceptable current gain and suppressing the forward-bias tunneling current. We obtained a 30-GHz cutoff frequency for a 35-nm base width.

Acknowledgment

We thank the staff of the Atsugi and Kawasaki semiconductor processing faculties for their invaluable aid and advice.

References

[1] T. Sugii, T. Yamazaki, and T. Ito, in Extended Abstracts of the 22nd Conf. SSDM Conf.SSDM, pp. 817-820, 1990.

[2] T. Sugii, T. Aoyama, Y. Furumura, and T. Ito, in Proceedings of the First Topical Symp. on Si Based Heterostructures (part of the American Vacuum Society 37th National Symposium), pp.124-129, 1990.

[3] M. Sugiyama, H. Takemura, C. Ogawa, T. Tashiro, T. Morikawa, and M. Nakamae, in IEDM Tech. Dig., pp. 221-224, 1989.

[4] T. Yamazaki, I. Namura, H. Goto, A. Tahara, and T. Ito, ibid.pp. 309-312, 1990.

[5] T. Sugii, T. Ito, Y. Furumura, M. Doki, F. Mieno, and M. Maeda, J. Electrochem. Soc., vol. 134, no. 10, pp. 2545-2549, Oct. 1987.

[6] T. Sugii, T. Ito, Y. Furumura, M. Doki, F. Mieno, and M. Maeda, IEEE Electron Device Lett., vol. 9, no. 2, pp. 87-89, Feb. 1988.

[7] T.Eshita,T.Suzuki, T. Hara, F. Mieno, Y. Furumura, M. Maeda, T. Sugii, and T. Ito, in Proceedings of Mat. Res. Soc. Symp., pp. 357-362, 1988.

[8] S. Nishino, J. Anthony Powell, and Herbert A. Will, Appl. Phys. Lett., vol. 42, no. 5, pp. 460-462, Mar. 1983.

[9] S. Kaneda, Y. Sakamoto, T. Mihara, and T. Tanaka, J. Cryst. Growth., vol. 81, pp. 536-542, 1987.

[10] H. Matsunami, in Extended Abstracts of the 1991 International Conf. SSDM, pp. 138-140, 1991.

[11] T. Sugii, T. Yamazaki, and T. Ito, IEEE Trans. on Electron Devices, vol. 37 no, 11, pp. 2331-235, Nov. 1990.

[12] H. Matsumura and S. Furukawa, in Proceedings 12th Conf. Solid-State Devices, pp. 275-280, 1980.

[13] T. Sugii, T. Yamazaki, T. Fukano, and T. Ito, IEEE Electron Devices Lett., vol. EDL-8, no. 11, pp. 528-530, Nov. 1987.

[14] T. Yamazaki, T. Ito, and H. Ishikawa, in VLSI Symp. Dig. Tech. Papers, pp. 56-57, 1984.

[15] T. Sugii, T. Yamazaki, K. Suzuki, T. Fukano, and T. Ito, in IEDM Tech. Dig., pp. 659-662, 1989.

[16] T. Sugii, T. Yamazaki, and T. Ito, Jpn. J. Appl. Phys., vol. 30, no. 6A, pp. L970-L972, June, 1991.

[17] L. J. Giacoletto, IEEE Trans. on Electron Devices, vol. 19, pp. 692-693, May, 1972.

[18] T. Yamazaki, T. Sugii, T. Ito, I. Namura, H. Goto, a. Tahara, in BCTM Tech. Dig., pp. 70-73, 1991.

A New Application of a-SiC Films for Realizing High Current Gain Si Heterojunction Bipolar Transistors

K. Sasaki, T. Miyajima, and S. Furukawa

Graduate School of Science and Engineering, Tokyo Institute of Technology, 4259 Nagatsuda, Midori-ku, Yokohama 227, Japan

Device characteristics of Si-HBT with μc-Si/a-SiC multilayered emitter structure are investigated. From the temperature dependence of the base current, current flow mechanism at the heterojunction and the energy bandgap of emitter region are discussed. Current gain as high as 523 is achieved by using a-SiC buffer layer, while for not using it, the current gain is only 22.

1. Introduction

In the study of Si based amorphous materials to Si-HBT, it has been reported that the position of the heterojunction should be coincident with the position of the pn junction and the μc-Si should be also deposited at low enough temperature, since the diffusion of phosphorus heavily doped in μc-Si must be prevented. However, the interface structure was not sufficiently investigated so far. Recently, we found that μc-Si had a serious problem, that is, μc-Si is easily crystallized near the interface between noncrystalline emitter materials and crystalline Si substrate when it is deposited directly on c-Si substrates by plasma CVD with an oil free evacuation system[1]. Due to this phenomenon, an abrupt and uniform heterojunction between μc-Si and c-Si is difficult to obtain. And, Si homojunction usually has been formed, resulting in a low current gain. In order to realize a true amorphous/crystalline heterojunction, the unintentional crystallization have to be suppressed. We propose to utilize a-SiC as a buffer layer in order to do this.

In this paper, the focus of our discussion is electrical characteristics of the Si based HBTs with the amorphous heteroemitter. The detailed investigation of the amorphous/crystalline Si heterojunction is given in another paper[2].

2. Experiments

μc-Si films were prepared using inductively coupled RF plasma CVD equipment with a SiH_4-PH_3-H_2-Ar gas mixture. The key technique for producing the microcrystalline phase is the addition of Ar to source gases. For preparing a-SiC layers, CH_4 gas was added to the gas mixture as the carbon source. In this experiment, the CH_4 fraction f_C, i.e., $f_C=CH_4/(SiH_4+CH_4)$, and the layer thickness t_{SiC} which was defined by the deposition rate were used as preparation parameters of the a-SiC buffer layers.

Si-HBTs with the μc-Si/a-SiC multilayered emitter structure were fabricated by using a conventional Si process. The current-voltage characteristics of the devices were measured as a function of operation temperature.

Springer Proceedings in Physics, Vol. 71 Amorphous and Crystalline Silicon Carbide IV
Editors: C.Y. Yang · M.M. Rahman · G.L. Harris

3. Results and Discussion

μc-Si emitter Si-HBTs with the a-SiC buffer layer as denoted in Table 1 were fabricated. Also, the device without the buffer layer was fabricated for reference. Characters and letters used in the following figures are denoted in Table 1. Typical cases of temperature dependence of the base current are shown in Fig. 1. In the lower current region, excess current components such as recombination current and tunneling current dominate the base current, while a parasitic resistance limits it in the higher current region, so that the characteristic difference is unclear between devices(a) and (b). In order to obtain more information from the data measured, the voltage drops by the parasitic resistances were corrected by assuming that the collector current obeys the exponential factor of qV/kT even in the resistance limited region and the same voltage drops appear in the base and collector current. The results corrected by such way are shown by bold lines in the figure. Temperature dependence of the slope is clearly different. For the device without a-SiC buffer layer, the slope become gentl with increasing the temperature, while with the buffer layer, the slopes show no temperature dependence[3].

Those temperature dependences were characterized by the exponential factors of qV/nkT and ξV as shown in Fig. 2. For the devices (a) and (c), n-values were almost unity, suggesting that the base currents are dominated by the hole diffusion current from the base to the emitter. Therefore, energy bandgaps of the emitters can be determined by Arrenius plots of the reverse base saturation current, which is determined by extrapolating the base current region having n-value of unity to the zero applied voltage. The plots well fit on lines over 6 decades as shown in Fig. 3. This fact shows the validity of the correcting way of the data and the assumptions used. It is noted that the derived values of 1.17

		$CH_4/(SiH_4+CH_4)$		
		0	0.2	0.5
a-SiC Thickness (nm)	0	○		
	2		—	△
	5		□	▽

Table 1.
Notations of the a-SiC buffer layer forming conditions used for Si-HBTs fabrication.

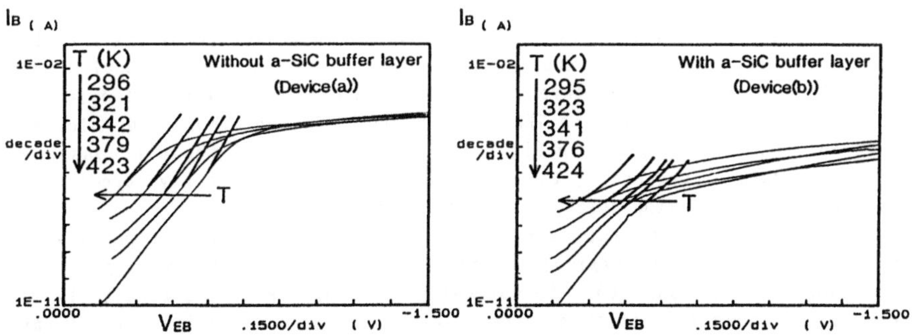

Fig. 1.
Temperature dependence of base current of devices (a) and (b).

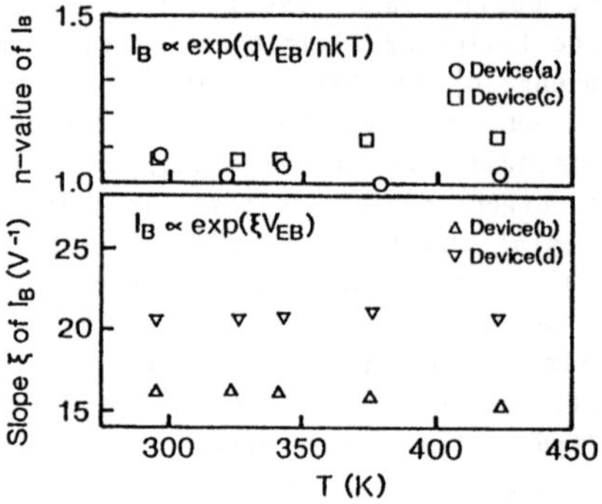

Fig. 2.
Temperature dependence of n-value and slope ξ of base current.

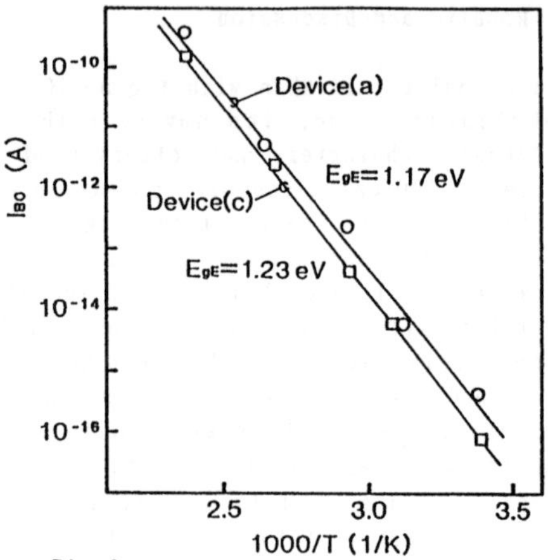

Fig. 3.
Arrenius plots of the reverse base saturation current.

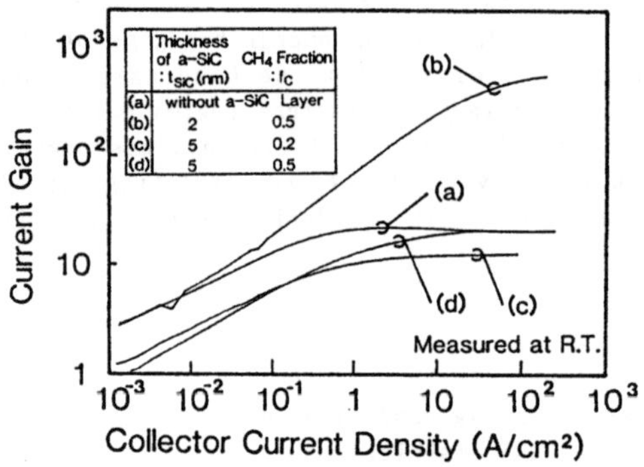

Fig. 4.
Characteristics of current gain for collector current density.
Maximum current gain:
(a) 22, (b) 523, (c) 13, (d) 20

and 1.23eV are much smaller than the optical bandgap of 1.75eV for μc-Si measured on glass substrates and are very close to the bandgap of c-Si. This means that the deposited layers have become poly or single c-Si. A little difference of the energy bandgap may be due to effects of the a-SiC buffer layer of the device (c). It can be said that a-SiC buffer layer formed by lower carbon content have little effect to suppress the crystallization. These results show that nearly homojunction structure is formed and are coincident with the results of photoresponse measurement and TEM observation[2]. On the other hand, for the devices (b) and (d), as shown in Fig. 2, the slope ξ is independent constants for the temperature, suggesting that tunneling current which we have predicted[4] is principal and diffusion current is suppressed.

Current gain characteristics of Si-HBTs for collector current density are shown in Fig. 4, where the maximum emitter-base applied voltage was kept -1.5V. For the devices (a) and (c), the current gain saturates at the relatively low value because the homojunction have been formed and hole diffusion current have flowed. On the other hand, for device (b), the current gain continues to increase with the collector current, the maximum current gain of as high as 523 is achieved, while current gain is only 22 for the device(a) of which the emitter has been fabricated without the buffer layer. From these results, it is considered that high current gain is achieved by suppressing the crystallization of μc-Si and forming a genuine heterojunction. By the way, because optimization of a-SiC buffer layer have not been done, a large number of localized state is supposed to present in a-SiC buffer layer, which may degrade the electrical interface property. Nevertheless, it is interesting that the high current gain have been obtained. Further higher current gain is expected by improving the a-SiC film quality and optimizing the buffer layer.

4. Conclusion

It is indicated that μc-Si is easy to be crystallized on the c-Si substrates, so that a homojunction is formed, which degrades current gain. The multilayered emitter structure such as a combination of a-SiC buffer layer and μc-Si film is effective to form a genuine and reliable heterojunction and also realize a high current gain.

References

1)K. Sasaki et al.: Ext. Abs. 22nd Int. Conf. SSDM, p.661(1990).
2)S. Furukawa et al.: Proc. of 4th ICACSC, (this issue).
3)K. Sasaki et al.: Ext. Abs. 21st Conf. SSDM, p.361(189).
4)K. Sasaki et al.: Proc. of 1st ICACSC, Springer Proc. in Phys. **34**, p.150(1989)

The Development of ECR-CVD SiC Coatings for X-Ray Mask Membranes

A.R. Shimkunas[1], *P.E. Mauger*[1], *L.P. Bourget*[2], *R.S. Post*[2], *L. Smith*[3], *R.F. Davis*[3], *G.M. Wells*[4], *F. Cerrina*[4], *J.E. Butler*[5], *and R.B. McIntosh*[6]

[1]Nanostructures, Inc., 1245 Pear Avenue, Mountain View, CA 94043, USA
[2]Applied Science and Technology, Inc., 35 Cabot Road, Woburn, MA 01801, USA
[3]Department of Materials Science and Engineering, North Carolina State University, Raleigh, NC 27695, USA
[4]Center for X-ray Lithography, University of Wisconsin, 3731 Schneider Drive, Stoughton, WI 53589, USA
[5]Naval Research Laboratory, Chemistry Division, Surface Chemistry Branch, Gas/Surface Dynamics Division, Code 6174, 4555 Overlook Avenue, S.W., Washington, DC 20375, USA
[6]Horizon Technology Group, Inc., Suite 1524, 4001 N. Ninth Street, Arlington, VA 22203, USA

High-quality SiC membranes for x-ray lithography masks have been fabricated from SiC coatings prepared by electron cyclotron resonance chemical vapor deposition (ECR-CVD). Stoichiometric SiC coatings were deposited on 100-mm-diameter silicon wafers at a susceptor temperature of 930 °C in a single-wafer ECR-CVD system at deposition rates of 200 Å/min with better than ±5% thickness uniformity and ±1% wafer-to-wafer reproducibility. The coatings were deposited under tensile stress, with a wafer-to-wafer stress repeatability of ±15%. The stress increased with coating density, which depended on the susceptor temperature and ion current density. The coatings were extremely smooth, and XRD and TEM studies showed that they were amorphous with scattered submicron crystalline inclusions. One- and 2-µm thick membranes were made in 30- and 50-mm diameters at yields of 80%. A peak transmittance of 65% was measured at 633 nm for 1-µm-thick SiC membranes. The bi-axial elastic modulus, $E/(1-\nu)$, of the SiC membranes was 4-6 x 10^{12} dyn/cm^2, about three times greater than that of (100) Si membranes. Hydrogen was undetectible in the form of either C-H or Si-H bonds to within the experimental detection limit of 0.04%. After an absorbed dose of 40 MJ/cm^3 during exposure at the Center for X-ray Lithography beamline at the University of Wisconsin's Aladdin synchrotron, there was no evidence of radiation damage.

Springer Proceedings in Physics, Vol. 71 Amorphous and Crystalline Silicon Carbide IV
Editors: C.Y. Yang · M.M. Rahman · G.L. Harris

Recent modifications to the ECR-CVD equipment have been made to increase substrate temperature uniformity, temperature limit, and deposition rate, and to reduce defect density. The modifications included the installation of a re-designed process chamber and a radiant substrate heater. New results based on the modified equipment are presented.

This work was supported by DARPA through the Naval Research Laboratory under Contract No. N00014-89-C-2241 to Horizon Technology Group, Inc., Arlington, VA.

Pattern Etching of Crystalline SiC by KrF Excimer Laser

K. Shirakawa[1] *and M. Murahara*[2]

[1]Hamamatsu Photonics K.K., 1126-1 Ichino-cho, Hamamatsu-shi, Shizuoka 435, Japan

[2]Faculty of Engineering, Tokai University, 1117 Kitakaname, Hiratsuka-shi, Kanagawa 259-12, Japan

Abstract. SiC exhibits a high resistance to heat and X–rays. With its characteristics, it is expected as a semiconductor material with which IC devices for car and space electronics can be developed. Minute pattern etching of SiC is, however, very difficult chemically or physically. Here, we propose a new photochemical pattern etching method, in which the crystalline SiC surface in an atmosphere of gaseous ClF_3 was etched by patterned KrF excimer laser light. The KrF laser beams were irradiated to the SiC substrate: one parallel and the other perpendicular to the substrate surface. The parallel beam was used to dissociate ClF_3 gas in the proximity of the substrate, and the perpendicular beam which was projected circuit pattern was used to locally excite the substrate. The line and space of the single crystalline SiC (3C–type) was 1 µm; etching depth, 1000 Å.

1. Introduction

SiC has a melting point of 2830 ℃ and has a property of X–rays resistance. And its band gap is wide as 2.3 eV. For these reasons, it has been expected to be used for semiconductor devices which are operative under extreme conditions such as a high temprature and X–rays. But, SiC is chemically high stable and is physically as hard as diamond. These characteristics make it very difficult not only to be etched by conventional chemical method in manufacturing semiconductor devices, but also to be made minute pattern etching by a mechanical method. Few paper on etching of SiC, therefore, has been presented. The study on etching of SiC by Ar ion laser was reported by R.M. Osgood [1]. In his study, SiC was placed in a HF solution of 2.5 % which an electric current of 1 to 2 µA/cm² was sent into and was then etched by Ar ion laser (second higher harmonics 257 nm) beam scanning. However, a circuit pattern etching by a lithography method used in semiconductor manufacturing process has never been conducted. We have proposed for selective area photo etching of SiC by using KrF excimer laser beam and ClF_3 gas as an

Springer Proceedings in Physics, Vol. 71 Amorphous and Crystalline Silicon Carbide IV
Editors: C.Y. Yang · M.M. Rahman · G.L. Harris

etchant [2][3]. In this paper, we report that 1 μm (line and space) pattern etching of the single–crystalline SiC was successfully achieved. Findings are reported below.

2. Principle of Photochemical Reactions of Etching of SiC

In order to run effective photo–chemical reactions, following two conditions must be simultaneously satisfied [4]:

1) The material must directly absorbs the laser light.
2) The photon energy of the laser light is higher than the bonding energy of constituent molecules.

Amorphous and crystalline SiC are different in absorption spectrum, as shown in Fig. 1. Absorption of amorphous SiC is uniformly 60 % in the range from 190 to 400 nm, while that of crystalline SiC is approximately 30 % at the wavelength of 193 nm (ArF) and 50 % at 249 nm (KrF). The bonding energy of Si–C is 104 kcal/mol; the photon energies of KrF and ArF laser beams are 114 kcal and 147 kcal, respectively. The amorphous SiC bond, consequently, can be dissociated directly by means of KrF as well as ArF laser beams. However, the crystalline SiC bond can be dissociated by KrF laser beam easier than by ArF laser beam. This SiC surface does not change when irradiating with only KrF laser perpendicular to the substrate because the Si and C atoms which have been dissociated from SiC by KrF laser beam recombine immediately. Then, we considered that if photo dissociation of SiC were caused in the presence of very active halogen radicals, each atom would form a halogen compound to be stable. The most active halogen gas, ClF_3, was accordingly chosen by us as an etchant.

As exhibited in Fig. 2, ClF_3 gas has an absorption band between 190 and 400 nm. Its bonding energy is low as 60 kcal/mol. This chemical bonding, therefore, can be easily dissociated by XeF

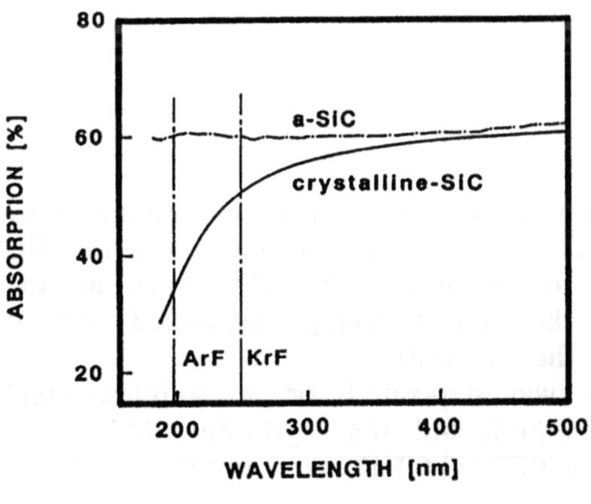

Fig.1
Absorption for SiC as a function of wavelength

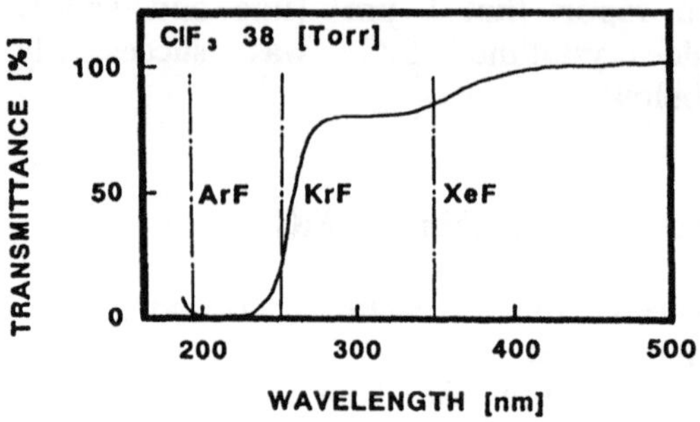

Fig.2
UV transmittance
of gaseous ClF_3

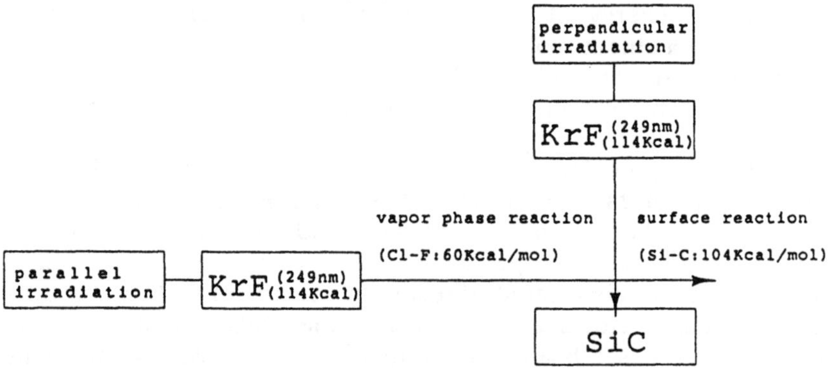

Fig.3 Fundamentals of the photoetching process for SiC by vapor phase

(wavelength 350 nm and photon energy 81 kcal), KrF or ArF excimer laser beam, and halogen radicals are produced. The fundamentals of the photochemical reactions is given in Fig. 3. The ClF_3 gas is excited by the KrF laser beam parallel to the SiC substrate. At the same time, the SiC is dissociated by the perpendicular laser beam. By the synergistic effect of surface reaction and gas phase reaction, a pattern etching of SiC is demonstrated.

3. Experimental Set–up

As in Fig. 4, the KrF laser beam is separated into two directions by a half mirror: one parallel and the other perpendicular to the SiC surface. The former is used to dissociate the ClF_3 gas in the proximity of the substrate; the latter being projected circuit pattern is used to locally excite the substrate.

The amorphous SiC, which was deposited on a stainless steel substrate by a plasma CVD method, and the crystalline SiC, which was grown on a Si substrate by a thermal CVD, were used.

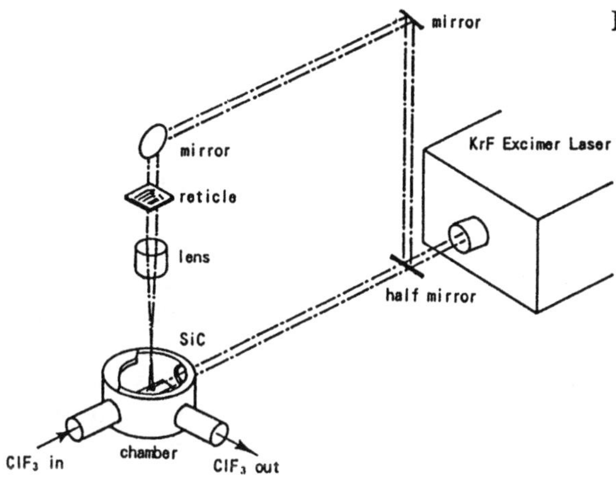

Fig.4 Schematic diagram of etching system

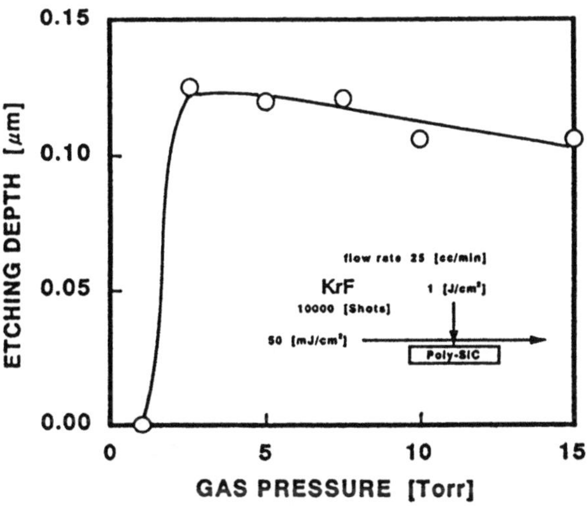

Fig.5 Dependence of gas pressure ClF with etching effect and surface

As mentioned earlier, the ClF_3 gas is essential for photo chemical etching of SiC. Fig. 5 illustrates that the etching of SiC can not observed at all when the gas pressure of ClF_3 is 0 torr, but can be observed when it is higher than 1 torr and above. Also, it shows that the etching rate slightly decreases when the gas pressure is 3 torr or more. This results from the absorption of the perpendicular beam by ClF_3 gas. Namely, the higher the gas pressure becomes, the lower the transmittance of light becomes. For this reason, the gas pressure of ClF_3 was set at 5 to 10 torr.

4. Results

4.1 Etching of amorphous SiC

The correlation of the perpendicular laser fluence and the etching depth on amorphous SiC is indicated in Fig. 6. The experimental conditions were as follows: the gas pressure of ClF_3 was 10 torr; the gas flow, 25 ccm; the fluence of the parallel laser beam, 50 mJ/cm^2 ;the number of shots, 200. This graph shows that the etching rate is drastically elevated when the fluence of the perpendicular laser beam reaches approximately 180 mJ/cm^2 . A clean etched pattern, however, could not be obtained because carbon was produced and adhered to the surface. According to our experiences, in order to achieve a fine patterned etching, it is desirable to etch at a laser fluence of the order of 160 mJ/cm^2 .

Fig. 7 exhibits the photo etched pattern which was reduced in size to 1/4 of the reticle pattern by projection. The line width of the

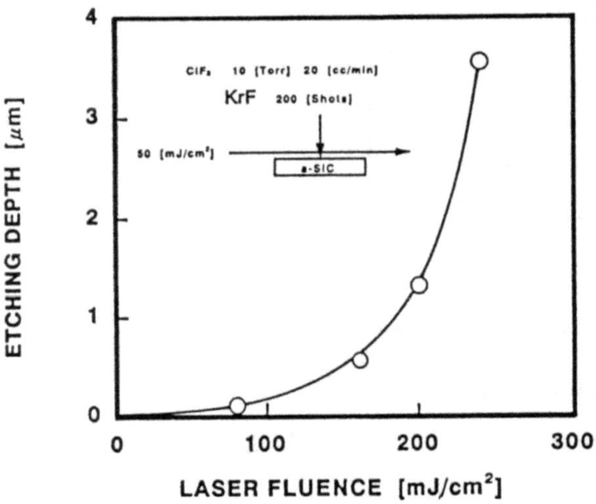

Fig.6
Correlation of perpendicular laser fluence and etching depth for amorphous SiC

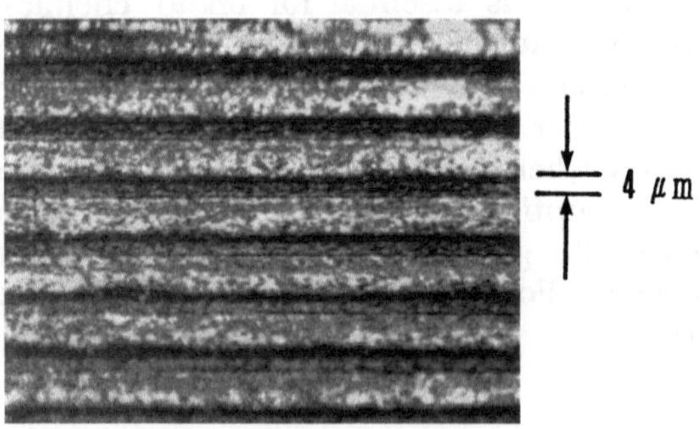

Fig.7
Etching feature of reticle pattern by reductive projection for amorphous SiC

amorphous SiC was 4 µm; the etching depth, 1000 Å. The experimental conditions were as follows: the gas pressure of ClF_3 was 10 torr; the gas flow, 10 ccm; the fluence of the parallel laser beam was 40 mJ/cm² and of the perpendicular, 160 mJ/cm² ; the number of shots, 100.

4.2 Etching of crystalline SiC

Compared with an amorphous SiC, a crystalline SiC is very difficult to etch. The most difficult one among all kinds of crystalline SiC was a single–crystalline SiC. Fig. 8 illustrates the relation between etching depth and perpendicular laser fluence on crystalline SiC. The conditions as follows: the gas pressure of ClF_3 was 5 torr; the gas flow, 10 ccm; the laser fluence of parallel laser beam, 50 mJ/cm² ; the number of shots, 10000. As shown in Fig. 8, in order to etch on a substrate the depth of 1000 Å, poly–crystalline SiC requires the laser fluence to be 1J/cm² as single–crystalline SiC does it to be 2J/cm².

The photo etched patterns of poly– and single–crystalline are shown in Fig. 9 and Fig. 10, respectively. First, the line and space of poly–crystalline SiC was 1.5 µm and the etching depth, 1000 Å as in Fig. 9. The experimental conditions as follows: the ClF_3 gas pressure was 5 torr; the gas flow, 10 ccm; the parallel laser fluence, 50 mJ/cm² ; the perpendicular laser fluence, 1 J/cm² ; the number of shots, 10000. Second, the line and space of single–crystalline SiC was 1 µm and the etching depth, 800 Å as in Fig. 10. The experimental conditions as folllows: the ClF_3 gas pressure was 5 torr; the gas flow, 5 ccm; the parallel laser beam fluence, 50 mJ/cm² ; the perpendicular laser fluence, 1.2 J/cm² ; the number of shots, 12000.

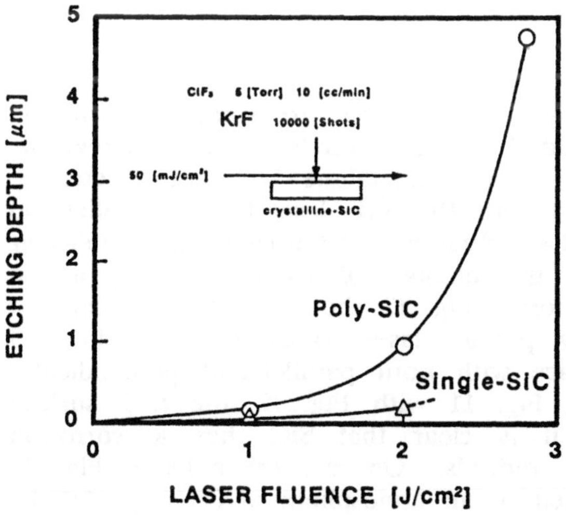

Fig.8
Correlation of perpendicular laser fluence and etching depth for crystalline SiC

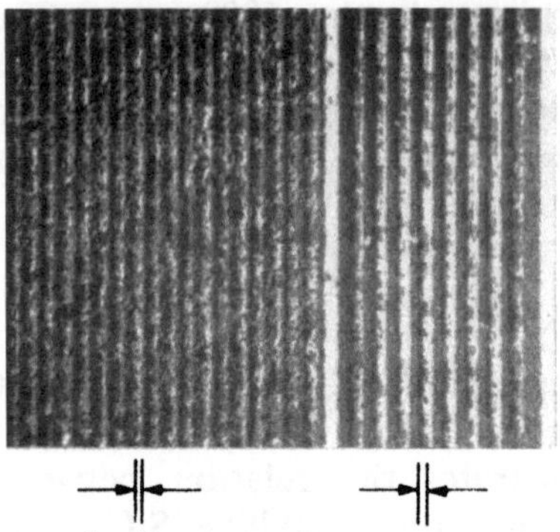

Fig.9 Etching feature of reticle pattern by reductive projection for poly–crystalline SiC

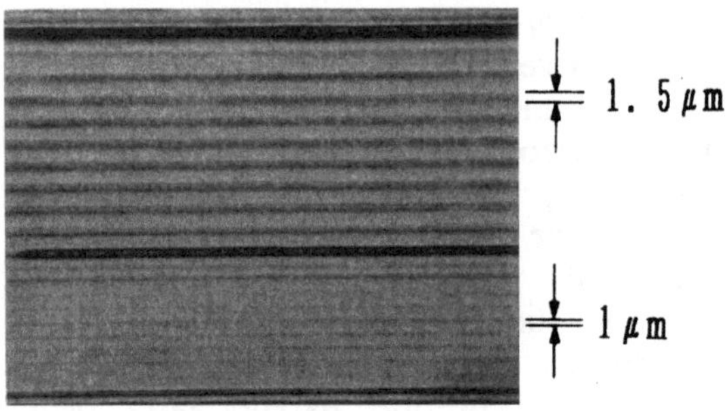

Fig.10 Etching feature of reticle pattern by reductive projection for single–crystalline SiC

4.3 Evaluation of STM

The SiC surface was observed by a STM to study how it was etched by the laser light and ClF_3 gas. Firstly, Fig. 11 shows the STM image of the surface of poly–crystalline SiC before etching.

The sample was irradiated with the KrF laser beams parallel and perpendicular to it (parallel laser fluence of 50 mJ/cm^2 ,perpendicular laser fluence of 1 J/cm^2 , the number of shots of 10000) in an atmosphere of ClF_3 gas (5 torr). Fig. 12 shows the STM image of the SiC surface with only parallel laser irradiation, and Fig. 13 shows that of the SiC surface with both parallel and perpendicular laser irradiation. Comparing Fig. 11 with Fig. 12, the SiC surface hardly changes; therefore, it is clear that SiC has a corrosion resistance ClF_3 gas and F radicals. On the other hand, Fig. 13 indicate that the SiC is etched in an atmosphere of ClF_3 gas by the

Fig.11 STM image of the poly–SiC surface before etching

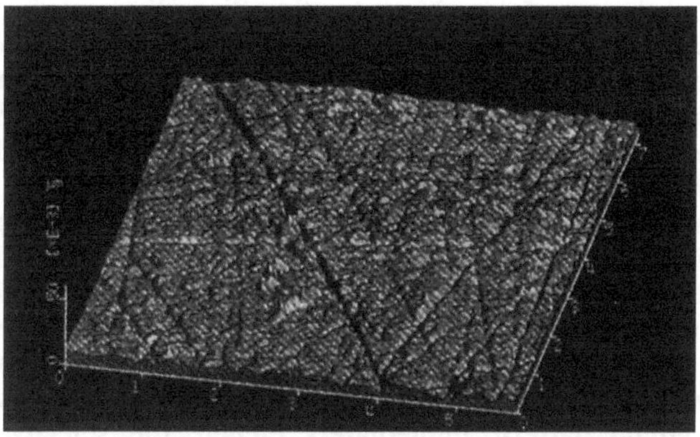

Fig.12 STM image of the poly–SiC surface without laser irradiation in an atmosphere of ClF_3 gas

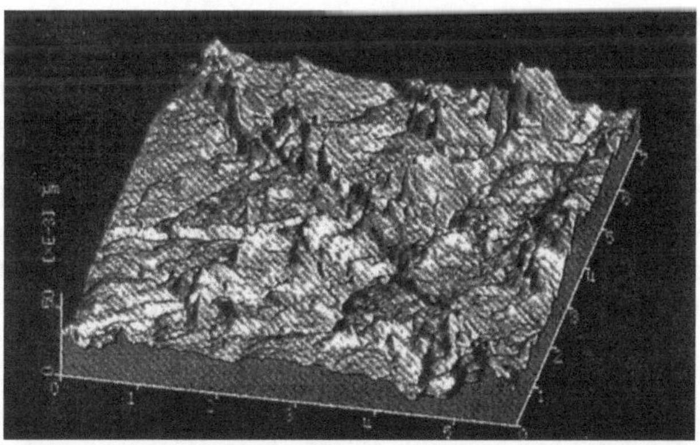

Fig.13 STM image of the poly–SiC surface with laser irradiation in an atmosphere of ClF_3 gas

parallel and perpendicular KrF laser irradiation. The surface looks uneven, for etching of crystalline SiC proceeds from a relatively frail part of crystal structure. The difference of altitude was 200 Å in average by STM evaluation.

5. Discussion

A purpose of this study is to achieve a fine pattern etching of SiC. If an extremely high fluence laser light is irradiated on SiC, it can be etched by abration. However, carbon is liberated on and adhered to the etched substrate surface. Once this carbon is adhered, the etching efficiency falls, because a laser irradiation is obstructed by it from the next time. It was a problem to clear the carbon to obtain a distinct surface. Therefore, we developed a new method to etch SiC in the presence of superfluous F radicals to avoid an isolation of carbon for clear surface. Namely, the Si and C atoms, photo–dissociated from SiC by the KrF laser, promptly react with the F radicals which are dissociated from ClF_3 gas. F has in general higher electronegativity than Cl. The bond energies of C–F and C–Cl are 128 and 95 kcal/mol, respectively; those of Si–F and Si–Cl are 128 and 107 kcal/mol, respectively. For the above reasons, it can be foreseen that the SiC photo–reacts with the F radicals to form SiF_4 and CF_4 . For the evaporation points of SiF_4 and CF_4 are low, being −107.2 ℃ and −143.6 ℃, respectively, the chemical compounds are evaporated without substrate heating. When the parallel laser beam dissociated a great deal of F radicals from ClF_3 in the proximity of the substrate, the perpendicular beam which was projected circuit pattern immediately etched the selective part of SiC irradiated. As it is clear from the STM images, the SiC surface which was not irradiated with the perpendicular laser beam hardly changed; however, the area irradiated with the laser beam was etched in an atmosphere of ClF_3 gas.

As the results, it was found that the etching of a single–crystalline SiC in all kinds of SiC required the highest laser fluence of KrF perpendicular to the substrate. We consider that this is due to the correlation of crystal structure and laser fluence, but the cause is still unclear. Also, the etching surface of poly–crystalline SiC was uneven. We think that etching proceeds from a relatively frail part of crystal structure. This is a question to be solved in order to achieve a clear, even surface at present.

6. Conclusion

ClF_3 gas is essential for photo–chemical etching of SiC. With the synergistic effect of the ClF_3 gas phase reaction and SiC surface reaction, a resistless pattern etching of SiC is demonstrated by the KrF laser beam. It is also revealed that an etching rate depends on the crystalline substance. The conditions (KrF laser) for etching depth of 1000 Å were as follows: 150 mJ/cm^2 x 100 shot for amorphous SiC, 1 J/cm^2 x 10000 shots for poly–crystalline SiC, 2 J/cm^2 x 10000 shots for single–crystalline SiC.
And, the minumum line and space of each SiC was as follows: amorphous SiC was 4 µm; poly–crystalline SiC, 1.5 µm; single–crystalline SiC, 1 µm.

References

1. J.S. Shor, X. Zhang, M.N. Ruberto, D.V. Podlesik, R.M. Osgood, CLEO'90, CWH2 (1990)
2. M. Murahara, H. Arai, T. Matumura, M.R.S. Symposium Proceeding 129,315 (1988)
3. M. Murahara, M. Yonekawa, K. Shirakawa, M.R.S. Symposium Proceeding 158, 295 (1989)
4. M. Murahara, K. Toyoda, Springer Series in Chemical Physics 39, 252 (1984)

Electrical Characterization of PiN Diode Structures in 6H-SiC

C.T. Gardner[1], *J.A. Cooper, Jr.*[1], *M.R. Melloch*[1], *J.W. Palmour*[2], *and C.H. Carter, Jr.*[2]

[1]School of Electrical Engineering, Purdue University, West Lafayette, IN 47907, USA
[2]Cree Research, Inc., 2810 Meridian Parkway, Suite 176, Durham, NC 27713, USA

1. Abstract

In this study, we report the first measurements of capacitance recovery times in 6H-SiC n-i-p-i-n diodes as a function of temperature, perimeter to area ratio, and pulse voltage, and show how this provides information on generation processes in the material. Two lots of wafers differing primarily in the oxidation conditions relating to the SiO_2 passivation are considered. Recovery times at 160 °C are about 100 seconds for lot A and 2.53 hours for lot B. Extensive measurements are reported on lot A, while the long recovery times seen in lot B have made it difficult to analyze these devices.

2. Introduction

In recent years there has been increased interest in electronic devices capable of prolonged high temperature operation. A limiting factor in many semiconductor devices is reverse leakage current due to thermal generation of carriers in the depletion regions. Reverse leakage current is expected to decrease exponentially as the band gap of the semiconductor increases. 6H-SiC gives promise for high-temperature applications because of its large bandgap, 2.9 eV. This work was undertaken to study capacitance recovery times and generation processes in 6H-SiC n-i-p-i-n devices.

3. Fabrication

A cross section of the structures studied is shown in Fig. 1. They consisted of a series of circular n-i-p-i-n diodes of diameters 1.5, 3, and 4 mils on 25.4 mm diameter single crystal 6H-SiC substrates. The n-i-p-i-n layers were grown on the (0001) Si-face 3.5° off axis toward the $\langle 11\bar{2}0 \rangle$ direction using a proprietary epitaxial

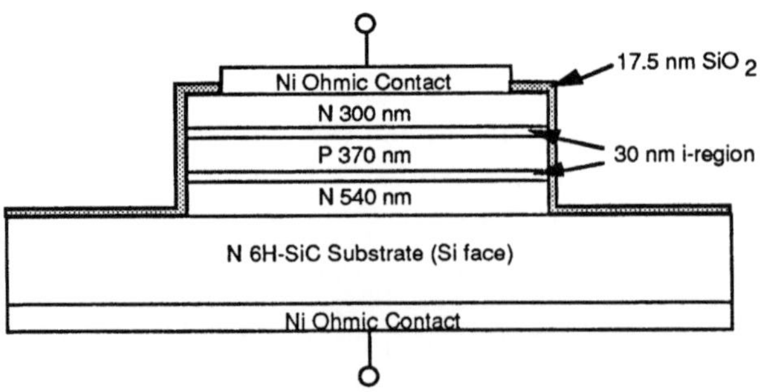

Figure 1. Cross section of devices studied.

Springer Proceedings in Physics, Vol. 71 **Amorphous and Crystalline Silicon Carbide IV**
Editors: C.Y. Yang · M.M. Rahman · G.L. Harris

growth process. The n and p-type dopants were nitrogen and aluminum, respectively, and were introduced in-situ during epitaxy. The 30 nm thick "i" layers were actually unintentionally doped n-type 6H-SiC with a carrier concentration in the range of $3\text{-}8\times10^{15}$ cm^{-3}. Doping of the n and p epilayers for the different wafers ranged from $1.3\text{-}6\times10^{18}$ cm^{-3}.

The circular mesas were defined by reactive ion etching in NF_3 [1], and the surfaces were then passivated by growth of SiO_2 via thermal oxidation. Two lots of wafers were fabricated at different times under similar conditions with the only variable being the method of oxide passivation. The wafers from lot A were oxidized in dry O_2 at 1200 °C for 30 mins, resulting in an SiO_2 thickness of 17.5 nm. The wafers from lot B were oxidized in wet O_2 at 1100 °C for 40 mins, which also yielded 17.5 nm of SiO_2. Ohmic contact was made to the top and bottom of the wafers with annealed nickel.

4. Electrical Characterization

P-i-n capacitors have been studied extensively for use in one-transistor dynamic memories in GaAs [2-4]. Similar concepts can be applied to SiC. When no charge is stored on an n-i-p-i-n capacitor, the p region is at ground and the p-n junctions are in equilibrium. When a positive bias is applied to the top n region, the top junction is reverse biased and the bottom is forward biased. Holes flow out of the p region through the forward biased junction, but can not be replenished by the reverse biased junction. This removal of holes charges the p region to a negative potential. Now, if the bias is suddenly removed, the p region will remain at a negative potential since there is no source of holes. Thus both p-n junctions become reverse biased.

The capacitor will eventually return to equilibrium as a result of electron-hole generation in the depletion regions. The generation current has both bulk and perimeter components. To first order, the leakage current can be modeled as [2,3,5]:

$$J_R = J_{R\,bulk} + J_{R\,perimeter} = qW_G G_B + qW_G \frac{P}{A} G_P \tag{1}$$

where G_B is the bulk generation rate ($cm^{-3}s^{-1}$), G_P is the effective perimeter generation rate ($cm^{-2}s^{-1}$), and P/A is the device perimeter-to-area ratio. W_G is the generation width, which is approximately equal to the reverse bias depletion width minus the equilibrium depletion width.

The generation current of a device is characterized by observing the capacitance transient in the dark following a positive or negative pulse. The capacitance recovery is exponential to first order and the recovery time constant τ_s can be related to the generation rates by [2]:

$$\tau_s = \frac{N_A N_D}{N_A + N_D}\left[\frac{1}{G_B + \frac{P}{A} G_P}\right] \tag{2}$$

The bulk and perimeter generation rates can be extracted from a plot of $1/\tau_s$ versus P/A ratio.

5. Experimental Results

5.1 Lot A

The recovery time constant τ_S is defined as the time required for the capacitance to recover to within 1/e of its zero-bias equilibrium value. The recovery times for several 1.5, 3, and 4 mil diameter mesa devices are shown in Fig. 2. The measurements were made at 160 °C using a 5.2 V pulse. In spite of the variation among devices, a general trend is apparent. As area decreases (and P/A ratio increases), the recovery time is reduced, in keeping with (2).

The capacitance recovery is thermally activated as can be seen in Fig. 3. Here we plot recovery time versus 1/T for several 3 mil devices on four different wafers. The plot is of the form:

$$\tau_S = C\exp\left[\frac{E_A}{kT}\right] \tag{3}$$

where C is a temperature independent proportionality constant and E_A is the activation energy. The activation energies from Fig. 3 range from 0.62 to 0.81 eV. Measurements of τ_S versus 1/T performed on 1.5 and 4 mil mesa devices produced activation energies within this same range. The origin of the less than half bandgap activation energy is not well understood at this time.

A plot of $1/\tau_S$ versus P/A ratio for four different temperatures is shown in Fig. 4. As indicated by (2), the slope of these curves corresponds to the perimeter generation rate (G_P) and the y-intercept corresponds to the bulk generation rate (G_B). The nearly zero y-intercept of all curves indicates that the bulk component of generation is very small compared to the perimeter generation rate. (It is seen that for two of the curves a negative y-intercept is obtained. Negative values for generation rate have no physical meaning - these less than zero intercepts are due to scatter in the data.)

Information about the thermal activation of the perimeter component can be found with (2) and the slopes of Fig. 4. The perimeter generation (G_P), shown in Fig. 5, exhibits an activation energy of 0.84 eV, which is just slightly above the range of activation energies given earlier for the recovery times of the devices.

Figure 6 shows the dependence of recovery time on the pulse magnitude and polarity. A possible explanation for the decrease in storage time with pulse magnitude is field-enhanced barrier lowering. When a large electric field is present, the shape of the Coulombic potential around a generation site is altered, effectively lowering the potential barrier that carriers must overcome to become mobile. A simple one-dimensional model for barrier lowering was derived by Frenkel [6,7]:

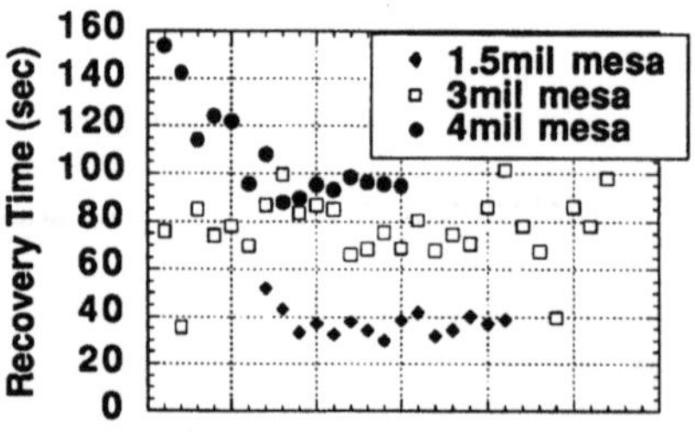

Figure 2. Lot A, recovery time at 160 °C

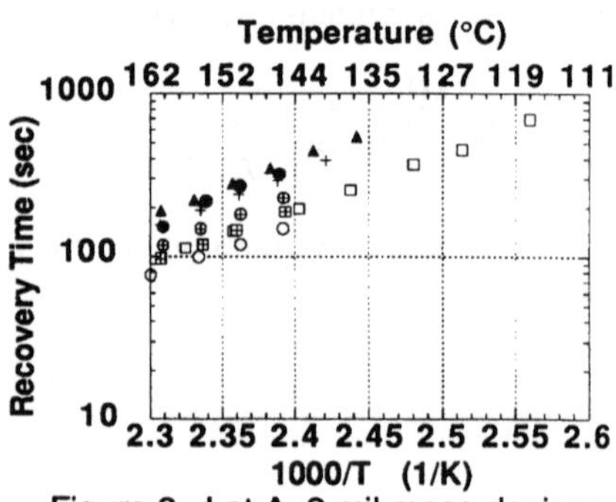

Figure 3. Lot A, 3 mil mesa devices

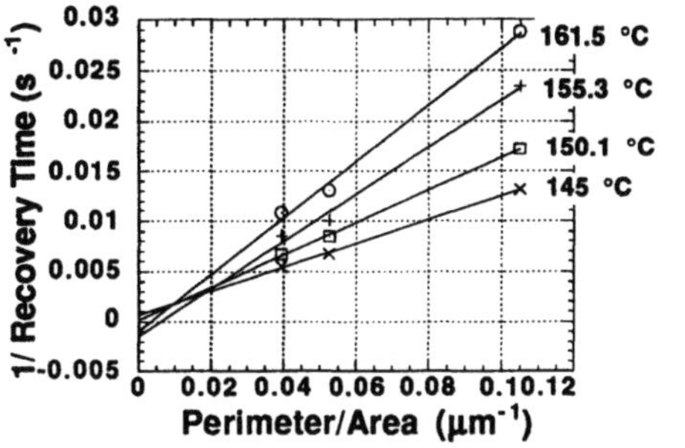

Figure 4. Lot A, $1/\tau_S$ versus Perimeter/Area

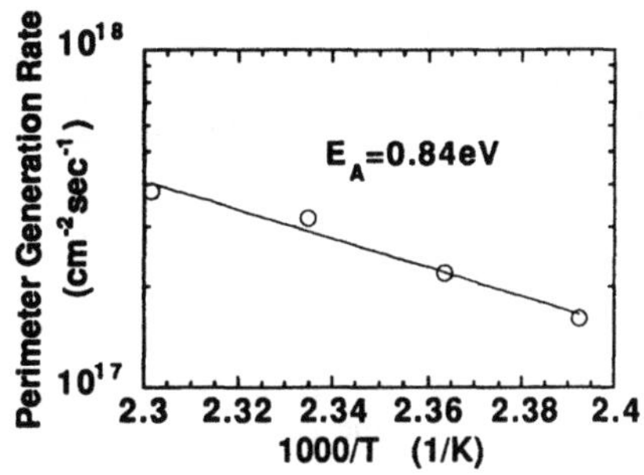

Figure 5. Lot A, Perimeter Generation

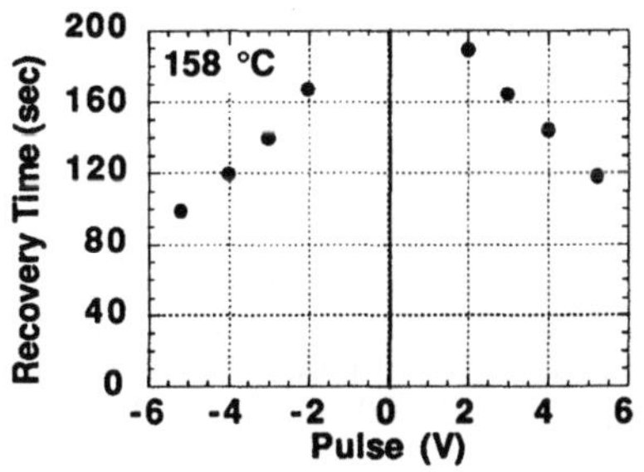

Figure 6. Lot A, τ_S versus pulse

$$\Delta E_A = -\left[\frac{q\mathcal{E}}{\pi\epsilon_s}\right]^{\frac{1}{2}} \qquad (4)$$

where ΔE_A is the barrier lowering in eV, $\mathcal{E}$ the electric field in V/cm, and ϵ_s is the semiconductor dielectric constant in F/cm. For the device shown in Fig. 6, as the pulse magnitude increases the amount of barrier lowering increases, decreasing the storage time. More evidence of field-enhanced generation in these devices is a decrease in activation energy with pulse magnitude. The activation energy measured for a 3 mil device using a 5.2 V pulse was 0.664 eV, while that obtained using a 2.5 V pulse was 0.696 eV. The difference of 0.032 eV is in reasonable agreement with the expected difference of 0.025 eV that is obtained using (4). While it seems that field-enhanced barrier lowering plays a role in these devices, this effect alone is not enough to account for the less than half bandgap activation energies. Equation (4) predicts a reduction of approximately 0.24 eV in the activation energy for a 5 V pulse, while half the bandgap of 6H-SiC is approximately 1.45 eV.

5.2 Lot B

As seen in Fig. 4, the devices in lot A are dominated by perimeter generation. This suggests that the recovery time may be very sensitive to thermal oxidation conditions during formation of the passivating oxide. To test this hypothesis, a new set of devices was fabricated and passivated using wet oxidation (lot B) instead of dry oxidation (lot A). Wet oxidation has been observed to yield better MOS electrical characteristics on (100) β-SiC than dry oxidation [8]. An exponential fit to the C-t transient of a 3 mil device, shown in Fig. 7, gives a recovery time of 2.53 hours at 160 °C. At this time we are unable to perform more extensive

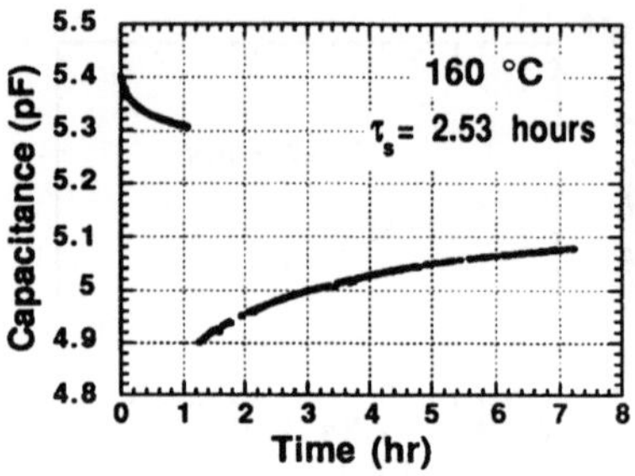

Figure 7. Lot B, C-t transient

measurements on lot B due to the extremely long recovery times and equipment limitations.

5.3 Comparison to GaAs and AlGaAs

Figure 8 shows a comparison of recovery times for SiC devices with GaAs and AlGaAs samples of similar structure [3]. Recovery times for SiC, lot A, at 144 °C are about 250 sec, increasing to about 500 sec at 125 °C. For comparison, GaAs samples exhibit recovery times of 200 msec and 700 msec at these two temperatures and AlGaAs samples have recovery times of 300 msec and 1.5 sec, respectively. In the temperature range measured, the lot A SiC devices show recovery times of nearly three orders of magnitude larger than GaAs or AlGaAs, while the measurement from lot B is five orders of magnitude larger.

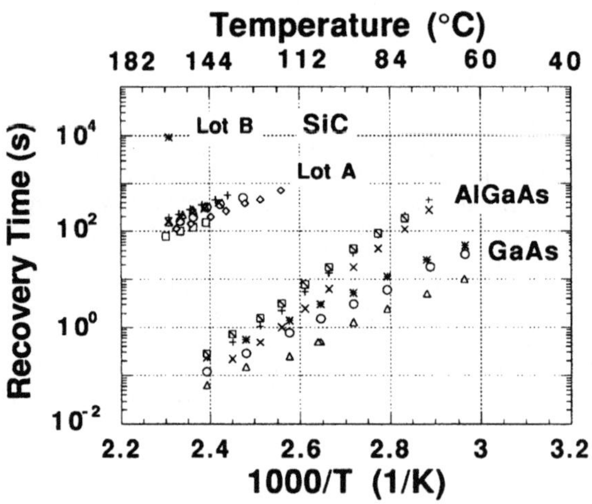

Figure 8. SiC, GaAs, and AlGaAs

6. Conclusion

In conclusion, we have reported the first measurements of capacitance recovery times in n-i-p-i-n SiC devices and compared them with similar GaAs and AlGaAs structures. For the devices in lot A, perimeter generation is the dominant source of leakage current. It appears that field-enhanced barrier lowering plays a role in these devices, but can not fully account for the less than half bandgap activation energies measured.

7. Acknowledgements

This work was supported by SDIO/IST under grants N00014-88-0527 and N00014-90-0038, administered by ONR.

8. References

1. J.W. Palmour, R.F. Davis, P. Astell-Burt, and P. Blackborow, Ceramic Transactions,
   2, 491 (1989).
2. T.E. Dungan, P.G. Neudeck, M.R. Melloch, and J.A. Cooper, Jr., IEEE Trans. Electron Devices, 37, 1599 (1990).
3. P.G. Neudeck, Ph.D thesis, Purdue University, 1991.
4. S.M. Bedair, B.T. McDermott, K.G. Reid, P.G. Neudeck, J.A. Cooper, Jr., M.R. Melloch, IEEE Electron Device Lett., 11, 261 (1990)
5. P.G. Neudeck, J.A. Cooper, Jr., and M.R. Melloch, Appl. Phys. Letts., 58, 1175 (1991).
6. J. Frenkel, Tech. Phys. USSSR, 5, 685 (1938).
7. J. Frenkel, Phys. Rev., 54, 647 (1938).
8. S.M. Tang, W.B. Berry, R. Kwor, M.V. Zeller and L.G. Matus, J. Electrochem. Soc., 137, 221 (1990).

High-Temperature Rectifiers, UV Photodiodes, and Blue LEDs in 6H-SiC

J.A. Edmond, H.-S. Kong, and C.H. Carter, Jr.

Cree Research, Inc., 2810 Meridian Parkway, Durham, NC 27713, USA

Abstract. Single junction devices in silicon carbide have been developed for use as high temperature rectifiers, UV photodiodes and blue LEDs. Rectifiers with blocking voltages from 15-1400 V and a forward current rating of 400 mA at ~3.0 V have been fabricated, packaged for 350°C operation and tested. The reverse bias breakdown appears to occur via avalanche multiplication processes exhibiting a sharp knee at breakdown. For a 710 V rectifier, the reverse bias leakage current density at 650 V is shown to increase from $\sim10^{-5}$ to $\sim10^{-4}$ A/cm^2 from 300 to 673K, respectively.

The energy bandgap of ~3.0 eV allows for inherently low dark currents and high quantum efficiencies for ultraviolet photodiode detectors made in 6H-SiC, even at high temperatures. Preliminary devices exhibit quantum efficiencies as high as ~100% with a tunable peak response from ~250-280 nm. The wavelength cut-off is at ~420 nm corresponding to the bandgap energy. The dark current density at -1.0 V and 473K is $\sim10^{-11}$ A/cm^2. This corresponds to an extrapolated room temperature current density of $\sim2\times10^{-17}$ A/cm^2 at -1.0 V.

As a light emitter, 6H-SiC junctions can be tailored to emit light across the visible spectrum. Currently, only blue LEDs are commercially available in this material. Over the past year, the quantum efficiency of the Cree blue LED has been increased significantly. The devices emit light with a peak wavelength of 470 nm with a spectral halfwidth of ~70 nm. The optical power output is typically between 12 and 17 μW at a forward current of 20 mA and 3 V. This represents a power efficiency of ~0.020-0.028%, the highest for any commercially available blue LED in the world today.

1. Introduction

Silicon carbide has been known for many years to have excellent physical and electronic properties which allows production of electronic devices that can 1) operate at significantly higher temperatures than devices produced from Si or GaAs, 2) absorb radiation below ~3.0 eV thus photogenerate power in the UV range and 3) emit radiation throughout the visible spectrum, in particular blue light emission. In comparison to competing wide bandgap materials for these applications, SiC can be readily doped either p- or n-type via ion implantation or epitaxial processes, thermally passivated and fabricated into viable solid state devices. Its chemical inertness and low dopant diffusivities show promise for long term, stable operation with low leakage currents and minimal degradation effects.

Several researchers have investigated p-n junction devices in 6H-SiC. Van Muench and Pfaffeneder [1] fabricated abrupt n^+-p rectifiers which exhibited a linear dependence between breakdown voltage and background p doping from 6×10^{17} to

Springer Proceedings in Physics, Vol. 71 Amorphous and Crystalline Silicon Carbide IV
Editors: C.Y. Yang · M.M. Rahman · G.L. Harris

$2x10^{18}$ cm^{-3}. This corresponded to a breakdown voltage of ~110 to ~30 V, respectively. In the carrier range 1-$6x10^{16}$ cm^{-3}, the breakdown voltage remained constant at ~110 V. This unexpected result was attributed to breakdown dominated by edge effects around the mesa periphery and local crystal imperfections. More recently, Anikin, et. al. [2,3] fabricated p^+-n mesa structures showing abrupt junction characteristics. The background doping of the diodes ranged from $5x10^{16}$ to $2x10^{17}$ cm^{-3}. The n-layer thickness was 1-3 μm. The top p-layers were formed by ion implantation of Al atoms or sublimation epitaxy. The resulting devices exhibited a built-in junction potential of 2.5-2.6 V and a maximum reverse bias breakdown voltage of 300-500 V.

The earliest published research on SiC UV photodiodes utilized diffusion of Al into n-type substrates [4]. This process resulted in high leakage devices with corresponding low quantum efficiency. Glasow, et. al. [5] improved upon this initial design by utilizing N ion implantation to form a very shallow (0.5 μm) n^+-p junction in a 5 μm p-type epitaxial layer grown on a p-type substrate. This research resulted in devices with a maximum quantum efficiency of 75% at a peak wavelength (λ_p) of 280 nm at room temperature. However, the reverse bias dark current density was excessive, on the order of 10^{-5} A/cm^2 at -10 V and room temperature.

Blue LEDs made in 6H-SiC has long been a major research effort at Siemens Research Laboratory, various labs in the Soviet Union and a number of Japanese labs, in particular Sanyo Electric Co., Ltd. Several of these groups have released blue LEDs as prototype products. In the laboratory, the highest external quantum efficiency (η_{ext}) reported has been achieved by Hoffmann, et. al. [6] (Siemens) using a transparent p-type substrate with contacts made to topside p^+ and compensated n-type epilayers. This nontraditional LED device design resulted in a maximum η_{ext} of ~$2x10^{-4}$. However, typical commercial devices have exhibited much lower values, η_{ext}~0.3-$1x10^{-4}$ with excessive forward voltage operation (3.5-4.5 V).

The present paper reports various characteristics of the state-of-the-art in rectifiers, UV photodiodes and blue LEDs fabricated in 6H-SiC.

2. 6H-SiC Substrates and Epitaxy

Wafers of n and p-type 6H-SiC having a diameter of 25.4 mm with typical resistivities in the range of 0.02-0.04 and 1-10 Ω-cm, respectively, were used as substrates for epitaxial growth. For the rectifier and LED device, an epitaxial junction was produced by first growing a predominantly nitrogen-doped n-type layer followed by a predominantly aluminum-doped p-type layer on an n-type substrate. The doping in the lower doped, or background layer, was varied from ~10^{16}-10^{19} cm^{-3} for the rectifiers and 10^{17}-10^{18} cm^{-3} for the LEDs. For the UV photodiode, an epitaxial junction was produced by first growing a predominantly aluminum-doped p-type layer followed by a predominantly nitrogen-doped n-type layer. The doping in the background layer was controlled between 1-$5x10^{16}$ cm^{-3} and the n-layer was heavily doped to ~10^{19} cm^{-3}. The thickness of the p-type layer was varied from 0.4-3.0 μm in order to alter the device response. All devices were fabricated using a mesa geometry. The junctions were passivated with thermally grown SiO_2. Ohmic contact materials for the p-side and n-side were sintered Al and Ni, respectively. Rectifier chips were encapsulated in a DO-204AM, axial lead, voidless and hermetically sealed glass package. The LED and photodiode devices were packaged into suitable lamps and TO-headers, respectively.

3. Device Characterization

Electrical characterization was performed using forward and reverse bias I-V measurements. For the low current measurements an electrometer was employed with 10 femtoamp resolution and a maximum current sensing capability of 1 mA. For the rectifier, I-V measurements were made at 300, 373, 473, 548 and 623K. High forward bias current (I_f) measurements up to 500 mA were performed at 300, 473 and 623K. For the UV photodiode, the reverse bias dark current versus voltage was measured at 473, 523, 573 and 623K. The responsivity and corresponding quantum efficiency as a function wavelength was also measured using a NIST calibrated deuterium UV source, ten calibrated narrow band (10 nm) filters from 205.8-380.0 nm and a Keithley Model 237 Electrometer. The blue LED emission spectra was measured with a Photoresearch Model PR703A Spectrascan System from 390-730 nm. The output power of the chip was determined with the use of an integrating sphere attachment. The external power efficiency as a function of drive current was measured.

3.1 Rectifier Characterization

Low current forward and reverse bias testing was performed for a 710 V diode up to 623K. The current scale was converted to current density in A/cm^2 denoted as J_f. Figure 1a shows the relationship between log J_f and linear V for the forward bias condition. The electrometer used to measure the current was limited to 1 A/cm^2.

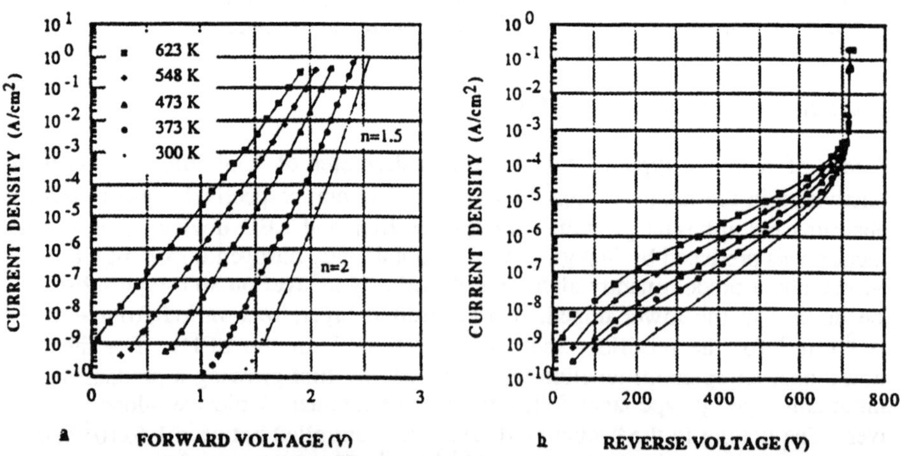

Fig. 1. Logarithmic current density-linear voltage characteristics of a 6H-SiC diode as a function of temperature in a) forward bias and b) reverse bias. The letter n in the 300K curve denotes the ideality factor of the diode in that particular region.

To determine the forward current conduction mechanisms operating in these devices, the slope of the linear regions was analyzed. For the 300K plot two linear regions are revealed. Up to approximately 1.5 V, only noise ($<10^{-10}$ A/cm^2) was observed due to the limitation in the measurement equipment. Past this voltage, the current increases rapidly. Thereafter, the plot can be separated into two regions, (a) low current region, from 1.5-2.2 V and (b) medium current region, from 2.2-2.55 V. In the low current region, the current increases linearly with an ideality factor, n, of 2.0. This indicates that in this region the current generation mechanism is due to generation-recombination processes at the junction. In the medium current region, the current increase is also linear with n~1.5 which is indicative of current from both diffusion and generation-recombination processes. This region can also be interpreted in terms of the modified Shockley-Noyce-Saah theory. This theory shows that when recombination in the depletion region takes place through multielectron centers having one deep and one shallow level, the ideality factor for the junction will be 1.5 and independent of temperature. Deep Level Transient Spectrometry is presently being performed to test for the presence of deep levels in these devices. As the temperature is increased, the current increases for a given voltage. However, the two linear regions can still be observed. For the highest temperature measurement, the diode exhibited ideality factors of n~1.9 and n~1.6 for the low and medium current regions, respectively.

Figure 1b shows the effect of temperature on the reverse bias current density (J_r) of the same diode shown in Fig. 1a. As shown, J_r increases with increasing voltage. This indicates that the current conduction mechanism may be due to generation current in the depletion region or perimeter leakage. Nevertheless, J_r is very low prior to the onset of avalanche breakdown at 710 V, even at 623K (J_r~$5x10^{-4}$ A/cm^2).

For forward currents higher than those shown in Fig. 1a, the I-V relationship becomes limited by the ohmic losses of the diode structure. This includes the series resistances from the ohmic contacts, epitaxial layers and substrate. Figure 2 shows the linear I_f-linear V relationship as a function of temperature for a typical device to 500 mA. At 300K, the diode current increases rapidly beyond about 2.7 V. The

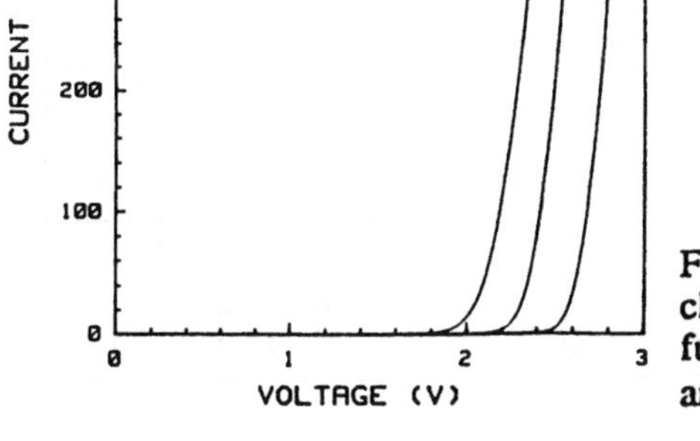

Fig. 2. Forward bias current-voltage characteristics of a 6H-SiC diode as a function of temperature. The junction area is $5.19x10^{-3}$ cm^2.

forward current of 500 mA is achieved at a forward bias of 2.9 V. As the temperature increases, the voltage at 500 mA decreases to 2.65 V and 2.49 V at 473K and 623K, respectively. From the slope of the linear regions of the curves, it is evident that the series resistance of the diode increases slightly with temperature. This is a result of the increase in resistivity of the n-type substrate above 300K. Low temperature measurements (not shown) were also performed. At 218K, the forward voltage at a forward current of 100 mA and 400 mA was 4.1 and 6.2 V, respectively.

3.2 UV Photodiode Characterization

Figure 3a shows the reverse bias dark current density versus reverse voltage as a function of temperature for a UV photodiode. The active junction is square with an area of 0.04 cm^2. For many applications, a photodiode is operated at a reverse bias of -1.0 V taking advantage of the fact that reverse current increases almost linearly with the increase in incident radiation. Since the sensitivity of the device depends on the level of background or dark current, it should be kept to a minimum for a high signal-to-noise level. As shown in this figure, the dark current is extremely low at -1.0 V, $\sim 10^{-11}$ A/cm^2, even at a temperature of 473K. At the highest voltage measured, -10 V, the dark current density increased from $\sim 10^{-9}$ to $\sim 3 \times 10^{-8}$ A/cm^2 when increasing the temperature from 473 to 623K. This relationship is plotted as a function of inverse temperature in Fig. 3b. Here, the dark current density at -10 V as a function of temperature is a straight line function with an activation energy of 0.60 eV. Data from previous researchers [5,7] is included in this figure for comparison purposes. As shown, the devices reported herein exhibit orders of magnitude lower leakage than any other published data in 6H-SiC. It is important to note that as low as the dark current is in these devices, it is still ~10-20 times higher than what has been measured in the rectifier devices (see Fig. 1b). We believe significant improvements can still be made in this parameter.

The responsivity and quantum efficiency of UV detectors as a function of wavelength is shown in Fig. 4. The p-layer thicknesses employed were 0.4, 1.2 and 3.0 μm with a junction depth of ~0.1 μm. For all the devices the peak quantum

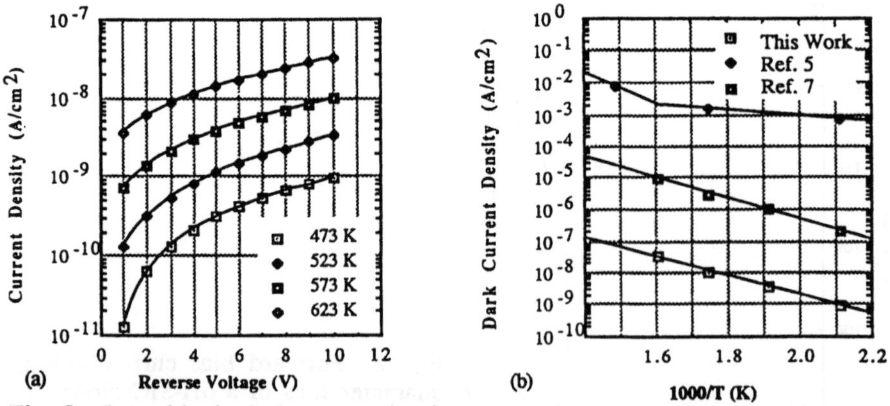

Fig. 3. Logarithmic dark current density versus a) reverse voltage and b) inverse temperature for 6H-SiC UV photodiodes.

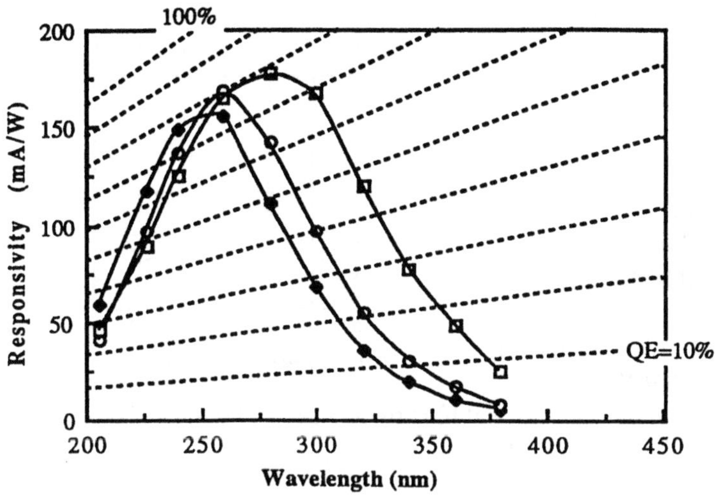

Fig. 4. Effect of p-layer thickness on the UV response of SiC photodetectors. The p-layer thicknesses are: ◆ 0.4 μm, O 1.2 μm and ❑ 3.0 μm.

efficiency was ~80%. The effect of p-layer thickness is to shift the peak response and overall long wavelength response. This is important for producing solar-blind UV detectors. For the 3 μm thick p-layer, the peak response occurs at ~280 nm. Decreasing the layer to 1.2 and 0.4 μm, shifts the peak response to ~260 nm and ~250 nm, respectively. In doing so, the long wavelength response is greatly reduced.

3.3 Blue LED characterization

The principle light generation mechanism in blue LEDs is phonon assisted donor-acceptor pair recombination between nitrogen donors and aluminum acceptors. This results in a device which emits light with a peak wavelength (λ_p) of ~470 nm with a spectral halfwidth ($\Delta\lambda_p$) of ~70 nm. The optical power output is typically between 12 and 17 μW at a forward current of 20 mA and 3 V. This represents an external power efficiency (η_{ext}) of ~0.020-0.028%, the highest commercially available blue LED in the world today. Figure 5a shows the relative spectral emission versus wavelength for a Cree blue LED operating at a forward current of 20 mA measured in an integrating sphere. As shown, the peak emission is ~470 nm with $\Delta\lambda_p$~69 nm and a dominant wavelength of 481.6 nm. The resultant spectral purity is 82.1 %. The radiant flux or optical power output of this device is 16 μW. With respect to photometric units, the die luminous intensity is 0.23 millicandela (mcd). The corresponding luminous intensity of a T1^3/4 packaged die (with an approximate viewing angle of 16°) is 15-20 mcd.

The effect of drive current to 50 mA on the radiant flux and corresponding value of h_{ext} is shown in Fig. 5b. As shown, the radiant flux increases in a sublinear

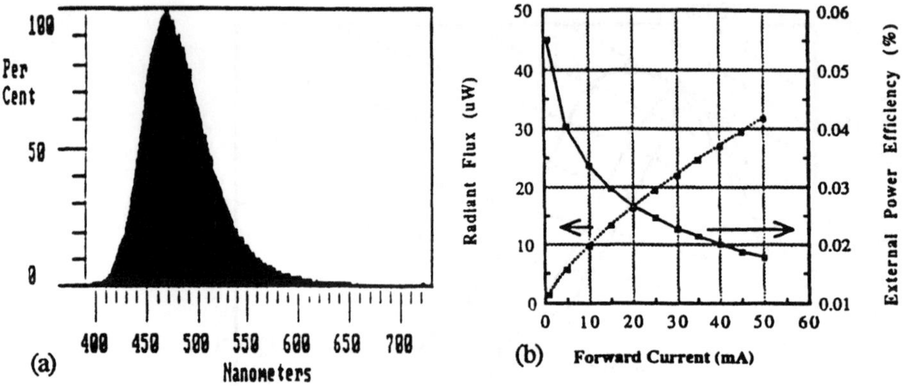

Fig. 5. Typical a) emission spectra and b) current dependent output of blue LEDs in 6H-SiC produced by Cree Research.

fashion with increasing current. This translates into a decreasing external power efficiency also shown in this figure. This sublinear behavior is most probably due to the onset of decreasing recombination of D-A pair or pair saturation and the increasing of lower efficiency recombination events at higher energy. By increasing the D-A pair density, it is believed that a more linear relationship between light output and drive current can be achieved as well as an overall increase in the efficiency of the device.

Another important aspect of SiC blue LEDs is their stability in light output over time. Using a stress current of 50 mA at room temperature, the typical 24 hour degradation is ~2.4%. After 10,000 hours the degradation is typically ~15%. Compared to GaP LEDs doped with N and Zn which show a typical degradation of 40%, SiC LEDs are much more stable.

4. Conclusion

State-of-the-art junction devices have been demonstrated in 6H-SiC. High voltage rectifiers with low leakage in reverse bias and low resistance in forward bias have been produced which operate from 218 to 623K. Ultraviolet sensitive photodiodes in this material exhibit high quantum efficiency and low dark current even at elevated temperatures. The peak response can be shifted by adjusting the low doped layer thickness. Pure blue LEDs with high brightness, 15-20 mcd and low degradation are now a commercially viable product in 6H-SiC.

References

1. W. von Muench and I. Pfaffeneder, J. Appl. Phys., **48**, 4831 (1977).
2. M.M. Anikin, A.A. Lebedev, I.V. Popov, V.P. Rastegaev, A.M. Strelchuk, A.L. Syrkin, Yu. M. Tairov, V.F. Tsvetkov and V.E. Chelnokov, Sov. Phys. Semicond., **22**, 181 (1988).
3. M.M. Anikin, A.A. Lebedev, A.L. Syrkin, and A.M. Strelchuk, to be published in *Amorphous and Crystalline SiC III*, edited by G.L. Harris and C.Y.-W. Yang (Springer-Verlag, Berlin, 1991).

4. R.B. Campbell and H.C. Chang, Sol. St. Electron., **63**, 949 (1967).
5. P. Glasow, G. Ziegler, W. Suttrop, G. Pensl and R. Helbig, SPIE **868**, 40 (1987).
6. L. Hoffman, G. Ziegler, D. Theis and C. Weyrich, J. Appl. Phys., **53**, 6962 (1982).
7. D.M. Brown, E. Downey, M. Ghezzo, J. Kretchmer, Y.S. Liu, J.A. Edmond, C.H. Carter, Jr., J. Palmour and J. M. Pimbley, Transactions of the First International High Temperature Electronics Conference, Albuquerque (1991).

Dependence of the Au-SiC(6H) Schottky Barriers Height on the SiC Surface Treatment

V.A. Dmitriev, K. Fekade, and M.G. Spencer*

Materials Science Research Center of Excellence,
School of Engineering, Howard University,
Washington, DC 20059, USA
*and A.F. Ioffe Institute, St. Petersburg, Russia

Abstract. We report the effect of surface preparation on the Schottky barrier potential for Au deposited on silicon carbide. Through repeated etchings, a profile of the barrier height as a function of distance into the material has been measured. For bulk substrates and sublimation epitaxial layers there is a significant variation in the barrier height near the surface. The final value of the barrier height was determined to be approximately 2eV, or 2/3 of the energy gap of 6H SiC. We believe that the variation in the previously reported values of barrier height (from 1.4 to 2.6 eV), are due to different SiC surface treatments.

1. Introduction

SiC is a wide bandgap semiconductor material which shows great promise for use in high-temperature and high-power applications. In order to optimize the performance of SiC devices, it is essential to understand the properties of contacts formed on SiC epitaxial films and substrates. We report the effect of surface preparation on Schottky barrier potentials. Previous authors have published values of 1.4 - 2.6 eV [1-3] for Au Schottky barriers on 6H SiC. We suggest that if attention is paid to surface preparation a constant value will be obtained for Au Schottky barrier. Further, we anticipate this to hold true for other metals.

2. Experimental procedure

Au films were vacuum deposited directly on SiC substrates (grown by the Lely technique) or epitaxial films through a metal shadow mask. The majority of the material was of the 6H polytype and the substrates were n type with the concentration of electrically active uncompensated donor concentration Nd - Na of 3×10^{18} cm^{-3}. Initially, the substrate surfaces were polished with diamond paste. The 6H SiC epitaxial films were grown by CVD at 1340°C on 6H SiC Lely substrates. 6H SiC epitaxial films were also grown on 6H SiC Lely by sublimation at ~2000°C. Ohmic contacts were deposited on the back side of the samples before Au deposition. The ohmic metalization consisted of 500 angstroms of Ni followed by 1500 angstroms of Au. The ohmic contacts were subsequently annealed at 600°C (for approximately 10 sec), in a rapid-thermal-annealing system. Barrier heights were determined by the capacitance method and a few data points were verified by photo-response techniques.

3. Experimental results

Schottky diodes formd by Au deposition on polished 6H SiC crystals had poor reverse bias characteristics with a breakdown voltage (reverse current 1 μA) about 0.5 V. Capacitance-voltage characteristics measurements near zero bias, indicate a zero bias barrier height of ~ 2.6 eV. Both ohmic and Schottky contacts were removed, prior to etching, in molten KOH at 470°C. The thickness of the etched region was measured with a surface profilometer (a mask of Al had previously been evaporated). Schottky diodes fabricated after removing a ~0.5μm layer of SiC, show slightly improved characteristics with a breakdown voltage of about 3 V and barrier height of 1.3 eV. Repeated etchings of the SiC surface yielded a

Springer Proceedings in Physics, Vol. 71 Amorphous and Crystalline Silicon Carbide IV
Editors: C.Y. Yang · M.M. Rahman · G.L. Harris

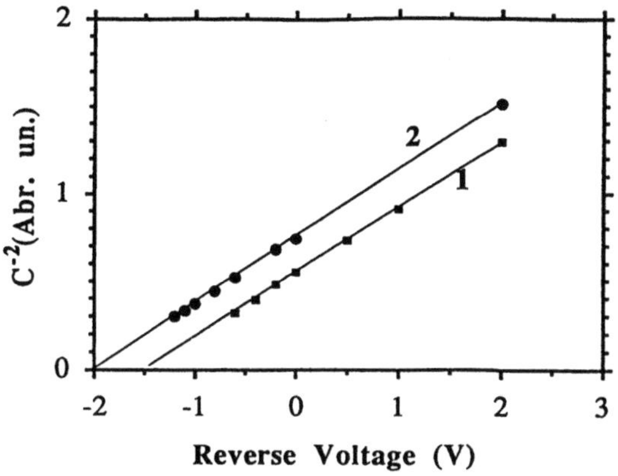

Fig. 1 Capacitance - voltage characteristics of Au Schottky barriers deposited on C-face of bulk 6H SiC crystal after removal layer of (1) 2 μm and (2) 8 μm.

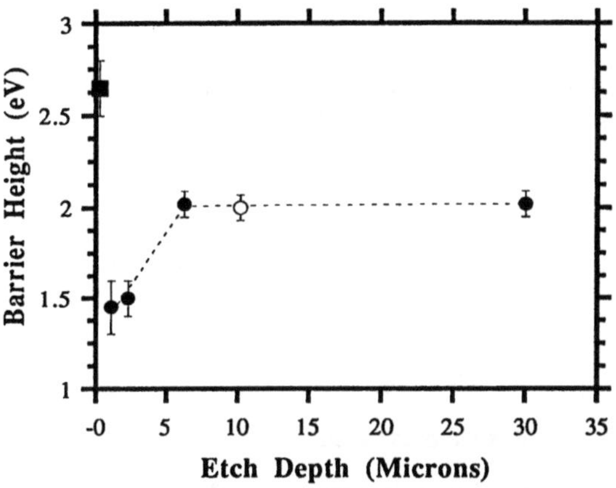

Fig. 2 Variation of Schottky barrier height measured at various depths from the original surface for a 6H substrate (grown by the Lely method). Data points represent average value of several measurements done at a particular depth. Error bars indicate the variation in the measured values at a particular etch depth.

profile of the barrier height as a function of distance into the material (Fig.1, 2, 3). Figure 2 shows the data for a substrate of Lely material, while Figure 3 displays data for an epitaxial film grown by sublimation. Closed circles and squares (unetched surface) indicate values determined the capacitance measurements. Open circles indicates datum measured by the photo-response technique. The data points represent the average value of measured barrier heights at a particular depth from the surface.

Several things are apparent from this data: 1) after sufficient etching, the final value of the barrier height is ~2eV; 2) both the substrate material and the sublimation epitaxial film showed variations of barrier height near the surface; 3) the sublimation epitaxial sample exhibits a value of 2eV with only 2 microns of etching as compared to almost 10 microns for the substrate. For Schottky barriers deposited as grown on CVD layers, 2eV barrier heights were measured. In Table I, we compare the literature data with our measurements.

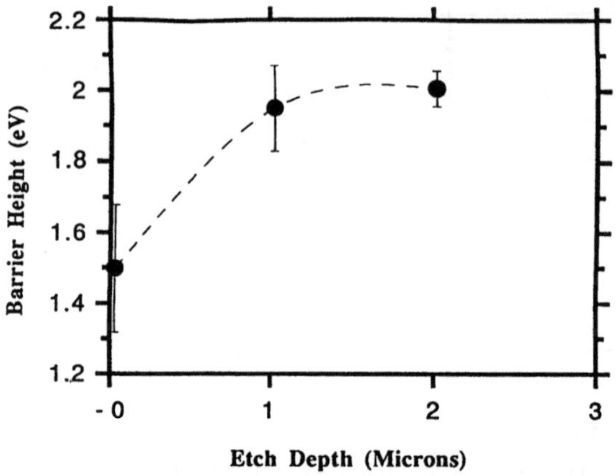

Fig. 3 Variation of Schottky barrier heights measured at various depths from the original surface of an epitaxial layer grown by the sublimation technique. Data points represent average value of several measurements done at a particular depth. Error bars indicate the variation in the measured value at a particular etch depth.

Table I Compilation of measured values of Schottky Barrier Height of Au and Cr metal deposited on 6H SiC substrates and epitaxial films.

Metal	Type	Polytype Orientation	Surface Prep.	Barrier Height (eV)	Ref.
Au	substrate	6H		1.95	1
Au	substrate	6H	Vac Cleaving & soda borax etch	1.45	2
Au	substrate	6H (0001) C	Oxidation & strip	1.4	3
Au	substrate	6H (0001) Si	KOH etching	2.0	5
Au	substrate	6H (0001) C	KOH etching	2.0	5
Au	subl. epy.	6H (0001) C	KOH etching	2.0	5
Au	CVD	6H (0001) Si	As grown	2.0	5

We believe that the variation in the reported results for Au, are due to differences in surface treatment. In the case of references 2 & 3, it is possible that the surface damage layer was not fully removed. In ref. 3, for a layer removal of 2-3 microns the barrier potential was 1.4 eV. This value agres well with our data for the same etch depth. Moreover, the current-voltage characteristics also continuously improved as a function of etch depth. A maximum
breakdown voltage of ~20 V is observed after 30 μm is removed from bulk crystals and more than 60 V is observed for sublimation epitaxial layers after 2-3 μm is etched. The

zero bias current density calculated from I-V characteristics decreased (for Au Schottky deposited on bulk crystals) from 10^{-11} A/cm^2 (after 1 μm etch) to 10^{-16} A/cm^2 (after 30 μm etch). Thus, both sublimation epitaxial layers and bulk substrates appears to have a surface layer with properties that differ from the bulk material. For substrates, this may be considerably polishing, For sublimation layers, it may be related to the high temperature vacuum anneal during the last stages of the epitaxial process.

4. Conclusion

Surface preparation is an important consideration in the formation of Au Schottky barriers on SiC. In particular, it is necessary to assure that all surfaces damaged is removed. It appears that for bulk crystals, this damaged layer can exceed 10 microns. After sufficient etching, the final value of the Au Schottky barrier height for both bulk material and epitaxial layers was determined to be approximately 2 eV, or 2/3 of the energy gap of 6H SiC.

We would like to thank A.M. Strelchuk from A.F.Ioffe Institute for photo-response measurements. We thank I. Jenkins of Howard University for CVD film of 6H SiC. We acknowledge support from the National Science Foundation MSRCE program. One of us (V.A. Dmitriev), acknowledge support from IBM Corporation and one of us (M.G. Spencer), acknowledge support from the National Science Foundation Presidential Young Investigator Program.

References

1. Mead, C.A. and Spitzer, W.G. Phys. Rev. v. 134, N 3A, pp. A713 - A716.

2. Hagen, S.H. J. of Appl. Phys. v. 39, pp. 1458 - 1461 (1968).

3. Wu, S.Y. and Campbell, R.B. Solid State Electron. v. 17, pp. 683 - 687 (1974).

4. Verenchikova, R.G., Sankin, V.I., and Radovanova, E.I. Sov. Phys. Semicond. v. 17, pp. 1123 - 1125 (1983).

5. This Work.

Photoelectrochemical Etching and Dopant Selective Etch-Stops in SiC

J.S. Shor[1,2], *X.G. Zhang*[1], *A.D. Kurtz*[2], *and R.M. Osgood, Jr.*[2]

[1]Columbia University, Microelectronics Sciences Laboratories, New York, NY 10027, USA

[2]Kulite Semiconductor Products, One Willow Tree Rd., Leonia, NJ 07605, USA

A new photoelectrochemical (PEC) etching technique has been developed which will enhance the processing capability of device structures in SiC. Using PEC etching, etch rates as high as 100 μm/min have been achieved, which are the highest reported thus far for any room temperature SiC etch. In addition, it is possible to implement electrochemical etch stops in SiC pn-junctions such that the n-type layer may be removed rapidly, while the p-type substrate underneath remains completely inert. PEC etching is also very easily controlled and results in high aspect ratio structures. This process is useful for the fabrication of electromechanical sensors, Mesa structures, and Bipolar and CMOS devices.

Introduction

SiC is a useful material for high temperature and high frequency device applications due to its unique properties [1], such as its wide band-gap (2.2-3.2 eV), high melting point (2830°C) and high breakdown electric field (4x10^{6}V/cm). However, SiC is chemically inert, making patterning devices a difficult problem. There are a number of molten metals and molten salts [1] that attack SiC, but these are impractical for device fabrication due to the high temperatures involved. Device structures have been selectively etched in SiC using Reactive Ion Etching (RIE) [2]. However, the etch rates reported for RIE are too slow for many applications and RIE shows no selectivity between different conductivity types. Currently, SiC is being investigated for sensor applications [3], which typically utilize conductivity selective etch stops to fabricate monolithic semiconductor diaphragm structures [4]. Dopant selective etch stops are also useful in the fabrication processes of CMOS, Bipolar and other devices. In this paper, a photoelectrochemical etching technique is reported which shows very high etch rates (up to 100 μm/min) and large selectivity between conductivity types. Although the electrochemistry of SiC has been studied to a limited extent and etching has been attempted [5-8], there have been no comprehensive studies which characterized electrochemical etching of SiC, outside those published by the current authors.

Experimental

Single-crystal ß-SiC n- and p-type epilayers as well as p-n junctions were grown on silicon substrates as previously described [9] at the NASA Lewis Research Center. To form ohmic contacts on the samples, aluminum was then evaporated onto the front of the SiC and patterned into contact regions. For the pn-junctions, the aluminum was deposited only on the n-type layer, which was the top layer. Semiconductor electrodes were formed by encapsulating the SiC samples in black wax, so that only the bare semiconductor surfaces were exposed. This SiC electrode was placed into a teflon electrochemical cell, along with a Pt wire counter electrode and a saturated calomel reference electrode. All voltages were measured with respect to the reference electrode (V_{sce} refers to the voltage with respect to the saturated calomel electrode). Dilute HF (2.5%in H_2O) was used as the electrolyte in all of the experiments. The light sources used for photoelectrochemical etching were a frequency doubled Ar^+ laser at 257 nm focused to a 2-3 μm spot and a diffuse Hg arc lamp with wavelengths above 400 nm blocked

Springer Proceedings in Physics, Vol. 71 Amorphous and Crystalline Silicon Carbide IV
Editors: C.Y. Yang · M.M. Rahman · G.L. Harris

out. Note that at the laser intensities used here along with the small laser spot, thermal heating in the focal region can be calculated to be <1°C [10]. For the Hg lamp, the electrolyte was pumped rapidly across the surface of the sample to eliminate heating effects.

Mechanism of the Process

The etch rates and process mechanism of SiC PEC etching have been described in an earlier publication [11]. Those results will be summarized here briefly for clarity. Fig. 1 is a diagram of the n-type ß-SiC/HF junction, which is similar in its charge transfer and energy band characteristics to a Schottky contact. At equilibrium, the Fermi level of the semiconductor lines up with the redox potential of the solution, which for n-type material results in upward surface band-bending. Under these conditions, photocarrier generation near the surface will result in hole confinement at the surface, while electrons will be driven into the bulk. The holes that reach the surface are transferred into the solution, causing anodic oxidation of the SiC by H_2O. The reaction products, SiO_2 and CO_2 are removed from the surface by HF dissolution and bubbling, respectively.

Due to the spectral absorption properties of SiC, visible wavelengths will not be absorbed efficiently at the surface; thus it is necessary to use deep UV light to promote fast etching. In addition, photoelectrochemical dissolution of n-type ß-SiC grown on Si requires a large degree of surface band-bending. Therefor, a potential of at least 0.5 V_{sce} must be applied to the semiconductor/liquid junction for etching to occur in dilute HF. This etch potential is 1 V. above the flat-band voltage of the n-SiC/HF system.

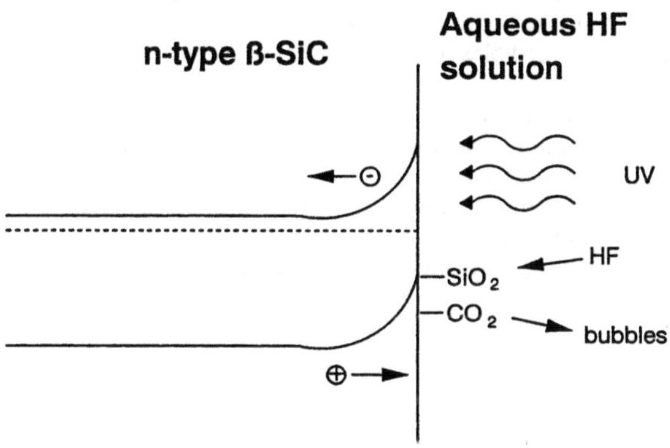

Fig. 1: n-type ß-SiC/HF(2.5%aq) junction. During etching the SiC is oxidized.

Etching

In order for PEC etching to be practical for device patterning, structures must be etched into the SiC using photolithographically defined masks. To accomplish this, Cr/Au was used as the masking material to prevent both light penetration into the semiconductor and surface electrolytic dissolution. Chromium reacts very slowly in the HF concentrations used and serves the dual purpose of being the ohmic contact to the SiC for potential control. Etching was conducted on several n-type ß-SiC samples (0.2 ohm-cm) at potentials of 0.8-0.9 V_{sce} and a Lamp intensity of ≈0.5 W/cm2. This resulted in etch rates of ≈ 530 Å/min with 1.5 μm and 5 μm features defined. A SEM micrograph of a snake resistor pattern etched into the SiC is shown in Fig. 2. The pattern definition was very good, despite a small undercut that occurred in the Cr etching. Based on the photocurrents, etch rates could be measured for both n-type ß-SiC and 6H-SiC. Using the Hg lamp set up described above, the etch rates were as high as

Fig. 2: SEM picture of a snake resistor etched into SiC using a Hg arc lamp.

6000 Å/min for both polytypes over a 1 cm^2 spot. These etch rates can be enhanced significantly by using a higher light intensity [11].

The morphology of the etched trenches exhibits a surface roughness on a 2000-3000 Å scale. This roughness is probably caused by nonuniform etching due to defect delineation. Si-C bonds are weaker at defect sites, which allow the bonds to be broken more easily. In addition, defects can also act as recombination centers for holes at the surface, preventing their transfer into the solution. Grooved structures and micro-spikes have been observed on etched surfaces along the <111> plane by high magnification SEM. It is known that a large density of stacking faults exist in the <111> direction for 3C-SiC grown on Si [9,12]. Thus, the grooves and spikes demonstrate that both etching enhancement and reduction are possible at defects and can cause etching nonuniformities. By using better quality SiC crystals, it may be possible to improve the morphology of the etching. Our preliminary results on 6H-SiC Lely crystals indicate that this is so. The morphology is also improved drastically when an etch stop barrier is used.

Etch-Stops

In reference # [11], the current authors characterized the photoelectrochemical etching of n-type ß-SiC under UV illumination in HF solutions and analyzed the electrochemistry involved in the process. C-V curves measured in that study suggested that the flat-band voltage of n-type ß-SiC in 2.5% HF(aq) is -0.55 V_{sce}. Since the flat-band voltage is fairly independent of frequency over the range 0.5-20 kHz, it is assumed that all of the potential drop takes place over the space charge layer of the semiconductor [13]. The position of the semiconductor fermi level with respect to the redox potential of the electrolyte and the position of the band edges at the surface with respect to the Fermi level will be the same for both n- and p-type materials. This positioning leads to a very different band bending for the two conductivity types. The band diagrams are depicted in Fig. 3, which shows the SiC/electrolyte interface of n- and p-type ß-SiC in dilute HF for an applied potential of 0 V_{sce}. At this potential, the bands of n-SiC bend upwards to the surface, while those of p-SiC extend downwards. From the values of the flat-band voltage of n-SiC in solution, the band-gap, and the position of the Fermi levels in n- and p-type SiC with respect to the band edges, it is possible to determine the degree of band-bending of p-SiC in Fig. 3b. Assuming E_c-E_f=100 meV for n-SiC and E_f-E_v≈50 meV for p-SiC, the band bending of p-SiC is ≈ 1.5 V at V=0 V_{sce} (e.g. Fig. 1), which suggests that the flat band voltage of p-SiC in this HF solution is ≈1.5 V_{sce}.

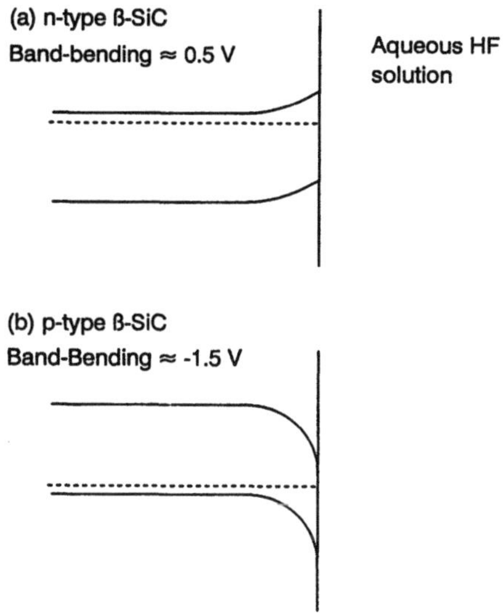

Fig. 3: SiC/HF(2.5%aq) junction for a) n-type 3C-SiC and b)p-type 3C-SiC.

The difference in the surface band-bending of n- and p-type SiC will affect the transport of holes across the semiconductor/liquid junction. These holes can be supplied from the bulk in p-type material, or they can be photo-generated by UV illumination in n- or p-type material. However, despite the source of carriers, the bands must be flat or bent upwards to allow the holes to be collected at the surface. Thus, in p-type semiconductors, etching will occur at much higher potentials than in n-type material. Therefore, it should be possible to select a potential at which the n-type material can be photoelectrochemically etched while the underlying p-type layer can act as an etch-stop. Fig. 3b suggests that below 1.5 V_{sce}, p-type material will not etch, while data in reference # [11] demonstrates that the minimum photo-etching potential for n-type ß-SiC is 0.5 V_{sce}. These potentials suggest that between 0.5-1.5 V_{sce}, n-type 3C-SiC can be removed from a p-type substrate selectively.

A 3C-SiC pn-junction with a 1 μm thick n-type layer on the surface was etched to demonstrate dopant selectivity. The evolution of the etching through the n-type material and the etch stopping at the p-type region is shown in Figs. 4 a-c. These features were fabricated at similar laser intensities, but at different laser dwell or etch times. The samples were placed into the SEM without any sputtered metal films on them so that n- and p-type materials would appear to have different gray shades. In Fig. 4a the n-type region has been partially etched at an etch rate of ≈25 μm/min, and the trench formed is similar to those found in bulk etching of ß-SiC [11]. In Fig. 4b the etch-stop layer is party visible, while in Fig. 4c the n-layer has been removed and etching has stopped completely at the p-type layer. Since almost all of the n-type layer has been removed, the morphology of the etched surface in Fig. 4c is as smooth as that of the top surface.

Fig. 3b suggests that above 1.5 V_{sce} p-type material will exhibit large dark currents. This has been independently confirmed by i-V measurements of p-type ß-SiC [14]. For lightly doped n-type material, dark currents begin only at potentials above 4 V_{sce}. The anodic dark currents in n-type material are due to tunneling of electrons from the solution into empty states in the conduction band, which cause the same dissolution reactions to occur as hole transfer from the valence band to the electrolyte does. Therefore, it should be possible to etch p-type material in the dark and stop at an n-type underlayer at potentials between 1.5-4.0 V_{sce}. However, any process that uses dark currents will not have the high anisotropy exhibited by photo-etching.

This PEC etch stop technique is equally applicable to other polytypes of SiC, since their wide band-gaps will cause a large difference in the etch potentials of p- and n-type material.

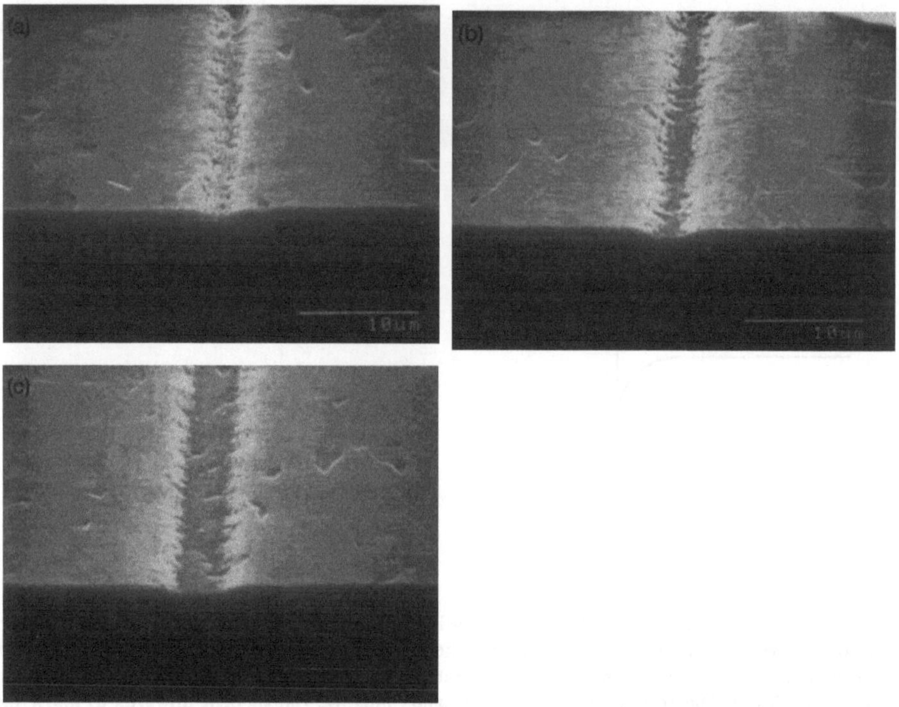

Fig. 4: Evolution of the etch stop. a) The n-type layer has been partially etched. b) The p-type layer is somewhat visible. and c) The n-type layer has been removed completely and etching has stopped at the p-type layer.

Therefor, as long as electrolytic dissolution takes place for a given polytype, etch-stopping should be possible. PEC etching has been demonstrated in 6H-SiC by us, and it is expected to work for most or all of the SiC polytypes.

Conclusion

We have demonstrated that by using electrochemistry, n-type ß-SiC can be selectively etched at high etch rates while p-type ß-SiC may be used as an etch-stop barrier and vice-versa. This process is unique, since conductivity selective etch stops are not currently possible in SiC by other means. The etch rates of this process are the highest demonstrated for SiC thus far, and the process can be controlled in situ by monitoring the photocurrents.

Acknowledgments

We would like to thank Tony Powell and Larry Matus of NASA Lewis for donating the ß-SiC samples used in this study and for their useful comments. We express thanks to X.G. Zhang of Cominco Alloys for his insights into semiconductor electrochemistry, and to D. Goldstein of Kulite Semiconductor for his support. One of us (R.M.O.) acknowledges DARPA/AFOSR support for a portion of the work described here.

References

1. " SiC 1973 " edited by R.C. Marshall, J.W. Faust and C.E. Ryan. Univ. of S. Carolina Press. p. 666.
2. J.W. Palmour, R.F. Davis, P. Astell-Burt, and P. Blackborow in " Science and Technology of Microfabrication " ed. by R.E. Howard, E.L. Hu, S. Namba and S.W. Pang (Mat. Res. Soc., Pittsburg, 1987) p. 185.
3. J.S. Shor, D. Goldstein, and A.D. Kurtz in "Transducers 91' Digest of Technical Papers" p. 912-916 (IEEE Press, 1991).
4. K.C. Lee: J. Electrochem. Soc. **137**, 2556 (1991).
5. M.M. Carrabba, J. Li, J.P. Hachey, R.D. Rauh and Y. Wang: Electrochem. Soc. Extended Abstracts **89-1** (1989) p. 727.
6. M. Glerria and R. Memming: J. Electroanal. Chem. **65**, 163 (1975).
7. H. Morisaki, H. Ono and K. Yazawa: J. Electrochem. Soc. **131**, 2081 (1984).
8. G.L. Harris, and K. Fekade: J. Electrochem. Soc. **135**, 405 (1988).
9. J.A. Powell, L.G. Matus, and M.A. Kuczmarski: J. Electrochem. Soc. **134**, 1558 (1987).
10. M. Lax: J. Appl. Phys. **48**, 3919 (1977).
11. J.S. Shor, X.G. Zhang and R.M. Osgood Jr.: to be published in J. Electrochem. Soc.
12. J.A. Powell, J.B. Petit, I.G. Jenkins, L.G. Matus, J.W. Yang, P. Pirouz, W.J. Choyke, L. Clemen, and M. Yoganathan: Appl. Phys. Lett. **59**, 333 (1991).
13. Y. Pleshkov and Y. Gurevich, "Semiconductor Photoelectrochemistry", Consultants Bureau, New York, 1986.
14. J.S. Shor, R.M. Osgood Jr. and A.D. Kurtz: to be published in Appl. Phys. Lett.

Fabrication and Electrical Properties of β-SiC/Si and Poly-SiC/Si Solar Cells

A. Solangi and M.I. Chaudhry

Department of Electrical and Computer Engineering,
Clarkson University, Postdam, NY 13699, USA

Abstract. Silicon heterojunction solar cells have been fabricated using polycrystalline and crystalline SiC films as window material. SiC films are grown on Si substrates by chemical vapor deposition technique. These films are *in situ* doped using phosphine and spin on dopant after growth. Typical results of the n-βSiC/p-Si heterojunction solar cells are V_{oc} = 492 mV; J_{sc} = 22.37 mA/cm^2; FF = 0.70; η = 7.70% at AM1 solar simulated radiation. The possible causes for this low efficiency are discussed.

1. Introduction

The first reported p-n junction silicon solar cell, in 1954, had 6% conversion efficiency. Since then a tremendous work in this field has resulted in 24% efficient silicon solar cells [1]. Achievement of the record efficiency has been possible by employing high efficiency schemes, such as, light trapping structure on front, long life times in base, reduced surface recombination velocity at front and rear, high open circuit voltage (V_{oc}) and high short circuit current (J_{sc}) [2]. Still there are two potential high efficiency schemes to work further to achieve limiting efficiency, 29.8%, for silicon solar cells [3]. These schemes are, reducing surface recombination velocity at front and reducing resistive losses in the cell to increase fill factor (FF).

If the recombination rate at the heterojunction is lower than the surface recombination rate of the Si homojunction solar cell, then Si heterojunction solar cell can be advantageous in reducing sheet resistance loses because the window effect allows the junction to be placed relatively deeper [4]. Investigating SiC-Si heterojunction solar cells, 8.04% efficiency has been demonstrated using a-SiC/a-Si system [5], 11.3% efficiency using a-SiC/c-Si system [6] and 15.4% efficiency using μc-SiC/poly-Si system [7]. Which indicates that efficiency can be further increased using β-SiC/Si system. Furthermore, it has been reported that recombination current at the β-SiC/Si interface is low [8-9].

β-SiC is almost transparent to the light of wavelength above 0.5 μm [10]. In a heteroface β-SiC/Si/Si solar cells, therefore, a relatively thicker window layer of β-SiC can be used to reduce lateral resistance of the cells without any appreciable power loss in the top window layer. The calculated data in Fig. 1 shows that thickness of top n-SiC layer in the n-βSiC/n-Si/p-Si heteroface solar cell does not considerably reduce photocurrent as compared to n-Si layer thickness in n-Si/p-Si homojunction solar cell.

Springer Proceedings in Physics, Vol. 71 Amorphous and Crystalline Silicon Carbide IV
Editors: C.Y. Yang · M.M. Rahman · G.L. Harris

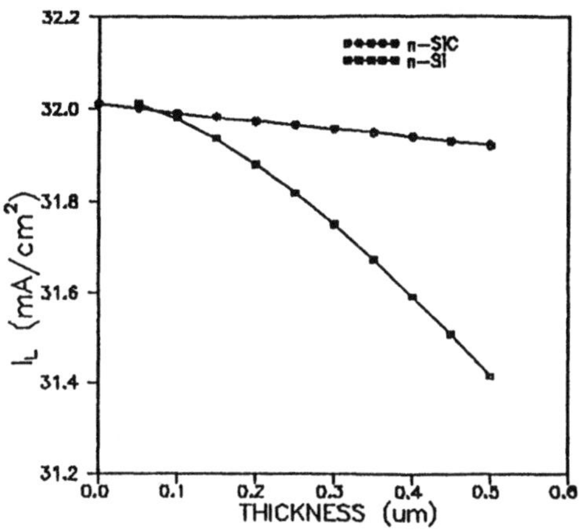

Figure 1
The variation of photocurrent with top layer thickness in Si homojunction and in an n-βSiC/n-Si/p-Si heteroface solar cells.

State-of-the-art silicon solar cells exhibit a poor quantum yield at short wavelengths, typical response dropping sharply below 0.5 μm. The extension of the spectral response to wavelengths as short as 0.3 μm will significantly improve the solar cell current. This may be achieved in the heteroface β-SiC/Si/Si solar cell, due to large bang gap (2.3 eV) [10] of β-SiC.

In this paper, we present and discuss the preliminary results of this new type solar cell for the first time .

2. Experimental

2.1 β-SiC film deposition system

β-SiC films are grown heteroepitaxially on (100) 15 Ω-cm, p-Si, substrates using a chemical vapor deposition (CVD) system described elsewhere [11]. The substrate surface is etched in hydrogen ambient at 1200°C for 5 min, to remove any native oxide. The β-SiC is grown using mixture of hydrogen (H_2), argon (Ar), silane (SiH_4) and propane (C_3H_8). Phosphine (PH_3) gas source is added to grow low resistivity n-type SiC films. The films are grown at atmospheric pressure, and the growth temperature ranges from 1000 to 1350°C. As confirmed by transmission electron microscopy [13-14], the films grown at and above 1150°C are single crystalline. The temperatures below 1150°C produce polycrystalline SiC films. The unintentionally doped SiC films are n-type with carrier concentration of approximately 10^{16} cm^{-3}.

The SiC films are also doped by diffusion of impurities. For this purpose, the Spin On Dopants (SOD), obtained from Filmtronics, Inc., are used.

2.2 Solar Cell Fabrication and Testing

The n-SiC/p-Si heterojunction solar cells are fabricated as follows. As the unintentional doping of SiC is low (10^{16} cm^{-3}), the intentional doping is carried out to reduce on series resistance of the solar cell. Two doping methods are employed:

phosphine gas is added during the growth of SiC films or SOD is used to dope the SiC films after they are grown.

The surface of grown SiC sample is prepared for applying SOD by soaking it in 4:1 H_2SO_4:H_2O_2 solution at 80°C for 15 min., rinsing with D.I. water and drying with acetone. The sample is then soaked for 5 min in 49% HF and rinsed with DI water before putting it on the spinner chuck. After a thin SOD film is applied, the sample is baked at 200°C for 20 min to harden the SOD film. The sample is then loaded in a diffusion furnace. The diffusion is carried in Argon atmosphere at 1100°C for 1 hr. After diffusion the sample is cleaned in HF to remove any oxide and rinsed with D.I. Water. In order to minimize the back contact resistance, the SOD is also used to diffuse Boron from the back Si substrate.

A 1x1 cm^2 piece is cut from the sample, front metal grid is formed with indium using shadow mask and back contact is formed by vacuum evaporation of aluminum. The front metal grid pattern covers 15 to 20% of the cell area. After metallization, the cells are annealed at ~600°C for 15 min.

2.3 I-V data collection

A home made solar simulator is used to characterize the solar cells. The simulator consists of a 300 W GE ELH projector lamp, a 7 cm long water filter (to reduce IR content of the light) and a stand for the solar cell under test. The distance between lamp and the stand is adjusted to give ~100 mW/cm^2 power (power in AM1 solar radiation) at the cell surface.

The I-V data are recorded using a Keithly 220 programmable current source, a Keithly 617 programmable electrometer and a microcomputer. The current is passed through the cell and voltage is measured across it in the presence of light. The parameters of the cell are calculated using the recorded fourth quadrant I-V data. The sheet resistance of the top SiC layer is measured by Van der Pauw technique.

3. Results and Discussion

The typical fourth quadrant I-V data of n-βSiC/p-Si heterojunction solar cell (#A) is shown in Fig. 2. The SiC film of this cell is grown at 1150 °C for 12 min, and is doped by adding phosphine gas (1% in H_2) during the growth. The flow rate of phosphine is 2 sccm. The considerable deviation of the data from squareness is due to a high series resistance. The high series resistance may be attributed to a high sheet resistance of top SiC layer (235 Ω/□), a poor back contact, and the high substrate resistivity (15 Ω-cm).

Fig. 3 shows the I-V characteristics of another n-βSiC/p-Si heterojunction solar cell (#B) with a low back contact resistance. The back contact is made by first diffusing boron from the back of the cell using SOD at 1050°C for 1 hr and then evaporating aluminum. The β-SiC film of this cell is also *in situ* doped using phosphine. In addition, it has been doped with phosphorous using SOD after growth, which further decreased the resistivity of the film. The reduction in the series resistance has almost doubled the fill factor and increased the efficiency. However,

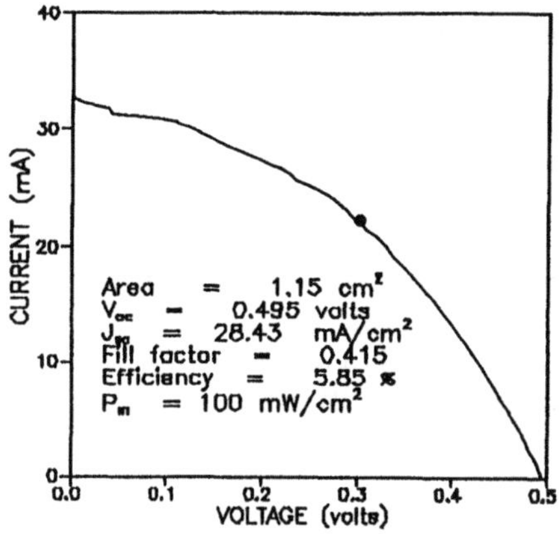

Figure 2
The I-V Characteristics of an n-βSiC/p-Si heterojunction solar cell.

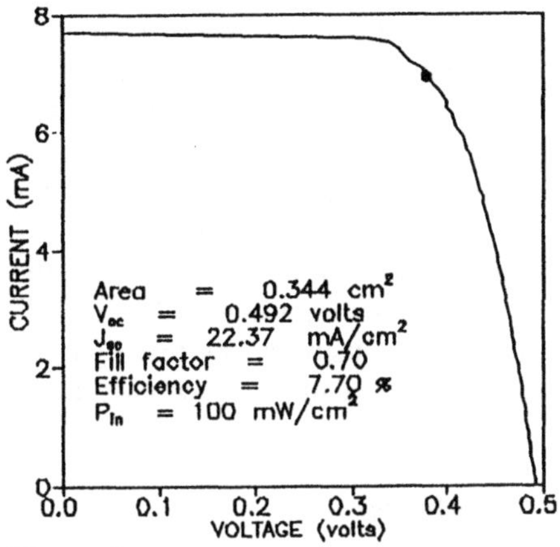

Figure 3
The I-V Characteristics of an n-βSiC/p-Si heterojunction solar cell with low back contact resistance.

the short circuit current is decreased. We speculate that the diffusion of phosphorous might have changed the optical properties of the SiC film and/or the quality of SiC/Si interface. The results of above two cells are summarized in the Table I. The X_j in the Table I is the thickness of top SiC layer.

We have found that it is very difficult to dope the top SiC layer by thermal diffusion of impurities. A poly-SiC/Si solar cell is fabricated by diffusing boron from the back Si substrate and phosphorous from the top SiC layer using SOD after SiC growth. The SiC layer of this cell is grown at 1000°C for 10 min without any *in situ* doping. The diffusion of the impurities is carried out at 1050°C for 1 hr. The results of this cell (#C) are shown in Table I. As can be seen, the fill factor and the efficiency are low due to the front contact resistance. As shown in Table I, the performance of this cell is poorer than that of β-SiC/Si cell, which could be due to a low quality

TABLE I. Results of n-βSiC/p-Si and n-poly-SiC/Si solar cells.

Cell #	X_j (μm)	Area (cm^2)	J_{sc} (mA/cm^2)	V_{oc} (mV)	FF	η
n-β-SiC/Si						
A	0.25	1.15	28.43	495	0.42	5.85
B	0.25	0.43	22.37	492	0.70	7.70
n-poly-SiC/p-Si						
C	0.2	1.80	15.56	486	0.50	3.80

poly-SiC/Si interface. Another possible reason may be a lower optical transmission of the polycrystalline SiC films.

This study shows that the doping of top β-SiC layers during the growth is essential to improve the overall performance of the SiC/Si solar cell. Moreover, a low resistivity substrate is essential for high efficiency. A low resistivity substrate will not only reduce the resistive losses in the substrate but also will yield a higher V_{oc} [16], which in turn will increase the efficiency of the cell. The optimum V_{oc} can be achieved using a highly doped, 10^{17} cm^{-3}, substrate [16]. We expect to improve efficiency by using a low resistivity Si substrate, a highly conductive SiC layer, a better design of the front contact, and a back surface field (BSF).

4. Conclusion

We have fabricated β-SiC/Si and poly-SiC/Si heterojunction cells using CVD techniques. The results of this study show that the diffusion of impurities may not be a suitable technique to fabricate SiC/Si solar cells. However, *in situ* doping is essential to reduce the series resistance. A comparison of β-SiC/Si and poly-SiC/Si solar cells shows that the single crystalline cells have higher efficiency than the poly crystalline SiC cells. This study shows that the β-SiC/Si heterojunction structure is a promising system for high efficiency Si-based solar cells.

Acknowledgement

This work has been supported in part by the New York State Science and Technology foundation and the Center for Advanced Materials Processing (CAMP) at Clarkson University.

References

[1]A. Wang, J. Zhao, and M. A. Green, Appl. Phys. Lett., **57**, 602 (1990).

[2]M. A. Green, *High Efficiency Silicon Solar Cells*, Trans. Tech. Publications, Aedermannsdorf, Switzerland 1987.

[3]Tom Tiedje, Eli Yablonvitch, George D. Cody, and Bonnie G. Brooks, IEEE Trans. Elec. Dev., **ED-31**, 711 (1984).

[4]A. G. Milnes and D. L. Feucht, *Heterojunctions and Metal-Semiconductor Junctions*, Academic Press, New York 1972.

[5]Yoshihisha Tawada, Masataka Kondo, Hiroaki Okamoto and Yoshihiro Hamakawa, Solar Energy Material, **6**, 299 (1982).

[6]M. M. Rahman and S. Furukawa, *Proc. 1st. Int. Conf. PVSEC (Kobe, Japan)*, (1984), p.172.

[7]Y. Matsumoto, G. Hirata, H. Takakura, H. Okamoto and Y. Hamakawa, J. Appl. Phys., **67**, 6538 (1990).

[8]T. Sugii, T. Ito, Y. Furumura, M. Doki, F. Mieno and M. Maeda, IEEE Trans. Elect. Dev. Lett., **9**, 87 (1988).

[9]T. Sugii, T. Yamakazi, and T. Ito, IEEE Trans. Elect. Dev., **ED-37**, 2331 (1990).
[10]A. Solangi and M. I. Chaudhry, J. Mater. Res., **7**, to appear in March 1992.
[11]M. I. Chaudhry, R. J. McCluskey and R. L. Wright, J. Crystal Growth, **113**, 120 (1991).
[12]A. Addamino and J. S. Sprague, App. Phys. Lett., **44**, 525 (1984).
[13]M. David, S. V. Babu, M. I. Chaudhry, and B.K. Flint, App. Phys. Lett., **57**, 1093 (1990).
[14]M. I. Chaudhry and R. L. Wright, J. Mater. Res., **5**, 1595 (1990).
[15]J. I. Pankove, *Optical Processes in Semiconductors*, Prentice Hall, Englewood Cliffs, New Jersey, 1971, p-103.
[16]M. A. Green, *Solar Cells: Operating Principles, Technology and Aystem Applications*, Printice Hall, Englewood Cliffs, New Jersey, 1982.

Simulations of Ge and C Implantations to Form $Si_{1-x}Ge_x$ BJT

A. Gupta, J.W. Waters, M.M. Rahman, and C.Y. Yang

Microelectronics Laboratory, Santa Clara University,
Santa Clara, CA 95053, USA

Abstract. Simulations of Ge and C implantations in Si were performed to study bandgap grading in the $Si_{1-x}Ge_x$ bipolar junction transistor. It was found that a wide-bandgap emitter and a narrow-bandgap base with no discontinuity and desirable bandgap grading were obtainable with implantation. Hence compared to epitaxially grown heterojunction bipolar transistors, these devices are likely to result in better device reliability.

1. Introduction

Materials with variable bandgaps are important candidates for high-speed bipolar devices [1]. Heterojunction bipolar transistors (HBTs) using silicon-Group IV alloys have been the focus of device research in recent years. Studies on the fabrication of SiGe/Si heterojunction bipolar transistors with epitaxially grown SiGe base have been reported [2-3]. Alternatively, SiGe can be formed by germanium ion implantation into silicon and subsequent solid phase epitaxy (SPE) [3-9]. Ion implantation is a technique convenient for selective SiGe growth, and is highly compatible with standard silicon technology. Hence it is advantageous over the various epitaxial processes. It has also been reported that carbon ion implantation into a germanium-implanted SiGe layer improves its crystallinity and the junction properties of p-n diodes fabricated in this layer [7-8]. The purpose of this study is to optimize the use of germanium and carbon for BJT fabrication.

2. Bandgap Profile Simulation

Being isoelectronic with both Si and Ge, substitutional C is not a dopant. Also, with a tetrahedral covalent radius of 0.77 Å, C is the only element in Group IV having a covalent radius smaller than that of Si (1.17 Å) [10]. This makes it ideal for compensation of lattice strain caused by implanted Ge (1.22 Å) in Si. It follows from empirical consideration that an alloy consisting of Si, Ge, and C can be formed with the Si lattice parameter. This alloy has an approximate Ge to C ratio

Springer Proceedings in Physics, Vol. 71 **Amorphous and Crystalline Silicon Carbide IV**
Editors: C.Y. Yang · M.M. Rahman · G.L. Harris

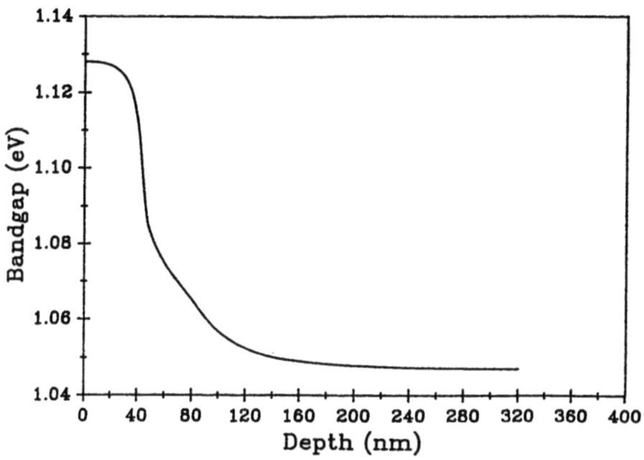

Figure 1. Desired bandgap profile with bandgap discontinuity across the emitter-base junction.

of 8:1 [10]. Furthermore, carbon in the diamond form has a much wider bandgap (5.6 eV) compared to both Si (1.12 eV) and Ge (0.66 eV). Thus it can be used to yield a wide-bandgap emitter in a HBT. To explore this possibility, simulations of Ge and C implantations into Si were performed and based on the results, bandgap profiles were obtained.

The aim of this study was to derive from simulations the implant parameters necessary to yield a bandgap versus depth profile as shown in Figure 1. Toward this end, different implant schemes, such as multiple Ge implants and implanting through a thick layer of oxide, were attempted. The simulation tool used was the commercially available program Profile Code [11]. The particle distributions in this code are calculated using up to 4 moments [12-14]. Based on empirical data, it also performs sputtering calculations for elemental solids [15]. To simulate SPE, the one-dimensional diffusion equation is solved assuming that the diffusion coefficient is temperature-dependent but independent of location or concentration i.e. $D = D_o \exp(-E_o/kT)$ [16]. D_o, E_o used for Ge and C were 6.26×10^5 cm^2/sec, 5.28 eV and 1.9 cm^2/sec, 3.1 eV, respectively.

3. Results

Figure 2 shows the SIMS profiles of the implanted Ge and C concentrations in a Si (100) wafer. Figure 3 shows the pre- and post-anneal simulated profiles using the same implant and anneal conditions. It has been reported that high dose Ge implantation on the order of 10^{16} cm^{-2} results in residual damage after the growth of the SiGe layer [7-8]. The simulations are unable to provide an estimate for the defect density due to this high dose implantation. However, the simulated results

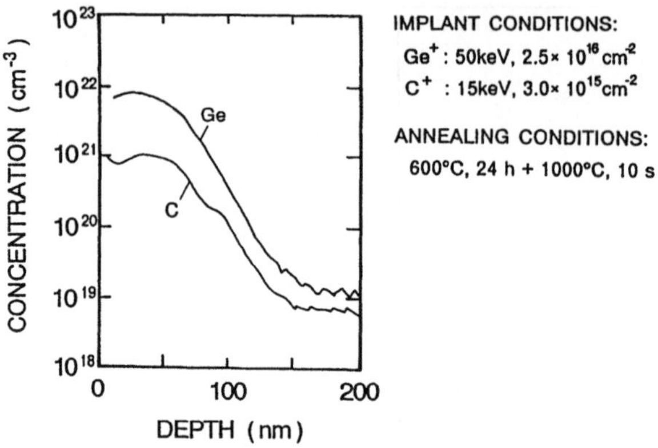

Figure 2. SIMS profiles of the implanted Ge and C concentrations in Si.

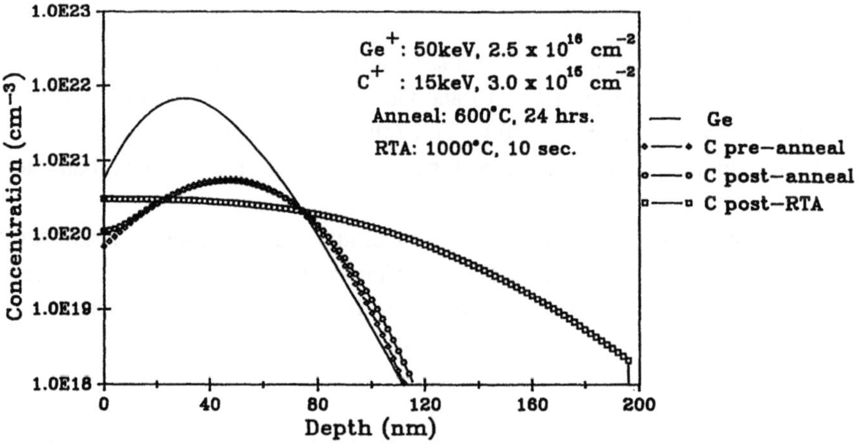

Figure 3. Simulated Ge and C profiles for comparison with the SIMS profiles.

indicate that the Ge profile is unaffected by both the ordinary anneal and the rapid thermal anneal (RTA) conditions of 600°C, 24 hrs and 1000°C, 10 s respectively. The C profile shows little change after ordinary anneal but spreads out considerably upon RTA. The projected ranges and the peak concentrations for the Ge and pre-RTA C profiles are similar to the SIMS profiles. The difference between the SIMS and the simulated post-RTA C profiles can be attributed to cubic SiC island formation in the region near the projected range during the first anneal cycle [10]. These islands, being made up of tightly bound Si-C bonds, are not affected by the subsequent RTA.

Figure 4 shows the optimized simulated Ge and C profiles [8]. These profiles were obtained from single Ge and C implants directly into Si as multiple Ge implantations resulted in large sputtering of the Si substrate. The implant energy

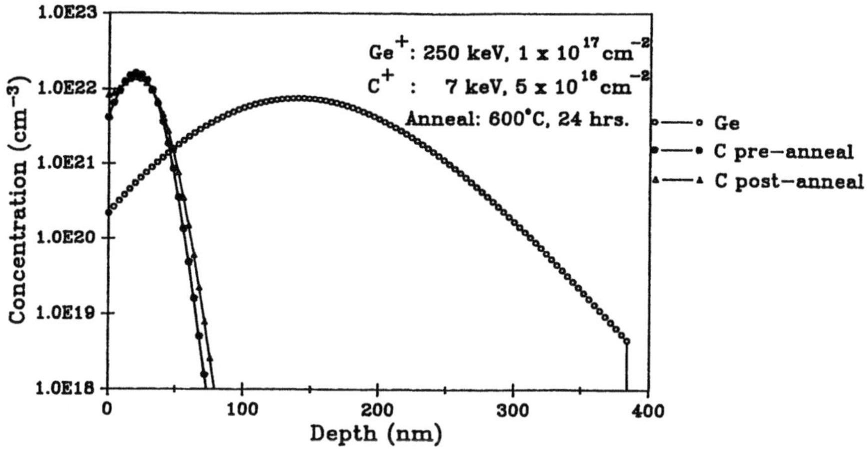

Figure 4. Optimized simulated Ge and C profiles.

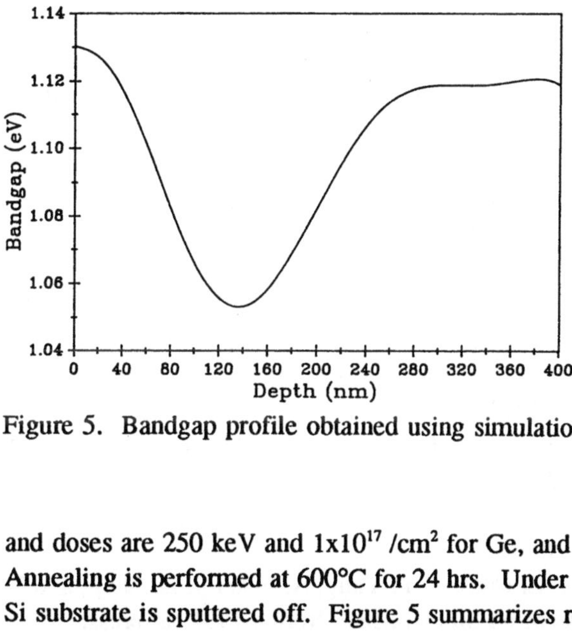

Figure 5. Bandgap profile obtained using simulation results.

and doses are 250 keV and $1x10^{17}$ /cm^2 for Ge, and 7 keV and $5x10^{16}$ /cm^2 for C. Annealing is performed at 600°C for 24 hrs. Under these conditions, 563 Å of the Si substrate is sputtered off. Figure 5 summarizes results of bandgap calculations based on the simulated profiles, assuming that all implanted ions are substitutional. The bandgap is calculated using the linear relationship

$$E_{g\,Si_{1-a-b}Ge_aC_b} = aE_{g\,Ge} + bE_{g\,\beta\text{-}SiC} + (1-a-2b)\,E_{g\,Si}$$

For an emitter width of 40 nm and a base width of 60 nm, the Eg profile reveals that a desirable bandgap grading exists in the base region. Also, no bandgap discontinuity results across the junctions.

4. Summary

Based on our simulation results, we have shown that C can be successfully used for bandgap engineering in the fabrication of SiGe BJT. The results indicate that low-energy, high-dose carbon implantation into SiGe layer can result in a wide-bandgap emitter region with no bandgap discontinuity across the junction. Such a device is likely to have better performance and reliability compared to epitaxially grown SiGe HBTs.

The authors wish to thank Dr. Takahiro Nagano, Hitachi Research Laboratory (HRL), Japan, for his valuable support and contributions throughout this work. Thanks are also due to Mr. Ken-ichi Shoji and Mr. Akira Fukami, also of HRL, for providing the SIMS data and for fruitful discussions.

References

[1] G.L. Patton, J.H. Comfort, B.S. Meyerson, E.F. Crabbe, G.J. Scilla, E.D. Fresart, J.M.C. Stork, J.Y.-C. Sun, D.L. Harame and J.N. Burghartz, IEEE Elect. Dev. Lett., vol. 11, no. 4, pp. 171-173, 1990.

[2] G.L. Patton, J.M.C. Stork, J.H. Comfort, E.F. Crabbe, B.S. Meyerson, D.L. Harame and J.Y.-C. Sun, IEDM Tech. Dig., pp. 13-16 (1990).

[3] D. Fathy, O.W. Holland, and C.W. White, Appl. Phys. Lett., vol. 51, no. 17, pp. 1337-1339 (1987).

[4] B.T. Chilton, B.J. Robinson, D.A. Thompson, T.E. Jackman, and J.-M. Baribeau, Appl. Phys. Lett., vol. 54, no. 1, pp. 42-44, 1989.

[5] D.C. Paine, D.J. Howard, N.G. Stoffel, and J.A. Horton, J. Mater. Res., vol. 5, no. 5, pp. 1023-1031, 1990.

[6] R.R. Hart, R.G. Hunsperger, H.L. Dunlap, and O.J. Marsh, Nucl. Instrum. Methods, vol. 190, pp. 70-74, 1981.

[7] A. Fukami, K. Shoji, T. Nagano and C.Y. Yang, Appl. Phys. Lett., vol. 57, no. 22, pp. 2345-2347, 1990.

[8] A. Gupta, J. W. Waters, C. Cook, C.Y. Yang, A. Fukami, K. Shoji, and T. Nagano, MRS Symposium on Silicon MBE, Spring 1991, in press.

[9] K. Shoji, A. Fukami, T. Nagano, T. Tokuyama, and C.Y. Yang, presented at ESSDERC '91, Switzerland.

[10] Y.T. Lee, N. Miyamoto and J. Nishizawa, J. Electrochem. Soc.: Solid-State Sci. and Tech., vol. 122, no. 4, pp. 530, 1975.

[11] S.N. Bunker and A.J. Armini, Nucl. Instrum. Methods Phys. Res., vol. B39, no. 1-4, pp. 7-10, 1989.

[12] J.F. Gibbons, W.S. Johnson and S.W. Mylroie, Projected Range Statistics: Semiconductors and Related Materials, Dowden, Hutchinson and Ross, Stroudsburg (1975).

[13] J.P. Biersack, The Calculation of Ion Ranges in Solids with Analytic Solutions, in "Ion Implantation Techniques", ibid.

[14] J.F. Ziegler, J.P. Biersack and U. Littmark, The Stopping and Range of Ions in Solids, Vol. I, Pergamon, Oxford (1985).

[15] N. Matsunami, Y. Yamamura, Y. Itikawa, N. Itoh, Y. Kazumata, S. Miyagawa, K. Morita, R. Shimizu and H. Tawara, At. Data and Nucl. Data Tables, Vol. 31, 1984.

[16] R. Ghez, Appl. Phys. Lett., vol. 45, pp. 881, 1984.

The Light Emitting Diodes on the Basis of Fast Electron Irradiated Silicon Carbide

Yu.A. Vodakov[1], *A.I. Girka*[2], *A.O. Konstantinov*[1], *E.N. Mokhov*[1], *A.D. Roenkov*[1], *S.V. Svirida*[2], *V.V. Semenov*[1], *V.I. Sokolov*[1], *and A.V. Shishkin*[2]

[1]A.F. Ioffe Physical-Technical Institute, Russian Academy of Sciences, Politechnicheskaya 26, 194021 St. Petersburg, Russia
[2]Moscow Engineering Physics Institute, Kashirskoe shosse 31, 115409 Moscow, Russia

Abstract. Green LEDs with λ_{max}=510÷530 nm have been obtained by means of fast electron irradiation of 6H-SiC diode structures, prepared by gaseous phase epitaxy. These LEDs are shown to have improved fasted-action and linearity of current-brightness characteristics compared with the analogous LEDs obtained by ion implantation methods.

1. Introduction.

High energy particle irradiation of SiC crystals stimulates the creation of irradiated recombination centers in them [1,2]. The most investigated is luminescence with D_1 spectrum [2]. For 6H-SiC polytype it consists of 3 phononless lines and a wide band with maximum in a green range (λ_{max}=530 nm). For 4H polytype D_1 spectrum includes 2 lines, and band maximum is in a blue range (λ_{max}=480 nm) [4].

D_1 center nature have not been established so far. It is assumed in some papers [2,3] that the divacancy is responsible for the D_1 formation. According to [5] D_1 is the center of (Si_i-V_C) type. It is confirmed by high luminescence efficiency of the SiC layers grown in vapour phase with silicon excess [6]. Some data testify [7] that D_1 contains nitrogen atom.

High luminescence efficiency with D_1 spectrum permits its use for the creation of effective LEDs. Such LEDs are usually formed by aluminium ion implantation into the nitrogen doped n-type 6H-SiC crystals [8,9]. After ion implantation high temperature annealing is carried out in order to form the p-n-junction and remove deep centres (that are mainly responsible for the radiationless recombination channels) from the

Springer Proceedings in Physics, Vol. 71 Amorphous and Crystalline Silicon Carbide IV
Editors: C.Y. Yang · M.M. Rahman · G.L. Harris

luminescence active layers. Unfortunately, because of small depth of p-n-junction occurence (≤ 0.1 µm) the p-layer evaporation is possible. This makes it difficult to fabricate LEDs of large area, used for multielement matrices.

In the present work, the D_1 type luminescence centers are injected into SiC samples by means of fast electron irradiation. Unlike [8,9], the p-n-junction was created before irradiation and its quality didn't depend on irradiation conditions and further, heat treatment. The paper gives the parameters of the obtained LEDs, and compares them with the known results.

2. Experimental results.

LED structures were formed on the 6H-SiC substrates doped by nitrogen impurities. N- and p-type conductivity epitaxial layers of 6H-SiC were grown step-by-step by "sublimation sandwich method" [10]. Epitaxial growth was carried out on (0001)C face at 1700-2300 °C. The layer thickness was 3-10 µm. N-type conductivity layer had concentration of non-compensated donors $N_D-N_A=(0.8 \div 5)\cdot 10^{17}$ cm^{-3}. P-type layer was doped during growth with Al at concentration of $\simeq 10^{20}$ cm^{-3}. Then the samples were irradiated with fast electrons. Irradiation dose varied within $(0.1 \div 5)\cdot 10^{18}$ cm^{-2} with a irradiation temperature being 300-1100K. The electron beam energy was above 0.4 MeV for these conditions.

After irradiation the samples were subjected (exposed) to heat treatment for removal of radiation induced deep centers from the luminescence-active zone (region). Such centers were revealed when studying the electron irradiated SiC samples by positron annihilation technique [11] or DLTS [12]. Aluminium contacts 0.4x0.4 mm^2 are precipitated on the p-region of diode structures. These contacts were the masks for a meza structure created by a chemical etching in an alkaline melt. The peripheral part of p-n-junction was protected by SiO_2 layer. Ohmic metal contact was deposited onto n-region.Plates were cut by individual chips 0.5x0.5 mm^2 which were fixed into metal frames by means of a conductive glue or Al heat compression. The p-regions were on the side of reflector. The light was lead out through the p-region and lateral faces of chip. LED structure was hermetically sealed by epoxy, that was poured out into semispherical polymer forms.

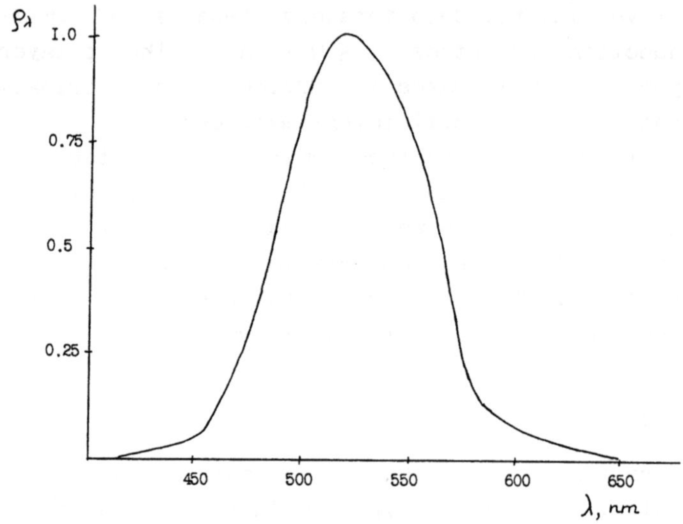

Fig. 1. Emission spectrum of 6H-SiC LED obtained by fast electron irradiation of p-n-junction.

Fig. 1 shows the typical spectrum of such prepared LED. It is seen that irradiation intensity maximum (λ_{max}=520 nm) is in a green region.FWHM of the spectrum is 70 nm. With the decrease of the diode current through p-n-junction the irradiation intensity maximum is shifted towards the long wave region.

Irradiation maximum depends also on the irradiation conditions and especially on the temperature of following annealing and it may vary between 505 - 535 nm. So it may conclude that the observed spectrum is conditioned by the overlap of two or more radiation bands. However, additional investigation is necessary to define the irradiation recombination nature more precisely.

LED irradiation intensity dependence on temperature is given in fig. 2. Irradiation intensity maximum is reached near room temperature. At high temperatures (T>500 K) activation extinguishing energy is about 280 meV. It should be noted that the diode structure preparing conditions influence strongly on luminescence extinguishing character. Thus, irradiation intensity maximums for different samples were observed at T=300 - 400 K, activation extinguishing varied within 180 - 280 meV.

Fig. 3 represents the current-brightness characteristics of LEDs, obtained by fast electron irradiation of p-n-junction (curves 1, 2). The analogous dependence is given there for

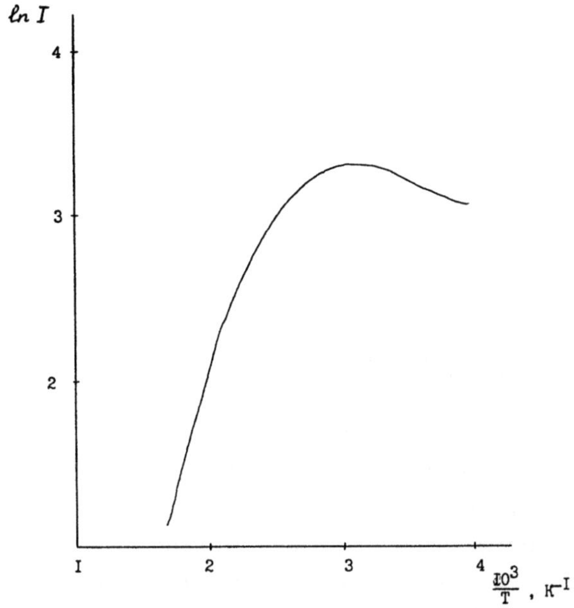

Fig. 2. Dependence of electroluminescence intensity of 6H-SiC LED, obtained by fast electron irradiation of p-n-junction, on temperature.

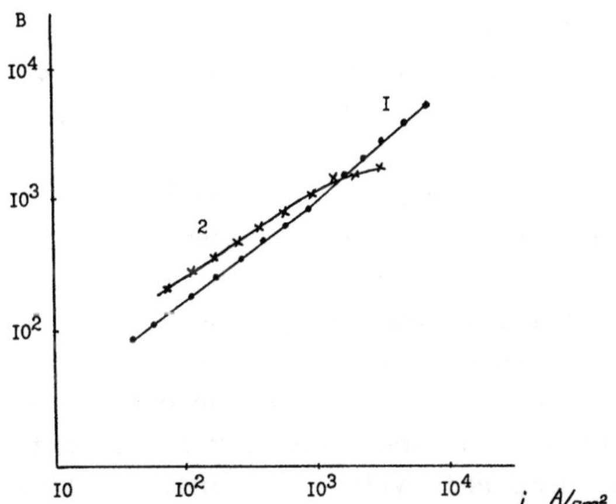

Fig. 3. Dependence of emission brightness of 6H-SiC LED, prepared by different techniques,on current density:
1 - fast electron irradiation;
2 - aluminium ion implantation.

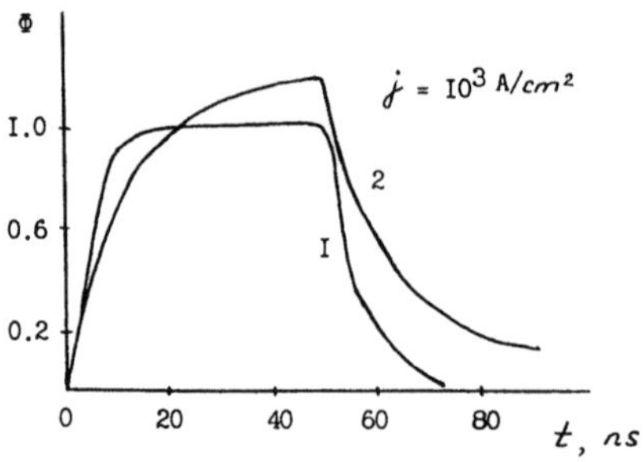

Fig. 4. Transient characteristics of 6H-SiC LED, prepared by different techniques:
1 - fast electron irradiation;
2 - aluminium ion implantation.

green LED obtained by traditional method (LED structure is formed by aluminium ion implantation into n-type 6H-SiC). It is seen that the current-brightness characteristics of LEDs obtained by fast electron irradiation are more linear. Luminescence brightness of such LEDs at low current density is lower than of ion implanted p-n structures. However, they have higher luminescence intensity at higher current (I>100 mA).

Transition characteristics of different kinds of LEDs are represented in fig. 4. LEDs obtained by fast electron irradiation have more short electroluminescence time-decay compared with the ion implanted diodes and have smaller mean charge and discharge times. Long-time discharge component of such LEDs is expressed considerably weakly, and charge time is less than 5 ns at high current densities $j>3\cdot 10^3$ A/cm^2.

Thus, one can obtain highly effective LEDs with the irradiation in green region of a spectrum (λ_{max}=520÷530 nm) by means of injection of radiative recombination centers into p-n junction obtained on 6H-SiC. These LEDs have the brightness of ≃10 mCd at 20 mA and fast-action and linearity of current-brightness characteristics compared with the ordinary LEDs obtained by means of aluminium ion implantation [8,9]. Studying of the cathodoluminescence spectrum of these fast electron irradiated 6H-SiC samples [13] shows, that after heat treatment at T≃1200 ^{o}C the D_1 spectrum is predominant, analogous to that observed in ion implanted SiC layers. At the same time,the fast electron irradiation diodes do not have the wide compensation band present in the ion implanted junction. Unlike ion implantation only point defects are injected into crystal at fast electron irradiation. These defects apparently bring relatively few deep centers stable up to high temperatures into forbidden gap (these centers are responsible for radiationless recombination), or traps with high electron trapping cross-section. This is confirmed by luminescence of SiC just after fast electron irradiation even at room temperature without preliminary annealing [12]. At the same time luminescence is effective only after heat treatment at T>1000°C on ion implantation. This appears to explain better linearity of the current-brightness characteristics and higher fast-action of electron irradiated LEDs.

In conclusion it should be noted that blue LEDs with λ_{max}=480 nm can be obtained similarly by means of electron irradiation of 4H-SiC crystals.

Reference

1. Makarov V.V., Petrov N.N. Sov. Solid State Physics, (1966) 5, 1602.
2. Patrick L., Choyke W.J. Phys. Rev. (1972) 8, 3253.
3. Andreyev A.P., Violin E.E., Demakov K.D., Tairov Yu.M., Udal'tsov V.E. Sov. Physics of Semiconductors, (1981) I, 200.
4. Makarov V.A. Sov. Solid State Physics, (1967) 2, 597.
5. Vodakov Yu.A., Lomakina G.A., Mokhov E.N., Ramm M.G., Sokolov V.I. Sov. Physics of Semiconductors,(1986) I2, 2153.
6. Vodakov Yu.A., Lomakina G.A., Mokhov E.N., Oding V.G., Ramm M.G., Sokolov V.I. Sov. Physics of Semiconductors, (1986) 8, 1433.
7. Suleimanov Yu.M., Grekhov A.M., Grekhov V.M. Sov. Physics of Semiconductors, (1983) 6, 1840.
8. Gusev V.M., Demakov K.D., Efimov V.M., Ionov V.I., Kasagonova M.G., Prokof'yeva I.K., Stolyarova V.G., Chekushkin Yu.N. Sov. Physics of Semiconductors, (1981) I2, 2430.
9. Gusev V.M., Demakov K.D., Kasagonova M.G.,Reifman M.B., Stolyarova V.G. Sov. Physics of Semiconductors (1975), 7, 1239.
10. Vodakov Yu.A., Mokhov E.N., Ramm M.G., Roenkov A.D. Krist. und Techn. (1979) 6, 729.

11. Girka A.I., Kuleshin V.A., Mokrushin A.D.,Mokhov E.N., Svirida S.V., Shishkin A.V. Sov. Physics of Semicond. (1989), I2, 2159.
12. Ballandovich V.S., Violina G.N. Cryst. Latt. Def. and Amorph. Mat. (1987) I3, 189.
13. Vavilov V.S., Vodakov Yu.A., Ivanov A.I., Mokhov E.N., Roenkov A.D., Chukichev M.V., Verenchikova R.G. Sov. Physics of Semiconductors (1991), 4, 762.

Graded-Gap and Quantum-Well Injection a-SiC:H p-i-n Light-Emitting Diodes

Jyh-Wong Hong[1], *Tean-Sen Jen*[1], *Sui-Liang Ning*[1], *Nerng-Fu Shin*[1], *and Chun-Yen Chang*[2]

[1]Department of Electrical Engineering, National Central University, Chungli 320, PR of China
[2]Institute of Electronics, National Chiao-Tung University, Hsinchu 300, Taiwan, PR of China

Abstract. As alternative approaches to improve the performance of hydrogenated amorphous silicon carbide (a–SiC:H) p–i–n thin–film light–emitting diodes (TFLED's), graded p–i junction and quantum–well–injection (QWI) structures have been incorporated into the i–layer of p–i–n TFLED's. The electroluminescence (EL) intensity of TFLED's with graded p–i junction is 100 times higher than that of basic p–i–n TFLED's and about 35 times lower than that of conventional green LED's at the same injection current density. The EL intensity of QWI TFLED's is also 20 times higher than that of basic p–i–n TFLED's. For graded–gap TFLED's, a red–orange light emission was experimentally observed by naked eyes, while for QWI TFLED's a yellow–like one was observed.

1. Introduction

The development of an a–SiC:H p–i–n TFLED has noticeable progress in recent years [1–6]. It has several advantages when compared with a conventional crystal LED, such as accessible to large–area flat–panel displays, tunable color using integrated multilayer structures and low cost. To meet the requirements of practical applications, it needs performance improvements, e.g. the increase of EL intensity and the enhanced emission of shorter wavelength light (green to blue). In this study, a graded–gap structure in the p–i interface is employed to improve the EL intensity. Additionally, quantum–well–injection (QWI) structures are added at the p–i and i–n interfaces to obtain a shorter wavelength light.

2. Device Fabrication and Operation

A. Substrate Treatment

To fabricate the TFLED's, indium–tin–oxide (ITO) coated Corning 7059 glass was used as a substrate. After a standard cleaning process, it was put into the ULVAC CPD–1108D plasma–enhanced chemical vapor deposition (PECVD) system. In order to improve the contact performance between ITO electrode and p–type layer, a CF_4–O_2 plasma dry–etching treatment was used to bombard the ITO film prior to the deposition of a p–type a–SiC:H layer [7,8].

B. Graded–gap TFLED

The schematic cross–section of the proposed graded–gap a–SiC:H TFLED is shown in Fig. 1(a). The graded–gap p–i junction was achieved by terminating the B_2H_6 dopant during continuous deposition of p^+– and i– type a–SiC:H films without RF power interruption. The thickness of graded region was approximately equal to 30 Å as estimated by using the equilibrium mass–transport continuity equation [9]. The deposition conditions for various layers of graded–gap a–SiC:H TFLED's are summarized in Table 1. Finally, the 5000 Å Al top electrode with a circular area = 1.41×10^{-2} cm^2 was deposited by employing thermal evaporation through a metal mask.

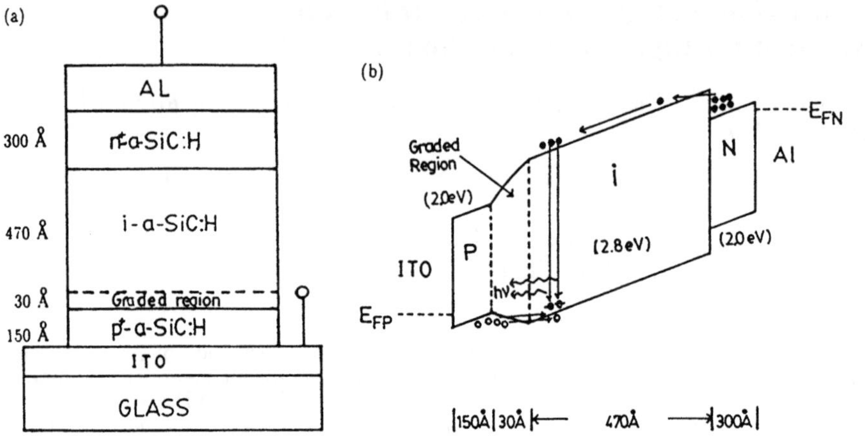

Fig. 1 a)Schematic cross–section of an a–SiC:H graded–gap p–i–n TFLED, b)energy band diagram of an a–SiC:H graded–gap p–i–n TFLED under forward bias.

Table 1. Deposition conditions for various layers of an a–SiC graded–gap p–i–n TFLED

Layer	Flow–rate ratio sccm	Pressure torr	RF power watt	Deposition rate Å/sec
p	$SiH_4^1:C_2H_2:B_2H_6^2$ = 200:6:38	0.3	20	1.6
i	$SiH_4:C_2H_2$=200:6	0.3	20	1.25
n	$SiH_4:CH_4:PH_3^3$ = 200:8:72	1.0	35	1.4

Substrate temperature: 180°C
1 : 96% H_2 + 4% SiH_4
2 : 99% H_2 + 1% B_2H_6
3 : 99% H_2 + 1% PH_3

Fig. 1(b) shows the schematic energy band diagram under forward bias of a graded–gap TFLED. The optical gap of the doped a–SiC:H layers is limited to 2.0 eV due to the valence–electron controllability [8], while the optical gap of the employed intrinsic layer is 2.8 eV [9]. Consequently, notch barriers exist at p–i and i–n interfaces if a p–i–n structure is used without graded–gap. The proposed graded–gap p–i junction is used to minimize the effect of notch barrier height on hole transport. The current transport of a p–i–n a–SiC:H TFLED is determined by carriers tunneling through triangular barriers at p–i and i–n interfaces and its emission of light is mainly based on the radiative recombination in the i–layer [5]. The nonradiative recombination centers at the previously used step p–i interface, which results from the interruption of RF power between p⁺– and i– layer depositions, degrade the EL intensity [1]. Hence, a graded–gap p–i junction could also be employed to obtain a better p–i interface property by continuous deposition process.

C. Quantum–Well–Injection TFLED

The schematic cross–section of a QWI TFLED is shown in Fig. 2(a), and the thickness of each layer is also indicated. In order to obtain a higher optical gap (3.1 eV) for the barrier–layers of QWI structure, the carbon source used was C_2H_2 rather than CH_4. The deposition conditions for various layers are summarized in Table 2.

The schematic energy band diagram under forward bias of a QWI TFLED is shown in Fig. 2(b). In this structure, the tunneling energy level could be increased by the quantum size effect, which results in a shorter–wavelength light emission due to the recombination of electron–hole pairs. Also, a large part of applied voltage drops at QWI barriers due to the larger energy–gap of barriers in this QWI structure, so the carrier tunneling–injection probability is increased [1].

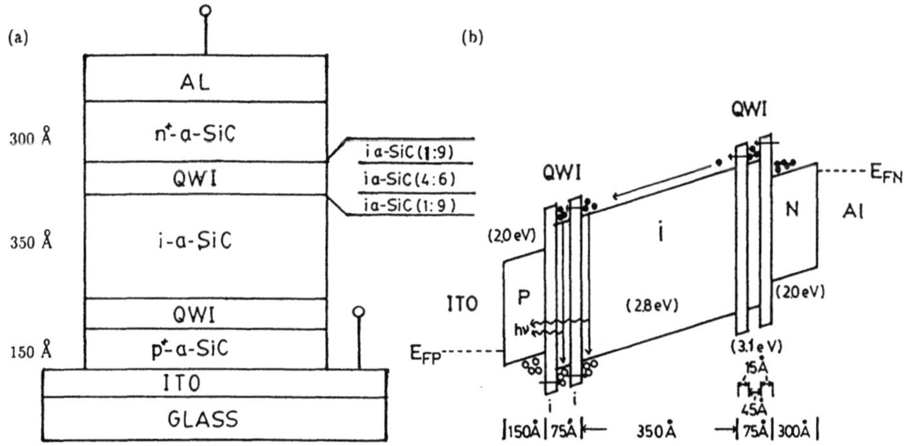

Fig. 2 a)Schematic cross–section of an a–SiC:H QWI TFLED, b) energy band diagram of an a–SiC:H QWI TFLED under forward bias.

Table 2. Deposition conditions for various layers of a QWI TFLED

Layer	Flow–rate ratio sccm	Pressure torr	RF power watt	Deposition rate Å/sec
p	SiH_4:CH_4:B_2H_6 = 200:12:38	1.0	35	1.2
i	SiH_4:C_2H_2=200:6	0.3	20	1.25
n	SiH_4:CH_4:PH_3 = 200:8:72	1.0	35	1.4
QWI(a–SiC:H):				
barrier	SiH_4:C_2H_2=1:9	0.3	20	2.2
well	SiH_4:C_2H_2=4:6	0.3	20	1.25

Substrate temperature: 180°C

3. Experimental Results and Discussion

Fig. 3 shows the EL spectrum of a graded–gap a–SiC:H p–i–n TFLED under pulse voltage bias (0–30 volt, 200 Hz, 50% duty cycle), measured by using a monochrometer (ORIEL 77200), a photomutiplier (ORIEL 7070) and a lock–in amplifier. The peak of the EL spectrum locates at 745 nm, which is in the

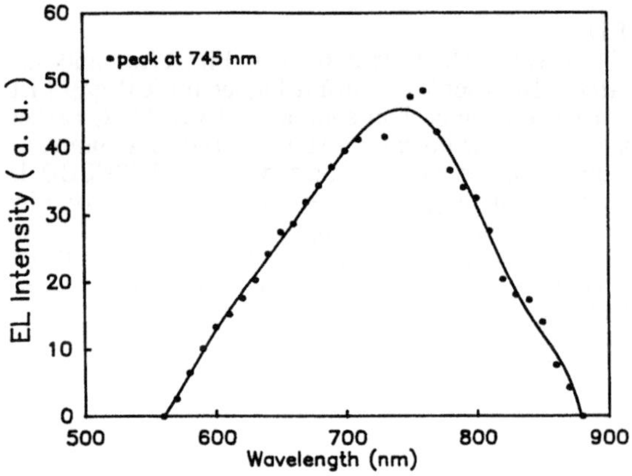

Fig. 3
Electroluminescence spectrum of an a–SiC:H graded–gap p–i–n TFLED.

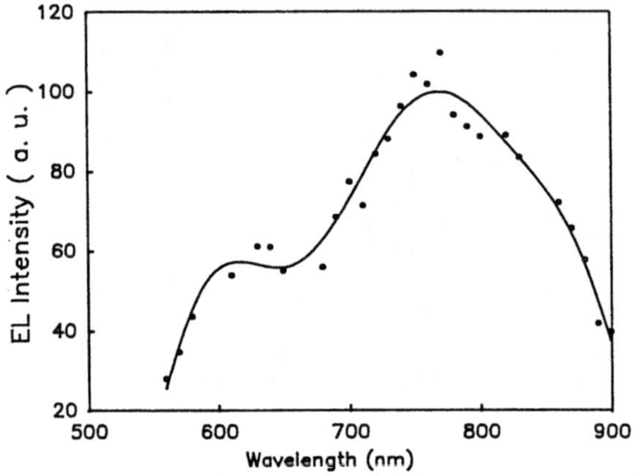

Fig. 4
Electroluminescence spectrum of an a–SiC:H QWI TFLED.

red–orange light region. It is evident that about one–half of the EL emission lays in the infrared region and is not observable by naked eyes. If the peak of the EL spectrum is moved to shorter wavelengths by adjusting the carbon content in the i–layer, the EL spectrum could be further improved. The abrupt p–i–n TFLED shows a red color (peak is around 750 nm and Full Width at Half Maximum (FWHM) is about 190 nm) while the HTI (Hot–carrier Tunneling Injection) TFLED shows an orange one (peak is around 700 nm and FWHM is about 200 nm) [7]. The HTI structure is achieved by inserting the thin (100 Å) higher–gap a–SiC:H layers at the p–i and i–n interfaces to improve the luminosity of the TFLED by enhancing the injection efficiencies of electrons and holes [6,7].

The EL spectrum of a QWI TFLED is shown in Fig. 4. We find that a second peak emerges at about 640 nm, as compare to the EL spectrum for a graded–gap p–i–n TFLED. The blue–shift of the EL spectrum could be ascribed to the quantum size effect which allows holes to be injected into the valence band of the i–layer with a higher energy. A yellow–like light emission was experimentally observed by naked eyes at 8.5–volt pulse bias.

Fig.5 shows the summary of EL intensities of an a–SiC:H p–i–n TFLED, an HTI TFLED [7], a QWI TFLED, a graded–gap p–i–n TFLED and a conventional green LED. The EL intensity of the TFLED was measured by placing the TFLED in front of a photomultiplier and was driven by an HP 4145B semiconductor parameter analyzer. As can be seen from the figure, the EL intensity of a graded–gap TFLED is

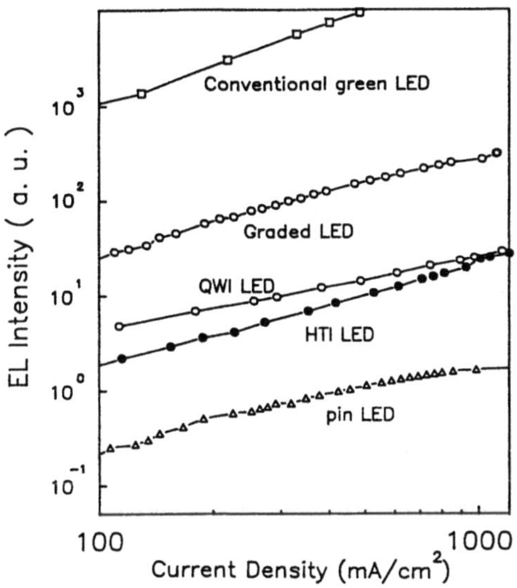

Fig. 5 Log–log plot of EL intensity versus injection current density for a p–i–n TFLED, an HTI TFLED [7], a graded–gap p–i–n TFLED, a QWI TFLED, and a conventional green LED.

100 times higher than that of an abrupt p–i–n TFLED and about 35 times lower than that of a conventional green LED (device area = 2.5×10^{-3} cm^2) under the same injection current density ranging from 100 to 1000 mA/cm^2. The QWI TFLED also has an EL intensity that is 20 times higher than that of a basic p–i–n TFLED and four times higher than that of an HTI TFLED.

It was also found that the graded–gap TFLED's showed a stable light emission even under a continuous dc bias, but the abrupt–gap TFLED's were occasionally burned–out.

4. Conclusion

The carrier injection of a graded–gap and QWI TFLED is mainly based on the injection mechanism of electrons and holes at n–i and p–i interfaces through the notch barriers. The EL intensity is dependent on the radiative recombination in the various layers. The EL intensity not only depends on carrier injection efficiency but also depends on the quality of film and interface properties. The poor interface properties of films result in nonradiative recombination of carriers, especially, at p–i junctions, which degrade the EL intensity of the TFLED. We have improved the p–i interface properties by a graded–gap structure. The obtained EL intensity of a graded–gap TFLED is about two orders of magnitude larger than that of an abrupt p–i–n TFLED and 35 times less than that of a conventional green LED at the same injection current density.

Another visible TFLED has been developed by inserting QWI structures at p–i and i–n interfaces, respectively. The emission light of the a–SiC:H QWI TFLED, as detected by human eye, is a yellow–like one . At present stage its EL intensity is still about 5 times lower than that of a graded–gap TFLED.

In this study, we found that the reliability of a TFLED could be improved because of the fact that the EL intensity could be increased. This is due to reducing the heat–generating nonradiative recombination of carriers which deteriorates the reliability of a TFLED. The application of a–SiC:H is foreseeable by employing technologies, such as ECR (Electron Cyclotron Resonance) CVD [8] to further reduce the nonradiative recombination.

Acknowledgement

This work was supported by National Science Council, R. O. C.

References

1. Kruangam, D., Deguchi, M., Toyama, T., Okamoto, H. and Hamakawa, Y.: IEEE Trans. on Electron Devices, ED–35, 957 (1988).
2. Matunami, H. and Yoshimoto, M.: J. Non–Cryst. Solids, 59, 569 (1983).
3. Munekata, H. and Kukimoto, H.: Appl. Phys. Lett., 42, 432 (1983).
4. Kruangam, D, Endo, D., Wei, G. P., Okamoto, H. and Hamakawa, Y.: Japan. J. of Appl. Phys., 24, L806 (1985).
5. Kruangam, D., Endo, T., Degauchi, M., Wei, G. P., Okamoto, H. and Hamakawa, Y.:Optoelectronics–Device and Technologies, 1, 67 (1986).
6. Paasche, S. M., Toyama, T., Okamoto, K. and Hamakawa, Y.: IEEE Trans. on Electron Devices, 36, 2895 (1989).
7. Hong, J. W., Jen, T. S., Shin, N. F., Lee, J. D. and Chang, C. Y.: to be appeared in Journal of The Chinese Institute of Engineers, Series A.
8. Harmakawa, Y., Kruangam, D., Toyama, T., Yoshimi, M., Paasche, S. and Okamoto, H.: Optoelectronics–Devices and Technologies, 4, 281 (1989).
9. Ning, S. L.: M. S. Thesis, NCU, Taiwan, R. O. C., (1991).

High-Temperature and High-Voltage Diamond Devices

H. Shiomi, Y. Nishibayashi, and N. Fujimori*

Itami Research Laboratories, Sumitomo Electric Industries, Ltd.,
1-1, Koya-kita, 1-chome, Itami, Hyogo 664, Japan
*Present address: Department of Material Science and Engineering,
Stanford University, Stanford, CA 94305-2205, USA

High-temperature and high-voltage devices were fabricated using a bi-layer structure consisting of undoped and boron-doped diamond layers. These layers were epitaxialy grown on a synthesized single-crystal diamond of (100) orientation by a microwave plasma-assisted chemical vapor deposition (CVD) using H_2, CH_4 and B_2H_6 as reactant gases.

High-voltage Schottky diodes had the structure: metal/undoped diamond/p^+ diamond. A p^+ layer of 1.8μm thickness was deposited with a reactant gas ratio of B_2H_6/CH_4=167ppm. Al Schottky electrodes and Ti ohmic electrodes were formed by a thermal evaporation method and an e-beam evaporation method, respectively. The forward current was independent of temperature because of the Fermi degeneracy in p^+-type diamond, while the performance of the conventional diamond Schottky diode with only a p layer instead of undoped/p^+ layers changed with increasing temperatures due to the ionization of the boron acceptor. The breakdown voltage of the diode with the undoped layer of 450nm thickness was 520V, while that of the conventional diode was as low as 100V.

Insulating diamond gate FET's were fabricated and operated at 300 C°. A boron-doped layer of 800nm thickness was deposited with a reactant gas of B_2H_6/CH_4=0.83ppm. The boron concentration was estimated to be $10^{18}cm^{-3}$ by SIMS. Tungsten was deposited for source and gate electrodes by an e-beam evaporation method. An undoped diamond layer of 1μm was selectively grown between the source and the drain electrode using tungsten electrodes as a mask. Al was thermally evaporated and patterned for a gate electrode on the undoped layer. The gate length and width were 30μm and 1.2mm, respectively. At 300°C, the transconductance was 0.6μS/mm at a drain voltage of 50V.

1. Introduction

Diamond is expected to be a good host material for electronic and optical devices under severe conditions[1)], such as high temperature and strong irradiation environment, because of its large energy gap (5.5 eV). Diamond is also expected to be an

Springer Proceedings in Physics, Vol. 71 Amorphous and Crystalline Silicon Carbide IV
Editors: C.Y. Yang · M.M. Rahman · G.L. Harris

attractive material for high-frequency and high-power devices, due to its high breakdown voltage (10^6-10^7 Vcm^{-1}),[2,3] high electron and hole mobilities (2000 $cm^2V^{-1}s^{-1}$ and 1800 $cm^2V^{-1}s^{-1}$, respectively)[4], high electron saturation velocity (2 x $10^7 cm\ s^{-1}$)[5] and low dielectric constant (5.7).

We have reported previously that smooth diamond epitaxial films without a graphitic component can be grown on synthetic Ib diamonds by a plasma assisted chemical-vapor-deposition (CVD) method[6] and have fabricated Schottky diodes and MESFET's on boron-doped diamond epitaxial films.[7,8] However, high temperature operation was not observed.

In this paper, we will report on the fabrication and the operation of novel Schottky diodes and FET's with a insulating diamond layer.

2. Experimental Procedure

A. Epitaxial Growth

The epitaxial films were grown by plasma-assisted chemical vapor deposition, as reported previously[6]. H_2 and CH_4 were used for source gases at a CH_4 concentration of CH_4/H_2=6%. B_2H_6 (10 ppm, diluted by H_2) was used for the doping gas at B_2H_6 concentration of B_2H_6/CH_4(=b)=0.83ppm, 8.3ppm and 167ppm. Epitaxial diamond films were obtained on (100) surfaces of synthesized single crystal Ib diamond substrates, 1.5 X 2.0 mm^2 in area and 0.3 mm thick. The resistivity of these substrates was 10^{12} - 10^{14} Ωcm. H_2 gas was introduced into a reactor at the rate of 100 sccm and 2.45GHz microwave radiation of 300 W was supplied at the same time. The plasma was spontaneously turned on when the gas pressure reached a few Torr. The gas pressure was finally adjusted and maintained at 40 Torr. The substrates were heated for 5 min in a H_2 plasma and then CH_4 and B_2H_6 were mixed with the H_2 feed. The substrate temperature was 830°C, measured by an optical pyrometer.

B. Shottky Diode Fabrication

A cross section of a high-voltage Schottky diode is shown in Fig. 1. Following the growth of a 1.8μm-thick p^+-type layer grown at b=167ppm, each sample was removed from the reactor and then reinserted in another reactor to avoid unintentional doping. An 75nm-thick or 450nm-thick undoped layer was grown on the p^+ layer. Al Schottky electrodes and Ti ohmic electrodes were formed by a thermal evaporation method and an e-beam evaporation method, respectively. Al electrodes were 500 μm diam dots, which were between two rectangular Ti electrodes 250 μm wide and 2.0 mm long.

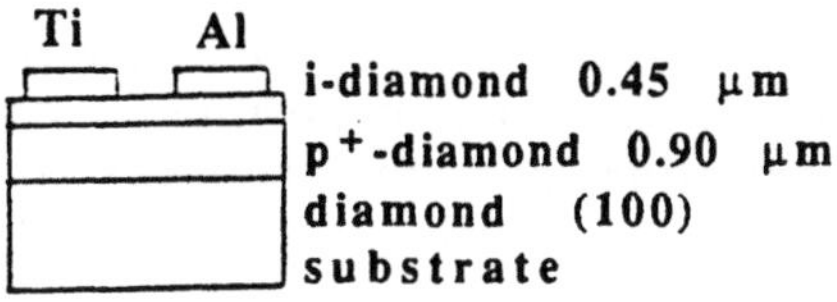

Fig. 1. Schematic cross-sectional view of the diode structure.

C. Fabrication of FET

The structure of the FET in this paper is shown in Fig. 2. The procedure of FET fabrication is shown in Fig. 3.

(a) A boron-doped film of 0.8μm-thick was grown at b=0.83ppm on an insulating diamond substrate.
(b) Tungsten was deposited by electron beam evaporation on the region of the source and the drain using a metal mask.
(c) A undoped epitaxial film of 1μm-thick was grown selectively except in the region of the source and drain.
(d) Annealing in air at 480°C for 1min was carried out.
(e) Aluminum film was deposited by thermal evaporation at the substrate temperature of 200°C.
(f) The aluminum film was patterned in the shape of gate electrodes.

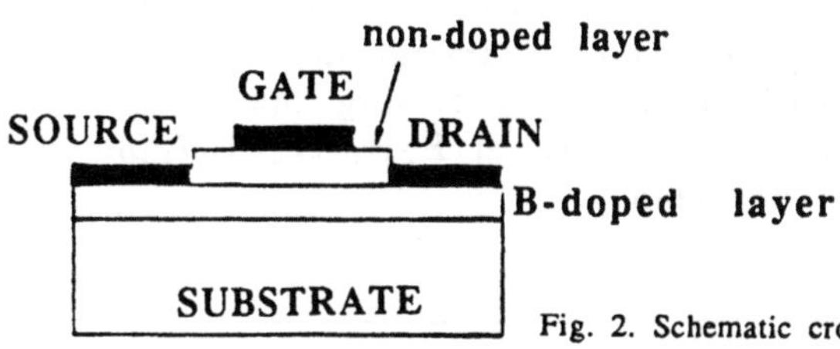

Fig. 2. Schematic cross-sectional view of the FET.

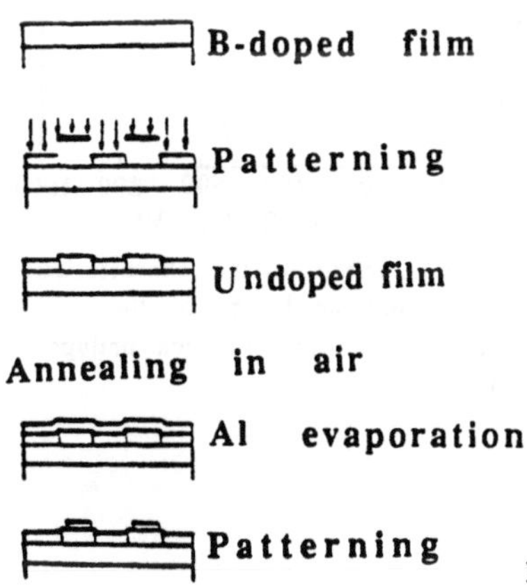

Fig. 3. Procedure of fabrication of FET

In order to minimize unintentional doping in the undoped layer, reactors and holders were separated for each deposition. As reported previously, the boron concentration and the mobility of boron-doped epitaxial film are estimated to be 10^{18} cm^{-3} and 100 $cm^2/V/s$, respectively[9]. The length between the source and the drain electrode, the gate width and the gate length were 90μm, 1.2mm and 30μm, respectively.

3. Results and Discussion

A. Diode Performance

Figure 4 shows the current-voltage characteristics of the Schottky diodes using p^+-type layers with and without intrinsic layers. Intrinsic layers were necessary to obtain rectifying properties on p^+ layers. The breakdown-voltage of the diodes with 450nm-thick intrinsic layers was found to be as high as 520V, as shown in Fig. 5, while that of the convetional diode was as low as 100V.

Current-voltage characteristics of the Schottky diode at 27°C and 300 °C are shown in Fig. 6. The upper temperature was limited by the measurement system rather than by the diode performance. The forward branch of the current-voltage characteristics was controlled by the series resistance of the diamond film. The resistivity of p-type diamond layers grown at B_2H_6/H_2=8.3ppm was actually reduced by raising the temperature while that of p^+-type diamond layers grown at B_2H_6/H_2=167ppm did not change because of Fermi degeneracy.

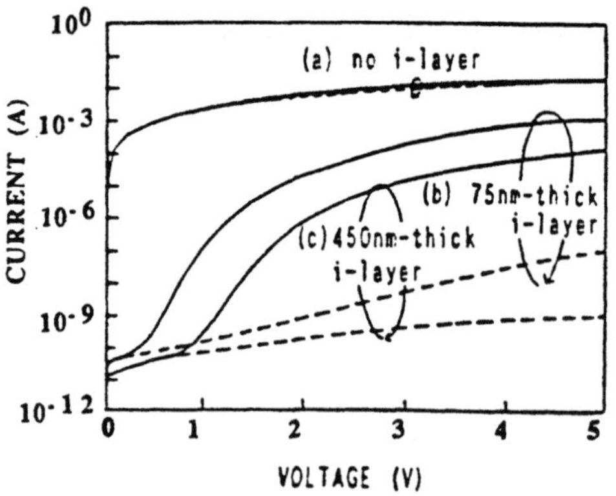

Fig. 4. Current-voltage characteristics of Schottky diodes (a) without an undoped layer, (b) with a 75 nm-thick undoped layer and (c) with a 450 nm-thick undoped layer.

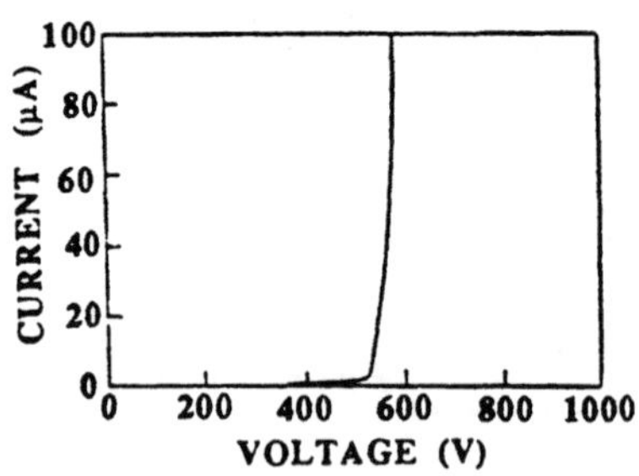

Fig. 5. Current-voltage characteristics at high reverse biases of the diode with a 450 nm-thick undoped layer.

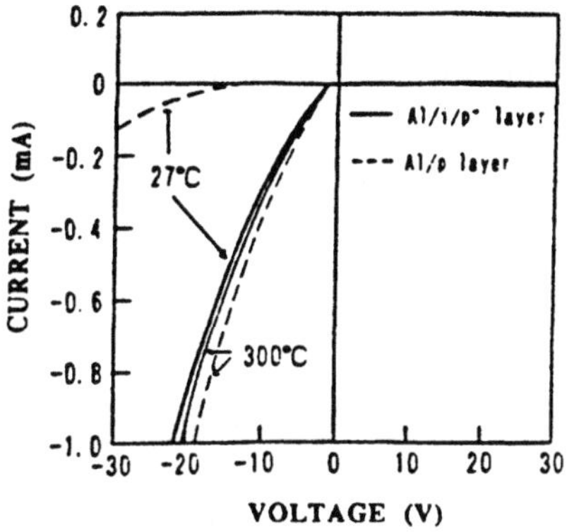

Fig. 6. Current-voltage characteristics of Schottky diodes at 27 °C and 300 °C. Solid lines are for Al/450nm-thick intrinsic layer/1.8μm p^+-type layer. Dashed lines are Al/1.8μm p-type layer.

B. FET Performance

The current-voltage characteristics of the FET at 26°C and 300°C are shown in Figs. 7 (a) and (b), respectively. These show that the drain current is controlled by the gate voltages. The drain current of Fig. 7(a) have a tendency toward saturation at higher voltage. The transconductance is calculated to be 0.02 S/mm. At 300°C, the drain current is not saturated and the transconductance is calculated to be 0.6 S/mm. The increase of the space charge in the depletion layer is considered to prevent saturation at 300°C.

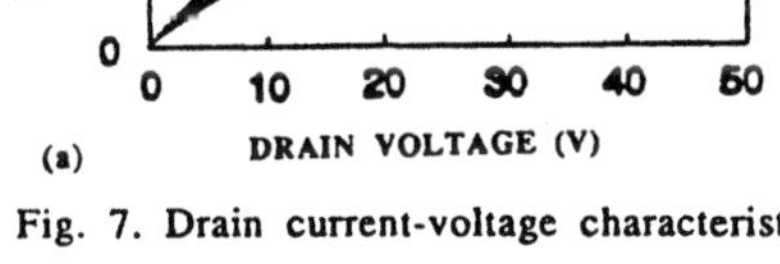

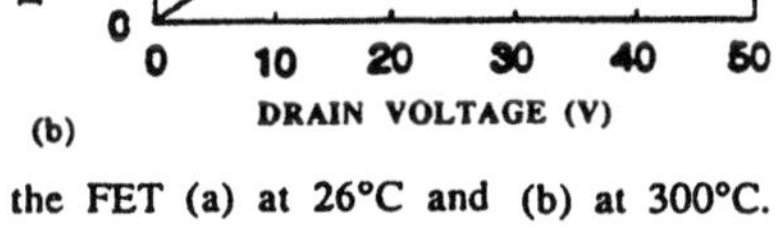

Fig. 7. Drain current-voltage characteristics of the FET (a) at 26°C and (b) at 300°C.

4. Conclusions

High-voltage Schottky diodes were fabricated on bi-layers consisting of intrinsic and p^+-type diamond epitaxial layers, using Al for Schottky contacts and Ti for ohmic contacts. The forward branch of current-voltage characteristics was independent of the temperature because of Fermi degeneracy in p^+-type layers. The breakdown voltage was found to be as high as 520V.

Insulating diamond gate FET's were fabricated using a selective epitaxial growth. Tungsten was used for the source and the drain electrodes, and also used for the metal mask of selective epitaxial growth of the non-doped layer. Aluminum was used for the gate electrode. The FET's operated at 300°C.

References

1) M.W.Geis, D.D.Rathman, D.J.Ehrlich, E.A.Murphy and W.T.Lindley : IEEE Electron Device Lett. 8 (1987) 341

2) A.V.Bogdanov, I.M.Vikulin and T.V.Bogdanova : Sov. Phys.-Semicond. 16 (1982) 720.

3) E.A.Konoirova, Yu.A.Kuznetsov, V.F.Sergienko, S.D.Tkachenko, A.V.Tsikanov, A.V.Spitsyn and Yu.Z.Danyushevski : Sov. Phys. - Semicond. 17 (1983)146.

4) V.K.Bazhenov, I.M.Viokulin and A.G.Gonar : Sov.Phys. - Semicond. 19 (1985) 829.

5) L.Reggiani, S.Bosi, C.Canali, F.Nava and S.F.Kozlov : Solid State Commun. 30 (1979) 333.

6) H.Shiomi, K.Tanabe, Y.Nishibayashi and N. Fujimori : Jpn. J. Appl. Phys. 29 (1990) 34.

7) H.Shiomi, H.Nakahata, T. Imai, Y. Nishibayashi and N. Fujimori: Jpn. J. Appl. Phys. : 28 (1989) 758.

8) H.Shiomi, Y.Nishibayashi and N. Fujimori : Jpn. J. Appl. Phys. 12 (1989) L2153.

9) H.Shiomi, Y.Nishibayashi and N.Fujimori : Jpn. J. Appl. Phys. 30 (1991) 1363.

Gamma-Ray Irradiation Effects on Cubic Silicon Carbide Metal-Oxide-Semiconductor Structure

M. Yoshikawa[1], *Y. Morita*[1], *H. Itoh*[1], *I. Nashiyama*[2], *S. Misawa*[2], *H. Okumura*[2], *and S. Yoshida*[2]

[1]Japan Atomic Energy Research Institute,
1233 Watanuki, Takasaki, Gunma 370-12, Japan
[2]Electrotechnical Laboratory,
1-1-4 Umezono, Tukuba, Ibaraki 305, Japan

Abstract. Gamma-ray irradiation effects on cubic silicon carbide (3C-SiC) metal-oxide-semiconductor (MOS) structures have been studied with high frequency capacitance-voltage measurements. Using Termann analysis to determine the densities of interface traps and oxide traps, we found a 2/3 power law dependence of the radiation-induced traps on the absorbed dose. The generation of these traps was found to depend on bias during irradiation, similar to those of Si MOS devices. We also compared the radiation response of dry and pyrogenic oxides on 3C-SiC with that of dry oxides of Si, and showed less trapping in pyrogenic 3C-SiC oxides.

1.Introduction

The large-area single crystals of cubic silicon carbide (3C-SiC) have been reproducibly obtained by recent technical progress in heteroepitaxial growth on Si substrates[1,2]. Insulators on 3C-SiC can be easily made by means of oxidation of 3C-SiC surfaces, being similar to the oxidation processes of Si. These have brought interests in 3C-SiC devices with planer-structures used in high temperature and/or radiation environments such as outer space and nuclear facilities.

For the application to the planer structures of 3C-SiC devices used in radiation fields, it is important to clarify irradiation effects on metal-oxide-semiconductor (MOS) structures. Recently, Nashiyama *et al.* have reported the effects of ^{60}Co gamma-ray irradiation on 3C-SiC MOS structures using metal-oxide-semiconductor field effect transistors (MOSFETs), and showed that these devices have a significant radiation tolerance up to 10 kGy(SiO_2)[3]. However, only a little has been reported about the irradiation effects near the 3C-SiC/SiO_2 interfaces in the MOS structures.

In this paper, capacitance-voltage (C-V) measurements have been carried out for 3C-SiC MOS capacitors irradiated by ^{60}Co gamma-rays under various biased conditions. We show the effects of the irradiation on the C-V characteristics, and analyze radiation-induced traps on the 3C-SiC/SiO_2 interfaces using Termann method. The absorbed dose dependence of these traps is examined for MOS capacitors with dry and pyrogenic oxides. We also investigate the bias dependence of the generation of the traps, and discuss the irradiation effects on the C-V characteristics of 3C-SiC MOS structures in comparison with those of Si MOS structures with dry oxide.

2.Experiments

Single crystals of 3C-SiC were grown epitaxially on Si(100) substrates at 1400°C by way of chemical vapor deposition using a C_3H_8/H_2-SiH_4/H_2-H_2 reaction-gas system(gas flow rate: 150sccm, 240sccm, and 8000sccm, respectively). The growth procedure has been reported elsewhere[4]. The unintentionally doped epilayers showed n-type conduction and were about 10 μm in thicknesses.

Immediately after the crystal growth, surfaces of the 3C-SiC epilayers were oxidized by dry and pyrogenic oxidation processes. Dry oxidation was carried out in pure O_2 gas at 1100°C for 4 hours to make the oxide layers of 480 Å thicknesses. Pyrogenic oxidation, where pure water vapor is made by direct reaction between H_2 and O_2 gases with a flow rate ratio of 1:2 in a furnace tube, was performed at 1100°C for 4 hours to obtain the pyrogenic oxide layers of 600Å thicknesses.

Thin aluminum films were evaporated onto the oxide layers to form gate electrodes of 0.3mm in diameter. After the oxide layers near the gate electrode were scratched, aluminum films were deposited on the exposed surfaces to serve as a ohmic contact. No post annealing for alloying was performed.

The 3C-SiC MOS capacitors obtained were irradiated with ^{60}Co gamma-rays in argon atmosphere at room temperature. The dose rate was 8.8 kGy(SiO_2)/h*. The capacitors were placed between aluminum plates of 2mm thickness to satisfy the secondary-electron equilibrium condition[5].

The *C-V* characteristics of 3C-SiC MOS capacitors were measured at a frequency of 1 MHz under dark condition at room temperature. The gate voltage was swept from -15V to +10V after a holding time of 30s at -15V, to accumulate minority carriers. Pulsed *C-V* characteristics were also measured with a response time of 100μs at a frequency of 1MHz to obtain the doping concentration of donors in unirradiated epilayers.

3.Results

Figure 1 shows the change in the high frequency *C-V* characteristics of 3C-SiC MOS capacitors with dry and pyrogenic oxides irradiated under open-circuited condition. The notations, C_{fb} and C_{mg}, indicate the capacitances corresponding to flatband and midgap conditions, respectively. The capacitances at positive voltages above +5V are equivalent to the oxide capacitances, C_{ox}. These values were in good agreement with an oxide capacitance derived from an ideal *C-V* curve, assuming a uniform doping level of donors of 1.8×10^{17} cm^{-3}. The donor concentrations were calculated from the slope of $1/C^2$ versus gate voltage plots obtained with the pulsed *C-V* measurements.

As the generation rate of minority carriers is extremely low because of the wide bandgap of 3C-SiC, the *C-V* curves in the Fig.1 did not show a clear

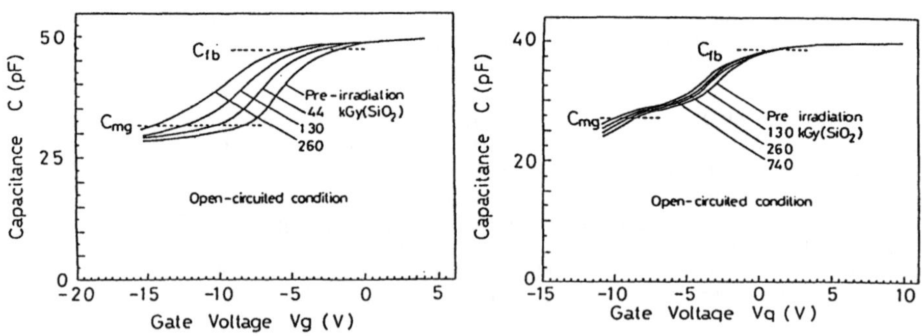

(a) Dry oxide. C_{ox} = 50.3 pF. (b) Pyrogenic oxide. C_{ox} = 39.9 pF.

FIG.1 Change in high-frequency C-V characteristics of 3C-SiC MOS capacitors with dry and pyrogenic oxides irradiated by ^{60}Co gamma-rays under open-circuited condition.

* 1Gy = 100 rad

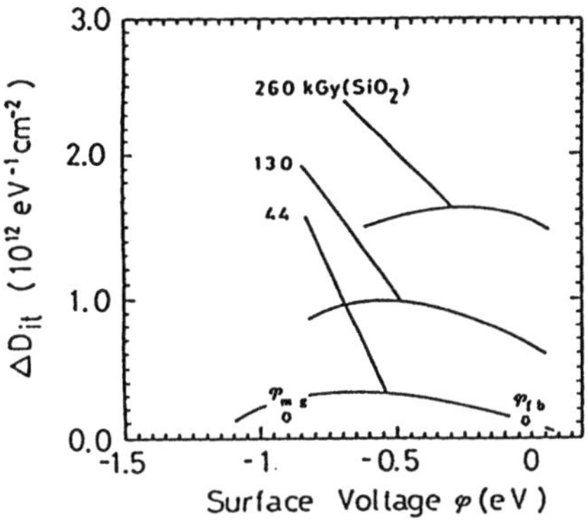

FIG.2 Energy profile of radiation induced interface trap densities, ΔD_{it}, in the band gap for a 3C-SiC MOS capacitor with a dry oxide.

inversion characteristic, as was reported by Shibahara *et al.* [6]. Interface traps densities, D_{it}, and oxide traps, N_{ot}, were calculated using Termann method[7]. The values of D_{it} and N_{ot} in the MOS capacitors with dry oxides were from 2×10^{11} to 1×10^{12} traps/(eVcm^2), and about 10^{12} traps/cm^2, respectively. For the MOS capacitors with pyrogenic oxides, those were from 3×10^{11} to 2×10^{12} traps/(eVcm^2), and about 10^{12} traps/cm^2, respectively.

By gamma-ray irradiation, the C-V curve of the 3C-SiC MOS capacitor with dry oxide shifted largely towards negative voltage and showed clear increase in stretchout along the gate voltage axis. The shift of C_{mg} level was about -6.3V for the irradiation of 260 kGy(SiO_2) in Fig.1(a). The increase in the stretchout between C_{fb} and C_{mg} levels was approximately -2.9V for the same dose. The C-V curve for the MOS capacitor with the pyrogenic oxide, on the other hand, shifted slightly towards negative voltage, and obvious stretchout was not observed in the curve. The gate voltage shift at C_{mg} level was approximately -2.0V for the irradiation of 740 kGy(SiO_2) in Fig.1(b).

The distribution of the radiation induced interface traps densities, ΔD_{it}, in the bandgap was estimated from the results[7]. Figure 2 shows ΔD_{it} profiles calculated from the C-V curves before and after irradiation corresponding to Fig.1(a). The flatband and midgap potentials are indicated by φ_{fb} and φ_{mg}, respectively. It was found that ΔD_{it} exhibits a broad peak distribution in the upper-half of the bandgap, and the peak height depends strongly on the absorbed dose. The same calculations were performed for the MOS capacitor with a pyrogenic oxide. The result is shown in Fig.1(b). However, a clear peak was not observed in the bandgap in the case of a pyrogenic oxide.

4.Discussion

4.1 Radiation tolerance of 3C-SiC MOS structures

Total interface traps induced by irradiation, ΔN_{it}, is an important measure of irradiation effects on MOS structures. To evaluate the absorbed dose dependence of ΔD_{it}, ΔN_{it} was calculated by integrating the ΔD_{it} profiles along the surface potential axis. For the comparison between these results and that of Si MOS structures, we fabricated Si MOS capacitors with Al/SiO_2/Si structures using dry oxidation process at 1100°C, in which no rad-hard treatment was applied. Figure 3 shows logarithmic plots of ΔN_{it} against the absorbed dose in 3C-SiC and Si MOS capacitors irradiated under open-circuited condition. The values of ΔN_{it} in 3C-SiC and Si MOS capacitors lie on the parallel straight lines, as is explained in section 4.2. In the region below 10^{11} cm^{-2}, the values deviate from the straight lines

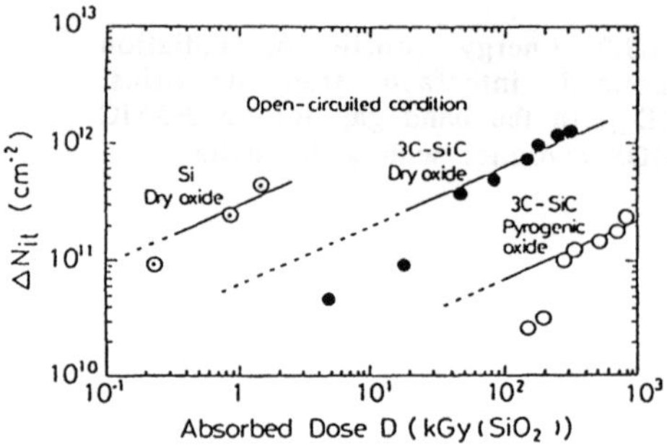

FIG.3 Logarithmic absorbed dose dependencies of radiation-induced interface traps, ΔN_{it}, in 3C-SiC and Si MOS capacitors irradiated under open-circuited condition.

because of a round off error in the calculation of the amount of ΔN_{it}'s. The values of ΔN_{it} in 3C-SiC MOS capacitors with a dry oxide are much smaller than those in Si MOS capacitors. Furthermore, those in 3C-SiC MOS capacitors with a pyrogenic oxide are smaller than those in 3C-SiC MOS capacitors with a dry oxide. These results indicate that 3C-SiC MOS structures, especially for those with pyrogenic oxides, are hard to generate interface traps by gamma-ray irradiation compared with those of Si MOS structures.

These results also support reasonably the fact that 3C-SiC MOSFETs have a tolerance to gamma-ray irradiation rather than Si MOS devices[3]. Oxygen deficient centers are known to be some precursors of the positive charge traps in Si MOS structures. The oxidation of 3C-SiC crystal is accompanied with a decarbonization process, which differs from the case of the oxidation of Si. The residual carbon atoms exist in 3C-SiC/SiO_2 interfaces[6] are considered to play some role in reducing the radiation-induced traps formation. However, the reason for the radiation tolerance of 3C-SiC MOS structures is not understood fully at present. For a better understanding of these results, we have to study microstructures of oxides and 3C-SiC/SiO_2 interfaces in 3C-SiC MOS structures in the future.

4.2 Irradiation effects on the 3C-SiC/SiO_2 interface

The bias effects were studied on 3C-SiC MOS capacitors with dry oxides. Figure 4(a) shows the logarithmic plots of ΔN_{it} as a function of the absorbed dose. The result shows that ΔN_{it} follows a power law dependence on the absorbed dose with an exponent of 2/3 for the irradiations under open-circuit and positive biased

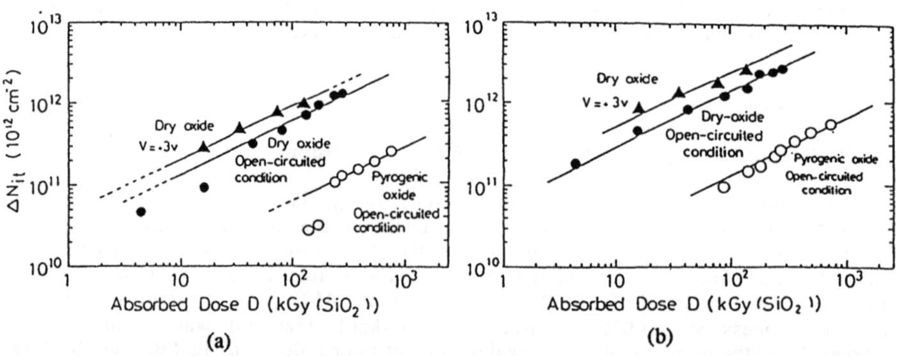

FIG.4 Logarithmic absorbed dose dependencies of radiation-induced traps, ΔN_{it}, and oxide traps, ΔN_{ot}, in 3C-SiC MOS capacitors with dry and pyrogenic oxides irradiated under various various condition.

(+3V) conditions, whereas a clear power law dependence was not observed under the negative biased (-3V) condition because of the suppression of the generation of ΔN_{it}.

Figure 4(b) shows logarithmic plots of oxide traps induced by the irradiation, ΔN_{ot}, in 3C-SiC MOS capacitors, which were calculated from the midgap shift on the *C-V* curve by irradiation[8]. In the calculations, we assumed no shift of midgap-voltage on the *C-V* curve owing to the generation of interface traps. This assumption is considered to be reasonable because the radiation responses of 3C-SiC MOS capacitors are quite similar to those of Si MOS capacitors[9]. The values of ΔN_{ot} were also proportional to the absorbed dose with an exponent of about 2/3 except for the negative biased condition in which a clear power low dependence was not observed. The bias dependence of ΔN_{ot} is quite similar to those of ΔN_{it} shown in Fig.4(a).

From these experimental results, the values of ΔN_{it} and ΔN_{ot} can be written by following equations,

$$\Delta N_{it} = k_{it} D^{\frac{2}{3}}, \quad \Delta N_{ot} = k_{ot} D^{\frac{2}{3}}, \tag{1}$$

where D is absorbed dose of the SiO_2 layers, k_{it}, and k_{ot} are proportional factors involving a term depending on applied electric field intensity, respectively. Since D is defined as absorbed energies per unit weight[10], the mean number of hole-electron pairs per unit volume induced by irradiation in SiO_2 layers, N_{h-e}, can be written using D as follows,

$$N_{h-e} = \frac{\rho D}{\varepsilon}, \quad \text{or} \quad D = \frac{\varepsilon N_{h-e}}{\rho} \tag{2}$$

where ρ is the density of SiO_2, ε the mean energy for generating a hole-electron pair in SiO_2 layers. These equations show that the radiation induced traps, ΔN_{it} and ΔN_{ot}, are relation to the mean number of hole-electron pairs per unit volume generated in the SiO_2 layers.

For silicon MOS structures, it is well known that generation of the traps near the Si/SiO_2 interface are related to holes escaping from initial recombinations of hole-electron pairs induced by irradiation. Furthermore, the 2/3 power law dependence has been also reported for Si/SiO_2 MOS structures[11,12,13]. The similarity between the absorbed dose dependence of Si/SiO_2 and $3C\text{-}SiC/SiO_2$ interfaces suggests that positive charges (holes and/or H^+ ions, probably) transfer to the $3C\text{-}SiC/SiO_2$ interface by positive bias applied to the oxide layer, and interact with precursors of the interface traps, as in the case of Si MOS structures[14,15].

5. Conclusions

Gamma-ray irradiation effects on 3C-SiC MOS structures were studied with *C-V* measurements. We estimated the radiation-induced interface traps, ΔN_{it}, and oxide traps, ΔN_{ot}, near the $3C\text{-}SiC/SiO_2$ interfaces for both dry and pyrogenic oxides on 3C-SiC epilayers. 3C-SiC MOS capacitors with pyrogenic oxides have a significant tolerance to gamma-ray irradiation. The generation of radiation-induced traps under positive biased or open-circuited condition was higher than that under negative biased condition. The values of ΔN_{it} and ΔN_{ot} follow a power law dependence on the absorbed dose with an exponent of approximately 2/3, which agrees well with those of Si MOS structures.

References

1 S.Nishino, J.A.Powell, and H.A.Will, Appl.Phys.Lett., **42**,460(1983).
2 M.Shinohara, M.Yamanaka, H.Daimon, E.Sakuma, H.Okumura, S.Misawa, K.Endo, and S.Yoshida, Jpn.J.Appl.Phys., **27**, L434(1988).
3 I.Nashiyama, M.Shinohara, T.Matsumoto, K.Endo, E.Sakuma, S.Misawa, H.Okumura, and S.Yoshida, *Amorphous and Crystalline Silicon Carbide* (Springer-Verlag, Berlin, 1989), p123.
4 M.Yamanaka, H.Daimon, E.Sakuma, S.Misawa, and S.Yoshida, J.Appl.Phys., **61**, 599(1987).
5 F.H.Attix, *Introduction to Radiological Physics and Radiation Dosimetry.* (Wiley-Interscience, New York, 1984), Vol 11, Chap.4 p61.
6 K.Shibahara, S.Nishino, and H.Matsunami, Jpn.J.Appl.Phys., **23**, L862(1984).
7 I.M.Termann, Solid-State Electron. **5**, 285(1962).
8 P.J.McWhorter and P.S.Winokur, Appl.Phys.Lett., **48**, 133(1986).
9 P.M.Lenahan and P.V.Dressendorfer, J.Appl.Phys., **55**, 3495(1984).
10 J.R.Srour,*Tutorial Short Course IEEE 1983 Nuclear and Space Radiation Effects Conference* , (Gatlinburg, Tennessee, 1983), Chap. 2, p13
11 F.B.McLean, IEEE.Trans.Nucl.Sci., **NS-27**, 1651(1980).
12 P.S.Winokur, H.E.Boesch.Jr., J.M.MaGarrity, and F.B.McLean. IEEE.Trans.Nucl.Sci., **NS-24**, 2113(1977).
13 F.B.McLean, IEEE Trans.Nucl.Sci. **NS-27**, 1651(1980).
14 H.S.Witham and P.M.Lenahan, IEEE.Trans.Nucl.Sci., **NS-34**, 1147(1987).
15 K.Naruke, M.Yoshida, K.Maeguchi and H.Tango, IEEE.Trans.Nucl.Sci., **NS-30**, 4054(1983).

Formation of SiGe/Si Heterostructures by Low-Temperature Germanium Ion Implantation

K. Shoji[1], *A. Fukami*[1], *T. Nagano*[1], *T. Tokuyama*[2], *A. Gupta*[3], *and C.Y. Yang*[3]

[1]Hitachi Reserach Laboratory, Hitachi, Ltd.,
4026 Kuji, Hitachi-shi, Ibaraki 319-12, Japan
[2]Institute of Applied Physics, University of Tsukuba,
Tsukuba-shi, Ibaraki 305, Japan
[3]Microelectronics Laboratory, Santa Clara University,
Santa Clara, CA 95053, USA

Abstract. $Si_{1-x}Ge_x$ alloys were formed by high-dose (on the order of 10^{16} cm^{-2}) germanium ion implantation into Si. It was found that the crystalline quality of the SiGe layer was improved by maintaining the substrate at low temperature during implantation. Cross-sectional transmission electron micrographs indicated a considerable reduction in the end-of-range defects. This improvement was further confirmed by electrical characterization of p-n junctions formed in the SiGe layer.

1. Introduction

Much interest is currently being focused on SiGe alloys for use in high-performance heterojunction bipolar transistors. SiGe/Si heterostructures have been formed by using techniques such as molecular beam epitaxy and ultra-high vacuum chemical vapor deposition [1-6]. Alternatively, they can be fabricated by high-dose Ge^+ implantation [7-11]. Low-dose Ge^+ implantation is a well-established technique to form shallow p-n junctions in pre-amorphized Si substrates [12], and is inherently compatible with standard silicon integrated circuit fabrication. However, high-dose Ge^+ implantation introduces residual defects, which significantly damage the crystalline quality [11]. There are two kinds of defects in the recrystallized SiGe layer formed by Ge^+ implantation and subsequent thermal annealing. One consists of misfit dislocations near the surface, while the other consists of dislocations due to residual damage at the interface between the recrystallized SiGe layer and Si substrate. The latter is known as "end-of-range (EOR) defects". Reduction of EOR defect density has been achieved by lowering the substrate temperature during ion implantation [13-15]. In this paper, we present experimental results on Ge^+ implantation with doses on the order of 10^{16} cm^{-2} and at substrate temperatures as low as −100 °C. Our results show improvements in the SiGe crystallinity and the p-n junctions characteristics.

Editors: C.Y. Yang · M.M. Rahman · G.L. Harris

2. SiGe "Growth" and Device Fabrication

Cross-sectional transmission electron microscopy (XTEM) was used to estimate the SiGe layer crystallinity. Germanium ions were implanted into a p-type Si(100) wafer through a thermally grown 10-nm oxide at an acceleration energy of 160 keV with a dose of 5×10^{16} cm^{-2}. The peak germanium concentration was 6×10^{21} cm^{-3}, which corresponds to a germanium content of 12 at.%. The Si substrate temperatures during ion implantation were maintained at –100 °C and 27 °C with liquid nitrogen and water as coolants, respectively. The beam current density was kept below 0.44 µA/cm^2 during implantation to eliminate the possibility of crystal regrowth due to beam heating. A tilt angle of 7° was used for implantation. Implanted samples were then annealed in N_2 at 700 °C for 8 hours.

To evaluate the SiGe crystalline quality, a planar p-n structure was fabricated using a standard bipolar process for collector junction formation. This structure is shown schematically in Figure 1. Antimony was diffused into a p-type Si(100) substrate to form the n^+-region using a two-step process with predeposition followed by drive-in. An undoped Si epitaxial layer was then grown, followed by phosphorus ion implantation to form the n-region. A SiO_2 overlayer was deposited, and diffusion windows were opened. Arsenic ions were implanted and a furnace anneal at 950 °C for 30 min was carried out to form the n^{++}-region for metal contact. To form the p-region, boron ions were implanted, followed by a furnace anneal at 900 °C for 10 min. A SiO_2 overlayer was added again, and the Ge^+ implantation window was opened. Germanium ions were implanted through a thermally grown 10-nm oxide with the same conditions as those for the samples used for XTEM observation. Subsequently, a two-step anneal (700 °C for 8 hours and 900 °C for 10 min) was carried out.

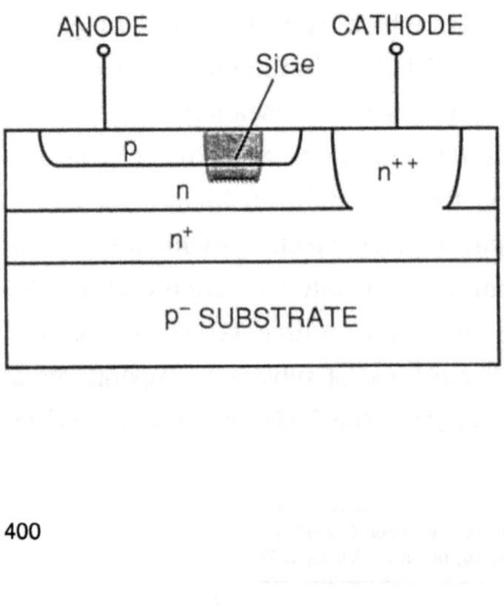

Figure 1. Schematic of SiGe diode.

3. Characterization Results

Figure 2 shows the concentration profile of germanium obtained from secondary ion mass spectroscopy (SIMS) and those of the dopants (antimony, phosphorus, and boron) obtained from process simulation. Comparing the SIMS data, the germanium profile of the –100 °C implanted sample is found identical to the 27 °C implanted one. The depth of the p-n junction is 150 nm, and the boron and phosphorus concentrations at the junction are about 4×10^{16} cm^{-3}. The thickness of the n-layer is 450 nm, with an average doping concentration of around 10^{16} cm^{-3}. The germanium content at the p-n junction is approximately 12 at.%, and the germanium distribution extends to the center of the n-layer.

Figures 3(a) and (b) show XTEM photographs for the 700 °C annealed samples with implantations at –100 °C and 27 °C, respectively. Two types of defects are observed. One is located near the surface and the other at the EOR. The surface defects extend down to about 100 nm in both cases, while the EOR defects are located at about 310 nm and 280 nm from the surface in the –100 °C and 27 °C implanted samples, respectively. The surface region in the –100 °C implanted sample shows slightly fewer defects compared to the 27 °C implanted sample, while the EOR defects decrease in number substantially. The recrystallized layer for the –100 °C implanted sample is thicker than the 27 °C implanted one by about 30 nm. The germanium concentration at the interface between the recrystallized layer and the Si substrate for the –100 °C implanted sample is 5×10^{18} cm^{-3}, whereas that for the 27 °C implanted one is around 10^{19} cm^{-3}. Thus, the residual damage resulting from the germanium ion bombardment for the

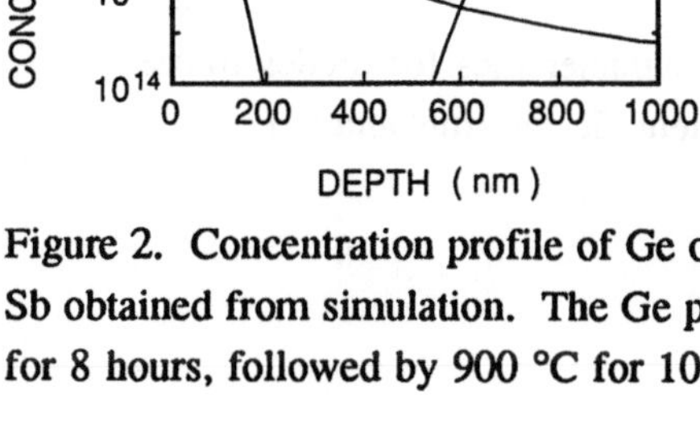

Figure 2. Concentration profile of Ge obtained from SIMS and those of B, P, and Sb obtained from simulation. The Ge profile is from a sample annealed at 700 °C for 8 hours, followed by 900 °C for 10 min.

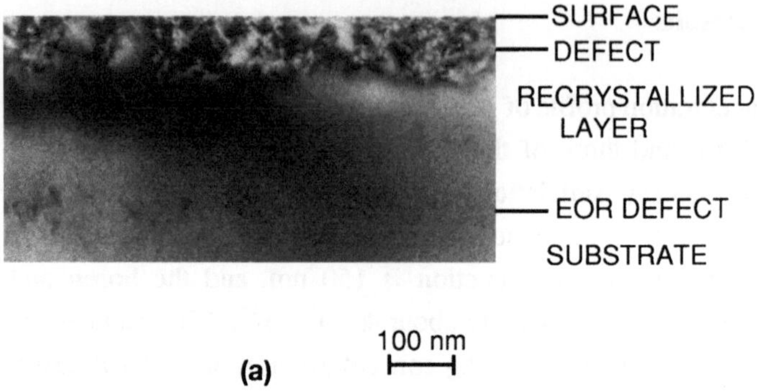

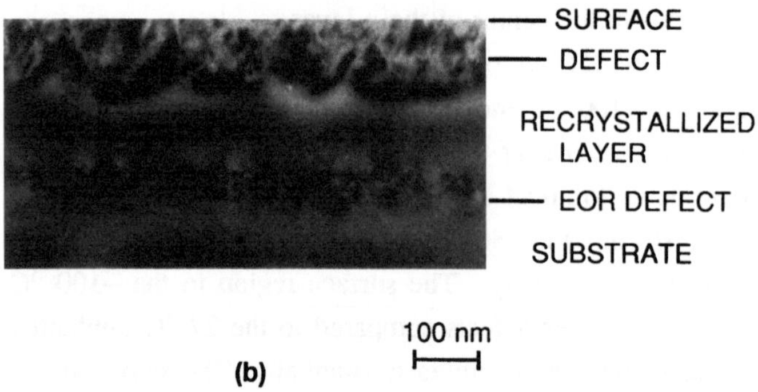

Figure 3. XTEM images of SiGe/Si heterostructures annealed at 700 °C for 8 hours: (a) –100 °C Ge$^+$ implantation, (b) 27 °C Ge$^+$ implantation.

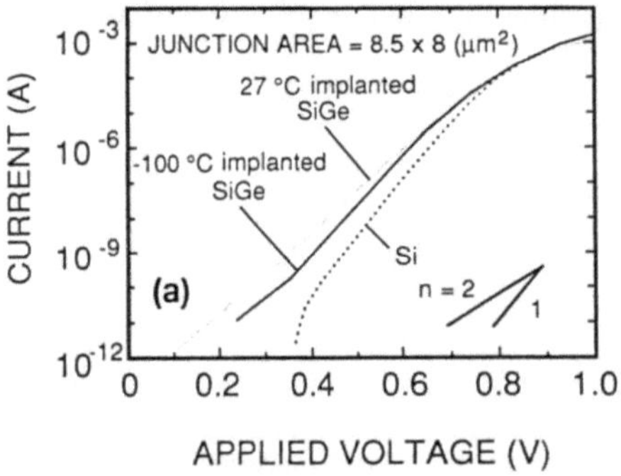

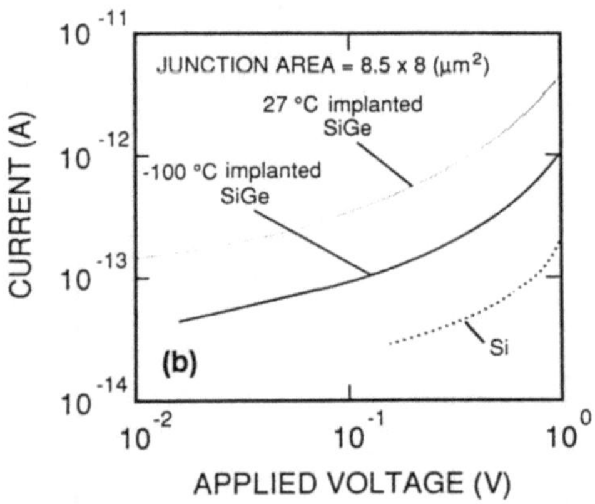

Figure 4. Comparisons of current-voltage characteristics for diodes formed with –100 and 27 °C Ge$^+$ implantations and without Ge ion implantation: (a) forward bias, (b) reverse bias.

−100 °C implanted sample is less than that for the 27 °C implanted one. This in turn results in fewer EOR defects for the −100 °C implanted sample.

Figure 4 summarizes the results of the room-temperature current-voltage measurements for the three p-n junction diodes. In figure 4(a), the SiGe heterojunction diodes exhibit larger currents and smaller slopes than the Si control diode. This is attributed to carrier recombination associated with the EOR defects across the p-n junction. However, the current and ideality factor for the −100 °C implanted diode are smaller than those for the 27 °C implanted one. The ideality factor of the former is 1.3, while that of the latter is 1.5. This confirms that reduction of EOR defects is achieved by lowering the substrate temperature during germanium ion implantation. Figure 4(b) compares reverse current-voltage characteristics for the three diodes. The reverse current of the 27 °C implanted diode is about one order of magnitude larger than that of the Si control diode. However, the reverse current of the −100 °C implanted diode is about half of the 27 °C implanted one. This further confirms the improvement in crystalline quality in the SiGe layer and near the SiGe/Si substrate interface. The larger reverse current for the −100 °C implanted diode may also be partially attributed to a narrower bandgap compared to that of Si.

4. Summary

It has been demonstrated that high-dose Ge^+ implantation into Si at a substrate temperature of −100 °C results in a significant reduction of EOR defects compared to 27 °C implanted samples. Direct observations of this improvement using XTEM are confirmed by measured current-voltage characteristics of p-n junctions formed in the SiGe layer.

The authors would like to thank Dr. N. Momma for his encouragement throughout this work. Thanks are also due to Messrs. S. Kawakami, K. Sukegawa, and the technical staff of Hitachi Research Laboratory for their contributions in the device fabrication.

REFERENCES

[1] J.W. Matthews and A.E. Blakeslee, J. Cryst. Growth, 27, 118 (1974).

[2] J.W. Matthews, J. Vac. Sci. Technol., 12, 126 (1975).

[3] J.H. van der Merme, J. Appl. Phys., 34, 123 (1963).

[4] B.S. Meyerson, Appl. Phys. Lett., 48, 797 (1986).
[5] B.S. Meyerson, F.K. LeGoues, T.N. Nguyen, and D.L. Harame, Appl. Phys. Lett., 50, 113 (1987).
[6] B.S. Meyerson, K.J. Uram, and F.K. LeGoues, Appl. Phys. Lett., 53, 2555 (1988).
[7] O.W. Holland, C.W. White, and D. Fathy, Appl. Phys. Lett., 51, 520 (1987).
[8] D. Fathy, O.W. Holland, and C.W. White, Appl. Phys. Lett., 51, 1337 (1987).
[9] K. Ohta, J. Sakano, and S. Furukawa, Ext. Abs. of 21st Conf. Solid State Devices and Materials, Tokyo, 555 (1989).
[10] A.R. Srivasta, S. Sharan, O.W. Holland, and J. Narayan, J. Appl. Phys., 65, 4028 (1989).
[11] A. Fukami, K. Shoji, T. Nagano, and C.Y. Yang, Appl. Phys. Lett., 57, 2345 (1990).
[12] M.C. Ozturk, J.J. Wortman, C.M. Osburn, A. Ajmera, G.A. Rozgonyi, E. Frey, W. Chu, and C. Lee, IEEE Trans. Electron Devices, 35, 659 (1988).
[13] K.S. Jones and D. Venables, J. Appl. Phys., 69, 2931 (1991).
[14] S. Prussin and K.S. Jones, J. Electrochem. Soc., 137, 1912 (1990).
[15] T. Suzuki, H. Yamaguchi, S. Ohzono, and N. Natsuaki, Ext. Abs. 22nd Conf. Solid State Devices and Materials, Sendai, 1163 (1990).

Characterization of Ge and C Implanted Si Diodes

C. Cook, A. Gupta, and C.Y. Yang
Microelectronics Laboratory, Santa Clara University,
Santa Clara, CA 95053, USA

Electrical characterization of diodes formed by Ge and C implantations in Si was performed. The results were compared with those obtained for a corresponding Si control device. Current-voltage measurements confirm that carbon doping improves the crystalline quality of the Ge-implanted SiGe layer. However, capacitance-voltage measurements indicate that carbon implantation causes significant dopant deactivation.

1. Introduction

SiGe is a strong candidate to meet future high-speed requirements of VLSI circuits due to its narrow bandgap relative to Si. Studies on the use of SiGe alloys to form high-performance heterojunction bipolar transistors (HBTs) have been reported [1-2]. These studies employed such fabrication techniques as ultra high vacuum chemical vapor deposition (UHV-CVD) and molecular beam epitaxy (MBE). An alternate approach to form the SiGe layer is by germanium ion implantation into silicon and subsequent solid-phase epitaxy (SPE) [3-8]. The implantation process is advantageous over the various epitaxial processes as it can be readily integrated into standard silicon processing and it is also convenient for selective SiGe growth. Furthermore, it has been reported that carbon ion implantation into the SiGe implanted layer can reduce surface defects, thus improving the crystallinity of the layer, leading to better junction characteristics [3]. This paper presents the results of electrical characterization of SiGe and SiGeC diodes formed by Ge and C implantations. The results are compared with those for Si diodes fabricated with the same process but without the Ge and C implantations.

2. Device Fabrication

The SiGe layer was formed by Ge implantation into p-type Si(100) at an energy of 120 keV with a dose of $5\text{x}10^{16}$ cm^{-2}. Carbon implantation was subsequently performed in some samples at 33 keV with a dose of either $6.25\text{x}10^{15}$ cm^{-2} or $1.67\text{x}10^{16}$ cm^{-2}. These samples were then annealed in N_2 ambient at 700 °C for 4~16 hours to grow the SiGe and SiGeC layers. B^+ ions were then implanted at 15 keV with a dose of $3.0\text{x}10^{13}$ cm^{-2} followed by an As implant through a 200 nm layer of poly-Si at 80 keV and a dose of $1.85\text{x}10^{16}$ cm^{-2}. The samples were then annealed at 900 °C for 20 minutes followed by a second anneal of 950 °C for 10 minutes to form the planar n^{++}- p^+ diodes. The device structure is shown in figure 1.

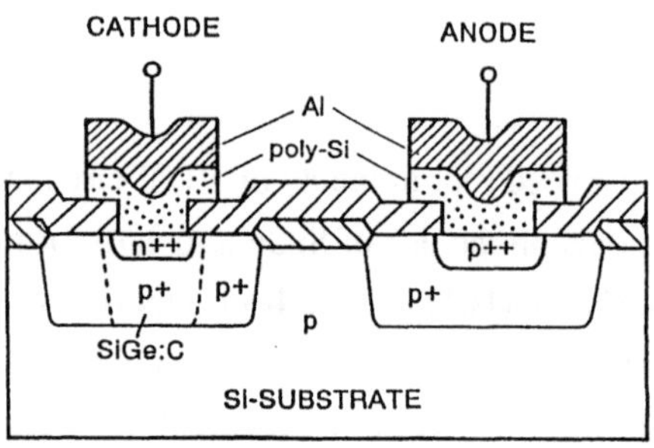

Figure 1. Schematic cross section of a SiGe n^{++}- p^+ diodes.

3. Results and Discussion

Current-voltage (I-V) and capacitance-voltage (C-V) measurements were performed on samples fabricated with different anneal times and carbon doses. Measurements were also performed on diodes having different device areas.

Figure 2 shows the comparisons of I-V characteristics of the SiGe, SiGeC, and Si diodes. A comparison of the reverse-biased I-V characteristics shows that the leakage current of the SiGe diode is about two orders of magnitude larger than that of the Si diode. This is attributed to the generation current induced by surface defects. However, the leakage current of the SiGeC diode is about one order of magnitude smaller than that of the SiGe diode. The decrease in leakage current confirms that C doping reduces the surface defects introduced by the Ge

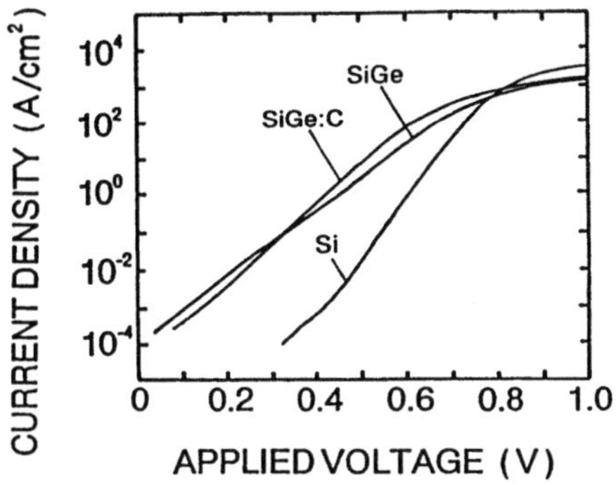

Figure 2a. Comparisons of current-voltage characteristics: forward bias.

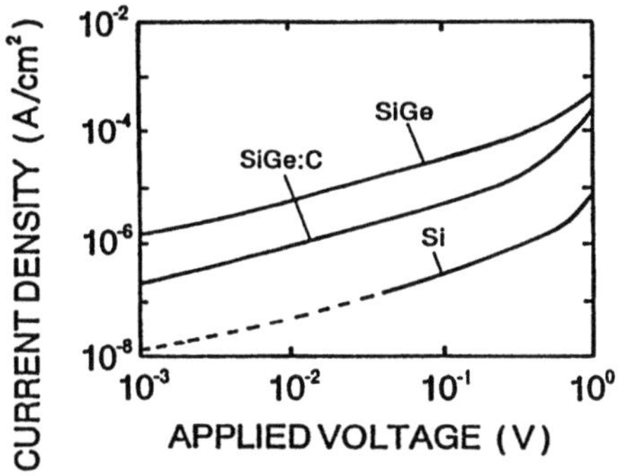

Figure 2b. Comparisons of current-voltage characteristics: reverse bias.

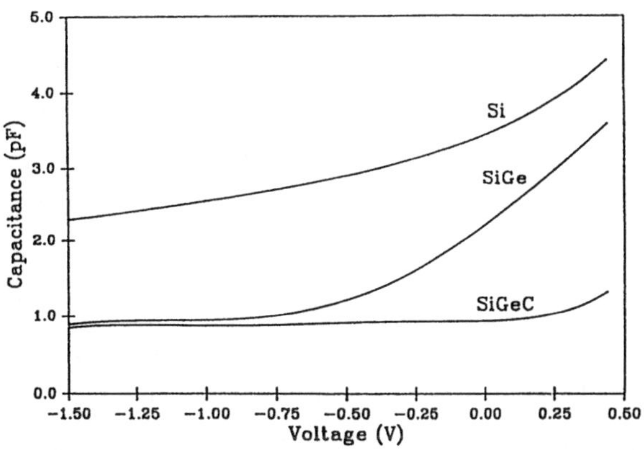

Figure 3. Comparison of capacitance-voltage characteristics at 100 kHz. Capacitor area = $30x30um^2$.

implantation, thereby improving the crystalline quality of the SiGe layer. In forward bias both the SiGe and the SiGeC diodes exhibit larger currents and smaller slopes than the Si diode. The larger currents are consistent with the smaller bandgaps of SiGe and SiGeC compared to that of Si. The estimated ideality factors (at a forward bias of 0.5 V) are 1.7, 1.5, and 1.09 for SiGe, SiGeC, and Si diodes respectively. This improvement of ideality factor for the SiGeC case is attributed to a reduction of recombination/generation centers due to the additional C implant [3,9].

C-V measurements of the diodes were performed at frequencies of 10 kHz, 100 kHz, and 1 MHz. Figure 3 shows the junction capacitance characteristics of SiGe,

SiGeC, and Si diodes measured at 100 kHz. The SiGe diode displays similar qualitative capacitance characteristics as the Si diode while the SiGeC diode shows an almost invariant junction capacitance with bias voltage. The capacitance of the SiGe diodes is considerably smaller than that of pure Si diode, represented by a parallel downshift of the C-V curve. At a reverse bias voltage of 0.5 V, the SiGe diode has a capacitance value only 40% that of the Si diode. At the same bias voltage, the SiGeC diode has a capacitance value 35% that of the Si diode. Assuming that the differences in permittivities are small, the decrease in capacitance will imply that the depletion widths of the SiGe and SiGeC diodes are larger than that of the Si diode. Moreover, the depletion width is a function of the doping level, mainly of the lightly doped side, i.e. the boron doping of the p^+ region in this case. Hence, the increase in depletion width suggests that the net doping of boron is reduced in both SiGe and SiGeC diodes. This reduction can be attributed to dopant deactivation caused by defects related to the Ge and C implantations. It can also be caused by boron out-diffusion and/or dopant compensation by donor complexes created by the defects. Using the step-junction model, an estimation of the number of electrically active B atoms in the SiGe diode indicates that it is 15 times less than that in the Si diode. The effect is particularly prominent in the SiGeC diode where the capacitance becomes very small and exhibits weak dependence on applied voltage. The capacitance curves for the SiGe and SiGeC merge at large reverse bias, consistent with the fact that the C implant range is less than that of the Ge. Therefore, at large reverse bias as the depletion width grows wider, the influence of C ions on the junction capacitance diminishes.

The phenomenon of further dopant deactivation due to C doping is also revealed by the forward-bias I-V characteristics. At forward bias voltages greater than 0.4V, the current of the SiGeC diode is larger than that of the SiGe diode. Since the diffusion current is inversely proportional to the activated dopant concentration, the larger current indicates a lower doping level. Therefore, the I-V results correlate with the C-V analysis.

Other factors which may influence the junction characteristics of these diodes were studied. No frequency dependence of the junction capacitance characteristics was observed for reverse bias and small forward bias. At large forward bias, junction capacitance cannot be measured correctly due to large current in this region of operation. The anneal time and carbon dose variations in the ranges stated above also had little effect on the observed junction characteristics.

4. Summary

Characterization of SiGe diodes formed by Ge implantation reveals that Ge implant creates additional generation/recombination centers as well as reduces the net boron doping. Although the addition of carbon doping seems to reduce the concentration of generation/recombination centers created by Ge implantation, C-V data indicate that the carbon implant results in further deactivation of boron in the device, thus limiting the effective use of carbon in the fabrication of SiGe/Si heterojunction bipolar transistors.

Acknowledgment

The authors wish to express their deepest appreciation to Mr. Akira Fukami, Mr. Ken-ichi Shoji, and Dr. Takahiro Nagano of Hitachi Research Labarotory, Japan for providing the samples and for their valuable suggestions and support throughout the project.

References

1. G.L. Patton, J.H. Comfort, B.S. Meyerson, E.F. Crabbe, G.J. Scilla, E.D. Fresart, J.M.C. Stork, J.Y.-C. Sun, D.L. Harame, and J.N. Burghartz, "75-GHz f_T SiGe-base heterojunction bipolar transistor," *IEEE Electron Device Lett.*, vol. 11, no.4, pp. 171-173, 1990.

2. G.L. Patton, J.M.C. Stork, J.H. Comfort, E.F. Crabbe, B.S. Meyerson, D.L. Harame, and J.Y.-C. Sun, "SiGe-base heterojunction bipolar transistors: physics and design issues," *IEDM Tech. Dig.*, pp. 13-16, 1990.

3. A. Fukami, K. Shoji, T. Nagano, and C.Y. Yang, "Characterization of SiGe/Si heterostructures formed by Ge^+ andC^+ implantations," *Appl. Phys. Lett.*, vol.57, no. 22, p. 2345-2347, 1990.

4. D. Fathy, O.W. Holland, and C.W. White, "Formation of epitaxial layers of Ge on Si substrates by Ge implantation and oxidation," *Appl. Phys. Lett.*, vol. 51, no. 17, pp. 1337-1339, 1987.

5. K. Ohta, J. Sakano, and S. Furukawa, "Formation of non-strained single crystalline SiGeB/Si heterostructure by Ge and B ions implantaation," *Extended Abstracts of 21st Conf. SSDM*, pp. 555-556, 1989.

6. B.T. Chilton, B.J. Robinson, D.A. Thompson, T.E. Jackman, and J.-M. Baribeau, "Solid phase epitaxial regrowth of $Si_{1-x}Ge_x$/Si strained-layer structures amorphized by ion implantation," *Appl. Phys. Lett.*, vol. 54, no. 1, pp. 42-44, 1989.

7. D.C. Paine, D.J. Howard, N.G. Stoffel, and J.A. Horton, "The growth of strained $Si_{1-x}Ge_x$ alloys on <001> silicon using solid phase epitaxy," *J. Mater. Res.*, vol. 5, no. 5, pp. 1023-1031, 1990.

8. A. Fukami, K. Shoji, T. Nagano, T. Tokuyama, and C.Y. Yang, "Graded bandgap SiGe bipolar transistor fabricated with germanium ion implantation," presented at ESSDERC '91, Switzerland.

9. A. Gupta, J.W. Waters, C. Cook, C.Y. Yang, A. Fukami, K. Shoji, and T. Nagano, "Simulations of Ge^+ and C^+ implantations to form SiGe/Si HBT and characterization of SiGe and SiGeC diodes," MRS Symposium on Silicon MBE, Spring 1991, in press.

Selective Growth of SiC and Application to Heterojunction Devices

S. Nishino, H. Tanaka, K. Takahashi, and J. Saraie

Department of Electronics and Information Science, Faculty of Engineering and Design, Kyoto Institute of Technology, Matsugasaki, Sakyo-ku, Kyoto 606, Japan

Single crystal cubic SiC films were selectively grown on a pattered Si substrate by chemical vapor deposition (CVD) using Si_2H_6, C_2H_2 and HCl at normal pressure at a substrate temperature of 1100 °C. A small amount of HCl was necessary to grow a SiC film selectively. By applying selective growth, a heterodiode and heterobipolar transistor(HBT) were made using a patterned SiO_2 mask on the Si substrate. The Heterojunction diode showed good rectifying characteristics(ideality factor:1.03, leakage current:1pA). The HBT showed a current gain of 20.

1. Introduction

Single crystalline cubic silicon carbide has a large band gap (2.2 eV), high electron mobility (1000 $cm^2/V/s$)[1], and high saturation velocity ($2.7x10^7$ cm/s, calculated at $2.5x10^5$ V/cm)[2], so active devices using SiC are expected. However, for a long time, epitaxial growth of cubic SiC on Si was difficult because of large lattice mismatch. Recently, techniques for epitaxial growth of SiC have been established and many device characteristics have been reported [3-9]. The temperature of crystal growth, however, is high, being around 1350 °C. When we consider a device combining SiC and Si, the epitaxial growth temperature should be lowered to minimize autodoping, the solid-solid diffusion effect and dislocation density [10]. Several papers using cubic SiC material in heterodiode and heterojunction bipolar transistors are reported by achieving low temperature growth.[11-20]

Furumura et al. reported epitaxial growth of cubic SiC on a Si substrate at 1000 °C using the $SiHCl_3$ + C_3H_8 + H_2 system in a low pressure CVD and showed good heterodiode characteristics.[11,12] Ohoshita et al.[10] reported selective growth of cubic SiC on a Si substrate at 1000 °C using the SiH_2Cl_2 + C_3H_8 + H_2 + HCl system in a low pressure CVD.[13] However, the grown layer is highly oriented polycrystalline SiC. Sugii et al.[14] reported an HBT with high gain using single crystalline SiC as a wide bandgap emitter.[13] Recently he also reported improved characteristics of the HBT using a fluorine doped SiC emitter.[15,16] Kamimura reported characteristics of an HBT using polycrystalline SiC as an emitter material.[17]

We previously reported low temperature growth of cubic SiC using the Si_2H_6 +C_2H_2 + H_2 + HCl system [18], and successful selective growth of cubic SiC films on a Si substrate using an SiO_2 mask.[19] There are many problems to be solved for devices using cubic SiC on Si. In this paper, selective growth and device fabrication are described. Single crystalline epitaxial films are formed only on the exposed area of the Si substrate. By using the selective growth of cubic SiC, a heterojunction diode(n-SiC/p-Si) and a heterojuction bipolar transistor(HBT), n-type cubic SiC/p-type Si/n-type Si, were fabricated. The heterodiode showed good characteristics and the HBT also showed good characteristics and current gain.

2.Experimental

Cubic SiC films were grown on Si(100) and Si(111) substrates by atmospheric pressure CVD using Si_2H_6 and C_2H_2 as souı ce materials. Crystal growth of cubic SiC consisted of two steps: etching and CVD growth. The substrate, placed on the SiC coated graphite susceptor, was heated by an RF generator. Crystal growth was carried out at a substrate temperature of 1100 °C with the standard gas composition: H_2=3 slm, Si_2H_6=0.15 sccm, C_2H_2=0.15 sccm. The gas ratio of Si/C was kept at 1. Details of the crystal growth are described elsewhere.[18,19] At this time, the n-type epilayer was aimed to be an emitter material for the HBT. And the emitter is used for an injector of electrons. A thick film is not necessary for this purpose. So the main effort was to get thin SiC films epitaxially on the selected area of the Si substrate. Selective growth was confirmed by observing the surface with an optical microscope. The thickness of the epilayer was estimated by interference spectra obtained in the reflectance measurement in the visible-light range and the refractive index.

The surface morphology of the film was observed by an optical microscope (Nomarski interference contrast). To study the crystal quality of the grown layers, the reflection high energy electron diffraction (RHEED) method was used.

To reduce the series resistance of the diode, a Si(100) substrate with a p/p+ layer was used. Resistivity of p-layer was 5 ohm.cm. And to minimize the leakage current of the diode, selective growth was used. Silicon dioxide was deposited by CVD using SiH_4+O_2 at 800 °C. The SiO_2 thickness was about 5000 A. And by using photolithography, an emitter mask pattern was made. the base area was 0.02 mm^2. The hetero-diode was made by growing cubic SiC on the selected area of the substrate. Phosphine (PH_3) was used to decrease the resistivity of the epilayer. The electrodes of the diode were made by evaporating Al in vacuum for n-type SiC and the p-type Si substrate. After annealing the diode at 400 °C for 5 min in N_2 atmosphere, Ohmic contact was formed.

For fabricating the HBT, four mask patterns were used for the photolithography. To reduce the series resistance of the transistor, a Si substrate with n/n+ was used. The resistivity of Si epilayer (n-layer:thickness about 5 µm) on the n+ substrate was 2 ohm.cm. The base region was made by diffusing boron at 1100 °C in the diffusion furnace using a solid source of BN. Then the emitter was made by depositing a SiC film on the base region. To prevent a diffusion of phosphorus into the base region, an introduction of PH_3 into the reaction gas was delayed for 5 minutes after undoped SiC was grown. Subsequently, the thickness of the undoped SiC and the Phosphorus-doped layers were 50 and 600 A, respectively. The collector is backside of the Si substrate. Electrodes for emitter, base, and collector were made by evaporating Al in vacuum and thermal annealing was done at 400 °C in N_2 ambient. In this manner, the HBT consisted of an n-SiC/p-Si/n-Si structure. Transistor characteristics were measured by a curve tracer(Tektronics 576) and a semiconductor parameter analyzer(HP 4145B).

3.Results and Discussion

Selective growth was carried out at flow rates of 0.15 sccm , 0.15 sccm, and 15 sccm, for the Si_2H_6, C_2H_2 and HCl respectively. Without an HCl addition, polycrystalline SiC was observed on the SiO_2 mask (triangular shape in the figure) as shown in Fig.1(a). By adding HCl to the reaction gases, the SiC growth rate was depressed. As the HCl flow rate was increased, the growth rate decreased and the selectivity became better. Perfect selective growth was obtained at 15 sccm of HCl as shown in Fig.1(b), where polycrystalline deposition was not observed on the SiO_2 mask. At an HCl flow rate of 15 sccm, no film was grown on the SiO_2 masked

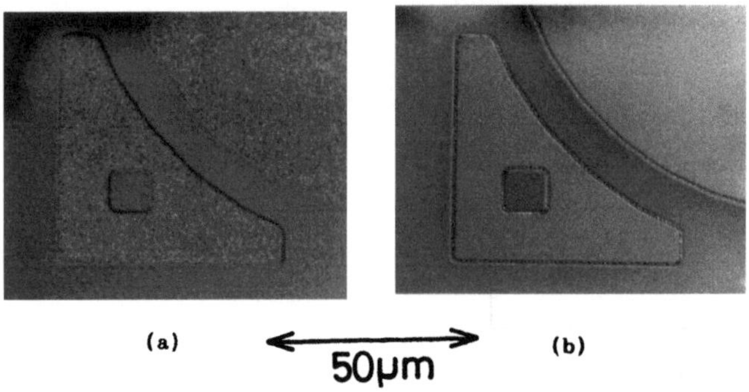

Fig.1 Selective growth of cubic SiC on patterened Si substrate. Square area was exposed to Si and surrounding triangle was SiO_2 masked area. (a) without HCl addition , (b) with HCl addition.

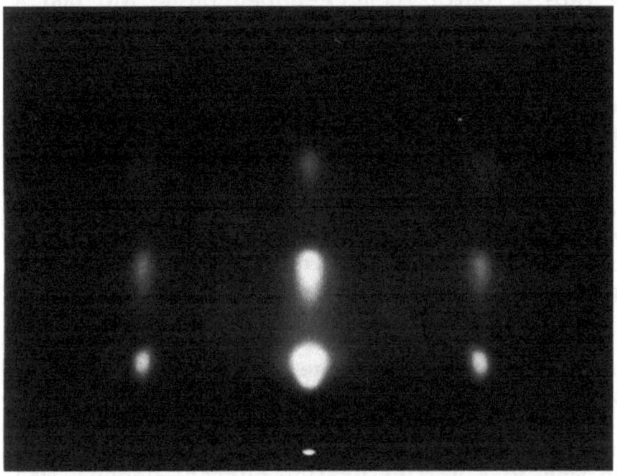

Fig.2 RHEED pattern of selectively grown SiC layer.

area during a 60 min deposition. The SiC film grew always on the selected area of the Si substrate. Similar results were reported for selective deposition of Si in the SiH_2Cl_2-HCl-H_2 system.[21] The mechanism of selective growth of cubic SiC on Si substrates is similar to selective growth of Si. By introduction of HCl in the Si_2H_6-C_2H_2-H_2 system, etching of Si adatoms and the SiO_2 mask by HCl occurs and the nucleation density of SiC is different for the SiO_2 mask and Si substrate. At the very beginning of the SiC formation, the Si surface is covered with C atoms and Si-C bonding is formed. However, on the SiO_2 mask the bonding energy of Si-C will be lower than the case of SiC on Si. Consequently, nucleation on the SiO_2 mask decreases drastically, and selective growth occurred. Without adding HCl, the nucleation density was almost the same on both materials. The diffraction pattern of selectively grown cubic SiC films showed single crystalline pattern as shown in Fig.2. Epitaxial films were grown only on the exposed area of the Si substrate.

To obtain a low resistive SiC epilayer on Si substrate (p-type), PH_3 was added to the reaction gas. When the flow rate of PH_3 was increased, the resistivity of the epilayer was decreased. At a flow rate of 0.005, and 0.03 sccm, the resistivity of the film was $1.6x10^{-2}$ and $8.5x10^{-3}$ ohm.cm,

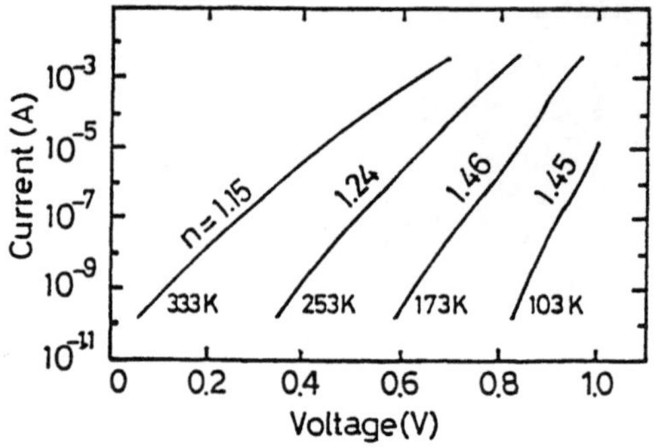

Fig.3 Temperature dependence of I-V characteristics of heterojunction diode.

respectively. The decrease of the resistivity varied with PH_3 flow rate and almost saturated at a PH_3 flow rate of 0.015 sccm. P-doped cubic SiC 600 A thick was grown on the B-doped Si epitaxial layer. The impurity distribution of SiC and underneath Si substrate was measured with SIMS (Cameca IMS-4F) using Cs^+ as a primary ion. Phosphorus did not penetrate into the Si epilayer, so the heterojunction is in the same plane as the pn junction. Similar results were already reported by several researchers.[12,14]

The diodes showed good rectification. Diode characteristics were improved when a low resistive SiC was used. Build-in voltage was slightly increased when the resistivity of the epilayer was decreased. This brings a Fermi level shift by the increase of carrier density of the epilayer. Current(I)-voltage(V) relations measured at room temperature were expressed by the relation $I \propto \exp(eV/nkT)$, where the value of n (ideality factor) was 1.03, which indicates that the current is predominately diffusion. Capacitance(C)-voltage(V) relations were expressed as the relation $V \propto 1/C^2$, which indicates that the junction was abrupt. The diffusion voltage obtained from the C-V curve was about 0.6 volt, and this value was almost the same value as that obtained from the I-V characteristics. Breakdown voltage of the diode was 30 V and saturation current density was 5×10^{-9} A/cm^2.

To understand the transport properties of electrons in the interface between SiC and Si, I-V characteristics depending on temperature were measured as shown in Fig.3. After mounting on transistor headers, the devices were placed in a vacuum vessel; the temperature could be varied from 103 K to 333 K. In the low voltage region, ideality factor n was 1.14, which indicates that current was transported by diffusion and recombination. The tunneling current which is insensitive to temperature was not observed in the diode and the hetero-interface was confirmed to be good in spite of large lattice mismatch.

On the basis of the heterojunction diode properties, we made a heterojunction bipolar transistor following the process as shown in Fig.4 , where top view and cross sections of the each process are shown. The emitter size was 0.02 mm^2. The thickness of the SiC was 600 A. The thickness of the base region was 2 µm which was determined by the spherical drilling method and stain etching. The carrier concentrations of emitter and base were 3×10^{19} and 5×10^{16} cm^{-3}, respectively. The I-V characteristics of the emitter-base junction was measured at room temperature. Ideality factor was 1.14 in the forward bias range of 0.43 - 0.59 V, however, excess current was observed in the lower voltage range (< 0.4 V). Leakage current in the backward direction was as low as 10^{-10} A and the breakdown voltage was 30 V. The ideality factor of 1.14 indicates that diffusion current is dominant in this region. The rectification ratio at 1.0 V is over 10^8,

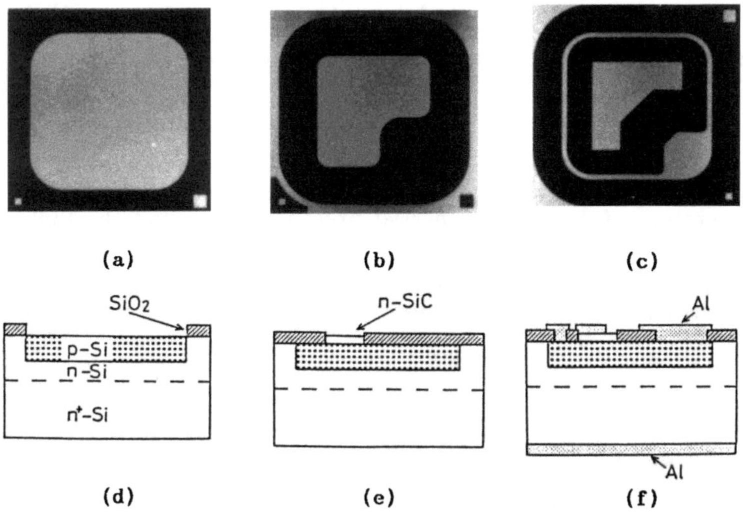

Fig.4 Fabrication of the transistor. Top view of masks for (a) base, (b) emitter and (c) metalization. Cross sections of (d) base, (e) emitter and (f) metallization.

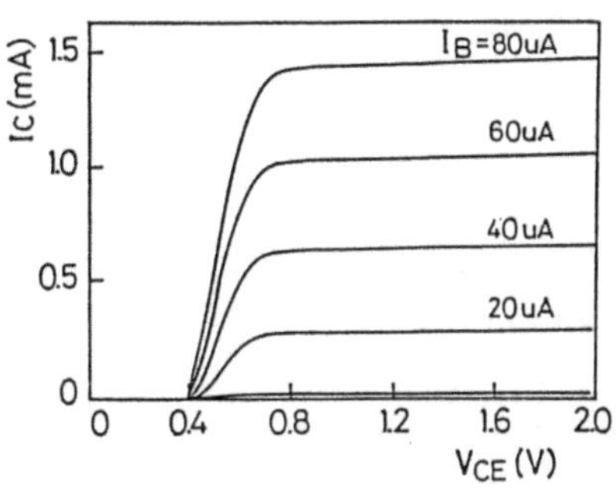

Fig.5 Common emitter characteristics of the HBT.

which shows good rectifying characteristics. Junction profile were determined by measuring C-V relation, and resulted abrupt junction.

Fig.5 shows the characteristics in the common-emitter configuration. Good transistor characteristics were observed. The current gain of the HBT was 20. Our process temperature (epitaxialgrowth and diffusion) was not as low as previously reported[12,14-16], however, rather high gain was obtained. A small off-set voltage was necessary to obtain the collector current. In the lower voltage, negative collector current increased as the base current increased. These data indicate that leakage current exists in both the collector and emitter junction. As far as the diode characteristics, the current across the junction was dominantly diffusion in the wide forward bias range. So it seems no problem in the emitter-base junction. However, the I-V characteristics of the emitter-base junction in the HBT show there was a small excess current in the lower voltage. This leakage current is related to the fabrication process of the HBT or carrier concentration of the base(10^{17} cm^{-3}). There is some possibility that surface of the base region is inverted to n-type, and subsequently a current path was formed for base to emitter and collector. If this current path can be cut, characteristics of HBT will be greatly improved.

4. Conclusion

Selective growth of cubic SiC on a patterned Si substrate was carried out at the low temperature of 1100 °C in normal pressure CVD using the Si_2H_6-C_2H_2-HCl-H_2 system. HCl addition during film growth plays an important role in low temperature epitaxy and selective growth. A heterojunction of n-SiC/p-Si was made on the patterned substrate and good rectification was observed. A heterojunction bipolar transistor was made using 4 masks and good characteristics with current gain of 20 were obtained.

References

1) W.E.Nelson,F.A.Halden, and A.Rosengreen:J.Appl.Phys.**37**,33(1966).
2) D.K.Ferry: Phys.Rev.**B12**,2361(1975).
3) S.Nishino,J.A.Powell and H.Will: Appl.Phys.Lett.**42**,460(1983).
4) S.Nishino,H.Suhara,H.Ono and H.Matsunami: J.Appl.Phys.**61**,4889(1987)
5) K.Shibahara,S.Nishino and H.Matsunami: Appl.Phys.Lett.**50**, 1888(1987).
6) K.Shibahara, T.Saito,S.Nishino and H.Matsunami: IEEE Electron Devices: **EDL-7**,692(1986).
7) P.Liaw and R.F.Davis: J.Elecrochem.Soc.**132**,642(1985).
8) G.L.Harris,K.H.Jackson,G.L.Felton,K.R.Osborne,K.Fekade and M.G.Spencer: Materials Lett.**4**,77(1986).
9) J.A.Powell,L.G.Matus and M.A.Kuczmarski: J.Electrochem.Soc.**134**,1558(1987).
10) H.P.Liaw and R.F.Davis,J.Electrochem.Soc.:**131**,3014(1984).
11) Y.Furumura,M.Doki,F.Mieno, and M.Maeda; Trans.IECEJ,**J69-C**,705(1986).
12) Y.Furumura,M.Doki,F.Mieno,T.Eshita,T.Sugii and M.Maeda: J.Electrochem.Soc.**135**,1255(1988)
13) Y.Ohoshita and A.Ishitani: J.Appl.Phys.**66**,4535(1989).
14) T.Sugii,T.Ito,Y.Furumura,M.Doki,F.Mieno, and M.Maeda: J.Electrochem.Soc.**134**,2545(1987).
15) T.Sugii,T.Yamazaki, and T.Ito: IEEE Trans.Electron Devices,**37**,2331(1990).
16) T.Sugii,T.Yamazaki, and T.Ito: Jpn.J.Appl.Phys.**30**,L970(1991).
17) K.Kamimura,Y.Nishibe, and Y.Onuma: Springer Proceedings in Physics,**43**,207(1989).
18) S.Nishino and J.Saraie: Springer Proceedings in Physics,vol **43**,8(1989).
19) S.Nishino,K.Takahashi,and J.Saraie:Proceedings of the 3rd International Conference on Amorphous and Crystalline Silicon Carbide and Related Materials (Springer-Verlag,in press).
20) M.I.Chaudhry and R.L.Wright: Appl.Phys.Lett.**59**,51(1991).
21) K.Tanno, N.Endo, H.Kitajima, Y.Kurogi and H.Tsuya: Jpn.J.Appl.Phys.**21**,L564(1982).

Electrical Characterization of Epitaxial Titanium Contacts to Alpha (6H) Silicone Carbide

L.M. Spellman[1], *R.C. Glass*[1], *R.F. Davis*[1], *T.P. Humphreys*[2], *R.J. Nemanich*[2], *K. Das*[3], *and S. Chevacharoenkul*[4]

[1]Department of Materials Science and Engineering, North Carolina State University, Raleigh, NC 27695-7907, USA
[2]Department of Physics, North Carolina State University, Raleigh, NC 27695-8202, USA
[3]Kobe Steel Research Laboratories (USA), Electronic Materials Center, P.O. Box 13608, Research Triangle Park, NC 27709, USA
[4]MCNC, Center for Microelectronics, P.O. Box 12889, Research Triangle Park, NC 27709, USA

Abstract. Titanium circular contacts grown epitaxially via electron beam MBE on unheated n-type 6H-SiC (0001) substrates were rectifying with low ideality factors (1.02 - 1.09) and leakage currents as low as 9 x 10^{-8} A/cm^2 at -10V. The contacts have remained stable after annealing for one hour at 700°C in ultra-high vacuum. Barrier heights for as-deposited and annealed contacts have been estimated from capacitance - voltage measurements to be 0.88 eV and 1.04 eV, respectively. A 1x1 ordered LEED pattern similar to that of the SiC surface was observed for all Ti coverages. Cross-sectional TEM diffraction patterns and images confirmed the deposition of crystalline Ti and revealed threading dislocations which relieve the strain caused by the 4% lattice mismatch between the film and the substrate.

1. Introduction

The thermal and electronic properties of silicon carbide make this wide bandgap semiconductor attractive for use in high-power, high-temperature, and radiation-hard applications. The future development of SiC device technology depends on, and may in fact be limited by, the ability to form good ohmic and Schottky contacts. For an ideal, abrupt junction the barrier height is defined by the Schottky -Mott limit, or the difference between the metal work function and the electron affinity of the semiconductor. However, interface states between the metal and semiconductor and traps common to wide bandgap semiconductors cause deviations from ideality. In fact, Pelletier *et al.* [1] have reported Fermi level pinning in 6H-SiC due to intrinsic surface states, indicating little dependence of barrier height on the work function of the metal. On the other hand, Waldrop *et al.* [2] have reported strong work function dependence for metal/β-SiC barrier heights, giving encouragement for the ability to control barrier heights of metals on 6H-SiC.

The ability to grow epitaxial metal layers on semiconductors presents opportunities to grow novel two- and three-dimensional device structures, extremely high quality heterostructures, and excellent electrical contacts [3]. This report focuses on electrical characteristics of epitaxially grown titanium contacts on alpha (6H) silicon carbide and the effect on these properties of annealing at 700°C for increasing periods of time.

2. Experimental

Vicinal single crystal, nitrogen-doped, n-type (10^{16} - 10^{18} cm^{-3}) substrates of 6H-SiC (0001) containing 0.5-0.8 μm thick, nitrogen-doped (10^{16} cm^{-3}) homoepitaxial films were provided by Cree Research, Inc. The Si-terminated (0001) surface, tilted 3°-4° towards $(11\bar{2}0)$, was used for all depositions and analyses.

Springer Proceedings in Physics, Vol. 71 Amorphous and Crystalline Silicon Carbide IV
Editors: C.Y. Yang · M.M. Rahman · G.L. Harris

The substrates were cleaned in sequence using a 10 min. dip in an ethanol / hydrofluoric acid / deionized water (10:1:1) solution and a thermal desorption in ultra-high vacuum (UHV) ($1 - 5 \times 10^{-10}$ Torr). A resistive graphite heater was used to heat the substrates at 700°C for 15 min. X-ray photoelectron spectroscopy (XPS) and low energy electron diffraction (LEED), both accessible by UHV transfer from the heating station and deposition chamber, were used to monitor surface chemistry and structure, respectively. The XPS system consisted of a Riber Mac2 semi-dispersive electron energy analyzer and a Riber pulse counter. The LEED studies were conducted with a Physical Electronics Instruments 11-020 electron gun. Titanium was deposited onto unheated substrates by electron beam evaporation (base pressure $< 2 \times 10^{-10}$ Torr). The first 10 nm was deposited at a rate of 1 nm/min. The deposition rate was then increased to 2 - 3 nm/min. to give a total thickness of 100 nm. For electrical characterization vertical contact structures consisting of 500 μm and 750 μm diameter Ti circular contacts were deposited through a Mo mask in contact with the (0001) SiC epitaxial layer, leaving a patterned metal film. Conductive liquid Ag served as the large area back contact. All subsequent annealing was done in UHV. Current-voltage (I-V) measurements were taken with a Rucker & Kolls Model 260 probe station in conjunction with an HP 4145A Semiconductor Parameter Analyzer.

Capacitance-voltage (C-V) measurements were taken with a Keithley Model 5956 Package 82 Simultaneous CV System in conjunction with an HP vector PC-308. The contact structures were the same as above. Measurements were taken at a frequency of 1 MHz.

Samples were prepared in cross-section for TEM analysis. Images were taken with both a Hitachi 800 and a Philips 430 microscope. High resolution images were taken with a JEOL 200 CX microscope.

3. Results

After chemically cleaning the substrates in a solution of ethanol, hydrofluoric acid, and deionized water for 10 min., XPS results showed a Si 2p peak at approximately 101eV, indicative of Si to C bonding in SiC [4-6]. No SiO_2, which has been reported at 102.1-102.5 eV [4], was detected in the signal. Other than a slight shift attributed to sample charging, the Si peak remained essentially the same after a 700°C desorption for 15 min. On the other hand, a C 1s peak at 285 eV present initially was removed after the desorption. This change indicates the desorption of adventitious C (i.e. hydrocarbons), while the major peak remained at 283.5 eV, as expected for C to Si bonding in SiC [4-7].

Surface structure before and after desorption also has been monitored. LEED patterns showed a 1x1 hexagonal pattern in all cases. Figure 1 shows a representative pattern of a (0001) 6H-SiC epitaxial layer which had undergone the described chemical cleaning procedure and thermal desorption. A LEED pattern similar to that of the SiC substrate was found for the Ti film; this result was the first indication of epitaxial growth.

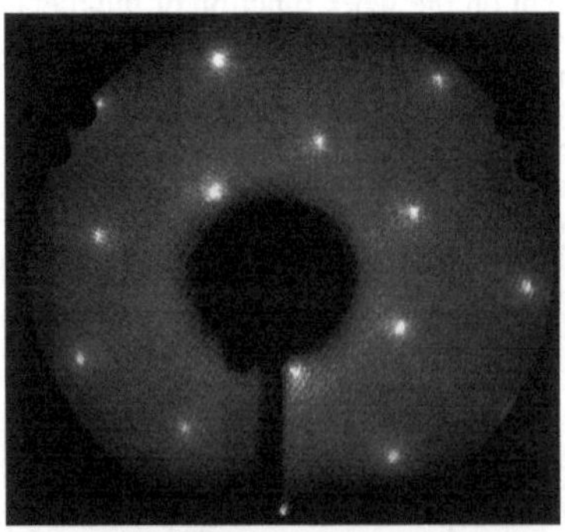

Figure 1. LEED pattern of (0001) 6H-SiC (120 V).

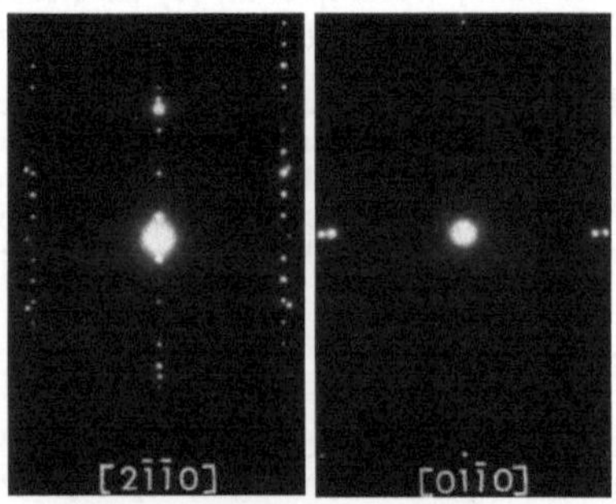

Figure 2. Selected area diffraction patterns of Ti/SiC interface region; Ti deposited at 400°C.

Figure 3. XTEM micrograph of Ti/SiC.

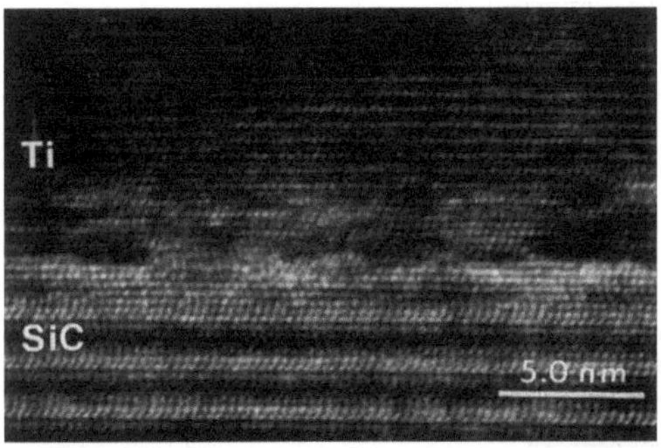

Figure 4. HRTEM micrograph of Ti/SiC.

Samples were prepared in cross-section for subsequent TEM analysis. Selected area diffraction (SAD) patterns of a Ti/SiC interface region are shown in Figure 2. It should be pointed out that both Ti (a = 2.95 Å, c = 4.68 Å) and 6H-SiC (a = 3.08 Å, c = 15.11 Å) have hexagonal crystal structures, corresponding to a 4% lattice mismatch in the (0001) basal plane. The diffraction patterns show a superposition of the spot pattern for the Ti film on that for the substrate. For each orientation of the 6H-SiC [8], the Ti spots lie outside the SiC spots due to the smaller lattice parameter of Ti. It is therefore shown that the Ti film is in the same orientation as the SiC and that a true lattice mismatch has occurred. This mismatch in lattice parameters is acheived through the creation of misfit and threading dislocations (Fig 3).

Actual lattice imaging of the interface region can be seen from high resolution images (Fig. 4). The 6-layer periodicity in the SiC meets the 2-layer periodicity of the Ti at a non-atomically smooth interface. Close examination of the interface reveals that there appear to be regions of lattice matching. Annealing this assembly at 700°C for 20 minutes resulted in the formation of a reacted layer 25 nm thick. The nature of this layer is currently under investigation.

Electrical characteristics of both as-deposited and annealed Ti contacts to n-type SiC (1×10^{16} cm^{-3}) were measured. Current-voltage characteristics of as-deposited Ti (curve (a) in Fig. 5) were rectifying with typical leakage currents at -10 V of 6 nA (1×10^{-7} A/cm^2); however, values as low as 0.4 nA (9×10^{-8} A/cm^2) were observed. Hard breakdown was not observed up to a voltage of -100 V; instead, the leakage current continuously increased, reaching 30 μA at –100V. This soft breakdown is thought to be due to sharp-edge effects from the contact geometries. After annealing in UHV for 20 min. at 700°C, the characteristics degraded (curve (b) in Fig. 5). However, the characteristics improved again with each subsequent 20 min. anneal (curves (c) and (d) in Fig. 5).

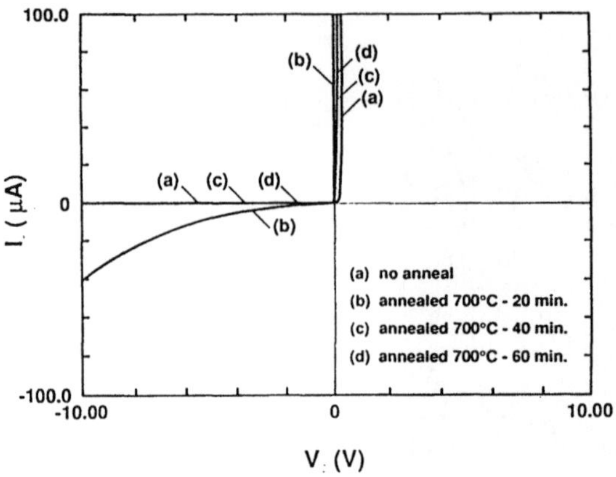

Figure 5. Current - voltage characteristics of Ti deposited on 6H-SiC (500 µm contact).

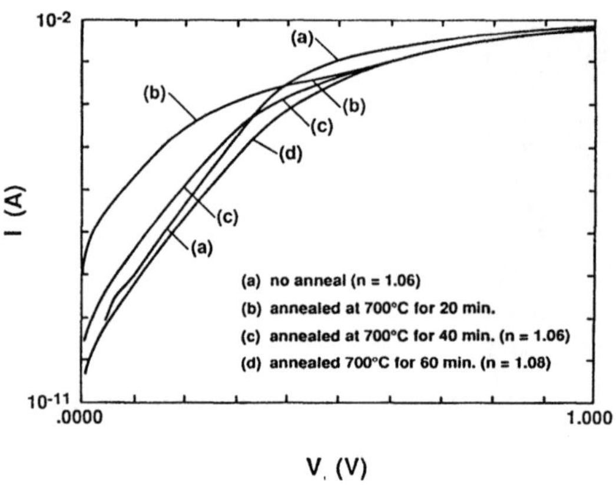

Figure 6. Log I vs. V of Ti deposited on 6H-SiC (500 µm contact).

Notably low ideality factors for silicon carbide were calculated from plots of log I vs. V for all but the first anneal (Fig. 6). For diodes in which the dominant current transport mechanism is thermionic emission and values of voltage V are greater than 3kT/q, the diode equation can be written as $J = Jo\ exp(qV/nkT)$ [9], where J is the current density and n is defined as the ideality factor. Plots of log I vs. V showed linearity over 3 decades of current yielding values of n below 1.1. No ideality factor could be calculated for the first anneal as the linear regime was too small.

Although the calculated ideality factors are low, thermionic emission may may not be the dominant current transport mechanism; therefore values for barrier height cannot be directly determined from current-voltage measurements (e.g. see Ref. [10]). Plots of log I vs. log V in the low voltage regime indicate that current transport is probably dominated by space charge limited (SCL) current, a mechanism which frequently occurs in wide bandgap semiconductors [11].

On the other hand, barrier height values were calculated from differential capacitance measurements as a function of reverse voltage. A plot of $1/C^2$ vs. V should give a straight line such that

$$\Phi_B = V_I + \xi + kT/q \qquad (2)$$

where Φ_B is the barrier height, V_I is the extrapolated intercept at $1/C^2 = 0$, and ξ is the energy

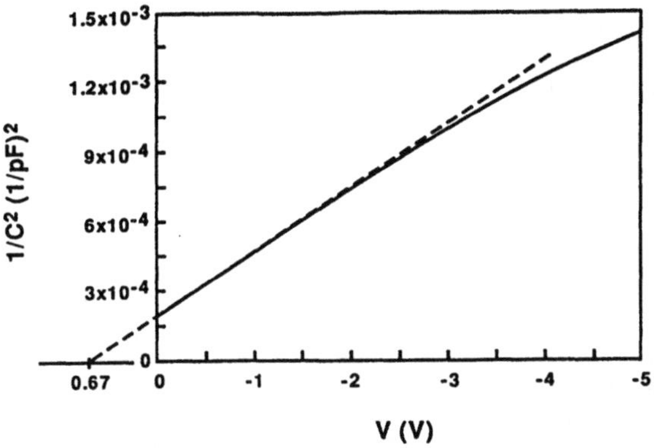

Figure 7. $1/C^2$ vs. V of as-deposited Ti/SiC (contact area = 2.0 x 10^{-3} cm^2) $\Phi_B \cong 0.88$ V.

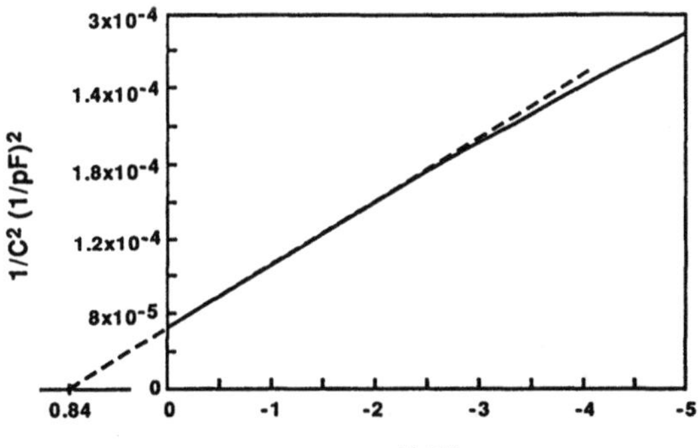

Figure 8. $1/C^2$ vs. V of annealed Ti/SiC (contact area = 4.6 x 10^{-3} cm^2). $\Phi_B \cong 1.04$ V.

difference between the Fermi level and the bottom of the conduction band [9]. Figures 7 and 8 show $1/C^2$ vs. V plots for an as-deposited and an annealed Ti contact, respectively. These curves deviate slightly from linearity but appear to be linear over 1 - 2V. By extrapolating the linear regions of the curves, the voltage intercepts were found to be 0.67 V for the unannealed contact and 0.84 Vfor the annealed contact. For a doping level of 10^{16} cm^{-3}, these intercepts correspond to a barrier height of 0.88V for the as-deposited Ti/SiC and 1.04V for the annealed Ti/SiC.

4. Discussion

By subtracting the electron affinity of the semiconductor from the workfunction of the metal, a theoretical, ideal barrier height can be calculated. Pelletier et. al. [1] have reported the Fermi level to be pinned near midgap in 6H-SiC with a workfunction of approximately 4.8 eV (intrinsic). Using a calculated value of 1.44 eV for the intrinsic Fermi level (measured from the the top of the valence band), the electron affinity (measured from the bottom of the conduction band to vacuum level) is estimated to be 3.38 eV. Since the workfunction of Ti is 4.33 eV, a barrier height of 0.95 eV is predicted. This corresponds to a 7% difference from the 0.88 eV value calculated from C-V measurements (Fig. 7).

The reaction chemistry on annealing directly affects the electrical characteristics (e.g. barrier height) because the characteristics become dependent upon the new interface formed with SiC. A reaction path in Ti/SiC diffusion couples at 1200°C in which TiC_{1-x} particles form in a matrix of Ti_5Si_3 has been reported [12]. The reported workfunction value of 3.71 eV for Ti_5Si_3 [13]

would indicate a smaller barrier height of 0.33 eV. This contradicts the experimental data (Fig. 8) in which a larger barrier height than that of the as-deposited Ti/SiC was calculated. With these issues in mind, more research on barrier heights and phase chemistry in the Ti/SiC system is anticipated.

5. Conclusions

Epitaxial growth of Ti contacts on (0001) 6H-SiC at room temperature has been confirmed by LEED and TEM. The 4% lattice mismatch strain is at least partly relieved by the creation of threading dislocations. Log I vs. log V plots in the low voltage regime indicate that space charge limited current is the dominant transport mechanism. Because thermionic emission may not predominate, barrier heights were not determined from current - voltage measurements. However, values were obtained from differential capacitance - voltage measurements. The measured value of 0.88 V for the as-deposited case agrees quite well with the predicted value of 0.95 V. The calculated barrier height increased to 1.04 V for the contact annealed at 700°C for 60 minutes. The identification of phases formed on annealing and additional methods of measuring barrier heights have been identified as key issues and are currently being investigated.

Acknowledgements

The authors gratefully acknowledge the support of the Office of Naval Research via Contract N00014-88-K-0341.

References

[1] J. Pelletier, D. Gervais, and C. Pomot, J. Appl. Phys. **55**, 994 (1984).
[2] J.R. Waldrop and R.W. Grant, Appl. Phys. Lett. **56**, 557 (1990).
[3] P.S. Peercy, J. Mater. Res., **5** (4), 852 (1990).
[4] M.N. Rahaman, Y. Bioteau, and L.C. De Jonghe, Am. Ceram. Soc. Bull. **65**, 1171 (1986).
[5] M.N. Rahaman and L.C. De Jonghe, Am. Ceram. Soc. Bull. **66**, 782 (1987).
[6] T.N. Taylor. J. Mater. Res. **4**, 189 (1989).
[7] C.D. Wagner, W.M.Riggs, L.E. Davis, J.F. Moulder, and G.E. Muilenberg, Handbook of X-ray Photoelectron Spectroscopy (Physical Electronics Division, Perkin-Elmer Corp., Eden Prairie, MN, 1979), p. 9.
[8] J.E. Lane, C.H. Carter, K.L. More, and R.F. Davis, J. Mater. Res. **1**, 737 (1986).
[9] E.H. Rhoderick and R.H. Williams in Metal-Semiconductor Contacts, 2nd Ed. (Oxford University, New York, 1988).
[10] S.M. Sze, Physics of Semiconductor Devices (John Wiley & Sons, New York, 1981).
[11] M.A. Lampert and P. Mark, Current Injection in Solids, (Academic Press, New York, 1970).
[12] M. Backhaus-Ricoult, Acta-Scripta Met. Proc., **4**, 79 (1989).
[13] G.V. Samsonov, L.N. Okhremchuk, N.F. Podgrushko, I.A. Podchernyaeva, and V.S. Fomenko, Inorganic Materials, **12**, 720 (1976).

Effect of H_2 Additive on Reactive Ion Etching of β-SiC in CHF_3/O_2 Plasma

A.J. Steckl and P.H. Yih

Nanoelectronics Laboratory, Department of Electrical and Computer Engineering, University of Cincinnati, 899 Rhodes Hall, ML 30, Cincinnati, OH 45221-0030, USA

Abstract. The effect of a low-level (~10%) addition of H_2 gas on the reactive ion etching of ß-SiC films on Si (100) substrates in a CHF_3/O_2 plasma has been investigated. The H_2 additive was found to prevent the formation of residues in the etched regions, while reducing the ß-SiC etch rate but only slightly affecting the SiC/Si etch rate ratio. For example, RIE etching at a pressure of 25 mtorr and an RF power of 200 W (0.4 W/cm^2) with 20 sccm of CHF_3/60%O_2 and 2 sccm of H_2 results in a SiC etch rate of 350 Å/min, a SiC/Si etch rate ratio of ~1.6, and *no residues*. The mechanisms involved in the effects of the H_2 additive are discussed.

1. Introduction

Silicon carbide technology [1] is making rapid progress toward electronic and optoelectronic device implementation, where its superior materials properties (e.g. wide band-gap, high melting point, high electric field breakdown, high thermal dissipation capacity) will open up new applications [2]. One of the critical steps in device fabrication is the etching of the semiconductor and related materials to form the appropriate device structure. This is particularly true for SiC because of its imperviousness to most widely used acids and other liquid etchants, an otherwise valuable property from a device stability and ruggedness point-of-view. Plasma-based etching in fluorinated compounds has been shown [3] to produce usable SiC etch rates (~10-100 nm/min) at, or near, room temperature. The basic mechanisms for the SiC etching process in fluorinated plasmas have been reported

Springer Proceedings in Physics, Vol. 71 Amorphous and Crystalline Silicon Carbide IV
Editors: C.Y. Yang · M.M. Rahman · G.L. Harris

[4]. They include a combination of chemical (by reaction with F and O) and physical (by energetic ions) removal processes. The specific desired characteristics for the plasma etching process depend on the device type and structure to be implemented [5]. For example, the fabrication of a SiC-Si heterojunction transistor would greatly benefit from a SiC/Si etch rate ratio greater than unity, whereas in a SiC homojunction device this would not be a factor. General considerations which would apply for most device structures include selectivity to masking layer, etching anisotropy and presence or absence of residues in the etched region.

2. Experimental Procedures

The plasma-based etching of SiC was performed in the reactive ion etching (RIE) mode, in other words at a relatively low operating pressure (~10^{-3} - 10^{-2} torr) and with the sample located on the RF-driven electrode. The system utilized was a parallel plate reactor (Plasma Therm PK1241) with Al electrodes. A computer-controlled optical spectrometer monitored the emission from the plasma during the process. Noble gas actinometry was employed, using 0.6 sccm Ar added to the feed gas, to obtain the relation between measured emission and actual concentration in the plasma for various species. In these experiments, several parameters were generally kept constant: the RF power - 200 W (0.4 W/cm^2), pressure - 20 mtorr, and total gas flow rate - 20 sccm. The variable parameters were the O_2% in the CHF_3 gas feed, the H_2 additive flow rate and the etching time. When hydrogen additive was used, the working pressure was increased from 20 to 25 mtorr to accommodate the increase in total gas flow rate.

To be able to accurately measure the etch rate and observe the presence or absence of residues it is preferable to use thicker films which can be etched for longer periods. 6 to 8 μm n-type β-SiC films grown on p-Si (100) substrates by chemical vapor deposition were provided by the NASA Lewis Research Center. The samples were first prepared by successively polishing with smaller grit (from 3 to 0.25 μm) diamond paste to achieve a surface roughness of 100 Å or less. Next, thermal oxidation in dry oxygen was performed to remove the polishing-induced surface damage layer. Finally, the samples were etched in HF to remove the oxide layer. The resulting SiC films, with mirror-like and damage-free surfaces, were used for all experiments. A standard etching pattern, prepared by basic photolithography and a lift-off process, was used for etch rate measurements. A thin layer of Au sputtered on the SiC sample was used as a thin film etch mask. The sample size was small enough such that the area ratio of sample-to-(power) electrode was less than 1%. Therefore, the sample did not disturb the plasma conditions during the processing. This is very important for accurate actinometry measurements. After RIE, the Au mask was removed by chemical etching, using a KI/I_2 solution, prior to further analysis.

3. Results and Discussion

We have previously reported [4] that a SiC/Si etch rate selectivity greater than unity (~1.5-2) can be obtained in CHF_3 with high percentage of O_2. This is appealing for SiC-Si heterostructure device fabrication. However, we have found that reactive ion etching through SiC films with thickness of 1 μm or greater using $CHF_3/90\%O_2$, which requires at least 20-30 min etching time, results in an array of residues being formed in the etched region. A representative example is shown in Fig. 1, where residues present after 20 min RIE under standard conditions

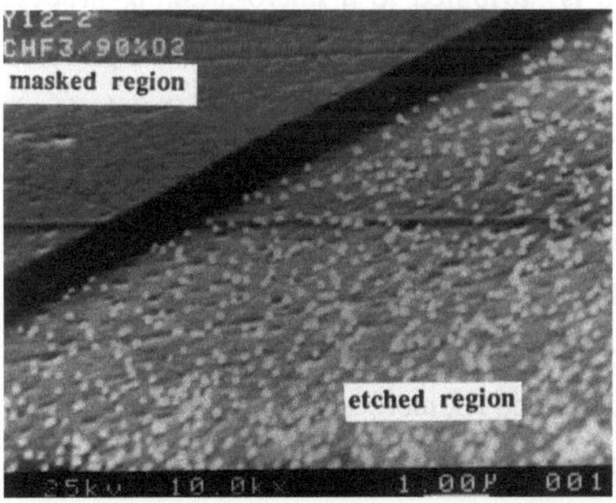

Fig. 1 Photomicrograph of SiC surface after RIE for 20 min with 20 sccm $CHF_3/90\%O_2$ at 200 W and 20 mtorr: 10 kx magnification.

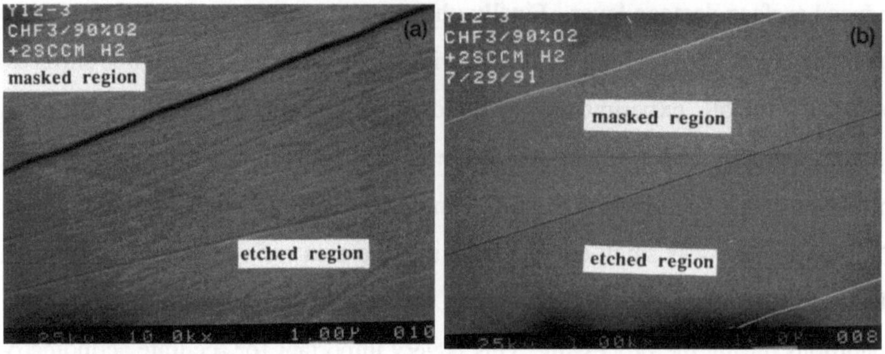

Fig. 2 Photomicrographs of SiC surface after RIE for 20 min with 20 sccm $CHF_3/90\%O_2$ and 2 sccm H_2 at 200 W and 25 mtorr: (a) 10 kx magnification; (b) 1 kx magnification.

(as defined above) uniformly cover the etched SiC surface. Similar residues have been reported to occur during reactive ion etching of SiC in CF_4 [6] and NF_3 [7].

Incorporating a low-level of H_2 additive to the CHF_3/O_2 gas feed results in the complete prevention of residues. This is illustrated by comparing the photomicrographs at the same magnification (10 kx) contained in Fig. 1, where no H_2 was used, and in Fig. 2a, where 2 sccm of H_2 was added. In the latter case, the surface morphology, before and after etching, is essentially identical. The only observable features, even over the larger area shown in the 1 kx photomicrograph contained in Fig. 2b, are striations believed due to the polishing process.

The formation of these residues is attributed to a micro-masking process initiated by the sputter deposition of microscopic Al particles from the powered electrode onto the SiC film. Since the etch rate of Al in a CHF_3/O_2 plasma is known to be extremely low, it can protect the SiC region directly underneath, hence resulting in the micro-masking effect.

Several mechanisms can be responsible for the suppression of the columnar residues: prevention of Al redeposition via the formation of volatile Al-H compounds, enhanced etch rate of redeposited Al, more isotropic etching leading to undercutting and removal of the Al micro-masks. It is significant to point out that for lower O_2 percentage in CHF_3 (0 - 50%), no residues are formed even in the absence of H_2 additive.

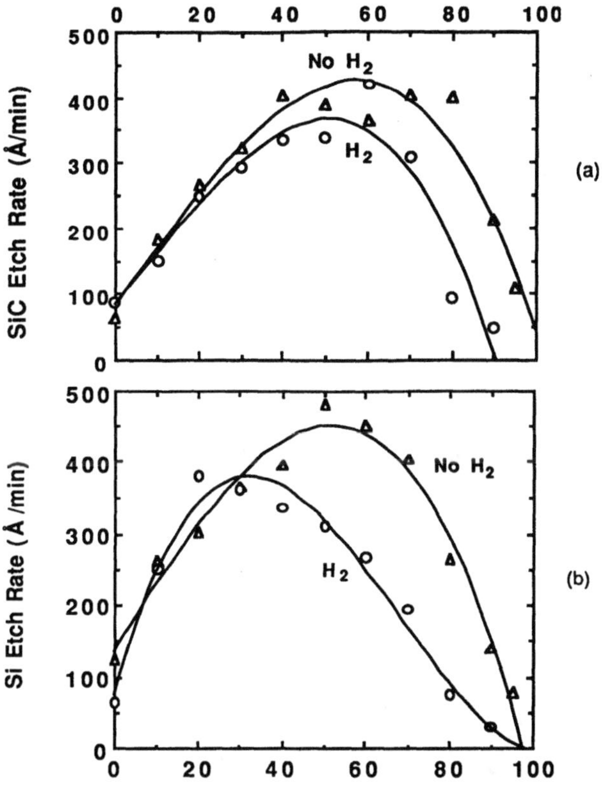

Fig. 3 Etch rate as a function of O_2% in CHF_3 with and without H_2 additive: (a) SiC; (b) Si.

Since the H_2 additive was successful in preventing residue formation, we investigated its effect on the SiC and Si etch rates. As can be seen in Fig. 3, the effect of H_2 at high O_2 percentage is to reduce the etch rate of both SiC and Si. However, as shown in Fig. 4 a SiC/Si etch rate selectivity in excess of unity is still maintained at high O_2 percentage.

To understand the effect of H_2 additive on the etch rate we have measured the plasma concentrations of [F], [O], and [H] as a function of O_2% in CHF_3. The [F] concentration at high O_2 percentage, Fig. 5a, is significantly reduced by the H_2 additive, presumably through HF formation. The reduction of [F] concentration leads to a reduced Si etch rate, while the combination of reduced [F] and [O] concentrations cause a drop in the SiC etch rate.

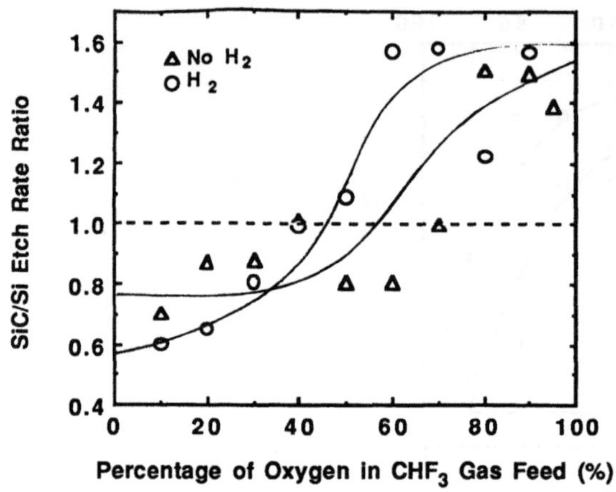

Fig. 4 SiC-to-Si etch rate ratio in $CHF_3/\%O_2$ with and without H_2 additive.

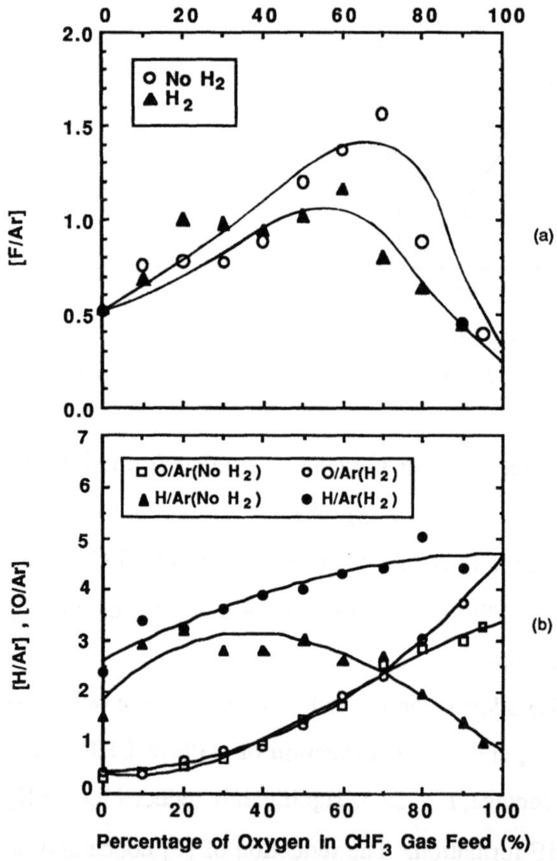

Fig. 5 Relative concentration of plasma species during RIE in $CHF_3/\%O_2$ with and without H_2 additive: (a) [F] at 703 nm; (b) [O] at 777 nm and [H] at 486 nm.

4. Summary and Acknowledgements

In summary, we have shown that a low-level H_2 additive to CHF_3/O_2 plasma prevents residues formation during SiC reactive ion etching. The H_2 additive reduces the SiC and Si etch rates, but the SiC/Si etch rate selectivity is maintained.

We gratefully acknowledge J. A. Powell of NASA Lewis Research Center for providing the SiC samples. We also acknowledge the support of this work by the Edison Materials Technology Center and the support provided by WPAFB and monitored by NASA LeRC.

References

[1] R. F. Davis, Z. Sitar, B. E. Williams, H. S. Kong, H. J. Kim, J. W. Palmour, J.A. Edmond, J. Ryu, J. T. Glass and C. H. Carter, Materials Science and Engineering, **B1**, 77 (1988).

[2] J. A. Powell and L. G. Mattus, in **Amorphous and Crystalline Silicon Carbide**, G.L. Harris and C. W.- Y. Wang, eds., Springer Proceedings in Physics, **34**, 2 (1989).

[3] J. Sugiura, W.-J. Lu, K. C. Cadien, and A. J. Steckl, J. Vac. Sci. Technol. B, **4**, 349 (1986).

[4] W.-S. Pan and A. J. Steckl, J. Electrochem. Soc., **137**, 213 (1990).

[5] W.-S. Pan and A. J. Steckl, in **Amorphous and Crystalline Silicon Carbide**, G.L. Harris and C. W.- Y. Wang, eds., Springer Proceedings in Physics, **34**, 192 (1989).

[6] J. W. Palmour, R. F. Davis, T. M. Wallett and K. B. Bhasin, J. Vac. Sci. Tech. A, **4**, 590 (1986).

[7] J. W. Palmour, R. F. Davis, P. Astell-Burt, P. Blackborrow, Mat'l. Res. Soc. Proc., **76**, 185 (1987).

Index of Contributors

Springer Proceedings in Physics

Managing Editor: H. K. V. Lotsch

1 *Fluctuations and Sensitivity in Nonequilibrium Systems*
Editors: W. Horsthemke and D. K. Kondepudi

2 *EXAFS and Near Edge Structure III*
Editors: K. O. Hodgson, B. Hedman, and J. E. Penner-Hahn

3 *Nonlinear Phenomena in Physics* Editor: F. Claro

4 *Time-Resolved Vibrational Spectroscopy*
Editors: A. Laubereau and M. Stockburger

5 *Physics of Finely Divided Matter*
Editors: N. Boccara and M. Daoud

6 *Aerogels* Editor: J. Fricke

7 *Nonlinear Optics: Materials and Devices*
Editors: C. Flytzanis and J. L. Oudar

8 *Optical Bistability III*
Editors: H. M. Gibbs, P. Mandel, N. Peyghambarian, and S. D. Smith

9 *Ion Formation from Organic Solids (IFOS III)*
Editor: A. Benninghoven

10 *Atomic Transport and Defects in Metals by Neutron Scattering*
Editors: C. Janot, W. Petry, D. Richter, and T. Springer

11 *Biophysical Effects of Steady Magnetic Fields*
Editors: G. Maret, J. Kiepenheuer, and N. Boccara

12 *Quantum Optics IV*
Editors: J. D. Harvey and D. F. Walls

13 *The Physics and Fabrication of Microstructures and Microdevices*
Editors: M. J. Kelly and C. Weisbuch

14 *Magnetic Properties of Low-Dimensional Systems*
Editors: L. M. Falicov and J. L. Morán-López

15 *Gas Flow and Chemical Lasers*
Editor: S. Rosenwaks

16 *Photons and Continuum States of Atoms and Molecules*
Editors: N. K. Rahman, C. Guidotti, and M. Allegrini

17 *Quantum Aspects of Molecular Motions in Solids*
Editors: A. Heidemann, A. Magerl, M. Prager, D. Richter, and T. Springer

18 *Electro-optic and Photorefractive Materials*
Editor: P. Günter

19 *Lasers and Synergetics*
Editors: R. Graham and A. Wunderlin

20 *Primary Processes in Photobiology*
Editor: T. Kobayashi

21 *Physics of Amphiphilic Layers*
Editors: J. Meunier, D. Langevin, and N. Boccara

22 *Semiconductor Interfaces: Formation and Properties*
Editors: G. Le Lay, J. Derrien, and N. Boccara

23 *Magnetic Excitations and Fluctuations II*
Editors: U. Balucani, S. W. Lovesey, M. G. Rasetti, and V. Tognetti

24 *Recent Topics in Theoretical Physics*
Editor: H. Takayama

25 *Excitons in Confined Systems*
Editors: R. Del Sole, A. D'Andrea, and A. Lapiccirella

26 *The Elementary Structure of Matter*
Editors: J.-M. Richard, E. Aslanides, and N. Boccara

27 *Competing Interactions and Microstructures: Statics and Dynamics*
Editors: R. LeSar, A. Bishop, and R. Heffner

28 *Anderson Localization*
Editors: T. Ando and H. Fukuyama

29 *Polymer Motion in Dense Systems*
Editors: D. Richter and T. Springer

30 *Short-Wavelength Lasers and Their Applications*
Editor: C. Yamanaka

31 *Quantum String Theory*
Editors: N. Kawamoto and T. Kugo

32 *Universalities in Condensed Matter*
Editors: R. Jullien, L. Peliti, R. Rammal, and N. Boccara

33 *Computer Simulation Studies in Condensed Matter Physics: Recent Developments*
Editors: D. P. Landau, K. K. Mon, and H.-B. Schüttler

34 *Amorphous and Crystalline Silicon Carbide and Related Materials*
Editors: G. L. Harris and C. Y.-W. Yang

35 *Polycrystalline Semiconductors: Grain Boundaries and Interfaces*
Editors: H. J. Möller, H. P. Strunk, and J. H. Werner

36 *Nonlinear Optics of Organics and Semiconductors*
Editor: T. Kobayashi

37 *Dynamics of Disordered Materials*
Editors: D. Richter, A. J. Dianoux, W. Petry, and J. Teixeira

38 *Electroluminescence*
Editors: S. Shionoya and H. Kobayashi

39 *Disorder and Nonlinearity*
Editors: A. R. Bishop, D. K. Campbell, and S. Pnevmatikos

40 *Static and Dynamic Properties of Liquids*
Editors: M. Davidović and A. K. Soper

41 *Quantum Optics V*
Editors: J. D. Harvey and D. F. Walls

42 *Molecular Basis of Polymer Networks*
Editors: A. Baumgärtner and C. E. Picot

43 *Amorphous and Crystalline Silicon Carbide II: Recent Developments*
Editors: M. M. Rahman, C. Y.-W. Yang, and G. L. Harris

MIX
Papier aus verantwortungsvollen Quellen
Paper from responsible sources
FSC® C105338

If you have any concerns about our products,
you can contact us on
ProductSafety@springernature.com

In case Publisher is established outside the EU,
the EU authorized representative is:
Springer Nature Customer Service Center GmbH
Europaplatz 3, 69115 Heidelberg, Germany

Printed by Libri Plureos GmbH
in Hamburg, Germany